AF608707

Quantum Mesoscopic Phenomena and Mesoscopic Devices in Microelectronics

NATO Science Series

A Series presenting the results of activities sponsored by the NATO Science Committee. The Series is published by IOS Press and Kluwer Academic Publishers, in conjunction with the NATO Scientific Affairs Division.

A. **Life Sciences**	IOS Press
B. **Physics**	Kluwer Academic Publishers
C. **Mathematical and Physical Sciences**	Kluwer Academic Publishers
D. **Behavioural and Social Sciences**	Kluwer Academic Publishers
E. **Applied Sciences**	Kluwer Academic Publishers
F. **Computer and Systems Sciences**	IOS Press
1. **Disarmament Technologies**	Kluwer Academic Publishers
2. **Environmental Security**	Kluwer Academic Publishers
3. **High Technology**	Kluwer Academic Publishers
4. **Science and Technology Policy**	IOS Press
5. **Computer Networking**	IOS Press

NATO-PCO-DATA BASE

The NATO Science Series continues the series of books published formerly in the NATO ASI Series. An electronic index to the NATO ASI Series provides full bibliographical references (with keywords and/or abstracts) to more than 50000 contributions from international scientists published in all sections of the NATO ASI Series.
Access to the NATO-PCO-DATA BASE is possible via CD-ROM "NATO-PCO-DATA BASE" with user-friendly retrieval software in English, French and German (WTV GmbH and DATAWARE Technologies Inc. 1989).

The CD-ROM of the NATO ASI Series can be ordered from: PCO, Overijse, Belgium

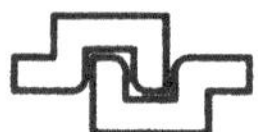

Series C: Mathematical and Physical Sciences – Vol. 559

Quantum Mesoscopic Phenomena and Mesoscopic Devices in Microelectronics

edited by

Igor O. Kulik

and

Recai Ellialtioğlu
Department of Physics,
Bilkent University,
Bilkent, Ankara

Springer-Science+Business Media, B.V.

Proceedings of the NATO Advanced Study Institute on
Quantum Mesoscopic Phenomena and Mesoscopic Devices in Microelectronics
Ankara, Turkey
13–25 June 1999

A C.I.P. Catalogue record for this book is available from the Library of Congress.

Library of Congress Cataloging-in-Publication Data

Quantum mesoscopic phenomena and mesoscopic devices in microelectronics / edited by Igor O. Kulik and Recai Ellialtioglu.
p. cm. -- (NATO science series. Series C, Mathematical and physical sciences ; v. 559)
Includes indexes.
ISBN 978-0-7923-6625-6 ISBN 978-94-011-4327-1 (eBook)
DOI 10.1007/978-94-011-4327-1

1. Mesoscopic phenomena (Physics)--Congresses. 2. Quantum solids--Congresses. 3. Quantum electronics--Congresses. I. Kulik, Igor O. II. Ellialtioglu, Recai. III. NATO Advanced Study Institute on Quantum Mesoscopic Phenomena and Mesoscopic Devices in Microelectronics (1999 : Ankara, Turkey, and Antalya, Turkey) IV. Series.

QC176.8.M46 .Q3 2000
537.5--dc21

00-064694

ISBN 978-0-7923-6625-6

Printed on acid-free paper

Contents

Part II QOULOMB BLOCKADE AND THE KONDO PROBLEM

Part V JOSEPHSON EFFECT

Part VI MESOSCOPIC SUPERCONDUCTIVITY

Part VIII NANO-ELECTRONICS

Preface

Mesoscopic world represents a field of physics one step up from the atomic (microscopic) level. Quantum mechanical laws, well documented at the level of a single or a few atoms and electrons, are extended to systems of size 1-100 nm containing 10^2 to 10^{10} electrons, still much smaller than in the usual "macroscopic" objects, but behaving in a manner similar to single atom. Besides the pure theoretical interest, such systems are challenging in the process of achieving the ultimate microelectronic applications. The objective of this book is to bring together various directions in meso-large quantum systems, including such intriguing phenomena as quantization of magnetic flux (the Aharonov-Bohm effect) and quantization of electric charge (the Coulomb blockade and Coulomb oscillation effects); quantization of electrical resistance in mesoscopic conductors and, in general, aspects of quantum transport in mesoscopic, nanoscopic and atomic contacts formed between the metallic, semiconducting and superconducting electrodes; Josephson effect and Andreev reflection in small tunneling junctions as well as manifestations of quantum coherence in normal-conducting metals (the persistent currents).

We tried to put material presented at the NATO Advanced Study Institute on Quantum Mesoscopic Phenomena and Mesoscopic Devices in Microelectronics (Ankara/Antalya, June 13-25, 1999) related to these issues in a systematic way by collecting major topics into *parts* in the book.

In *Part I*, atomic and ballistic contacts between metals are discussed as counterparts to more familiar tunneling contacts (tunneling junctions) which have been pioneers in the electronic applications. The recent advances in nanofabrication technology made possible the preparation of direct metallic constrictions with dimensions down to atomic sizes. The flow of current in a constriction is a regular process with a reduced shot noise and therefore appealing for low-noise circuits, and quantized conductance in units of a fundamental quantity $2e^2/h$, twice the fundamental conductance quantum e^2/h (a "klitzing") first appeared in the physics of Quantum Hall Effect.

Part II considers phenomena associated with the so called Coulomb blockade as well as Coulomb oscillation in small metallic islands. This direction of mesoscopic physics promises the fastest applications in microelectronics, and is already being used in the super-sensitive scientific instrumentation.

Part III addresses the fundamental issue of mesoscopic physics: which is the maximal spatial size, and the maximal temporal scale at which quantum coherence in a mesoscopic system is preserved? "Dephasing" is a factor which limits the existence of quantum behavior in a system comprising many linear, as well as nonlinear electronic components, and therefore establishes a limit for the advanced applications of quantum phenomena like for example the quantum computation.

In Part IV, the Aharonov-Bohm effect and the related phenomena of "persistent" currents in various mesoscopic geometries are reviewed.

Part V is devoted to Josephson effect with a new angle of thought related to mesoscopic objects and to new artificially fabricated states of matter like Bose-Einstein condensates.

Part VI is devoted to mesoscopic superconductivity and pairing effect in ultrasmall superconducting grains. The trend in recent mesoscopic development directs to solid-state realization of the ultimate goal in the macroscopic quantum physics: the quantum computation, which is considered in *Part VII* (while some aspects of optoelectronic physics selected to *Part VIII*). The multi-degree of freedom (multi-qubit) quantum system develops, through the unitary transformation of its global wave function, a process similar to multi-bit computation. This "parallel processing" may dramatically increase the speed of computers as compared to standard classical Von Neumann computers. The papers mentioned in *Part VIII* illustrate the progress so far achieved in the fulfillment of this ultimate goal.

We thank the lecturers and the speakers of this NATO-ASI for delivering their particular subjects with a special care for the global task: the understanding and developing of the quantum aspects of mesoscopic structures. Special thanks are addressed to the members of International and Local Organizing Committees: Antonio Barone, Joseph Imry, Konstantin Likharev, Cemal Yalabik and Bilal Tanatar. Their valuable advises determined the scope of the Meeting and its final success. We thank Bilkent University which served as a host of the Meeting in its Ankara period, and Ador Tourizm and Travel Agency for the excellent organization of the second period of the Meeting in a stimulating atmosphere of small suburb of Antalya at the Mediterranean coast of Turkey. It is our pleasure to acknowledge the generous grant by the Scientific Affairs Division of the North Atlantic Treaty Organization (NATO) which made our Meeting possible. We acknowledge the support by the Abdus Salam International Centre for Theoretical Physics (ICTP), as well as partial support by National Science Foundation (USA), the Centre Culturel et Linguistique (France), and Deutsche Forschungs Gemeinschaft (Germany) for the young scientists from the respective countries.

IGOR O. KULIK AND RECAI ELLIALTIOGLU

I

QUANTUM CONTACTS AND WIRES

Chapter 1

NONLINEAR PHENOMENA IN METALLIC CONTACTS

I. O. Kulik
Department of Physics, Bilkent University
Ankara 06533, Turkey

Abstract We review and extend theoretical approaches to nonlinear and nonequilibrium effects in metallic microcontacts ranging in their dimension from the atomic to macroscopic sizes. *Atomic* contacts are shown to quantize their conductance in units of $2e^2/h$ provided the charge redistributes near the constriction to establish the maximal electron transmittivity through the orifice. *Ballistic* semiclassical contacts are treated both from the Landauer point of view and from the Boltzmann transport theory. The J-V nonlinearity in contacts is related to the inelastic scattering near the narrowest part of the constriction and permits for spectroscopic investigation of phonons in solids (the point-contact spectroscopy).The effects of phonon emission and reabsorption in contacts are taken into consideration. Phonon relaxation is shown to determine the frequency dependence of the nonlinear contact conductivity. *Thermal* contacts develop specific nonequilibrium states with *hot spots* in the center of metallic constriction whose temperature is much in excess of the ambient contact temperature and is uniquely related to voltage.

1. INTRODUCTION

It is the aim of this paper to present a coherent approach to linear and nonlinear, as well as to equilibrium and nonequilibrium, phenomena in metallic contacts of diameter ranging from the atomic size to macroscopic size. Our understanding of these properties arises from the works of Landauer [1], Sharvin [2], Yanson [3], Holland groups [4, 5], and others [6], *etc.* Unlike tunneling junctions, direct metallic constrictions (or links) develop a number of peculiarities of which we mention the following.

I. O. Kulik and R. Ellialtioğlu (eds.),
Quantum Mesoscopic Phenomena and Mesoscopic Devices in Microelectronics, 3–26.

(1) Conductance of contact scales with the quantum of conductance

$$G_0 = \frac{2e^2}{h} = 1/12.9\text{k}\Omega \qquad (1.1)$$

in such a way that minimal conductance reaches a value G_0 before the contact breaks to the tunneling-type junction with a much smaller or zero conductance, and is even quantized in units of $G_0 = R_0^{-1}$ in a proper arrangement. In particular, this happens if contact size or shape is varied by applying a gate voltage to change the electron concentration (in semiconducting constrictions), or contacting electrodes are pulled away to increase the length (and possibly the contacting area), in metallic contacts. The typical dependence of the contact conductance on the pulling strength [7] is presented in Fig. 1.1.

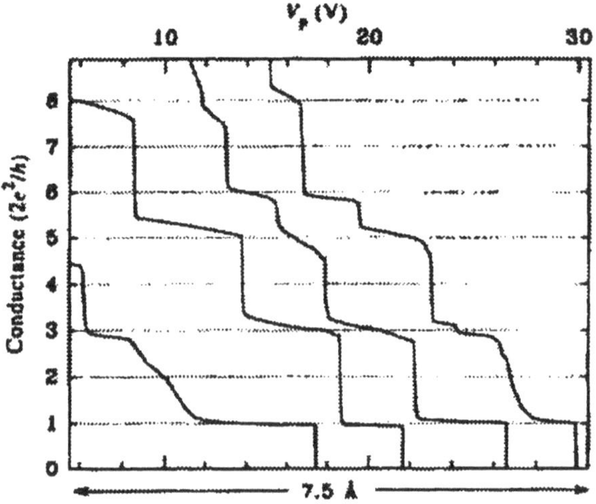

Figure 1.1 Conductance of sodium contact at 4.2 K as a function of stretch [7]. Measurements have been performed by pressing two pieces of metal and then pulling them away from one another with a piezoelectric sensor. Reproduced by permission from Ref. [7].

(2) The electron flow in a constriction is a regular quantum process (a kind of "nondemolition measurement") while the energy dissipation takes place away from its narrowest part. Because of this, the shot noise in direct metallic constriction reduces compared to its value in the tunneling junction of similar resistance [8]

$$S_V \sim 2eVR\frac{d}{l} \qquad (1.2)$$

where S_V is the shot noise power and l the phase-breaking electron mean free path assumed to be larger than the contact diameter d. Reduced shot noise in a metallic contacts was first observed in an experiment in 1984 [9] (see Fig. 1.2). (Further works are reviewed in [10].)

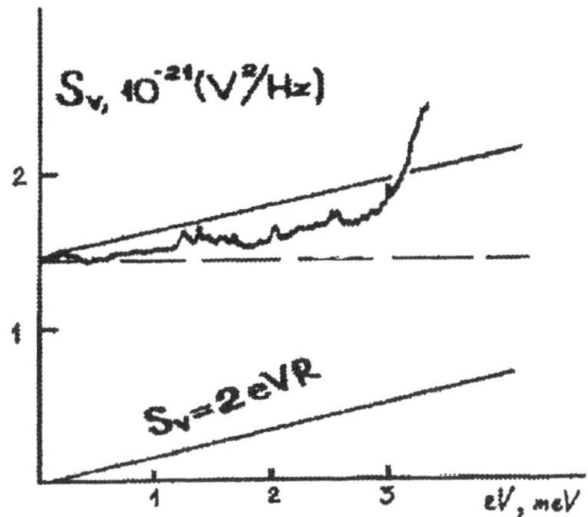

Figure 1.2 Current noise in a Na microcontact at $T = 1.7$ K [9]. Contact was produced by shortening a tunneling barrier between two metallic electrodes with an electric shock creating a small metallic bridge between the electrodes. Taken from Ref. [9].

(3) The superconducting properties of contacts with direct metallic conductivity are controlled by the Andreev reflection [11]. In short narrow constrictions ($d \ll \xi$ where ξ is a superconducting coherence length), the current-phase relation is nonsinusoidal [12]

$$J(\varphi) = G\frac{\pi\Delta}{e}\sin\frac{\varphi}{2}\tanh\frac{\Delta\cos\frac{\varphi}{2}}{2T} \tag{1.3}$$

unlike in the tunneling Josephson junctions, and larger in magnitude than the critical Josephson current at same conductance.

(4) Nonlinearity in the contact conductance arises due to inelastic processes of electron-phonon interaction (EPI) in the narrowest part of constriction where the drift velocity of electrons approaches the velocity of acoustic waves. The derivative of current with respect to voltage is proportional to the density of phonon states (and also to the frequency dependent matrix element of EPI)

$$\frac{dG}{dV}(V) \simeq F(\omega)|_{\omega=eV/\hbar} \tag{1.4}$$

thus providing for the spectroscopy of phonons with microcontacts [3, 4]. An example of the nonlinear current-voltage characteristics of microcontact [14] is shown in Fig. 1.3. Metallic contacts survive quite large voltage biases (say, $eV \sim 100$ mV) at which a small region of metal near the orifice enters into the extreme nonequilibrium, nonthermal state superimposed over the background of the cold lattice.

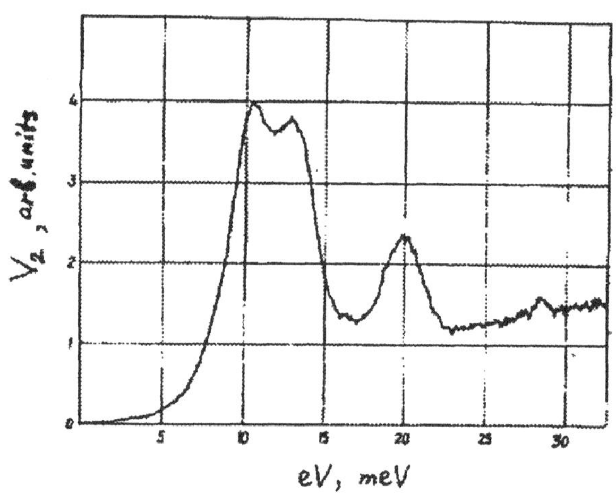

Figure 1.3 Point-contact spectrum of EPI in *Ag* needle-anvil contact at 1.6 K [14]. Second derivative of the $J-V$ characteristics was recorded by measuring the amplitude of the second harmonic, V_2, of the oscillating voltage versus the *d.c.* voltage on the contact, V. Taken from Ref. [28].

(5) In plastically deformed contacts, phonons emitted due to electron scattering reabsorb near the orifice. Since phonon relaxation rate is much slower than the electron relaxation, the nonlinear electron conductivity shows a dispersion at characteristics frequencies [15, 16]

$$\nu_{ph} \simeq \lambda \frac{s}{v_F} \omega_D \sim 10^{10}\ \mathrm{s}^{-1} \tag{1.5}$$

(6) Larger-size contacts, $d \geq 100$ nm, enter the non-ballistic regime of current transport in which ***hot spot*** is formed near the orifice with a high temperature uniquely related to voltage [17]

$$k_B T = 3.63 eV \tag{1.6}$$

resulting in a strong nonlinearity of its $J(V)$ dependence and the transistor effect [18].

The theoretical description of contacts divide them into three categories:

- *Atomic* contacts with the size of the order of few atoms. The mechanism of conduction is described as hopping between atomic sites similar to tight-binding approximation in the theory of solids.

- *Ballistic* microcontacts, those of size larger than the atomic size but smaller than the mean free path of electrons

$$a \ll d \ll l \tag{1.7}$$

Such contacts are treated in a semiclassical approximation using transport theories such as the Boltzmann kinetic equations.

- *Thermal* contacts ($d \gg l$) developing, due to a current concentration, "hot spots" of small size in a very cold steady state environment.

2. ATOMIC CONTACTS

The model of contact [19] assumes regular arrangement of atoms in its narrowest part in the form of two cone-shaped surfaces contacting over a plate with N_t atoms (and possibly making a bridge of length of L atoms), and connected through N_l leads to the thermal reservoirs specified with their respective temperatures (T_i), voltages (V_i) and phases of the order parameter (φ_i) (in case when the contact is formed between superconducting electrodes). Schematic presentation of contact is given in Fig. 1.4*a*.

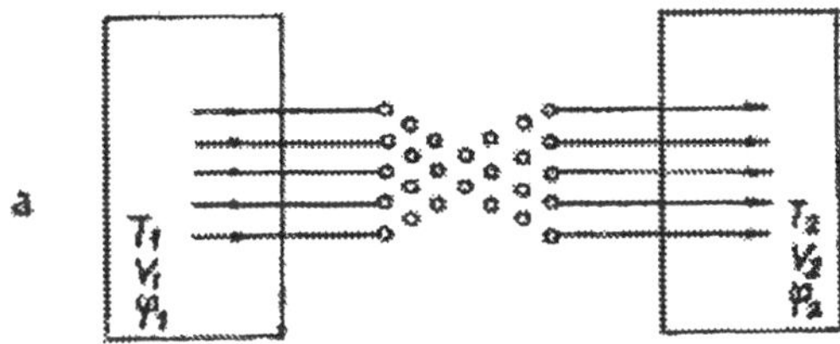

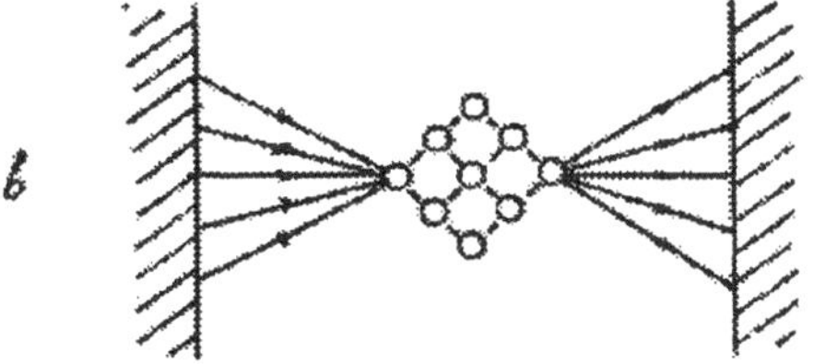

Figure 1.4 Models of the atomic contact with $N = 26$, $N_t = 2$, and $N_l = 5$ (*a*), and of the atomic link with $N = 9$, $N_l = 5$ (*b*).

Perfect contact geometry assumes that the number of leads, N_l, is much larger than the product of the number of the ***transition channels*** N_t to the number of ***conduction channels*** N_c. By the latter we mean, for example s, p_x, p_y, p_z, *etc.* electronic bands, or their hybridized bands. The channels are presented with their respective hopping amplitudes (transfer matrix elements) t_s and the positions of band centers ε_s, $s = 1, ..., N_c$. The Hamiltonian of the junction is

$$H = \sum_{s=1}^{N_c} \left(\varepsilon_s \sum_{i=1}^{N} c_{is}^{+} c_{is} - t_s \sum_{<i,j>} c_{is}^{+} c_{js} \right) + H_{lead} \tag{1.8}$$

where

$$H_{lead} = -t \sum_{k=1}^{N_l} \left(\sum_{n=1}^{\infty} a_{nk}^{+} a_{n+1,k} + \sum_{s=1}^{N_c} a_{1k}^{+} c_{ks} \right) + \text{h.c.}$$
$$-t \sum_{k=1}^{N_l} \left(\sum_{n=1}^{\infty} b_{nk}^{+} b_{n+1,k} + \sum_{s=1}^{N_c} b_{1k}^{+} c_{N-k+1,s} \right) + \text{h.c.} \tag{1.9}$$

The atoms in the central part of contact are numbered from 1 to N (the electron creation operators at atom sites are c_{is}^{+}, $i = 1, ..., N$, $s = 1, ..., N_c$ connected to the left and right leads with the creation operators a_{1k}^{+}, $k = 1, ..., N_l$, and b_{1k}^{+}, $k = N - N_l + 1, ..., N$, respectively). We assume that electrons arrive to the contact through the leads from the left reservoir independently from one another, and are transmitted to the right reservoir after passing the contact with the transit amplitudes $t_{ks,k's'}$. Then, according to Landauer [1] and Imry [20] the contact conductance at $T = 0$ may be expressed as

$$G = G_0 \sum_{k,k'=1}^{N_l} \sum_{s,s'=1}^{N_c} |t_{ks,k's'}|^2. \tag{1.10}$$

Calculation shows the dependence of the conductance on the occupation level (the Fermi energy μ) in metals. Typical dependences $G(\mu)$ are presented in Figs. 1.5 and 1.6. They show that the conductance, although in its magnitude of the order of the conductance quantum, is not exactly equal to or multiple of G_0. The non-monotonic behavior of conductance versus energy is an inevitable consequence of the scattering concept, and follows as a result of quantum reflection at the contact boundary.

Maximal conductance is proportional to the number of conducting channels N_c and also to the number of contacting points (the "transition channels") N_t in the narrowest part of metallic connection

$$G_{max} \leq N_c N_t G_0. \tag{1.11}$$

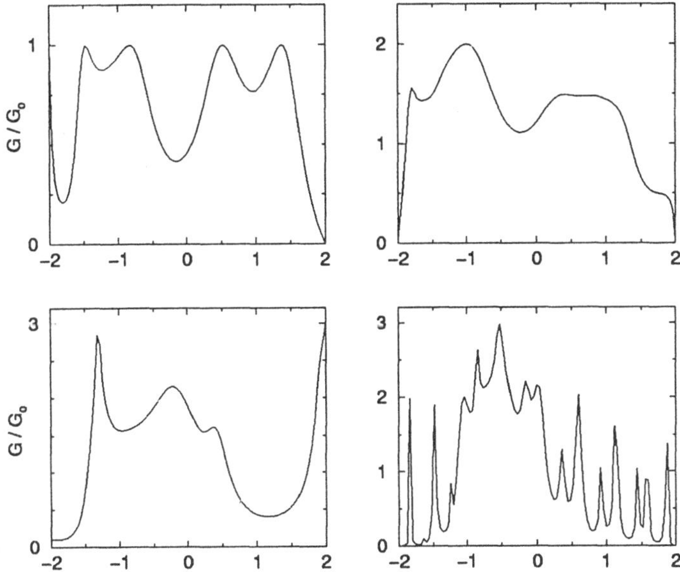

Figure 1.5 Examples of the calculated conductance versus Fermi energy dependences in the atomic contacts with $t = t_s = -1$ and $\varepsilon_s = 0$. Upper left panel: 2d contact with $N_t = 1$, $N_c = 1$, $N_l = 5$, $N = 29$, $L = 0$. Upper right panel: 2d contact with $N_t = 2$, $N_c = 1$, $N_l = 5$, $N = 26$, $L = 0$. Lower left panel: 3d contact with $N_t = 3$, $N_c = 1$, $N_l = 30$, $N = 57$, $L = 0$. Lower right panel: 3d contact with $N_t = 3$, $N_c = 1$, $N_l = 30$, $N = 81$, and the channel between the tips of length $L = 8$.

Conductance is independent of the number of the "lead channels" N_l provided that N_l is larger than $N_c N_t$. These are the conclusions derived from the "rigid" model of the contact which assumes that the electron distribution in the contact area is not subject to variations due to proximity with the bulk electrodes. There is a reason, however, to believe that such variations may take effect.

Consider in particular the contact in the form of a link presented in Fig. 1.4b. Conductance $G(\mu)$ displays sharp peaks (Fig. 1.6) which correspond to the transmittance resonances at the discrete levels in the link. Similar resonances also appear in $G(U)$ dependence where U is the energy shift added to the atoms at the inner sites. If we allow for charge to accumulate in the link, or to deplete from the atoms in the inner block, the Fermi level in the link will level off with one of such resonances with the result that the transmissivity between the left and right electrodes substantially increases which in turn will lower the total system energy. The spontaneous accumulation (depletion) of charge at the link is therefore energetically favorable. We may assume that contact may automatically adjust its Fermi level by accreting (or losing)

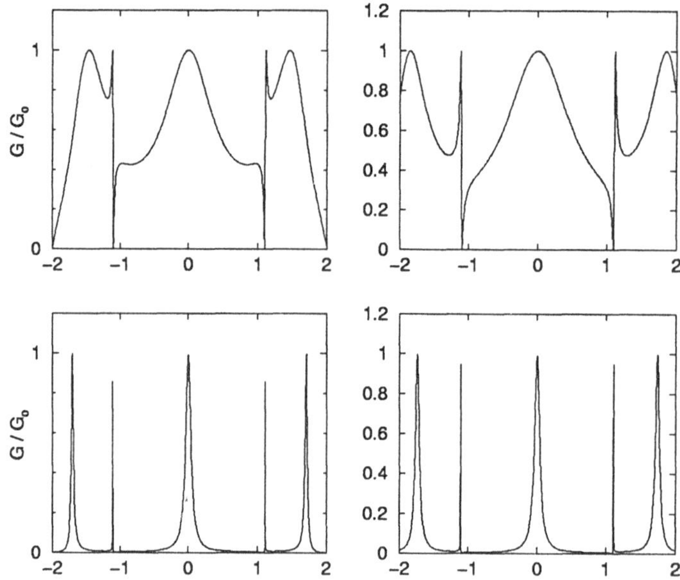

Figure 1.6 Conductance of atomic link with parameters $t = t_s = -1$, $N = 9$, $N_c = 1$ (see Fig. 1.4*b*). Upper panels correspond to $N_l = 3$ (left panel) and $N_l = 30$ (right panel), and show the dependence of conductance on the Fermi energy μ (in units of $|t|$). Lower panels correspond to same values of N_l, and show the dependence of conductance on the energy shift added to atoms in the link, U (in units of $|t|$) at $\mu = 0$.

some charge from bulk metals. Of course, this will cost some energy of charging the link, of the order of e^2/d, which however is less than the energy gain due to increased transmissivity (of order of t) provided that $d \gg a$ and assuming that $|t| \sim e^2/a$.

This is opposite to the Coulomb blockade situation [21] characteristic of weakly coupled granules to the banks ($|t| \ll e^2/a$) in which, because of small $|t|$, the metallic cohesion energy between the granule and the massive electrode is insignificant. We conclude therefore on the possibility of explaining the exact quantization of conductance in contacts which is often found in an experiment, in terms of the self-charging effect of atoms in constriction.

3. BALLISTIC MICROCONTACTS

Contacts with the size of the contact area d much larger than the interatomic spacing can be treated semiclassically by introducing the distribution of electrons in the momentum and coordinate space $f(\mathbf{p}, \mathbf{r})$ and solving for $f(\mathbf{p}, \mathbf{r})$ from the Boltzmann equation. Sharvin [2] as-

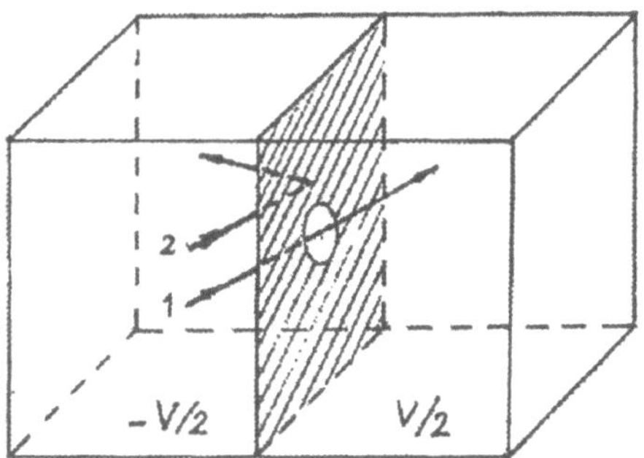

Figure 1.7 Sketch of contact in the form of an orifice in the nontransparent screen. 1) electron trajectory piercing through the orifice, 2) trajectory reflecting from the screen. At the fixed direction of the electron momentum **p**, probability of electron transition between the two boxes is equal to the ratio between surface of the orifice and the surface of screen.

sumed that contact conductance in this case is independent of the mean free path and may be estimated as

$$G \simeq \sigma \frac{d^2}{l} \sim \frac{ne^2 d^2}{p_F} \tag{1.12}$$

where σ is the bulk conductivity and l the mean free path of electron. Since product σl is independent of the mean free path, so the full conductance will be. The calculation of the Sharvin conductance can be achieved with the help of the Landauer formula (1.10), or by using directly the Boltzmann approach [13].

In the Landauer language, we may assume that the probability of electron traversing the impenetrable screen through the circular orifice of surface S in it (Fig. 1.7) is equal to the ratio of S to the total surface of the screen S_0,

$$T = |t|^2 = S/S_0 \ . \tag{1.13}$$

Summation over the states of electron in a box is semiclassically equivalent to integration over $dp_x dp_y$ with a factor $L_x L_y/(2\pi\hbar)^2$ where L_x, L_y are transverse dimensions of the quantization box, thus giving for the conductance

$$G = \frac{2e^2}{h}\frac{S}{S_0}\int_{p_x^2+p_y^2<p_F^2} \frac{L_x L_y dp_x dp_y}{(2\pi\hbar)^2} = \frac{2e^2}{h} N_\perp \tag{1.14}$$

where $N_\perp$ is defined as the number of transverse channels corresponding to the contact area S:

$$N_\perp = \frac{S k_F^2}{4\pi}. \tag{1.15}$$

This is the number of states at the Fermi energy per cross sectional area S. The expression (1.14) clearly complies with the Sharvin conductance (1.12). In the alternative derivation of contact conductance using the Boltzmann approach [13], we calculate the current at the orifice as

$$J = 2eS \int f(\mathbf{p}) v_z \frac{d^3 p}{(2\pi\hbar)^3} \tag{1.16}$$

where f is the distribution function at $z = 0$, by using the expression for the latter

$$f(\mathbf{p}) = f_0\left(\varepsilon_{\mathbf{p}} - \frac{eV}{2}\ \mathrm{sgn} v_z\right). \tag{1.17}$$

Such form is an immediate consequence of the energy conservation on the ballistic electron trajectory entering the orifice from the left box ($z = -\infty$) in case when the velocity of the electron at orifice is positive ($v_z > 0$), or from the right box ($z = +\infty$) if the velocity is negative ($v_z < 0$). f_0 is an equilibrium Fermi distribution $f_0(\varepsilon) = 1/[\exp{(\varepsilon - \mu)/T} + 1]$, V is the voltage difference between metals. Left box and the right box are the two "thermal reservoirs" since at any point inside the box, except at the immediate vicinity of the contact ($|\mathbf{r}| \sim d$), distribution of electrons is the equilibrium one. The electrons with z-component of velocity $v_z > 0$ at $z = 0$ are in equilibrium at $z = -\infty$ where the maximal energy of Fermi distribution equals to $\varepsilon_F + eV/2$ whereas the electrons having z-component of velocity $v_z < 0$ at $z = 0$, arrive from $z = +\infty$ where the maximal energy is $\varepsilon_F - eV/2$. Expanding f in Eq. (1.16) in powers of eV/ε_F, we receive at $V \to 0$ the current at the orifice

$$J = -\frac{2eS}{h^3}\frac{eV}{2}\int \frac{\partial f_0}{\partial \varepsilon_p}|v_z| d^3p = GV \tag{1.18}$$

with the conductance

$$G = \frac{e^2 S S_F}{2(2\pi\hbar)^3} \tag{1.19}$$

S_F is the surface of the Fermi sphere $4\pi p_F^2$. This formula is equivalent to the Landauer expression, Eq. (1.14).

According to the derivation presented, distribution of electrons at the orifice consists ot two electron "beams" moving in opposite directions with maximal energies at the truncated Fermi surface equal to $\varepsilon_F \pm eV/2$ (see Fig. 1.8*a*).

At any point $\mathbf{r}$ away from the orifice, the truncated Fermi surface has same energy shift between two parts, eV, but the parts are inequivalent in size. The electron distribution at point $\mathbf{r}$ equals to

$$f(\mathbf{p}, \mathbf{r}) = f_0[\varepsilon_{\mathbf{p}} + e\phi(\mathbf{r})\ \mathrm{sgn} v_z] \tag{1.20}$$

where $\phi(\mathbf{r})$ is the electrostatic potential at point $\mathbf{r}$, and $\Omega(\mathbf{r})$ is a solid angle showing orifice from point $\mathbf{r}$. By requiring that charge density remains unchanged (the condition of the local neutrality) at any point $\mathbf{r}$, we find the potential distribution

$$\phi(\mathbf{r}) = \frac{V}{2}\left[1 - \frac{\Omega(\mathbf{r})}{4\pi}\right]\ \mathrm{sgn} z. \tag{1.21}$$

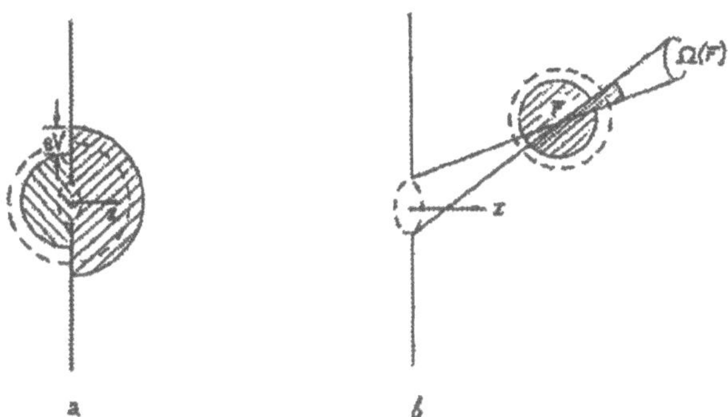

Figure 1.8 Distribution of electrons at the contact surface (a), and at point $\mathbf{r}$ outside the surface (a). $\Omega(\mathbf{r})$ is a solid angle at which the orifice is seen from $\mathbf{r}$. Fermi surface at each point is truncated along the line which is an image of the orifice to the Fermi sphere. The energy difference between two parts of truncated Fermi surface equals at each point to eV.

which along z axis becomes

$$\phi(z) = \frac{V}{2} \frac{z}{\sqrt{z^2 + d^2/4}} \tag{1.22}$$

The voltage continuously changes from $-V/2$ to $V/2$ at distances from the orifice of the order of its diameter d which is much smaller than the mean free path of electron l. Within the distances of order d near the orifice, a strongly nonequilibrium stationary state exists as long as a current is supplied through the contact. Since energy is conserved along the electron trajectory, Joule heat is not released inside the contact and is transferred to the lattice only at distances of order l much away from the orifice.

The Landauer calculation directly relates conductance G to the number of conducting channels inside the contact, $N_\perp$. It was then argued [22] that if the number of transverse channels changes discretely at the increasing contact diameter, so the conductance will do, *i.e.* G will be an integer multiple of the conductance quantum $2e^2/h$. It was assumed that in a smooth contact continuously changing its diameter from infinity to d in the narrowest part, discrete channels will open one by one thus resulting in a conductance quantization $G = nG_0$.

These considerations do not apply directly to the atomic contacts. Subsequent microscopic calculation of the waveguide modes in a finite-size contact of various geometry [23]-[26] showed oscillatory behavior (see Fig. 1.9) as a function of occupation, which however to our knowledge have been never observed. We suggest that the self-focusing behavior

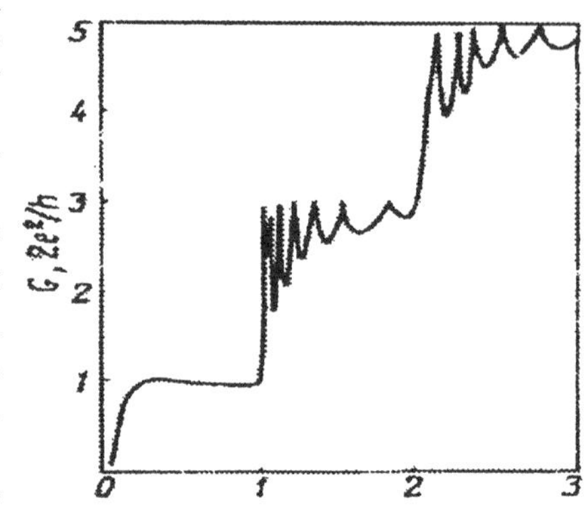

Figure 1.9 Conductance of ballistic contact in a form of cylinder of length L such that $k_F L = 40$, as a function of the parameter $k_F d/2$. Taken from Ref. [27].

of electron concentration near the contact "throat" discussed in page 9 may instead be relevant to the observed discrete G values.

4. INELASTIC SCATTERING AND J–V NONLINEARITY IN SEMICLASSICAL CONTACTS

According to Landauer or Boltzmann theory of ballistic contact conductance, its $J-V$ dependence is linear up to biases of the order of Fermi energy. Introduction of the inelastic scattering of electrons on phonons results in the nonlinearity of the current-voltage characteristics at energy of the order of typical phonon energies [6]. This nonlinearity serves as a tool of the phonon spectroscopy in metals [3, 4, 28, 29, 30] since the nonlinear dependence is directly related to the density of phonon states at voltage bias equal to phonon energy,

$$eV = \hbar\omega. \tag{1.23}$$

To find the nonlinear correction to the contact current, we need to calculate $f_{\mathbf{p}}$ from the Boltzmann equation

$$\frac{\partial f_{\mathbf{p}}}{\partial t} + \frac{\partial \varepsilon_p}{\partial \mathbf{p}}\frac{\partial f_{\mathbf{p}}}{\partial \mathbf{r}} - e\frac{\partial \phi}{\partial \mathbf{r}}\frac{\partial f_{\mathbf{p}}}{\partial \mathbf{p}} = I_{e-ph}\{f_{\mathbf{p}}, N_{\mathbf{q}}\} \tag{1.24}$$

and to find the phonon distribution $N_{\mathbf{q}}$ from

$$\frac{\partial N_{\mathbf{q}}}{\partial t} + \frac{\partial \omega_{\mathbf{q}}}{\partial \mathbf{q}}\frac{\partial N_{\mathbf{q}}}{\partial \mathbf{r}} = I_{ph-e}(N_{\mathbf{q}}, f_{\mathbf{p}}) \tag{1.25}$$

where I_{e-ph} and I_{ph-e} are the electron-phonon and phonon-electron collision integrals

$$I_{e-ph} = \sum_{\mathbf{q}} W_{\mathbf{q}}\{[f_{p+q}(1-f_p)(N_q+1) - f_p(1-f_{p+q})N_q]\delta(\varepsilon_{p+q} - \varepsilon_p - \omega_q)$$

$$+[f_{p-q}(1-f_p)N_q - f_p(1-f_{p-q})(N_q+1)]\delta(\varepsilon_{p-q} - \varepsilon_p + \omega_q)\} \quad (1.26)$$

and

$$I_{ph-e} = 2W_{\mathbf{q}}\sum_{\mathbf{p}}[f_{p+q}(1-f_p)(N_q+1) - f_p(1-f_{p+q})N_q]\delta(\varepsilon_{p+q} - \varepsilon_p - \omega_q) \quad (1.27)$$

$W_{\mathbf{q}} = (2\pi\hbar)|M_{\mathbf{q}}|^2$ where $M_{\mathbf{q}}$ is the matrix element of electron-phonon interaction.

To find the nonlinear correction to current, we solve Eqs. (1.24) and (1.25) to first order in the collision integral which are in effect the first corrections in the ballistic small parameters d/l_{e-ph} and d/l_{ph-e} where l_{e-ph} and l_{ph-e} are the electron-phonon and phonon-electron mean free paths, respectively. In the nonequilibrium state, the mean free paths are defined as

$$\frac{1}{l_{e-ph}(\varepsilon, T)} = \frac{2\pi}{v_F}\int_0^{\omega_m}(2N_\omega + 1 + f_{\varepsilon+\omega} - f_{\varepsilon-\omega})\alpha^2(\omega)F(\omega)d\omega \quad (1.28)$$

and

$$\frac{1}{l_{ph-e}(\omega, T)} = \frac{4\pi}{v_F}N(\varepsilon_F)\omega\alpha^2(\omega) \quad (1.29)$$

where $N(\varepsilon)$ and $F(\omega)$ are the electron and phonon densities of states, and $\alpha^2(\omega)$ is the square of the matrix element of electron-phonon interaction averaged over the Fermi surface. The product

$$g(\omega) = \alpha^2(\omega)F(\omega) \quad (1.30)$$

is known as a function of electron-phonon interaction (the Eliashberg function) and is defined as

$$g(\omega) = \frac{1}{(2\pi)^3}\int\frac{dS_p}{v_p}\frac{dS_{p'}}{v'_p}W_{\mathbf{p}-\mathbf{p}'}\delta(\omega - \omega_{\mathbf{p}-\mathbf{p}'})/\int\frac{dS_p}{v_p} \quad (1.31)$$

(integration is running over the Fermi surface, $v_{\mathbf{p}} = |\partial\varepsilon_p/\partial\mathbf{p}|$ is electron velocity at $\varepsilon = \varepsilon_F$). At $T = 0$ and at energy equal to the Debye energy, mean free paths can be estimated as

$$l_{e-ph} \sim l_{ph-e} \sim \frac{v_F}{\lambda\omega_D} \quad (1.32)$$

where λ is a dimensionless electron-phonon coupling constant

$$\lambda = 2\int_0^\infty g(\omega)\frac{d\omega}{\omega}. \quad (1.33)$$

Typically, $\lambda \sim 0.1-1$ in most metals, therefore both the electron-phonon and the phonon-electron mean free paths are of order of $10 - 100$ nm at

$\varepsilon \sim \hbar\omega_D$, whereas the electron-phonon and phonon-electron scattering frequencies differ by 3 order of magnitude:

$$\tau_{e-ph}^{-1} \sim 10^{13}\ \mathrm{s}^{-1}, \quad \tau_{ph-e}^{-1} \sim 10^{10}\ \mathrm{s}^{-1}. \tag{1.34}$$

Solving Eq. (1.24) perturbatively to first order in d/l_{e-ph}, we receive for the correction to the distribution function an expression

$$f_1 = e\phi_1 \frac{\partial f_0}{\partial \varepsilon_p} + \int_{-\infty}^{0} I_{e-ph}(\mathbf{p}(t), \mathbf{r}(t))dt \tag{1.35}$$

where ϕ_1 is a correction to the electrostatic potential. $\mathbf{p}(t)$ and $\mathbf{r}(t)$ are the momentum and the coordinate at electron trajectory at time t. At $eV \ll \varepsilon_F$, the trajectory is a straight line arriving at time $t = 0$ to point $\mathbf{r}$ from $-\infty$ or from $+\infty$ at $t = -\infty$, depending on the direction of the electron velocity $\mathbf{v}$. The potential can be found from the electroneutrality $< f_1 >= 0$. The first order correction to the current

$$J_1 = 2e \int dxdy \int \frac{d^3p}{h^3} v_z f_1(\mathbf{p}, \mathbf{r}). \tag{1.36}$$

is received finally in the form [6, 30]

$$J_1 = -\frac{2e\Omega_{eff}}{(2\pi)^6} \int_0^\infty d\omega L(\omega, eV, T) \int \frac{dS_p}{v_p} \int \frac{dS_{p'}}{v'_p} K(\mathbf{v}, \mathbf{v}') W_{\mathbf{p}-\mathbf{p}'} \delta(\omega - \omega_{\mathbf{p}-\mathbf{p}'}) \tag{1.37}$$

where

$$L(\omega, \varepsilon, T) = M(\omega, \varepsilon) - M(\omega, -\varepsilon), \quad M(\omega, \varepsilon) = \frac{(\omega - \varepsilon)(e^{\varepsilon/T} - 1)}{[1 - e^{(\varepsilon - \omega)/T}](e^{\omega/T} - 1)} \tag{1.38}$$

where $\Omega_{eff} = d^3/3$ is an effective volume near the orifice in which nonequilibrium phonons are emitted by "hot" electrons. Backscattering of electrons is the cause of such emission and serves to the decrease the of electron current.

At fixed phonon frequency, $J - V$ curve changes its slope at $eV = \hbar\omega$ (Fig. 1.10), while the second derivative of J with respect to V acquires a negative peak. For the continuous distribution of phonons on frequency, $F(\omega)$, the derivative of conductance with respect to voltage takes form

$$G^{-1}\frac{dG}{dV} = -\frac{8ed}{3\hbar v_F} \int_0^\infty g_c(\omega) \chi\left(\frac{\omega - eV}{T}\right) \frac{d\omega}{T} \tag{1.39}$$

at finite temperature T, and

$$G^{-1}\frac{dG}{dV} = -\frac{8ed}{3\hbar v_F} g_c(eV) \tag{1.40}$$

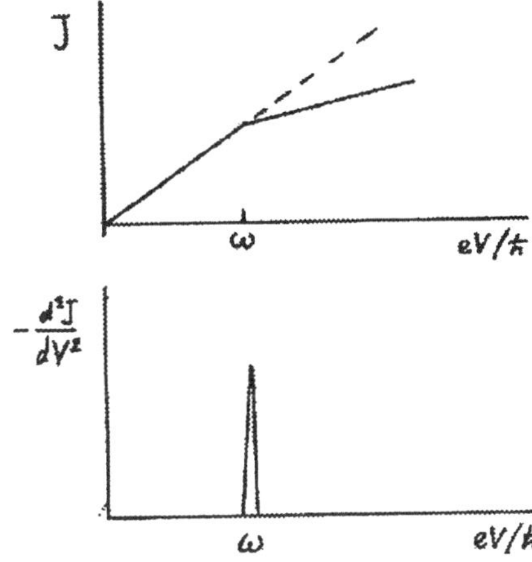

Figure 1.10 $J-V$ characteristics (*a*) and its second derivative (*b*) for a contact with fixed phonon frequency ω.

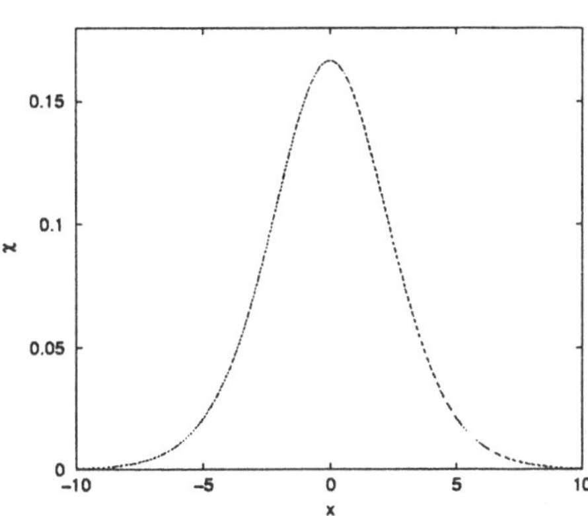

Figure 1.11 Temperature broadening of the phonon spectrum. The linewidth at the half height equals to 5.44.

at $T = 0$. $\chi(x)$ is the temperature broadening function (Fig. 1.11)

$$\chi(x) = \frac{d^2}{dx^2}\left(\frac{x}{e^x - 1}\right) \tag{1.41}$$

and $g_c(\omega)$ is the transport function of electron-phonon interaction

$$g_c(\omega) = \frac{1}{(2\pi)^3}\int \frac{dS_p}{v_p}\frac{dS_{p'}}{v'_p} K(\mathbf{v},\mathbf{v}')W_{\mathbf{p}-\mathbf{p}'}\delta(\omega - \omega_{\mathbf{p}-\mathbf{p}'})/\int \frac{dS_p}{v_p} \tag{1.42}$$

which differs from the Eliashberg function (1.31) in a an additional form factor, the so called K-factor, $K(\mathbf{v},\mathbf{v}')$, taking into consideration the kinematic restrictions on electron scattering at the orifice. In the case of circular orifice

$$K(\mathbf{v},\mathbf{v}') = \frac{4|v_z v_{z'}|\theta(-v_z v_{z'})}{|v_z\mathbf{v}' - v'_z\mathbf{v}|}. \tag{1.43}$$

where $\theta(x)$ is a step function, $\theta(x) = 1$ at $x = 0$ and $\theta(x) = 0$ at $x < 0$.

The function $K(\mathbf{v},\mathbf{v}')$ is singular at $\mathbf{v}' = -\mathbf{v}$ (for the reverse scattering) but since the singularity is an integrable one, it does not much affect the shape of $g_c(\omega)$ as compared to the isotropic EPI function $g(\omega)$ (mention that K in Eq. (1.43) is normalized to unity, $< K >_{FS} = 1$). For a spherical Fermi surface with the matrix element of EPI depending only on the transfer momentum $\mathbf{q} = \mathbf{p}' - \mathbf{p}$, the important is the dependence of K on the scattering angle θ

$$q = |\mathbf{p}' - \mathbf{p}| = 2p_F \sin\frac{\theta}{2}, \tag{1.44}$$

and on the angle γ between the direction of $\mathbf{q}$ and the normal $\mathbf{n}$ to metal surface

$$\gamma = \arccos(\mathbf{q}, \mathbf{n}).$$

The integration over the other angles gives

$$K(\theta,\gamma) = \frac{2}{\pi\sin\theta}\int_{\varphi_0}^{\pi/2} \frac{\sin^2\frac{\theta}{2}\cos^2\gamma - \cos^2\frac{\theta}{2}\sin^2\gamma\cos^2\varphi}{(\cos^2\varphi + \cos^2\gamma\sin^2\varphi)^{1/2}} d\varphi \tag{1.45}$$

where $\varphi_0 = \arccos(\tan\frac{\theta}{2}/\tan\gamma)\theta(\gamma - \theta/2)$ (in Fig. 1.12 we present a 3d plot of $K(\theta,\gamma)$). Some authors (see [31]) further integrate K over γ to receive

$$\bar{K}(\theta) = \frac{1}{2}\left(1 - \frac{\theta}{\tan\theta}\right). \tag{1.46}$$

The singularity mentioned above is at $\theta = \pi$.

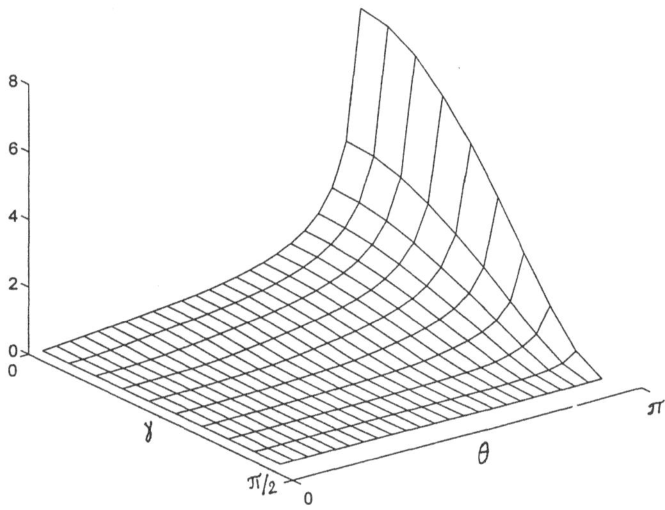

Figure 1.12 3d plot of K-factor $K(\theta,\gamma)$ in a circular orifice.

The above results have been generalized to the models of contact of various geometry (the orifice, the channel, *etc.*) [32] and to scattering conditions concerning elastic (impurity) scattering [33]. Such scattering in itself does not lead to the nonlinearity of the $J(V)$ but decreases the phonon nonlinear part of d^2J/dV^2 as compared to the perfect ballistic regime. In a ***diffusive*** contact,

$$l_i \ll d \ll l_{ph-e}, \tag{1.47}$$

scattering by phonons can be calculated as an effect proportional to $(l_i d)^{1/2}/l_{e-ph}$, where l_i is the elastic (impurity) scattering length of electron. The nonlinear part of $J(V)$ can in a diffusive contact be presented in the same form as in a ballistic contact, with an appropriate K-factor. In Table. 1.1, K-factors are listed for some contact geometries.

Table 1.1 K-factors of contacts with various geometries. d is the diameter of the orifice or cylindrical channel, L is the length of the channel. $\mathbf{n}$ is a unit vector in the direction of electron velocity $\mathbf{v}$, l_i is the elastic mean free path of electron, l_ε is the inelastic (electron-phonon) mean free path.

Contact geometry	K-factor	Parameters
Orifice in clean metal	$4\lvert n_z n'_z\rvert\theta(-n_z n'_z)/\lvert n_z\mathbf{n}' - n'_z\mathbf{n}\rvert$	$d \ll l_i$
Orifice in dirty metal	$3[(\mathbf{n}_\perp - \mathbf{n}'_\perp)^2 + 2(n_z - n'_z)^2]/8$	$l_i \ll d \ll \sqrt{l_i l_\varepsilon}$
Clean channel	$2\theta(-n_z n'_z)$	$L \gg d$
Dirty channel	$3(n_z - n'_z)^2/2$	$d, l_i \ll L \ll \sqrt{l_i l_\varepsilon}$

Another example of inelastic scattering is associated with the localized lattice defects, the so called two-level systems (TLS) [34, 35]. The general treatment of inelastic scattering by TLS is similar to that of phonons, except that the population of TLS is strongly eV- dependent and in itself contribute to the lineshape of the point-contact spectrum. Fig. 1.1 shows this nonsymmetric lineshape of the TLS's point-contact spectra at various temperatures.

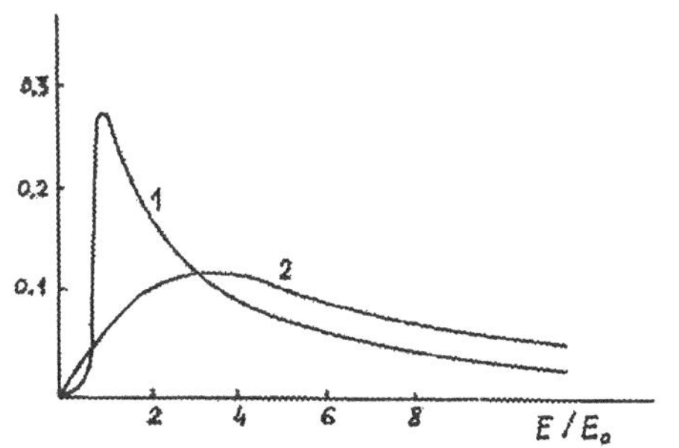

Figure 1.13 Two-level-system point-contact spectrum lineshape at $T/E_0 = 0.1$ (line 1) and at $T/E_0 = 0.5$ (line 2). Taken from Ref. [34].

5. PHONON TRAPPING AND RELAXATION

Inelastic events of electron-phonon interaction result in the phonon emission from the narrowest part of the contact. Since the phonon mean free path is much larger than the contact dimension, the phonons leave the contact and release their energy away from the nonequilibrium part of the junction. In plastically deformed constrictions, however, phonons can be scattered back and reabsorbed near the orifice. The nonequilibrium phonon gas with an effective temperature T^* much larger than the ambient temperature T is then formed near the region of nonequilibrium electrons.

The nonequilibrium phonons increase the electron scattering near the orifice and produce an additional nonlinearity in the $J-V$ curve, in particular the nonzero value of d^2J/dV^2 at voltage larger than the maximal phonon energy. Such ***background*** point-contact spectroscopy signals are often observed in microcontacts of the needle-anvil geometry [36].

The second derivative of the $J-V$ characteristics of microcontact in the phonon reabsorption regime can be presented in the form [15]

$$G^{-1}\frac{dG}{dV} = -\frac{8ed}{3\hbar v_F}[g_c(eV) + B(eV)] \tag{1.48}$$

where the background part, $B(eV)$, is presented as

$$B(\varepsilon) = 2\frac{d^2}{d\varepsilon^2}\left\{\varepsilon \int_0^\infty \frac{g(\omega)d\omega}{e^{\omega/T^*}-1}\right\}. \tag{1.49}$$

The effective temperature of the nonequilibrium phonons T^* is found from the equation of energy balance

$$\int_0^{eV}(eV-\omega)\tilde{g}(\omega)\omega d\omega = 2\int_0^\infty \frac{g(\omega)\omega^2 d\omega}{e^{\omega/T^*}-1} \tag{1.50}$$

in which the function $\tilde{g}(\omega)$ differs from the conventional contact EPI function $g(\omega)$ with an additional factor $\theta(-p_z p_z')$ corresponding to integration over the half of the Fermi sphere. As an approximation, we assume then that $\tilde{g}(\omega) \sim (1/2)g(\omega)$. By introducing a factor η such that

$$\int_0^{eV}(eV-\omega)\tilde{g}(\omega)\omega d\omega = \eta\int_0^\infty \frac{g(\omega)\omega^2 d\omega}{e^{\omega/T^*}-1} \tag{1.51}$$

we receive

$$T^* \sim eV/\eta \tag{1.52}$$

with $\eta \sim 4$. At the bias energy eV much above the phonon spectrum,

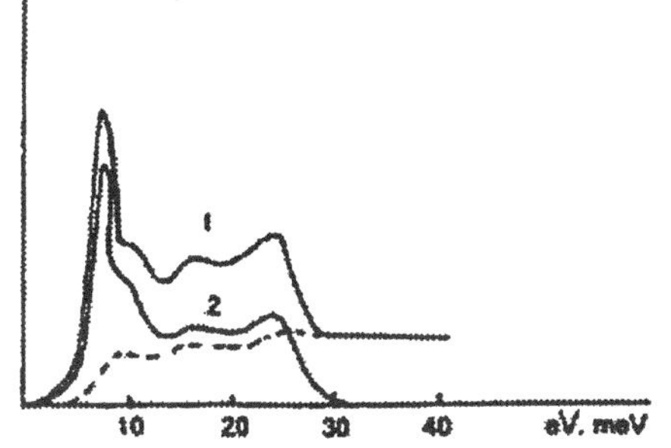

Figure 1.14 Point-contact spectrum of Zn contact [35] (line 1) decomposed into its spectral part (line 2) and to the background contribution (dashed line) [14].

$eV \gg \hbar\omega_D$, Eq. (1.49) gives

$$B(\infty) = \frac{4}{\eta}\int_0^{\infty} g(\omega)\frac{d\omega}{\omega} = \frac{2\lambda}{\eta}. \tag{1.53}$$

λ is the EPI coupling parameter, Eq. (1.33).

The Fig. 1.14 shows, as an example, the point contact spectrum (the derivative $-G^{-1}dG/dV(V)$) of a dirty Zn contact with a relatively large background [37], together with the EPI interaction function $g_c(\omega)$ received by inverting the integral equation (1.48). The latter have the unexpected property, namely, the strong frequency dependence with respect to frequency of current modulation used in the PC spectroscopy measurements [28]. The origin of frequency dependence is related to relaxation of nonequilibrium phonons. Trapped phonons have relaxation frequency of the order of the phonon-electron relaxation rate $\nu_{ph} \sim \tau_{ph-e}^{-1} \sim 10^{10}\text{s}^{-1}$. Trapping and desorbing of phonons is a relatively slow process as compared to the characteristic electron-phonon relaxation frequencies $\nu_{e-ph} \sim 10^{13}\text{s}^{-1}$.

Inelastic part of the current with trapped phonons is presented as

$$J_1(V) = \frac{8ed}{3\hbar v_F}G(0)\int_0^{\infty} d\omega\, g(\omega)\left[\frac{\omega+eV}{e^{(\omega+eV)/T}-1} - \frac{\omega-eV}{e^{(\omega-eV)/T}-1} - 2eVN(\omega)\right] \tag{1.54}$$

where the last term takes into account the effect of trapped phonons. At zero ambient temperature, the phonon distribution takes the simple form

$$N_\omega = \frac{eV-\omega}{\eta(\omega+\omega_0)}\theta(eV-\omega) \tag{1.55}$$

where ω_0 is the phonon escape frequency introduced in [15]. The last term in Eq. (1.54) (a PC background) is shown to depend on frequency of the external signal applied to the contact, $V = V_0 + V_1\cos\omega t$, as

$$< J_1 > = -\frac{2e^2V_1^2\eta}{\hbar v_F}G(0)\int_0^{\infty} g(\nu)\frac{\tau_{ph-e}^{-2}(\nu)}{\tau_{ph-e}^{-2}(\nu)+\omega^2}\frac{d\nu}{\nu} \tag{1.56}$$

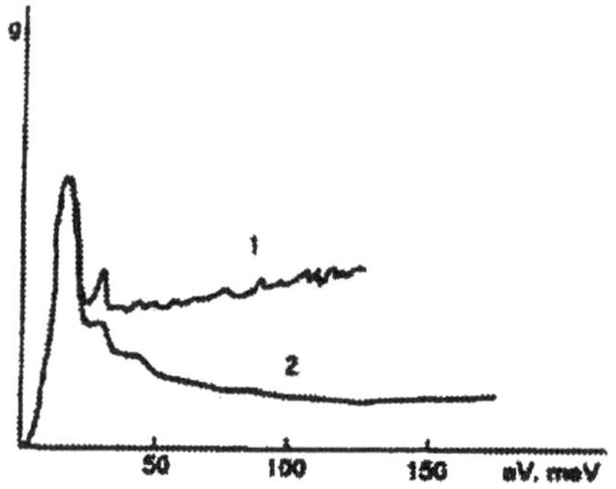

Figure 1.15 Point-contact spectra of Cu contact [15] measured at frequency f=3 kHz (line 1) and at frequency f=80 GHz (line 2). Taken from Ref. [15].

and shows the decrease above the cutoff ω_c of the order of the phonon-electron relaxation rate ν_{ph-e} [16] (Fig. 1.15).

6. THERMAL CONTACTS AND HOT SPOTS

In a thermal contact, electron and phonon mean free paths are smaller than the contact diameter d. Therefore, the lattice and electrons stay in equilibrium between themselves, but the temperature of this local equilibrium $T(\mathbf{r})$ is much higher than the ambient temperature of the environment. The distributions of temperature and electrostatic potential $\phi(\mathbf{r})$ are found from the equation of the energy balance

$$-\text{div}\mathbf{q} + \mathbf{jE} = 0, \quad \mathbf{q} = -\kappa\nabla T, \quad \mathbf{j} = \sigma\mathbf{E} \tag{1.57}$$

where the local thermal conductivity $\kappa = \kappa(T(\mathbf{r}))$ and electrical conductivity $\sigma = \sigma(T(\mathbf{r}))$. Equations (1.57) are solved in a circular orifice with transformation to the oblate spheroidal coordinates ($a = d/2$)

$$\begin{aligned} x &= a \sin u \cosh v \cos\varphi \\ y &= a \sin u \cosh v \sin\varphi \\ z &= a \cos u \sinh v\,. \end{aligned} \tag{1.58}$$

Eqs. (1.57) reduce to

$$\text{div}(\sigma\nabla\phi) = 0, \quad \frac{\pi^2 k_B^2}{6e^2}\text{div}[\sigma\nabla(T^2)] + \sigma(\nabla\phi)^2 = 0. \tag{1.59}$$

Assuming further the applicability of the Wiedemann-Franz law relating κ to σ,

$$\frac{\kappa}{\sigma T} = \frac{\pi^2 k_B^2}{3e^2} \tag{1.60}$$

and solving Eq. (1.59) in spheroidal coordinates, we receive potential and temperature distributions with no adjustable parameters

$$\phi(\mathbf{r}) = \frac{V}{2}\left(-1 + \frac{4}{\pi}\arctan e^{v}\right), \tag{1.61}$$

$$T(\mathbf{r}) = \frac{eV\sqrt{3}}{2\pi}\left[1 - \left(1 - \frac{4}{\pi}\arctan e^{v}\right)^{2}\right]^{1/2} \tag{1.62}$$

The temperature T_0 at the contact center, $v = 0$, is related to the applied voltage according to formula

$$eV = \frac{2\pi}{\sqrt{3}}k_B T_0 = 3.63\ k_B T_0 \tag{1.63}$$

and is very much larger than the temperature in bulk.

Current-voltage relationship is strongly nonlinear and takes universal form

$$J(V) = G(0)V\int_0^1 \sigma_{red}\left(\frac{eV\sqrt{3}}{2\pi}\sqrt{1-x^2}\right)dx \tag{1.64}$$

where $G(0)$ is a linear conductance $G(V \to 0)$ and $\sigma_{red}(T)$ is the reduced conductivity $\sigma_{red}(T) = \sigma(T)/\sigma(0)$. The current-carrying state in the thermal contact is of some interest, in particular with respect to possible application as a fast nonlinear switch or transistor [18, 38].

References

[1] R. Landauer, Electrical resistance of disordered one-dimensional lattices, Phil. Mag. **21**, 863 (1970).

[2] Yu. V. Sharvin, Possible method of Fermi surface investigation. Sov. Phys. JETP **21**, 655 (1965).

[3] I. K. Yanson, Nonlinear effects in the electrical conductance of point contacts and electron-phonon interaction in normal metals, Zh. Eksp. Teor. Fiz. **66**, 1035 (1974) [Sov. Phys. JETP **39**, 506 (1974)].

[4] A. G. M. Jansen, A. P. Van Gelder, and P. Wyder. Point-contact spectroscopy in metals. Journ. Phys. **F13**, 6073 (1980).

[5] B. J. Van Wees, H. Van Houten, C. W. J. Beenakker, J. G. Williamson, L. P. Kouwenhowen, D. Van Der Marel, and C. T. Foxon. Quantized conductance of point contacts in a two-dimensional electron gas. Phys. Rev. Lett. **60**, 848 (1988).

[6] I. O. Kulik, A. N. Omelyanchouk and R. I. Shekhter, Electrical conductivity of point microbridges and phonon and impurity spectroscopy in normal metals, Fiz. Nizk. Temp. **3** 1543 (1977) [Sov. J. Low Temp. Phys. **3**, 740 (1977)]; I. O. Kulik, R. I. Shekhter and A. N. Omelyanchouk, Electron-phonon coupling and phonon generation in normal metal microbridges, Sol. St. Commun., **23**, 301 (1977).

[7] J. M. Krans, J. M. Van Ruitenbeek, V. V. Fisun, I. K. Yanson and L. J. De Jongh, The signature of conductance quantization in metallic point contacts, Nature **375**, 767 (1995).

[8] I. O. Kulik and A. N. Omelyanchouk, Nonequilibrium fluctuations in normal-metal point contacts, Fiz. Nizk. Temp. **10** 305 (1984) [Sov. J. Low Temp. Phys. **10**, 158 (1984)].

[9] A. I. Akimenko, A. B. Verkin, and I. K. Yanson. Point-contact noise spectroscopy of phonons in metals. J. Low Temp. Phys. **54**, 247 (1984).

[10] M. J. M. De Jong, and C. W. J. Beenakker. Shot noise in mesoscopic systems, in: Mesoscopic Electron Transport, p.225. Eds. L. P. Sohn, L. P. Kouwenhowen, and G. Schoen. Kluwer, 1997.

[11] I. O. Kulik, Macroscopic quantization and the proximity effect in SNS junctions, Zh. Eksp. Teor. Fiz. **57**, 1745 (1969) [Sov. Phys. JETP **30**, 944 (1969)].

[12] I. O. Kulik, and A. N. Omelyanchouk, Contribution to the microscopic theory of the Josephson effect in superconducting bridges, Zh. Eksp. Teor. Fiz. Pis'ma **21**, 216 (1975) [JETP Lett. **21**, 96 (1975)]; I. O. Kulik and A. N. Omelyanchouk, Properties of superconducting microbridges in the pure limit, Fiz. Nizk. Temp. **3** 945 (1977) [Sov. J. Low Temp. Phys. **3**, 459 (1978)].

[13] A. N. Omelyanchouk, I. O. Kulik, R. I. Shekhter. Contribution to the theory of nonlinear effects in the electric conductivity of metallic junctions. JETP Lett, **35**, 437 (1977).

[14] Yu. G. Naydiuk, I. K. Yanson, A. A. Lysykh, O. I. Shklyarevskii. Electron-phonon interaction in microcontacts of Au and Ag, Fiz. Nizk. Temp. **8**, 922 (1982).

[15] I. O. Kulik, Nonequilibrium current-carrying states in metallic point contacts, Fiz. Nizk. Temp. **11** 937 (1985) [Sov. J. Low Temp. Phys. **11**, 516 (1985)].

[16] I. K. Yanson, O. P. Balkashin and Yu. A. Pilipenko, Relaxation of nonequilibrium phonons in metallic point contacts, Zh. Eksp. Teor. Fiz. Pis'ma **41**, 304 (1985) [JETP Lett. **41**, 372 (1985)].

[17] B. I. Verkin, I. K. Yanson, I. O. Kulik, O. I. Shklyarevskii, A. A. Lysykh, and Yu. G. Naydyuk. Singularities in d^2V/dI^2 dependences of point contacts between ferromagnetic metals. Sol. St. Commun. **30**, 215 (1979); I. O. Kulik. On the determination of $\alpha^2F(\omega)$ in metals by measuring $I-V$ characteristics of "wide" (non-ballistic) point-contact spectra. Phys. Lett. **106A**, 187 (1984).

[18] I. O. Kulik. Nonlinear four-terminal microstructures: A hot-spot transistor. J. Appl. Phys. **76**, 1920 (1994).

[19] J. C. Cuevas, A. Levy Yeyati, and A. Martin-Rodero. Microscopic origin of the conducting channels in metallic atomic-size contacts. Phys. Rev. Lett. **80**, 1066 (1998).

[20] Y. Imry. Physics of mesoscopic systems, in: Directions in Condensed Matter Physics, p.101. Eds. G. Grinstein and G. Mazenko. World Scientific, Singapore, 1986.

[21] I. O. Kulik and R. I. Shekhter, Kinetic phenomena and charge discretness effects in granulated media, Zh. Eksp. Teor. Fiz. **68**, 623 (1975) [Sov. Phys. JETP **41**, 308 (1975)].

[22] L. I. Glazman, G. B. Lesovik, D. E. Khmelnitskii, and R. I. Shekhter. Reflection-free quantum transport and the fundamental jumps of ballistic resistance in microconstrictions. JETP Lett. **48**, 238 (1988).

[23] A. Szafer, and A. D. Stone. Theory of quantum conduction through a constriction. Phys. Rev. Lett. **60**, 300 (1989).

[24] E. Tekman, and S. Ciraci. Theoretical study of transport through a quantum point contact. Phys. Rev. **B43**, 7145 (1991).

[25] A. G. Scherbakov, E. N. Bogachek, and U. Landman. Quantum electronic transport through three-dimensional microconstrictions with variable shapes. Phys. Rev. **B53**, 4054 (1996).

[26] E. N. Bogachek, A. M. Zagoskin and I. O. Kulik, Conductance jumps and magnetic flux quantization in ballistic point contacts, Fiz. Nizk. Temp. **16** 1404 (1990) [Sov. J. Low Temp. Phys. **16**, 796 (1990)].

[27] A. M. Zagoskin, and I. O. Kulik. Quantum oscillations of the electrical conductivity of two-dimensional ballistic contacts. Sov. J. Low Temp. Phys. **16**, 911 (1990).

[28] I.K. Yanson, and A. V. Khotkevich. Atlas of Point Contact Spectra of Electron- phonon Interaction in Metals (in Russian), Naukova Dumka, Kiev, 1986; Engl. transl.: Kluwer Acad. Publ., 1995.

[29] A. M. Duif, A. G. M.Jansen, and P. Wyder. Point-contact spectroscopy. J. Phys. **C1**, 3157 (1989).

[30] I. O. Kulik, Ballistic and non-ballistic regimes in point-contact spectroscopy, Fiz. Nizk. Temp. **18** 450 (1992) [Sov. J. Low Temp. Phys. **18**, 302 (1992)].

[31] A. P. Van Gelder, On the structure of the d^2I/dV^2 characteristics of point contacts between metals, Sol. St. Commun. **35**, 19 (1980).

[32] M. Ashraf, and J. C. Swihart, Calculated point contact spectra of sodium and potassium, Phys. Rev. B25, 2049 (1982).

[33] I. O. Kulik, R. I. Shekhter and A. G. Shkorbatov, Point-contact spectroscopy of electron-phonon coupling in metals with a small electron mean free path, Zh. Eksp. Teor. Fiz. **81**, 2126 (1981) [Sov. Phys. JETP **54**, 1130 (1981)].

[34] V. I. Kozub, and I. O. Kulik, Microcontact spectroscopy of population of two-level systems, Zh. Eksp. Teor. Fiz. **91**, 2243 (1981) [Sov. Phys. JETP **64**, 1332 (1986)].

[35] R. J. P. Keijsers, O. I. Shklyarevskii, and H. Van Kempen, Point-contact study of fast and slow two-level fluctuators in metallic glasses, Phys. Rev. Lett. **77**, 3411 (1996).

[36] I. K. Yanson, I. O. Kulik, A. G. Batrak, Point-contact spectroscopy of electron-phonon interaction in normal metal single crystals, J. Low Temp. **42**, 527 (1981).

[37] I. K. Yanson, Point-contact spectroscopy of electron-phonon interaction in Zn and Cd, Fiz. Nizk. Temp. **3** 1516 (1977) [Sov. J. Low Temp. Phys. **3**, 726 (1977)].

[38] R. Ellialtıoğlu and İ. İ. Kaya, Conductance in metallic submicron cross-junctions, Chapter 33 of this volume, p. 479.

Chapter 2

CONDUCTANCE CHANNELS OF GOLD ATOMIC-SIZE CONTACTS

E. Scheer
Physikalisches Institut, Universität Karlsruhe
D-76128 Karlsruhe, Germany

W. Belzig
Department of Applied Physics & DIMES, Delft University of Technology
Lorentzweg 1, NL-2628 CJ Delft, The Netherlands

M. H. Devoret, D. Esteve and C. Urbina
Service de Physique de l'Etat Condensé Commissariat à l'Energie Atomique
Saclay, F-91191 Gif-sur-Yvette Cedex, France

1. INTRODUCTION

Since the development of the scanning tunneling microscope (STM) [1] it is not only possible to see, but also to manipulate and to measure the transport properties of individual atoms on surfaces [2]. By energy dependent measurements of the differential conductance a certain chemical information can be achieved [3]. The challenging aim of building up electronic circuits atom by atom with tailor-made properties, however, would require the detailed knowledge of the relation between the physical and chemical properties of the respective atoms and their conduction properties, a problem which has been addressed by different methods during the last years [4, 5, 6]. The most simple system for all investigations - including the present - is a one-atom contact between two metallic banks of the same element.

Electrical transport through such contacts is suitably described by the Landauer formalism, which treats it as a wave scattering problem.

I. O. Kulik and R. Ellialtioğlu (eds.),
Quantum Mesoscopic Phenomena and Mesoscopic Devices in Microelectronics, 27–34.

The transport properties of the contact connected to the leads are described by a set of transmission coefficients $\{T_n\}$ with $T_n \leq 1$. *E.g.* the conductance G of a contact is $G = G_0 \sum_n T_n$. Here, $G_0 = 2e^2/h$ is the conductance quantum [7]. Since in few-atom metallic point contacts the structure size is of the same order as the Fermi wavelength, such a contact has only a small number N of channels. If the set $\{T_n\}$ is known, all further transport properties, as *e.g.* shot noise, thermopower or superconducting properties of the contact can be deduced.

In order to test basic concepts of the quantum mechanical transport theory, it is of particular interest to study metallic systems transmitting only one single channel with adjustable transmission coefficient. According to a quantum chemical model by Cuevas, Levy Yeyati and Martín-Rodero (CLM) [8, 9] the transmission coefficients of single-atom contacts are a function of the chemical properties of the metal and the atomic arrangement of the region around the central atom. Within their model only single atom contacts of monovalent metals as *e.g.* the alkali or noble metals should transmit one single channel. In particular for Au it has been predicted that the transmission coefficient of this single channel can achieve, in a perfectly ordered geometry of the central atom and its neighbors, a nearly saturated value of $T > 0.99$ [9]. We present here preliminary results of an experiment which allows to determine the transmission coefficients of this model substance.

2. DETERMINATION OF THE TRANSMISSION COEFFICIENTS

It has been shown [10] that the set $\{T_n\}$ of atomic-size constrictions is amenable to measurement using the strong non-linearities in the current-voltage (IV) characteristics of *superconducting* atomic contacts due to multiple Andreev reflections (MAR). The upper left inset of Fig. 2.1 shows the numerical predictions for a single channel with transmission T [11]-[14]. The $i(V,T)$ curves present a series of sharp current steps at voltage values $eV = 2\Delta/m$, where m is a positive integer and Δ is the superconducting gap. The order $m = 2, 3, ...$, of a step corresponds to the number of electronic charges transferred in the underlying MAR process. Energy conservation imposes a threshold $eV \geq 2\Delta/m$ for the process of order m. For low transmission, the contribution to the current arising from the process of order m scales as T^m. The analysis is performed by decomposing the experimental IVs through a contact into contributions of N independent terms with the $\{T_n\}$ (where $n = 1, ..., N$) as fitting parameters: $I(V) = \sum_{n=1}^{N} i(V, T_n)$.

Since Au is not superconducting, the described method of determining conduction channels is not directly applicable. However, due to the so-called proximity effect [15, 16], a finite piece of a non-superconducting metal in good contact with a superconducting metal adopts certain superconducting properties, *e.g.* it develops a "minigap" E_g in the quasiparticle excitation spectrum. As will be explained below, the appearance of the minigap is the basic property for determining the channel ensemble. A detailed analysis of the proximity effect in our mesoscopic samples is subject of a separate paper [17].

3. EXPERIMENTAL TECHNIQUES

Experimentally stable atomic-size contacts can be achieved with different techniques including STM [18, 19] and mechanically controllable breakjunctions (MCB) [20]. Using such methods it has been shown that smallest stable contacts of gold have most often a conductance close to G_0. We have produced a micro-fabricated breakjunction [21] of Al, where the center part of the constriction consists of Au. Using shadow evaporation through a suspended mask we evaporate perpendicular to the substrate surface two Al electrodes separated by a gap of width w with a thickness 300 nm. Without breaking the vacuum, two Au layers of thickness 10 nm are evaporated at two different angles $\pm$ 10° in order to fill the gap and to form a continuous film (see right inset of Fig. 2.1). After lifting off the mask the sample is dry etched to form a freestanding nanobridge. The bridge is broken at the constriction at very low temperatures and under cryogenic vacuum conditions by controlled bending of the elastic substrate mounted on a three-point bending mechanism. Details of the sample preparation and measuring setup are given in [10, 17, 22]. As found in previous experiments [20] the conductance G decreases in steps of the order of G_0, with smaller steps within a plateau (see Ref. [10, 23]).

Figure 2.1 shows typical IVs of an Al-Au-Al sample (circles) and an Al sample [24] (triangles) obtained at 50 mK for comparable conductances of about 0.7 G_0, determined by the slope of the IV at large voltages $V > 0.8$ mV. Both IVs exhibit the characteristic features of MAR and a supercurrent. However, the voltage values of the MAR steps are different for both samples. By analyzing IVs obtained in the tunnel regime (not shown) [10, 17] we deduce $\Delta/e = 185\ \mu$V for the Al sample and $E_g/e = 160\ \mu$V for the Al-Au-Al sample. Because of the small length and layer thickness of the normal metal of this particular Al-Au-Al sample its superconducting properties can be well described by the theory of MAR for BCS superconductors (the gap parameter is then the experimentally determined minigap E_g). The IV of the Al-Au-Al sample can

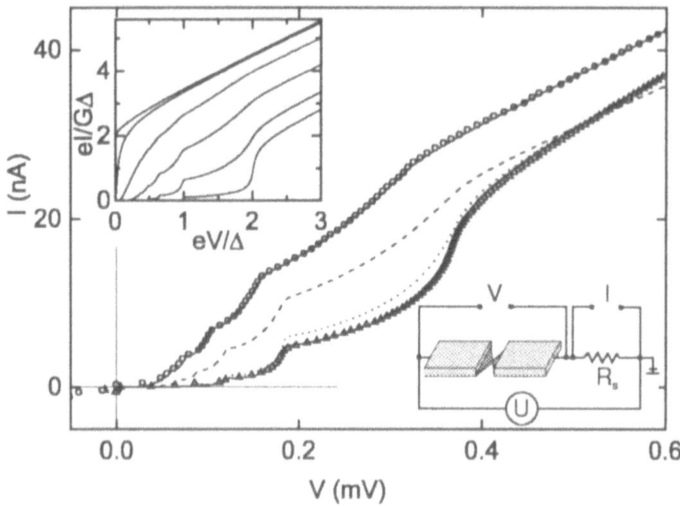

Figure 2.1 Measured IVs for few-atom contacts obtained with the Al-Au-Al sample (circles) and the Al sample (triangles) and best numerical fits (lines). The $\{T_n\}$ and total transmissions D obtained from the fits are: $T = D$=0.692, $E_g/e = 160\ \mu$V for the Al-Au-Al sample and $\{0.387, 0.246, 0.126\}$, D=0.759 $\Delta/e = 185\ \mu$V for the Al sample. The dashed (dotted) line is the best fit to the IV of the Al sample assuming one (two) channel(s). Not all measured data points are shown. Left inset: Theoretical IVs for a single channel superconducting contact for different values of T (from bottom to top: 0.1, 0.4, 0.7, 0.9, 0.99, 1) after [13]. Both axes are in reduced units. Right inset: Simplified sample geometry of the Al-Au-Al sample and measuring circuit. Dark grey regions: Au, light grey regions: Al.

be described by one single channel without any adjustable parameter, since the transmission of this channel is determined by the slope of the IV at high voltages $eV \gg 2\Delta$ (see line in Fig. 2.1), as predicted by CLM. According to their model, imperfect coupling of the constriction atom to its neighbors reduces the value of T [8, 9]. When pulling the two parts of the bridge apart, it is reasonable to assume that the local order in the central region is changed, explaining the fact that often the last plateau is not flat and its conductance is below G_0.

Although the conductance of the Al sample is also below G_0, three channels have to be taken into account (solid line). Best fits with one (dashed) or two (dotted) channels reveal systematic deviations from the measured IV. If the fitting procedure is performed with more than three channels, the quality of the fit is not improved any more and the transmissions of the supplementary channels turn out to be negligibly small (< 1 % of G).

Thus, we find for Au stable contacts, whose transport properties can be explained by taking into account only one conduction channel.

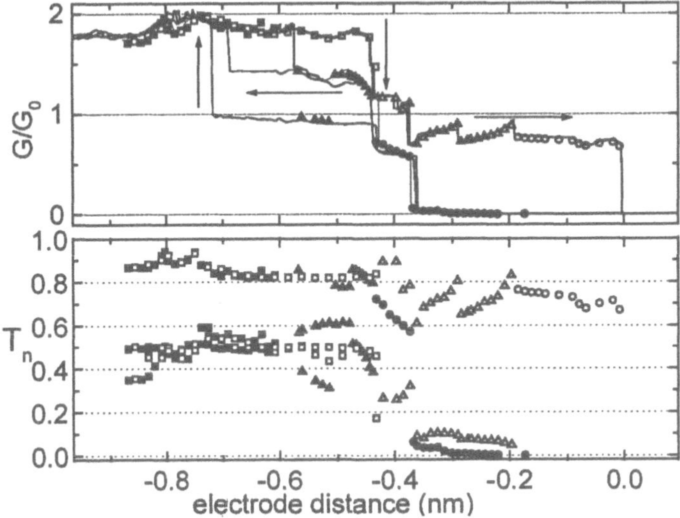

Figure 2.2 Upper panel: Conductance as a function of electrode distance of an Al-Au-Al sample (with E_g=125 μV) when repeatedly opened and closed. The distance axis has been scaled by the exponential dependence of G in the tunnel regime. Due to hysteresis in the mechanical setup, an varying offset of about 0.03 nm has been subtracted such that the origin of the distance axis is for all repetitions at the position where the contact breaks. Lines: continuous motion at $v \approx$ 10-20 pm/s and dc-conductance measurement with applied magnetic field of 10 mT for suppressing the superconductivity, symbols: sums of the transmission values when motion was stopped for recording IVs and determining the channel ensemble: Open symbols: opening, closed symbols: closing; circles (triangles, squares): one (two,three) channels. Not all data are shown. Lower panel: Transmission ensemble.

When further closing the contact to higher conductances, a decomposition analysis as for the Al sample taking into account more channels has to be performed. When the closing is stopped at conductances around 2 G_0, stable reproducible configurations can be achieved, which allow for repeated breaking and closing of the contact with the same conduction properties. As an example we plot in Fig. 2.2, a representative subset of conductance *vs.* distance curves of an Al-Au-Al sample, measured below 100 mK during two days when changing the configurations. The data shown as solid lines have been measured when continuously opening at a speed of 10-20 pm/s. For determining the channel ensemble (see lower panel of Fig. 2.2) two opening and closing cycles have been performed with stopping the motion at different positions and recording IV characteristics. Starting at $G \approx 2\ G_0$ from a three channel configuration with one well transmitted channel $T_1 \approx 0.8$ and two almost degenerate ones $T_2 \approx T_3 \approx 0.5$, the conductance jumps down to a short and tilted plateau at around 1.2 G_0, before a long last plateau with a reproducible substructure evolves. The sawtooth-like part of the plateau can be decomposed into the contributions of one widely open channel (which reveals the sawtooth behaviour) and a smaller second one. The last part of the plateau has a conductance around 0.8 G_0 and can be described by a single conductance channel. After breaking, the contact is closed by an almost exponential increase of the conductance to about $G \approx 0.07\ G_0$ and a sudden jump to $G \approx 0.7\ G_0$, still with a single channel. After closing the contact further by about 0.1 nm, the contact jumps to one of two two-channel configurations (with different transmissions than the two-channel situation when opening). These configurations appear to be metastable since the contact remains in those states for different length intervals, which differ from repetition to repetition. The analysis of the channel ensemble reveals that one of the channels has the same transmission $T_1 \approx 0.6$ for both configurations while the second jumps between two very different values (0.3 and 0.8). A possible interpretation of this behaviour would be one atom in a stable position and a second one alternating between two almost degenerate positions.

Finally the contact arrives at its initial stable plateau with three channels. The transmission values as well as the substructure of the plateau are reproducible within the accuracy of the conductance measurement in the continuous measurements and the determination of the channels ($\approx$ 3%). Although not all details of the substructure are observed in any of the $\approx$ 20 repetitions, the substructure of this stable plateau is used to distinguish the extrinsic hysteresis of the mechanical setup and the intrinsic hysteresis of the atomic motion. From the distance values we have subtracted varying offsets of the order of 0.1 nm such that the contacts break at zero. The mechanical hysteresis has been determined such that the curvatures of the stable three-channel plateaux when opening and closing the contact overlay. The remaining hysteresis of $\approx$ 0.4 nm is of the order of the lattice spacing of gold and is thus an intrinsic effect of the atomic contact [25]. The reproducible situation was kept until the contact was closed thoroughly. These observations support the assumption that the conductance of the whole system is determined by the central atom and its coupling to the nearest neighbors.

In conclusion, we have presented an investigation of the transport properties of gold tunnel and few-atom contacts with superconducting leads. We have analyzed quantitatively the IV characteristics of few-atom contacts containing up to three conduction channels. Single-atom gold contacts accommodate one conduction channel in accordance with theoretical predictions.

Acknowledgements

We thank Y. Naveh, J. M. van Ruitenbeek, G. Schön and C. Strunk for helpful discussions. We have enjoyed fruitful interaction with D. Averin, J. C. Cuevas and A. Levy Yeyati, and we thank them for providing us with their respective computer codes. E. S. thanks H. v. Löhneysen for continuous support. This work was partially supported by the Bureau National de la Métrologie and the Deutsche Forschungsgemeinschaft.

References

[1] G. Binnig, H. Rohrer, Ch. Gerber and E. Weibel, Tunneling through a controllable vacuum gap, Appl. Phys. Lett. **40**, 178-180 (1982).
[2] M. F. Crommie, C. P. Lutz and D. M. Eigler, Confinement of electrons to quantum corrals on a metal surface, Science **262**, 218-220 (1993).
[3] G. Binnig, H. Rohrer, Scanning tunneling microscopy, IBM J. Res. Dev. **30**, 355 (1986).
[4] N. D. Lang, Resistance of atomic wires, Phys. Rev. B **52**, 5335-5342 (1995).
[5] C. C. Wan, J.-L. Mozos, G. Taraschi, J. Wang and H. Guo, Quantum transport through atomic wires, Appl. Phys. Lett. **71**, 419-421 (1997).
[6] A. Yazdani, D. M. Eigler and N. D. Lang, Off-resonance conductance through atomic wires, Science **272**, 1921-1924 (1996).
[7] R. Landauer, Electrical resistance of disordered one-dimensional lattices. Philos. M. **21**, 863-867 (1970).
[8] J. C. Cuevas, A. Levy Yeyati and A. Martín-Rodero, Microscopic origin of the conducting channels in metallic atomic-size contacts, Phys. Rev. Lett. **80**, 1066-1069 (1998).
[9] A. Levy Yeyati, A. Martín-Rodero and F. Flores, Conductance quantization and electron resonances in sharp tips and atomic size contacts, Phys. Rev. B **56**, 10369-10372 (1997).
[10] E. Scheer, P. Joyez, D. Esteve, C. Urbina and M. H. Devoret, Conduction channel transmission of atomic-size aluminum contacts, Phys. Rev. Lett. **78**, 3535-3538 (1997).
[11] G. B. Arnold, Tunneling without the tunneling Hamiltonian II. Subgap harmonic structure, Journal of Low Temp. Phys. **68**, 1-27 (1987).
[12] D. Averin and A. Bardas, AC Josephson effect in a single quantum channel, Phys. Rev. Lett. **75**, 1831-1834 (1995).
[13] J. C. Cuevas, A. Martín-Rodero and A. Levy Yeyati, Hamiltonian approach to the transport properties of superconducting quantum point contacts, Phys. Rev. B **54**, 7366-7379 (1996).

[14] E. N. Bratus, V. S. Shumeiko, E. V. Bezuglyi and G. Wendin, dc-current transport and ac Josephson effect in quantum junctions at low voltage, Phys. Rev. B **55**, 12666-12677 (1997).
[15] D. Esteve, The proximity effect in mesoscopic diffusive conductors. In *Mesoscopic Electron Transport* (eds. L. L. Sohn, L. P. Kouwenhoven and G. Schön, Kluwer Academic Publishers, 1997) pp. 375-406, and references therein.
[16] W. Belzig, F. K. Wilhelm, C. Bruder, G. Schön and A. D. Zaikin, Quasiclassical Greens function approach to mesoscopic superconductivity, Superlattices and Microstructures **25**, 1251-1288 (1999).
[17] E. Scheer, W. Belzig, Y. Naveh, D. Esteve, M. H. Devoret and C. Urbina, Proximity effect and multiple Andreev reflections in gold atomic contacts, submitted for publication.
[18] L. Olesen, E. Lægsgaard, I. Stensgaard, F. Besenbacher, J. Schiøtz, P. Stoltze, K. W. Jacobsen and J. K. Nørskov, Quantized conductance in an atom-sized point contact, Phys. Rev. Lett. **72**, 2251-2254 (1994).
[19] N. Agraït, J. G. Rodrigo and S. Vieira, Conductance steps and quantization in atomic-size contacts, Phys. Rev. B **47**, 12345-12348 (1996).
[20] J. M. van Ruitenbeek, Quantum point contacts between metals; in *Mesoscopic Electron Transport* (eds. L. L. Sohn, L. P. Kouwenhoven and G. Schön, Kluwer Academic Publishers) pp. 549-579, and references therein.
[21] J. M. van Ruitenbeek, A. Alvarez, I. Piñeyro, C. Grahmann, P. Joyez, M. H. Devoret, D. Esteve and C. Urbina, Adjustable nanofabricated atomic size contacts Rev. Sci. Inst. **67**, 108-111 (1996).
[22] E. Scheer, P. Joyez, D. Esteve, C. Urbina and M. H. Devoret, Conduction channels of superconducting quantum point contacts, to appear in Physica B.
[23] E. Scheer, N. Agraït, J. C. Cuevas, A. Levy Yeyati, B. Ludoph, A. Martín-Rodero, G. Rubio Bollinger, J. M. van Ruitenbeek and C. Urbina, The signature of chemical valence in the conduction through a single-atom contact, Nature **394**, 154-157 (1998).
[24] The Al sample has been fabricated as explained in [10] within a single evaporation step through a mask without the bridge.
[25] C. Untiedt, G. Rubio, S. Vieira and N. Agraït, Fabrication and characterization of metallic nanowires, Phys. Rev. B **56**, 1251-1288 (1997).

Chapter 3

EXPERIMENTS ON CONDUCTANCE AT THE ATOMIC SCALE

J. M. van Ruitenbeek
Kamerlingh Onnes Laboratorium, Universiteit Leiden
Niels Bohrweg 2, NL-2333 CA Leiden, The Netherlands

1. INTRODUCTION

Quantization of the conductance may be observed when the width of a contact between two electron reservoirs becomes comparable to the Fermi wavelength. For a smooth (adiabatic) dependence of the confining potential along the current direction, the wave functions inside the constriction are described by modes with well-defined quantum numbers for the x- and y-coordinate dependence, and freely propagating in the z-direction (taken along the axis of the contact). In the absence of back-scattering from defects or the boundaries of the contact, each of the occupied modes contributes $2e^2/h$ to the conductance. Conductance quantization was first discovered in a semiconductor two-dimensional electron gas (2DEG) device, where the width of the contact can be adjusted using a gate potential ([1, 2], for reviews see [3, 4]). As a function of the contact width the conductance is observed to increase stepwise in units of $G_0 = 2e^2/h$.

When attempting to observe such effects in metals, we meet two important obstacles. First, there are no techniques available to continuously control the contact size by a gate electrode, as could be done in the semiconductor experiment. Second, when we adjust the contact size by mechanically adjusting a point contact, the atomic scale structure becomes visible. The diameter of an atom is comparable to the Fermi wavelength. Steps in the conductance can, therefore, result both from the atomic structure and from conductance quantization, and the two effects are difficult to separate, a priori. One should expect that plane waves form a poor starting point as a basis for describing the electron

I. O. Kulik and R. Ellialtioğlu (eds.),
Quantum Mesoscopic Phenomena and Mesoscopic Devices in Microelectronics, 35–50.

wave functions in atomic-size contacts, and atomic orbitals might be a more appropriate starting point. In the following we will address the questions as to what determines the conductance through a single atom. We will discuss how the modes can be uniquely defined and what is their nature. It will be shown that their number can be obtained from experiments. For this purpose measuring the linear conductance, G, of a contact is not adequate, as G gives only the sum of the conductance through all participating modes.

2. LANDAUER FORMALISM FOR ELECTRON TRANSPORT

Let us consider a general ballistic system with a narrow constriction, representing the contact. The connections between the ballistic system and the outside world are formed by electron reservoirs on each side of the contact, which are held at a potential difference eV by an external voltage source. When the leads connecting the reservoirs to the contact are straight wires of constant width, there is a well defined number of conducting modes in each of these wires, say N and M for the left and right lead, respectively. In a free electron gas model the modes are simply plane waves, which can propagate in the current direction (to the left and right) and are standing waves in the perpendicular directions. The modes can be labeled by an index corresponding to the number of nodes in the perpendicular direction. The numbers N and M are limited by the requirement that the energy of the modes is lower than the Fermi energy.

The conductance of the system can now be simply expressed as [3, 4],

$$G = \frac{2e^2}{h}\mathrm{Tr}(\mathbf{t}^\dagger \mathbf{t}) \; , \tag{3.1}$$

where e is the electron charge, h is Planck's constant, and $\mathbf{t}$ is an $N \times M$ matrix with matrix element t_{mn} giving the probability amplitude for an electron wave in mode n on the left to be transmitted into mode m on the right of the contact. It can be shown that the product matrix $\mathbf{t}^\dagger \mathbf{t}$ can always be diagonalised by going over to a new basis, consisting of linear combinations of the original modes in the leads. Further, the number N_c of non-zero diagonal elements is only determined by the number of modes at the narrowest cross section of the conductor [5, 6, 7]. Equation (3.1) thus simplifies to

$$G = \frac{2e^2}{h}\sum_{n=1}^{N_c} T_n \; , \tag{3.2}$$

where $T_n = |t_{nn}|^2$ and the index refers to the new basis. The new basis is the basis of "eigenchannels", which are associated with orthogonal wave functions around the narrowest part of the conductor. For a single-atom contact the eigenchannels can be associated with the valence orbitals of the atom [7]. As an example, a single-atom contact for a monovalent metal (Au, Na, *etc.*) has a single conductance channel. For *s-p*-metals and *s-d*-metals the number of channels is 3 and 5, respectively. We will discuss below how this can be tested experimentally.

3. ATOMIC-SIZE CONTACTS

Two intimately related techniques have been frequently used to study metallic point contacts in the quantum limit. As a very versatile instrument the Scanning Tunneling Microscope (STM) has been used in many experiments. The metal probe tip is driven into the sample surface to form a large contact, and it is subsequently pulled back, while the contact resistance is measured during pull-off. These cycles typically take 1 ms. The first experiment of this type was reported by Gimzewski and Möller [8], who observed an exponential dependence of the tunnel resistance with distance, which tended to saturate at close proximity. The transition to contact was seen to manifest it self as a distinct jump to contact. An important improvement to the experiment is obtained when cooling down to helium temperatures. At low temperatures the drift of the piezoelectric elements can be eliminated, and the violent thermal motion of atoms at the surface of the tip, the substrate and in the atmosphere is frozen. Using a low-temperature STM single-atom contacts can be held stable for very long time, and can be used to obtain detailed information on the conductance properties and the force in atomic size contacts [9, 10, 11].

In our group we introduced a slightly different method for the study of atomic-size contacts, which is a modified version of the break-junction technique developed by Moreland and Ekin [12], and which was baptised the Mechanically Controllable Break Junction (MCBJ) technique [13]. The principle of the technique is illustrated in Fig. 3.1. By breaking the metal, two clean fracture surfaces are exposed, which remain clean due to the cryo-pumping action of the low-temperature vacuum can. This method circumvents the problem of surface contamination of tip and sample in an STM experiment, where a UHV chamber with surface preparation and analysis facilities are required to obtain similar conditions. The fracture surfaces can be brought back into contact by relaxing the force on the elastic substrate, while a piezo-electric element is used for fine control. The roughness of the fracture surfaces results in

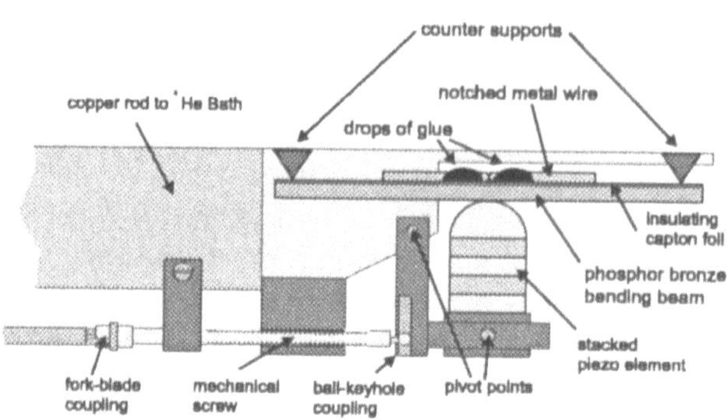

Figure 3.1 Schematic view of the mounting of a MCBJ, where the metal to be studied has the form of a notched wire, which is fixed onto an insulated elastic substrate, the bending beam, with two drops of epoxy adhesive very close to either side of the notch. The substrate is mounted in a three-point bending configuration between the top of a stacked piezo element and two fixed counter supports. This setup is mounted inside a vacuum can and cooled down to liquid helium temperatures. Then the substrate is bent by moving the piezoelement forward using the mechanical screw arrangement. The bending causes the top surface of the substrate to expand and the wire to break at the notch. After breaking, the contact fine-adjustment is done through the voltage over the piezoelement. Typical sizes are $L \simeq 20$ mm and $u \simeq 0.1$ mm.

a first contact at one point, and experiments usually give no evidence of multiple contacts. In addition to a clean surface, a second advantage of the method is the stability of the two electrodes with respect to each other. The stability results from the reduction of the mechanical loop which connects one contact side to the other, from centimetres, in the case of an STM scanner, to ~ 0.1 mm in the MCBJ.

Figure 3.2 shows some examples of the conductance measured during breaking of a gold contact at low temperatures, using an MCBJ device. The conductance decreases by sudden jumps, separated by plateaux, which have a negative slope, the higher the conductance the steeper. Some of the plateaux are remarkably close to multiples of the conductance quantum, G_0; in particular the last plateau before loosing contact is nearly flat and very close to 1 G_0. Closer inspection, however, shows that many plateaux cannot be identified with integer multiples of the quantum unit, and the structure of the steps is different for each new recording. Also, the height of the steps is of the order of the quantum unit, but can vary by more than a factor of 2, where both smaller and larger steps are found. Drawing a figure such as Fig. 3.2, with grid lines at multiples of G_0, guides the eye to the coincidences and may convey that the origin of the steps is in quantisation of the conductance. However, in evaluating the graphs, one should be aware that a plateau

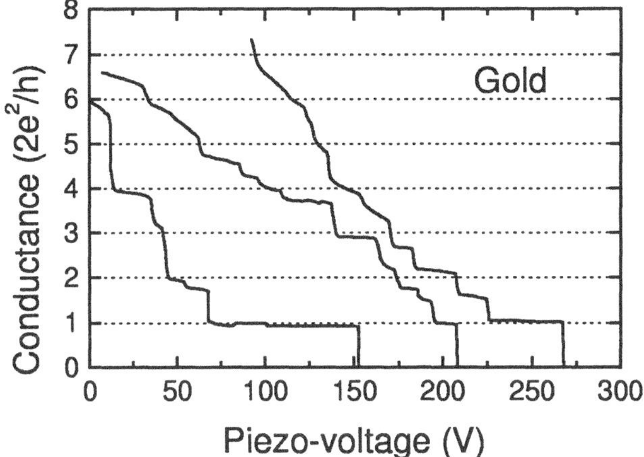

Figure 3.2 Three typical recordings of the conductance G measured in atomic size contacts for gold at helium temperatures, using the MCBJ technique. The electrodes are pulled apart by increasing the piezo-voltage. The corresponding displacement is about 0.1 nm per 25 V. After each recording the electrodes are pushed firmly together, and each trace has new structure. (After J. M. Krans [14])

cannot be farther away than one half from an integer value, and that a more objective analysis is required. Still, it is clear that we can use these fairly simple techniques to produce and study atomic-scale conductors, for which the conductance is dominated by quantum effects. Below we show how superconductivity, on the one hand, and shot noise in the current, on the other, can be exploited to elucidate the quantum nature of current transport in these contacts.

4. SUPERCONDUCTING SUBGAP STRUCTURE

We start by discussing the current-voltage (IV) characteristics of atomic-size junctions in the tunneling regime, *i.e.*, after the jump from the last conductance plateau into the regime where the resistance varies exponentially with distance. Fig. 3.3 shows several examples of IV characteristics for niobium junctions recorded in the tunneling regime, using voltage bias. They have been plotted on a semilogarithmic scale to make the steps at small voltages visible. The conductance of the junctions decreases from (a) to (d): $G/G_0 = 0.0707$, 0.0321, 0.0183 and 0.0133, respectively. The voltage scale is expressed in units of the superconducting gap, which is taken as $\Delta = 1.41\,\mathrm{meV}$ for these samples. The IV curves have the largest current step at $eV = 2\Delta$, as expected for

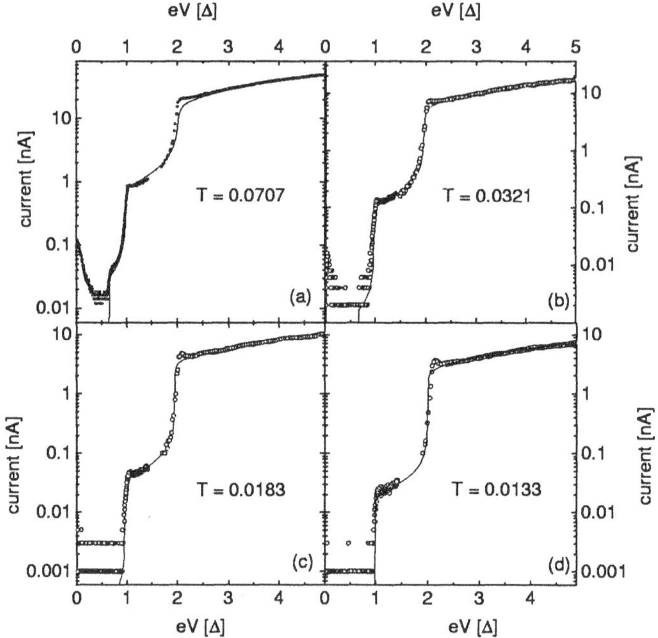

Figure 3.3 Tunneling curves for niobium MCB junctions recorded at 1.4 K. The transmission probabilities have been obtained from the normal state conductance of the junctions assuming a single channel dominates the tunneling, $T = G/G_0$. The curves have been generated from the theory of Bratus' *et al.* [15, 16, 17] without any adjustable parameters. (From: [18])

tunneling, and are linear above this value. Smaller current steps are seen at $2\Delta/2$ and in Fig. 3.3a even at $2\Delta/3$. At still lower currents the limitations of the digital resolution of the experiment become visible, which explains the discrete levels in the figure. The rise in the current for V approaching zero is a remnant of the Josephson current. At $V = 0$ we should expect to find the Josephson current due to the coherent tunneling of Cooper pairs across the junction, and which should ideally have a the Ambegaokar-Baratoff value $I_c = \pi\Delta/2eR$ [19]. The latter is of the order of the current just above the step at $eV = 2\Delta$ and the observed current is more than two orders of magnitude smaller. This strong suppression and the widening of the Josephson effect into a finite width anomaly can be attributed to the coupling of the junction to its electromagnetic environment, and experiments testing this coupling using a controlled environment are under way [20]. The remnant of the supercurrent is more strongly suppressed for weaker Josephson coupling (increasing resistance) as can be seen in Fig. 3.3, in agreement with [21].

In recent years the IV characteristics of superconducting junctions with a single conductance mode have been obtained from exact theoretical analysis [15, 17, 22, 23], for all values of the transmission probability, $0 \leq T \leq 1$. The curves through the data in Fig. 3.3 have been generated using this theory for a single conductance channel. The only parameters in the theory are the gap Δ, the temperature and the transmission probability T. The gap Δ=1.41 meV has been determined experimentally from tunneling curves with a large vacuum barrier, where subgap structure is absent. The temperature is fixed by the measured bath temperature, and the transmission probability is given by the normal state conductance of the junction through $T = G/G_0$, assuming a single conductance channel. The normal state conductance is determined from the *IV* curves at bias voltages several times the gap value. This makes the description entirely independent of freely adjustable parameters, and as one can observe in Fig. 3.3, the theory fits the experiment quite convincingly. The theory reproduces not only the relative height of the current steps at $2\Delta/n$, but also fits the shape of the curve in between the steps. Although the theory incorporates the supercurrent, the coupling to the environment is not included, and we have not attempted to describe this aspect of the experiment.

In the contact regime the conductance evolves in a series of steps and plateaux as the electrodes are pulled apart by increasing the piezo voltage, similar as for gold in Fig. 3.2. The last plateau for niobium has a conductance between 2 and 3 G_0 and is expected to consist of a single niobium atom. As an example we show in Fig. 3.4, a curve recorded at the beginning of the last plateau. Following Scheer *et al.* [24, 25] we assume that the total current can be decomposed into the sum of the individual channels contributing to the conductance,

$$I(V) = \sum_{j=1}^{N} i(T_j, V) \ ,$$

where $i(T_j, V)$ is the current of channel j. This procedure is justified, since multiple Andreev reflections do not mix the normal conduction channels. This property allows us to extract the mode composition of atomic size contacts from its *IV* characteristic, using the exact single channel functions $i(T_j, V)$ for BCS superconductors calculated in [17, 22, 23].

The recorded *IV* characteristic can be well described by the five channels with transmission probabilities listed in the figure. For comparison also fits using four, three and two channels are shown. Using four channels to fit the data results in a slightly lower quality fit, where it overestimates the current at voltages between Δ and 2Δ, and underestimates

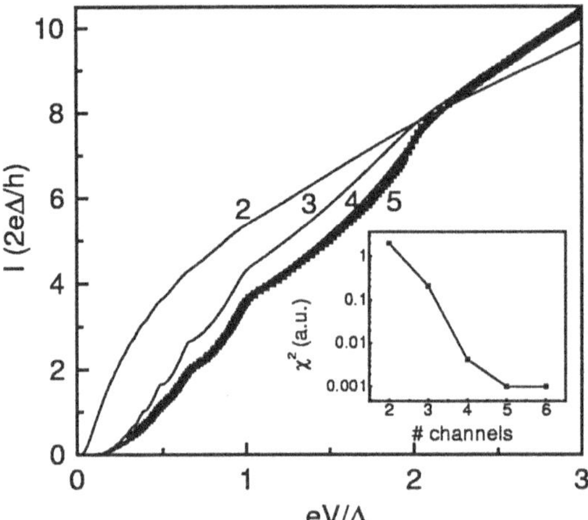

Figure 3.4 IV curve for a niobium contact measured on the last plateau before the jump to tunneling, together with best fits for two, three, four and five channels. The theoretical curves have been labeled corresponding to the number of channels used in the fit. For the gap Δ/e=1.41 mV was used taken from the vacuum tunneling data. The five channel fit has transmission probabilities T_1=0.811, T_2=0.669, T_3=0.627, T_4=0.327 and T_5=0.131. The total transmission probability is $T_{tot} = \sum_{j=1}^{5} T_j = 2.57$. The inset shows the error sum, χ^2, versus the number of channels used in the fits. (From: [18])

below Δ. The quality of the fit is expressed in a so called χ^2 factor, which is the sum of the square of the deviation of the measured current at all the points on the experimental curve from the current of the calculated curve at the same bias voltage, divided by the number of recorded points. Using three or two channels to reproduce the experimental data results in increasingly poorer fits and consequently in larger values for χ^2 (inset). Using six channels to fit the data does not result in any improvement with respect to five channels, and the value of the sixth channel remains near its insignificantly small initialisation value (T_6 = 0.0001). Hence, we find that five channels are necessary to reproduce the IV characteristics measured on the last plateau before breaking the contact.

From the measurements it can be concluded that the last plateau for niobium with a conductance usually between 2 and 3 G_0 is composed of 5 channels. This is in excellent agreement with the number of channels and the total conductance predicted by the theory of Cuevas *et al.* [7] using the orbital nature of a single atom as a starting point, and with similar observations for Pb, Al and Au [25].

5. SHOT NOISE

Although the use of the analysis of the superconducting subgap structure gives us a very powerful tool to study conductance at the atomic scale it has the disadvantage that a number of fitting parameters are required, for which it is not immediately evident that the results are independent of the fitting procedure. On closer inspection one can be convinced that the number of channels can indeed be determined uniquely. However, we have a second tool which allows us to make an independent test of the interpretation, and which does not require superconductivity, so that it can be directly applied to important non-superconducting metals, such as gold.

Shot noise is a non-equilibrium type of noise, directly resulting from the discreteness of electric charge: the passage of individual electrons causes the current to be a sum of random pulses. These intrinsic current fluctuations were already predicted in 1918 by Schottky, as an artifact of vacuum diodes [26]. Shot noise turns out to be present in all kinds of devices, including microscopic conductors. In the last decade, it has become clear that it can be used to obtain information on the electron transport mechanism, as is explained in Chapter 14 by M. Büttiker. For a general multichannel contact at zero temperature, the shot noise spectral density can be expressed in terms of the transmission probabilities T_n of all conducting channels [27, 28],

$$S_I = 2eV\frac{2e^2}{h}\sum_n T_n(1-T_n)\ . \tag{3.3}$$

For a contact with linear current-voltage characteristics, we see that the shot noise power is still linear in bias current. In a ballistic QPC in a two-dimensional electron gas (2DEG), the conductance, G, as a function of contact diameter shows a step-wise increase by integer multiples of the conductance quantum, $G_0 \equiv 2e^2/h$ [1, 2], and shot noise was indeed shown to be strongly suppressed at quantized conductance values, where the contributing conductance modes are fully transmitted [29, 30].

From the Landauer formula for conductance, $G = G_0\sum_n T_n$, we only know the sum of the transmission coefficients. However, from Eq. (3.3) we also know the sum of the transmission coefficients squared. Hence, the combination of conductance and shot noise measurements can provide us with new knowledge on electron transport properties of the smallest metallic contacts.

Stable atomic scale contact were fabricated using the MCBJ technique where its high degree of stability was even further improved by careful shielding from external electromagnetic, mechanical and acoustic vibrations. The effect of measuring at low temperature is that the thermal noise is reduced. However, the noise level of the preamplifiers is in general exceeding the shot noise we are interested in. Using two sets of preamplifiers in parallel and measuring the cross-correlation, this undesired noise is reduced. By subtracting the zero bias thermal noise from the current biased noise measurements, the preamplifier noise, present in both, is further eliminated. For currents up to 0.9 μA the shot noise level has the expected linear dependence on current. For further details on the measurement technique, we refer to [31].

First we discuss the results for the monovalent metal gold, for which a single atom contact is expected to transmit a single conductance mode. In Fig. 3.5 the experimental results for a number of conductance values are shown, where the measured shot noise is given relative to the classical shot noise value $2eI$. All data are strongly suppressed compared to the full shot noise value, with minima close to 1 and 2 times the conductance quantum. We compare our data to a model that assumes a certain evolution of the values T_n as a function of the total conductance. In the simplest case, the conductance is due to only fully transmitted modes ($T_n = 1$) plus a single partially transmitted mode (full curve). The model gives a measure for the deviation from this ideal case in terms of the contribution x of other partially open channels; the corresponding behaviour of the shot noise as a function of conductance is shown as the dashed curves in Fig. 3.5. This model has no physical basis but merely serves to illustrate the extent to which additional, partially open channels are required to describe the measured shot noise. For a more physical model fitting the data of Fig. 3.5 see [32].

We see that for $G < G_0$ the data are very close to the $x = 0\%$ curve, while for $G_0 < G < 2\,G_0$ the data are closer to the $x = 10\%$ curve. For $G > 2\,G_0$ the contribution of other partially open channels continues to grow. For all points measured on the last plateau before the transition to tunneling, which is expected to consist of a single atom (or a chain of single atoms [33]), we find that the results are well-described by a single conductance channel, in agreement with the fact that gold has only a single valence orbital.

When the experiment is repeated for aluminum contacts, we observe a different behaviour of the shot noise power as a function of conductance compared to gold. For contacts between $0.8\,G_0$ and $2.5\,G_0$ the shot noise values vary from 0.3 to 0.6 $(2eI)$, which is much higher than for gold (see Fig. 3.5). A systematic dependence of the shot noise power on the conductance seems to be absent.

From the two measured parameters, the conductance, G, and the shot noise, S_I, one cannot determine the full set of transmission probabilities.

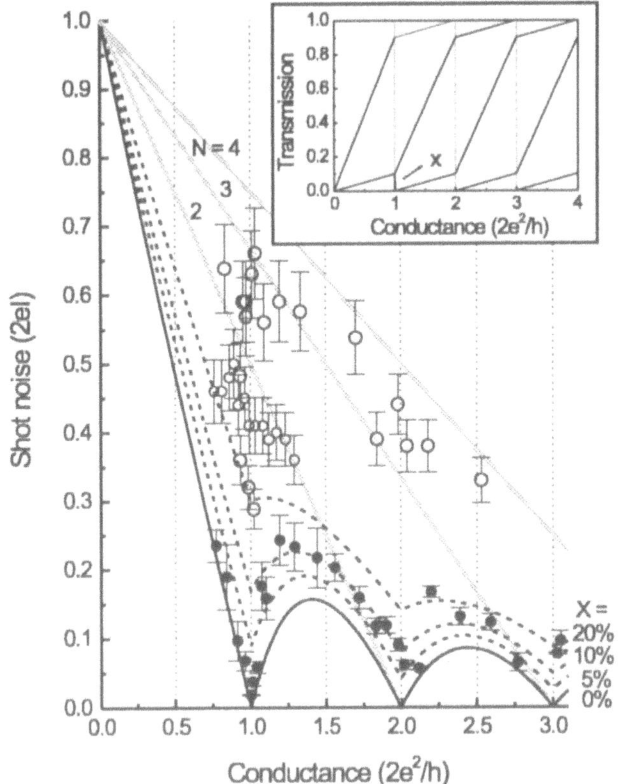

Figure 3.5 Measured shot noise values for gold (full circles) and aluminum (open circles) contacts at 4.2 K with a bias current of 0.9 μA. For gold, comparison is made with calculations described in the text and in the inset (full and dashed black curves). For aluminum, comparison is made with the maximum shot noise that can be produced by N modes (grey curves), as explained in the text. The minimum shot noise is given by the full black curve. Note that in the limit of zero conductance, the theoretical curves all converge to full shot noise. *Inset:* Model for visualising the effect of contributions of different modes to the conductance and shot noise. The model gives a measure for the deviation from the ideal case of channels opening one by one, by means of a fixed contribution $(1-T_{n-1})+T_{n+1} = x$ of the two neighboring modes. As an illustration we show the case of $x = 10\%$ contribution from neighboring modes. (From: [34])

However, the shot noise values found for aluminum, especially the ones at conductance values close to G_0, agree with Eq. (3.3) only if we assume that more than one mode is transmitted. The maximum shot noise generated by two, three or four modes as a function of conductance is plotted as the grey curves in Fig. 3.5; the minimum shot noise in all cases is given by the full black curve. Hence, for a contact with shot noise higher than indicated by the grey N-mode maximum shot noise curve, at least $N+1$ modes are contributing to the conductance. From this simple analysis we can see that for a considerable number of contacts with a conductance close to $1\,G_0$, the number of contributing modes is at least three. Again, this is consistent with the number of modes expected based on the number of valence orbitals, and with results of the subgap structure analysis [24, 25]. Note that the points below the line labeled $N=2$ should not be interpreted as corresponding to two channels: the noise level observed requires at least two channels, but there may be three, or more.

6. NEARLY-FREE-ELECTRON GAS METALS

We have argued that the conductance properties at the atomic scale are best described in terms of an atomic-orbitals basis. However, when we confine our attention to free-electron-like materials such as the alkali metals, then it turns out that plane wave models make very good predictions. Clear evidence for this was found in histograms of conductance values for sodium [35]. A similar histogram for potassium is shown in Fig. 3.6. The histograms are constructed from many individual curves of conductance versus distance, as in Fig. 3.2. The shape of each curve is different due to the lack of control over the atomic structure of the

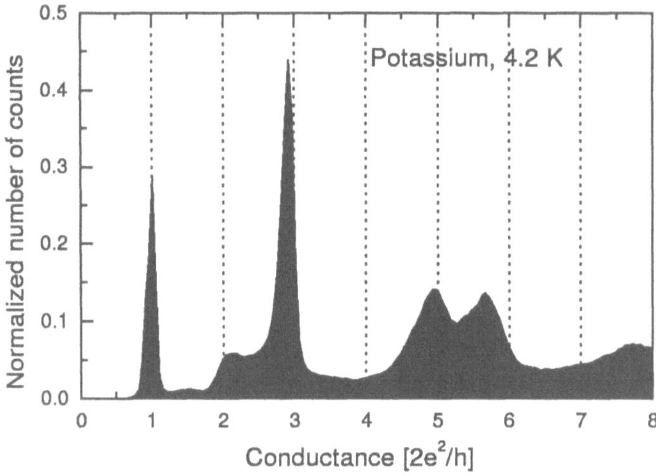

Figure 3.6 Histogram of conductance values, constructed from several thousands curves of conductance versus piezo voltage, measured for potassium at 4.2 K with an MCB device. The characteristic sequence of peaks ($G = 1, 3, 5, 6$) is regarded as a signature for conductance quantisation. (From [36])

contacts. The histogram may be regarded as an average over atomic configurations, and it shows a clear preference of the contacts for conductance values near 1, 3, 5 and 6 times the unit of conductance. The slight shift below exact multiples of G_0 may be attributed to defect scattering [37]. The characteristic series of conductance peaks,with notably peaks at 2 and 4 G_0 absent, can be explained in terms of the free-electron waves inside a cylindrical nanowire. The conductance modes for such a wire are described by Bessel functions, where modes 2 and 3, and modes 4 and 5 are degenerate [38, 39].

In a free-electron-like metal the conduction electrons are also responsible for the cohesion of the material. Several groups have considered a possible modulation of the cohesive energy of metallic nanowires as a result of the occupation of a discrete set of quantum modes [40, 41, 42]. This mechanism forms the basis for an interpretation of rich peak structure observed in the conductance histograms for sodium at higher temperatures [43].

7. CONCLUSIONS

With the developments of the last few years a microscopic understanding of atomic-scale electrical transport properties is beginning to grow. A coherent picture of conductance modes in a single atom derived from atomic valence orbitals is obtained, which finds strong support in the experimental observations for various metallic elements. Further evidence is obtained from conductance fluctuations [37], thermopower [44], and the strain dependence of the conductance on the last conductance plateau [11]. On the other hand, the predictions of simple free-electron-gas models can be observed in alkali metal contacts. The relative importance of the two pictures will be subject of discussion for some time to come and it will be interesting to investigate a possible cross-over between the two.

References

[1] B. J. van Wees, H. van Houten, C. W. J. Beenakker, J. G. Williamson, L. P. Kouwenhoven, D. van der Marel and C. T. Foxon, Quantised conductance of point contacts in a two-dimensional electron gas, Phys. Rev. Lett. **60**, 848–850 (1988).

[2] D. A. Wharam, T. J. Thornton, R. Newbury, M. Pepper, H. Ahmed, J. E. F. Frost, D. G. Hasko, D. C. Peacock, D. A. Ritchie and G. A. C. Jones, One-dimensional transport and the quantisation of the ballistic resistance, J. Phys. C **21**, L209–L214 (1988).

[3] C. W. J. Beenakker and H. van Houten, Josephson current through a superconducting quantum point contact shorter than the coherence length, Phys. Rev. Lett. **66**, 3056–3059 (1991).

[4] Y. Imry, Introduction to mesoscopic physics (Oxford University Press, Oxford, 1997).

[5] M. Büttiker, Coherent and sequential tunneling in series barriers, IBM J. Res. Dev. **32**, 63–75 (1988).

[6] M. Brandbyge, M. R. Sørensen and K. W. Jacobsen, Conductance eigenchannels in nanocontacts, Phys. Rev. B **56**, 14956–14959 (1997).

[7] J. C. Cuevas, A. Levy Yeyati and A. Martín-Rodero, Microscopic origin of conducting channels in metallic atomic-size contacts, Phys. Rev. Lett. **80**, 1066–1069 (1998).

[8] J. K. Gimzewski and R. Möller, Transition from the tunneling regime to point contact studied using scanning tunneling microscopy, Physica B **36**, 1284–1287 (1987).

[9] N. Agraït, J. G. Rodrigo and S. Vieira, Conductance steps and quantization in atomic-size contacts, Phys. Rev. B **47**, 12345–12348 (1993).

[10] R. Rubio, N. Agraït and S. Vieira, Atomic-sized metallic contacts: Mechanical properties and electronic transport, Phys. Rev. Lett. **76**, 2302–2305 (1996).

[11] J. C. Cuevas, A. Levy Yeyati, A. Martín-Rodero, G. Rubio Bollinger, C. Untiedt and N. Agraït, Evolution of conducting channels in metallic atomic contacts under elastic deformation, Phys. Rev. Lett. **81**, 2990–2993 (1998).

[12] J. Moreland and J. W. Ekin, Electron tunneling experiments using Nb-Sn 'break' junctions, J. Appl. Phys. **58**, 3888–3895 (1985).

[13] C. J. Muller, J. M. van Ruitenbeek and L. J. de Jongh, Conductance and supercurrent discontinuities in atomic-scale metallic constrictions of variable width, Phys. Rev. Lett. **69**, 140–143 (1992).

[14] J. M. Krans, Size effects in atomic-scale point contacts, (PhD thesis, Universiteit Leiden, The Netherlands, 1996).

[15] E. N. Bratus', V. S. Shumeiko and G. Wendin, Theory of subharmonic gap structure in superconducting mesoscopic tunnel contacts, Phys. Rev. Lett. **74**, 2110–2113 (1995).

[16] V. S. Shumeiko, E. N. Bratus' and G. Wendin, dc-current transport and ac josephson effect in quantum junctions at low voltage, Low Temp. Phys. **23**, 181–192 (1997).

[17] E. N. Bratus', V. S. Shumeiko, E. V. Bezuglyi and G. Wendin, dc-current transport and ac josephson effect in quantum junctions at low voltage, Phys. Rev. B **55**, 12666–12677 (1997).

[18] B. Ludoph, N. van der Post, E. N. Bratus', E. V. Bezuglyi, V. S. Shumeiko, G. Wendin and J. M. van Ruitenbeek, Multiple Andreev reflection in single atom niobium junctions, Phys. Rev. B **61**, 8561–8569 (2000).

[19] V. Ambegaokar and A. Baratoff, Tunneling between superconductors, Phys. Rev. Lett. **10**, 486–489 (1963).

[20] M. Goffman and C. Urbina, to be published.

[21] C. J. Muller, J. M. van Ruitenbeek and L. J. de Jongh, Experimental observation of the transition from weak link to tunnel junction, Physica C **191**, 485–504 (1992).

[22] D. Averin and A. Bardas, ac josephson effect in a single quantum channel, Phys. Rev. Lett. **75**, 1831–1834 (1995).

[23] J. C. Cuevas, A. Martín-Rodero and A. Levy Yeyati, Hamiltonian approach to the transport properties of superconducting quantum point contacts, Phys. Rev. B **54**, 7366–7379 (1996).

[24] E. Scheer, P. Joyez, D. Esteve, C. Urbina and M. H. Devoret, Conduction channel transmissions of atomic-size aluminum contacts, Phys. Rev. Lett. **78**, 3535–3538 (1997).

[25] E. Scheer, N. Agraït, J. C. Cuevas, A. Levy Yeyati, B. Ludoph, A. Martín-Rodero, G. Rubio Bollinger, J. M. van Ruitenbeek and C. Urbina, The signature of chemical valence in the electrical conduction through a single-atom contact, Nature **394**, 154–157 (1998).

[26] W. Schottky, Über spontane Stromschwankungen in verschiedenen Elektrizitätsleitern, Ann. Phys. (Leipzig) **57**, 541–567 (1918).

[27] G. B. Lesovik, Excess quantum noise in 2d ballistic point contacts, Sov. Phys. JETP Lett. **49**, 592–594 (1989), [Pis'ma Zh. Eksp. Teor. Fiz. **49**, 513 (1989)].

[28] M. Büttiker, Scattering theory of thermal and excess noise in open conductors, Phys. Rev. Lett. **65**, 2901–2904 (1990).

[29] M. Reznikov, M. Heiblum, H. Shtrikman and D. Mahalu, Temporal correlation of electrons: suppression of shot noise in a ballistic quantum point contact, Phys. Rev. Lett. **75**, 3340–3343 (1995).

[30] A. Kumar, L. Saminadayar, D. C. Glattli, Y. Jin and B. Etienne, Experimental test of the quantum shot noise reduction theory, Phys. Rev. Lett. **76**, 2778–2781 (1996).

[31] H. E. van den Brom and J. M. van Ruitenbeek, Quantum suppression of shot noise in atom-size metallic contacts, Phys. Rev. Lett. **82**, 1526–1529 (1999).

[32] J. Bürki and C. A. Stafford, Comment on "quantum suppression of shot noise in atom-size metallic contacts", Phys. Rev. Lett. **83**, 3342 (1999).

[33] A. I. Yanson, G. Rubio Bollinger, H. E. van den Brom, N. Agraït and J. M. van Ruitenbeek, Formation and manipulation of a metallic wire of single gold atoms, Nature **395**, 783–785 (1998).

[34] H. E. van den Brom, Fluctuation phenomena in atomic-size contacts, (PhD thesis, Universiteit Leiden, The Netherlands, 2000).

[35] J. M. Krans, J. M. van Ruitenbeek, V. V. Fisun, I. K. Yanson and L. J. de Jongh, The signature of conductance quantization in metallic point contacts, Nature **375**, 767–769 (1995).

[36] A. I. Yanson and J. M. van Ruitenbeek, to be published.

[37] B. Ludoph, M. H. Devoret, D. Esteve, C. Urbina and J. M. van Ruitenbeek, Evidence for saturation of channel transmission from conductance fluctuations in atomic-size point contacts, Phys. Rev. Lett. **82**, 1530–1533 (1999).

[38] E. N. Bogachek, A. N. Zagoskin and I. O. Kulik, Conductance jumps and magnetic flux quantization in ballistic point contacts, Sov. J. Low Temp. Phys. **16**, 796–800 (1990).

[39] J. A. Torres, J. I. Pascual and J. J. Sáenz, Theory of conduction through narrow constrictions in a three-dimensional electron gas, Phys. Rev. B **49**, 16581–16584 (1994).

[40] J. M. van Ruitenbeek, M. H. Devoret, D. Esteve and C. Urbina, Conductance quantization in metals: The influence of subband formation on the relative stability of specific contact diameters, Phys. Rev. B **56**, 12566–12572 (1997).

[41] C. A. Stafford, D. Baeriswyl and J. Bürki, Jellium model of metallic nanocohesion, Phys. Rev. Lett. **79**, 2863–2866 (1997).

[42] C. Yannouleas and U. Landman, On mesoscopic forces and quantized conductance in model metallic nanowires, J. Phys. Chem. B **101**, 5780–5783 (1997).

[43] A. I. Yanson, I. K. Yanson, and J. M. van Ruitenbeek, Observation of shell structure in sodium nanowires, Nature **400**, 144–146 (1999).

[44] B. Ludoph and J. M. van Ruitenbeek, Thermopower of atomic-size metallic contacts, Phys. Rev. B **59**, 12290–12293 (1999).

Chapter 4

WHY DOES A METAL–SUPERCONDUCTOR JUNCTION HAVE A RESISTANCE?

C. W. J. Beenakker
Instituut-Lorentz, Universiteit Leiden
P.O. Box 9506, 2300 RA Leiden, The Netherlands

Abstract The phenomenon of Andreev reflection is introduced as the electronic analogue of optical phase-conjugation. In the optical problem, a disordered medium backed by a phase-conjugating mirror can become completely transparent. Yet, a disordered metal connected to a superconductor has the same resistance as in the normal state. The resolution of this paradox teaches us a fundamental difference between phase conjugation of light and electrons.

1. INTRODUCTION

In the late sixties, Kulik used the mechanism of Andreev reflection [1] to explain how a metal can carry a dissipationless current between two superconductors over arbitrarily long length scales, provided the temperature is low enough [2]. One can say that the normal metal has become superconducting because of the proximity to a superconductor. This proximity effect exists even if the electrons in the normal metal have no interaction. At zero temperature the maximum supercurrent that the metal can carry decays only algebraically with the separation between the superconductors — rather than exponentially, as it does at higher temperatures.

The recent revival of interest in the proximity effect has produced a deeper understanding into how the proximity-induced superconductivity of non-interacting electrons differs from true superconductivity of electrons having a pairing interaction. Clearly, the proximity effect does not require two superconductors. One should be enough. Consider a junc-

I. O. Kulik and R. Ellialtioğlu (eds.),
Quantum Mesoscopic Phenomena and Mesoscopic Devices in Microelectronics, 51–60.

tion between a normal metal and a superconductor (an NS junction). Let the temperature be zero. What is the resistance of this junction? One might guess that it should be smaller than in the normal state, perhaps even zero. Isn't that what the proximity effect is all about?

The answer to this question has been in the literature since 1979 [3], but it has been appreciated only in the last few years. A recent review [4] gives a comprehensive discussion within the framework of the semiclassical theory of superconductivity. A different approach, using random-matrix theory, was reviewed by the author [5]. In this lecture we take a more pedestrian route, using the analogy between Andreev reflection and optical phase-conjugation [6, 7] to answer the question: Why does an NS junction have a resistance?

2. ANDREEV REFLECTION AND OPTICAL PHASE-CONJUGATION

It was first noted by Andreev in 1963 [1] that an electron is reflected from a superconductor in an unusual way. The differences between normal reflection and Andreev reflection are illustrated in Fig. 4.1. Let us discuss them separately.

- *Charge is conserved in normal reflection but not in Andreev reflection.* The reflected particle (the hole) has the opposite charge as the incident particle (the electron). This is not a violation of a fundamental conservation law. The missing charge of $2e$ is absorbed into the superconducting ground state as a Cooper pair. It is missing only with respect to the excitations.

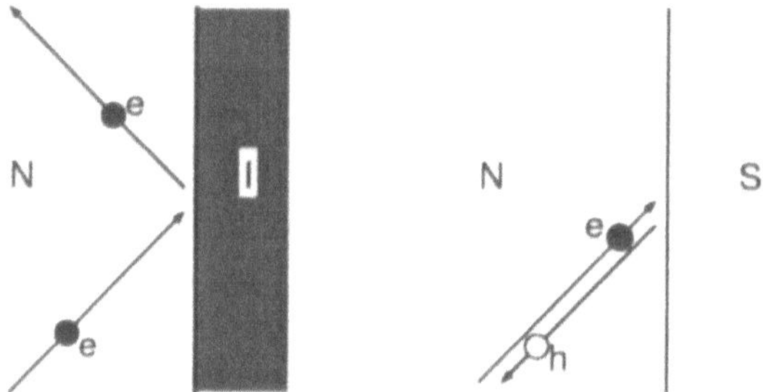

Figure 4.1 Normal reflection by an insulator (I) versus Andreev reflection by a superconductor (S) of an electron excitation in a normal metal (N) near the Fermi level. Normal reflection (left) conserves charge but does not conserve momentum. Andreev reflection (right) conserves momentum but does not conserve charge: The electron (e) is reflected as a hole (h) with the same momentum and opposite velocity. The missing charge of $2e$ is absorbed as a Cooper pair by the superconducting condensate.

- *Momentum is conserved in Andreev reflection but not in normal reflection.* The conservation of momentum is an approximation, valid if the superconducting excitation gap Δ is much smaller than the Fermi energy E_F of the normal metal. The explanation for the momentum conservation is that the superconductor can not exert a significant force on the incident electron, because Δ is too small compared to the kinetic energy E_F of the electron [8]. Still, the superconductor has to reflect the electron somehow, because there are no excited states within a range Δ from the Fermi level. It is the unmovable rock meeting the irresistible object. Faced with the challenge of having to reflect a particle without changing its momentum, the superconductor finds a way out by transforming the electron into a particle whose velocity is opposite to its momentum: a hole.

- *Energy is conserved in both normal and Andreev reflection.* The electron is at an energy ε above the Fermi level and the hole is at an energy ε below it. Both particles have the same excitation energy ε. Andreev reflection is therefore an *elastic* scattering process.

- *Spin is conserved in both normal and Andreev reflection.* To conserve spin, the hole should have the opposite spin as the electron. This spin-flip can be ignored if the scattering properties of the normal metal are spin-independent.

The NS junction has an optical analogue known as a phase-conjugating mirror [9]. Phase conjugation is the effect that an incoming wave $\propto \cos(kx - \omega t)$ is reflected as a wave $\propto \cos(-kx - \omega t)$, with opposite sign of the phase kx. Since $\cos(-kx-\omega t) = \cos(kx+\omega t)$, this is equivalent to reversing the sign of the time t, so that phase conjugation is sometimes called a time-reversal operation. The reflected wave has a wavevector precisely opposite to that of the incoming wave, and therefore propagates back along the incoming path. This is called retro-reflection. Phase conjugation of light was discovered in 1970 by Woerdman and by Stepanov, Ivakin, and Rubanov [10, 11].

A phase-conjugating mirror for light (see Fig. 4.2) consists of a cell containing a liquid or crystal with a large nonlinear susceptibility. The cell is pumped by two counter-propagating beams at frequency ω_0. A third beam is incident with a much smaller amplitude and a slightly different frequency $\omega_0 + \delta\omega$. The non-linear susceptibility leads to an amplification of the incident beam, which is transmitted through the cell, and to the generation of a fourth beam, which is reflected. This non-linear optical process is called "four-wave mixing". Two photons of

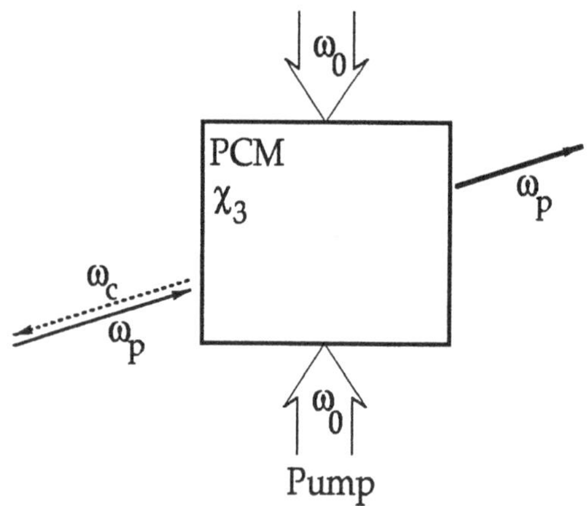

Figure 4.2 Schematic drawing of optical phase-conjugation by means of four-wave mixing. The phase-conjugating mirror (PCM) consists of a cell filled by a medium with a third-order non-linear susceptibility χ_3. (Examples are $BaTiO_3$ and CS_2.) The medium is pumped by two counter-propagating beams at frequency ω_0. A probe beam incident at frequency $\omega_p = \omega_0 + \delta\omega$ is then retro-reflected as a conjugate beam at frequency $\omega_c = \omega_0 - \delta\omega$. From Ref. [12].

the pump beams are converted into one photon for the transmitted beam and one for the reflected beam. Energy conservation dictates that the reflected beam has frequency $\omega_0 - \delta\omega$. Momentum conservation dictates that its wavevector is opposite to that of the incident beam. Comparing retro-reflection of light with Andreev reflection of electrons, we see that the Fermi energy E_F plays the role of the pump frequency ω_0, while the excitation energy ε corresponds to the frequency shift $\delta\omega$.

A phase-conjugating mirror can be used for wavefront reconstruction. Imagine an incoming plane wave, that is distorted by some inhomogeneity. When this distorted wave falls on the mirror, it is phase conjugated and retro-reflected. Due to the time-reversal effect, the inhomogeneity that had distorted the wave now changes it back to the original plane wave. An example is shown in Fig. 4.3. Complete wavefront reconstruction is possible only if the distorted wavefront remains approximately planar, since perfect time reversal upon reflection holds only in a narrow range of angles of incidence for realistic systems. This is an important, but not essential complication, that we will ignore in what follows.

3. THE RESISTANCE PARADOX

We have learned that a disordered medium (such as the frosted glass in Fig. 4.3) becomes transparent when it is backed by a phase-conjugating mirror. By analogy, one would expect that a disordered metal backed by a superconductor would become "transparent" too, meaning that its resistance should vanish (up to a small contact resistance that is present even without any disorder). This does not happen. Upon decreasing the temperature below the superconducting transition temperature, the resistance drops slightly but then rises again back to its high-temperature value. (A recent experiment is shown in Fig. 4.4, where the conduc-

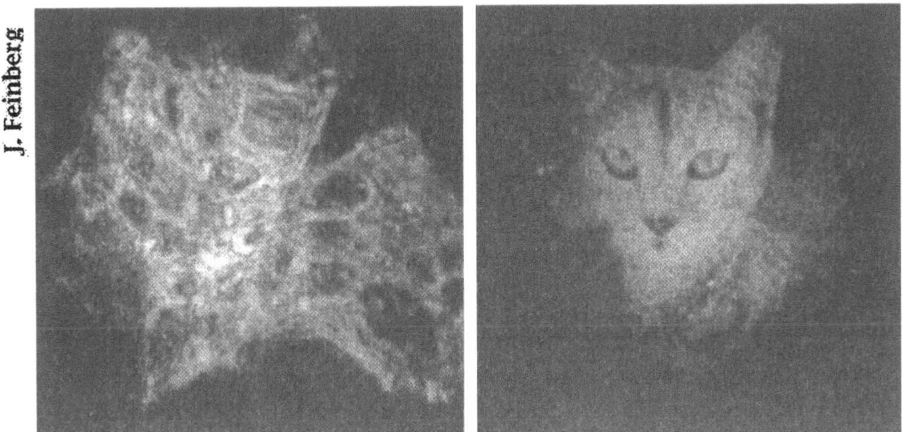

Figure 4.3 Example of wavefront reconstruction by optical phase-conjugation. In both photographs the image of a cat was distorted by transmitting it through a piece of frosted glass, and reflecting it back through the same piece of glass. This gives an unrecognizable image when reflected by an ordinary mirror (left panel) and the original image when reflected by a phase-conjugating mirror (right panel). From Ref. [13].

tance is plotted instead of the resistance.) This so-called "re-entrance effect" has been reviewed recently by Courtois *et al.* [4], and we refer to

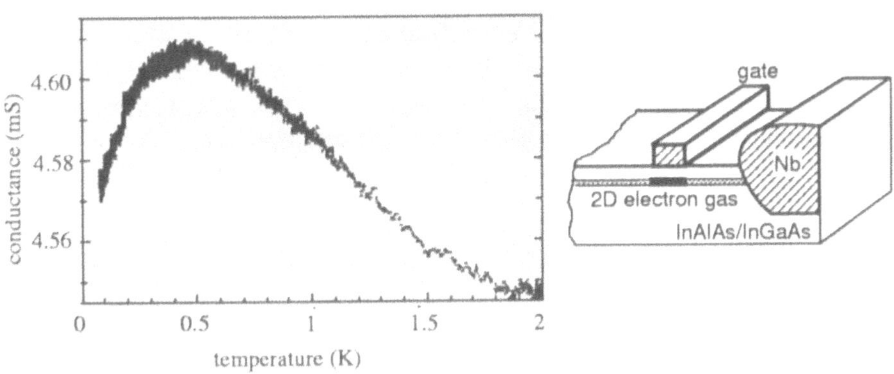

Figure 4.4 Temperature dependence of the conductance of an NS junction, showing the re-entrance effect. The superconductor is Nb, the normal metal is a two-dimensional electron gas. A gate creates a strongly disordered region in the 2D gas that dominates the conductance of the junction. Upon lowering the temperature the conductance first rises and then drops again. Under ideal circumstances the low- and high-temperature limits would be the same. From Ref. [16].

that review for an extensive list of references. The theoretical prediction [3, 14, 15] is that *at zero temperature the resistance of the normal-metal–superconductor junction is the same as in the normal state.* How can we reconcile this with the notion of Andreev reflection as a "time-reversing" process, analogous to optical phase-conjugation? To resolve this paradox, let us study the analogy more carefully, to see where it breaks down.

For a simple discussion it is convenient to replace the disordered medium by a tunnel barrier (or semi-transparent mirror) and consider the phase shift accumulated by an electron (or light wave) that bounces back and forth between the barrier and the superconductor (or phase-conjugating mirror). A periodic orbit (see Fig. 4.5) consists of two round-trips, one as an electron (or light at frequency $\omega_0 + \delta\omega$), the other as a hole (or light at frequency $\omega_0 - \delta\omega$). The miracle of phase conjugation is that phase shifts accumulated in the first round trip are cancelled in the second round trip. If this were the whole story, one would conclude that the net phase increment is zero, so all periodic orbits would interfere constructively and the tunnel barrier would become transparent because of resonant tunneling.

But it is not the whole story. There is an extra phase shift of $-\pi/2$ acquired upon Andreev reflection that destroys the resonance. Since the periodic orbit consists of two Andreev reflections, one from electron to hole and one from hole to electron, and both reflections have the same phase shift $-\pi/2$, the net phase increment of the periodic orbit is $-\pi$ and not zero. So subsequent periodic orbits interfere destructively, rather than constructively, and tunneling becomes suppressed rather than enhanced. In contrast, a phase-conjugating mirror adds a phase shift that alternates between $+\pi/2$ and $-\pi/2$ from one reflection to the next, so the net phase increment of a periodic orbit remains zero.

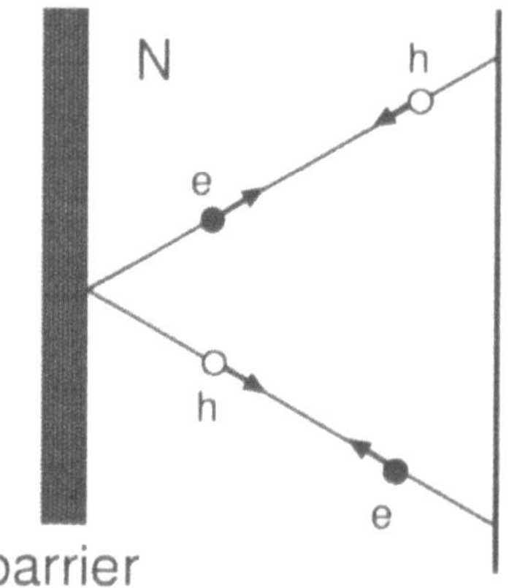

Figure 4.5 Periodic orbit consisting of two normal reflections and two retro-reflections. The net phase increment is zero in the optical case and $-\pi$ in the electronic case. Hence the periodic orbits interfere constructively for light and destructively for electrons. This explains why the barrier becomes transparent for light but not for electrons.

For a more quantitative description of the conductance we need to compute the probability R_{he} that an incident electron is reflected as a hole. The matrix of probability amplitudes r_{he} can be constructed as a geometric series of multiple reflections:

$$\begin{aligned} r_{\text{he}} &= t^{\dagger}\frac{1}{\mathrm{i}}t + t^{\dagger}\frac{1}{\mathrm{i}}r\frac{1}{\mathrm{i}}r^{\dagger}\frac{1}{\mathrm{i}}t + t^{\dagger}\frac{1}{\mathrm{i}}\left[r\frac{1}{\mathrm{i}}r^{\dagger}\frac{1}{\mathrm{i}}\right]^{2} t + \cdots \\ &= t^{\dagger}\frac{1}{\mathrm{i}}\left[1 - r\frac{1}{\mathrm{i}}r^{\dagger}\frac{1}{\mathrm{i}}\right]^{-1} t. \end{aligned} \tag{4.1}$$

Each factor $1/\mathrm{i} = \exp(-\mathrm{i}\pi/2)$ corresponds to an Andreev reflection. The matrices $t, t^{\dagger}$ and $r, r^{\dagger}$ are the $N \times N$ transmission and reflection matrices of the tunnel barrier, or more generally, of the disordered region in the normal metal. (The number N is related to the cross-sectional area A of the junction and the Fermi wavelength λ_{F} by $N \simeq A/\lambda_{\text{F}}^{2}$.) The matrices t, r pertain to the electron and the matrices $t^{\dagger}, r^{\dagger}$ to the hole. The resulting reflection probability $R_{\text{he}} = N^{-1}\,\text{Tr}\, r_{\text{he}} r_{\text{he}}^{\dagger}$ is given by [14]

$$R_{\text{he}} = \frac{1}{N}\,\text{Tr}\left(\frac{tt^{\dagger}}{1 + rr^{\dagger}}\right)^{2} = \frac{1}{N}\,\text{Tr}\left(\frac{tt^{\dagger}}{2 - tt^{\dagger}}\right)^{2}. \tag{4.2}$$

We have used the relationship $tt^{\dagger} + rr^{\dagger} = 1$, dictated by current conservation. The conductance G_{NS} of the NS junction is related to R_{he} by [17, 18]

$$G_{\text{NS}} = \frac{4e^{2}}{h} N R_{\text{he}}. \tag{4.3}$$

In the optical analogue one has the probability $R_{\pm}$ for an incident light wave with frequency $\omega_0 + \delta\omega$ to be reflected into a wave with frequency $\omega_0 - \delta\omega$. The matrix of probability amplitudes is given by the geometric series

$$\begin{aligned} r_{\pm} &= t^{\dagger}\frac{1}{\mathrm{i}}t + t^{\dagger}\frac{1}{\mathrm{i}}r\mathrm{i}r^{\dagger}\frac{1}{\mathrm{i}}t + t^{\dagger}\frac{1}{\mathrm{i}}\left[r\mathrm{i}r^{\dagger}\frac{1}{\mathrm{i}}\right]^{2} t + \cdots \\ &= t^{\dagger}\frac{1}{\mathrm{i}}\left[1 - r\mathrm{i}r^{\dagger}\frac{1}{\mathrm{i}}\right]^{-1} t. \end{aligned} \tag{4.4}$$

The only difference with Eq. (4.1) is the alternation of factors $1/\mathrm{i}$ and i, corresponding to the different phase shifts $\exp(\pm\mathrm{i}\pi/2)$ acquired at the phase-conjugating mirror. The reflection probability $R_{\pm} = N^{-1}\,\text{Tr}\, r_{\pm} r_{\pm}^{\dagger}$ now becomes independent of the disorder [19],

$$R_{\pm} = \frac{1}{N}\,\text{Tr}\left(\frac{tt^{\dagger}}{1 - rr^{\dagger}}\right)^{2} = 1. \tag{4.5}$$

The disordered medium has become completely transparent.

It is remarkable that a small difference in phase shifts has such far reaching consequences. Note that one needs to consider multiple reflections in order to see the difference: The first term in the series is the same in Eqs. (4.1) and (4.4). That is probably why this essential difference between Andreev reflection and optical phase-conjugation was not noticed prior to Ref. [19].

4. HOW BIG IS THE RESISTANCE?

Now that we understand why a disordered piece of metal connected to a superconductor does not become transparent, we would like to go one step further and ask whether the resistance (or conductance) is bigger or smaller than without the superconductor. To that end we compare, following Ref. [14], the expression for the conductance of the NS junction [obtained from Eqs. (4.2) and (4.3)],

$$G_{\mathrm{NS}} = \frac{4e^2}{h} \sum_{n=1}^{N} \frac{T_n^2}{(2-T_n)^2}, \tag{4.6}$$

with the Landauer formula for the normal-state conductance,

$$G_{\mathrm{N}} = \frac{2e^2}{h} \sum_{n=1}^{N} T_n. \tag{4.7}$$

The numbers $T_1, T_2, \ldots T_N$ are the eigenvalues of the matrix product $tt^\dagger$. These transmission eigenvalues are real numbers between 0 and 1 that depend only on the properties of the metal (regardless of the superconductor). Both formulas (4.6) and (4.7) hold at zero temperature, so we will be comparing the zero-temperature limits of G_{NS} and G_{N}.

Since $x^2/(2-x)^2 \leq x$ for $x \in [0,1]$, we can immediately conclude that $G_{\mathrm{NS}} \leq 2G_{\mathrm{N}}$. If there is no disorder, then all T_n's are equal to unity, hence G_{NS} reaches its maximum value of $2G_{\mathrm{N}}$. For a tunnel barrier all T_n's are $\ll 1$, hence G_{NS} drops far below G_{N}. A disordered metal will lie somewhere in between these two extremes, but where?

We have already alluded to the answer in the previous section, that $G_{\mathrm{NS}} = G_{\mathrm{N}}$ for a disordered metal in the zero-temperature limit. To derive this remarkable equality, we parameterize the transmission eigenvalue T_n in terms of the localization length ζ_n,

$$T_n = \frac{1}{\cosh^2(L/\zeta_n)}, \tag{4.8}$$

where L is the length of the disordered region. Substitution into Eqs. (4.6) and (4.7) gives the average conductances

$$\langle G_{\rm NS}\rangle_L = \frac{4e^2}{h}N\int_0^\infty d\zeta\, P_L(\zeta)\cosh^{-2}(2L/\zeta), \qquad (4.9)$$

$$\langle G_{\rm N}\rangle_L = \frac{2e^2}{h}N\int_0^\infty d\zeta\, P_L(\zeta)\cosh^{-2}(L/\zeta). \qquad (4.10)$$

(For Eq. (4.9) we have used that $2\cosh^2 x - 1 = \cosh 2x$.) The probability distribution $P_L(\zeta)$ of ζ is independent of L in a range of lengths between l and Nl [5]. It then follows immediately that

$$\langle G_{\rm NS}\rangle_L = 2\langle G_{\rm N}\rangle_{2L}. \qquad (4.11)$$

Since $G_{\rm N} \propto 1/L$, according to Ohm's law, we arrive at the equality of $G_{\rm NS}$ and $G_{\rm N}$.

The restriction to the range $l \ll L \ll Nl$ is the restriction to the regime of diffusive transport: For smaller L we enter the ballistic regime and $G_{\rm NS}$ rises to $2G_{\rm N}$; For larger L we enter the localized regime, where tunneling takes over from diffusion and $G_{\rm NS}$ becomes $\ll G_{\rm N}$.

5. CONCLUSION

We have learned a fundamental difference between Andreev reflection of electrons and phase-conjugation of light. While it is appealing to think of the Andreev reflected hole as the time reverse of the incident electron, this picture breaks down upon closer inspection. The phase shift of $-\pi/2$ acquired upon Andreev reflection spoils the time-reversing properties and explains why a disordered metal does not become transparent when connected to a superconductor.

Acknowledgements

The research on which this lecture is based was done in collaboration with J. C. J. Paasschens. It was supported by the "Stichting voor Fundamenteel Onderzoek der Materie" (FOM) and by the "Nederlandse organisatie voor Wetenschappelijk Onderzoek" (NWO).

References

[1] A. F. Andreev, Zh. Eksp. Teor. Fiz. **46**, 1823 (1964) [Sov. Phys. JETP **19**, 1228 (1964)].

[2] I. O. Kulik, Zh. Eksp. Teor. Fiz. **57**, 1745 (1969) [Sov. Phys. JETP **30**, 944 (1970)].

[3] S. N. Artemenko, A. F. Volkov, and A. V. Zaĭtsev, Solid State Comm. **30**, 771 (1979).

[4] H. Courtois, P. Charlat, Ph. Gandit, D. Mailly, and B. Pannetier, J. Low Temp. Phys. **116**, 187 (1999).
[5] C. W. J. Beenakker, Rev. Mod. Phys. **69**, 731 (1997).
[6] D. Lenstra, in: *Huygens Principle 1690–1990; Theory and Applications*, edited by H. Blok, H. A. Ferweda, and H. K. Kuiken (North-Holland, Amsterdam, 1990).
[7] H. van Houten and C. W. J. Beenakker, Physica B **175**, 187 (1991).
[8] A. A. Abrikosov, *Fundamentals of the Theory of Metals* (North-Holland, Amsterdam, 1988).
[9] D. M. Pepper, Sci. Am. **254** (1), 56 (1986).
[10] J. P. Woerdman, Optics Comm. **2**, 212 (1970).
[11] B. I. Stepanov, E. V. Ivakin, and A. S. Rubanov, Dokl. Akad. Nauk USSR **196**, 567 (1971) [Sov. Phys. Dokl. **16**, 46 (1971)].
[12] J. C. J. Paasschens, Ph.D. Thesis (Universiteit Leiden, 1997).
[13] J. Feinberg, Opt. Lett. **7**, 486 (1982).
[14] C. W. J. Beenakker, Phys. Rev. B **46**, 12841 (1992).
[15] Yu. V. Nazarov and T. H. Stoof, Phys. Rev. Lett. **76**, 823 (1996).
[16] E. Toyoda, H. Takayanagi, and H. Nakano, Phys. Rev. B **59**, 11653 (1999).
[17] G. E. Blonder, M. Tinkham, and T. M. Klapwijk, Phys. Rev. B **25**, 4515 (1982).
[18] Y. Takane and H. Ebisawa, J. Phys. Soc. Japan **61**, 3466 (1992).
[19] J. C. J. Paasschens, M. J. M. de Jong, P. W. Brouwer, and C. W. J. Beenakker, Phys. Rev. A **56**, 4216 (1997).

Chapter 5

POINT-CONTACT SPECTROSCOPY OF SUPERCONDUCTORS

I. K. Yanson
B. Verkin Institute for Low Temperature Physics and Engineering
National Academy of Sciences of Ukraine, 310164 Kharkiv, Ukraine

1. INTRODUCTION

In this review we describe the current-voltage spectroscopy of superconductors which, in contrast to the tunneling spectroscopy, considers contacts with direct conductivity, without any barrier. These contacts are usually called point contacts as their lateral dimension should be smaller than some microscopic lengths such as the coherence length in superconductors and the inelastic mean free path of conduction electrons in normal metals. Some of 3-dimensional metallic constrictions are schematically depicted in Fig. 5.1. A shorted thin-film tunneling junction (*a*) was historically the first structure in which the point-contact spectroscopy was discovered [1]. A much more sophisticated thin-film structure is displayed in (*b*) where the tiny controlled orifice was made in the insulating diaphragm of Si_3N_4 by means of ion-beam nanolithography [2]. Point-contacts with bulk electrodes are shown in Fig. 5.1(*c*-*e*) for different geometries. The simplest are the pressure-type contacts in which two sharp edges (*c*) [3] or a needle and an anvil (*d*) [4] are forced to touch each other gently. Most point-contact spectroscopic studies were made just with these very simple experimental devices. The mechanically controlled break-junction (MCB) (Fig. 5.1*e*) gives an experimentalist new advantages [5]. A small rode of a metal 1 is glued by the Staycast varnish 3 to the bending beam 4. The latter can be bent by a piezo driver 5 so that the notch 2 at the centre of the sample becomes stretched and the cross-section area of the constriction is gradually decreased. This device allows us to study the contact properties as a function of constriction size.

I. O. Kulik and R. Ellialtioğlu (eds.),
Quantum Mesoscopic Phenomena and Mesoscopic Devices in Microelectronics, 61–77.

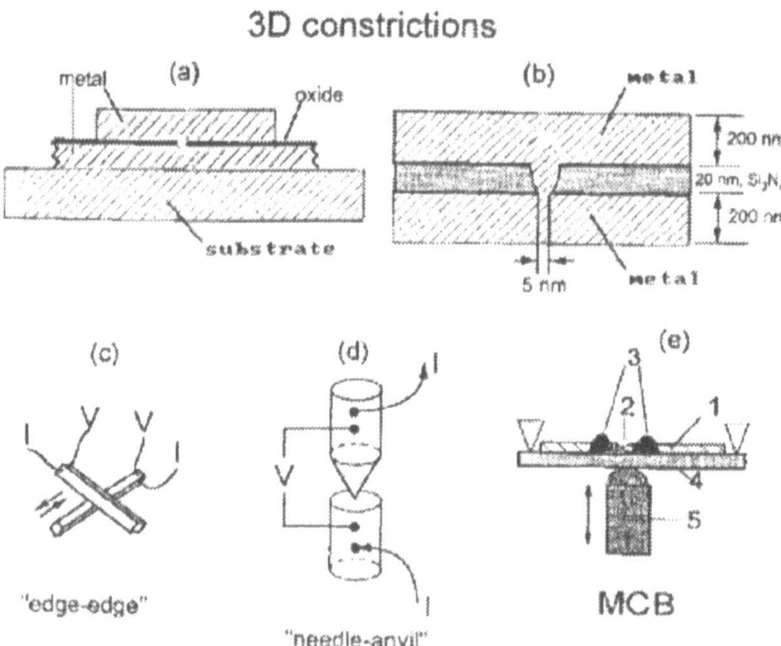

Figure 5.1 The three-dimensional constrictions made by different technique: (*a*) shorted thin-film tunneling structure [1]; (*b*) thin-film point contact made by nanolithography [2]; (*c*) contacted sharp edges of bulk metals; arrows show the movement of electrode to choose the proper spot for measuring [3]; (*d*) "spear-anvil" geometry of the bulk electrodes [4]; (*e*) mechanically controllable break-junction [5]; arrow shows the movement of piezo drive 5 in order to bend the substrate 4.

In this review we do not concern the so-called Andreev-reflection spectroscopy of superconducting energy gap. The reader can find this in Refs. [6, 7, 8] and later extensions [9, 10]. Our focus will be on the above-gap-energies where one could expect nonlinearities caused by the electron-boson interactions responsible for the formation of Cooper pairs. The problem unsolved yet is that in standard superconductors (lead, tin, Nb_3Sn and many others) the above-gap nonlinearities give the evidence about the electron-phonon-interaction mechanism of Cooper pairing [11], whilst the tunnel junctions of unconventional superconductors (high-T_c's, heavy fermions, organic superconductors, *etc.*) do not show any noticeable nonlinearities for energies larger than the superconductor energy gap [12, 13] (albeit see Ref. [14]). On the contrary, the point contacts do manifest strong nonlinearities at above-gap energies which in many cases correlate with the boson density of states known, for instance, from neutron study [15, 16].

We thus think that the study of nonlinearities of point contacts is the alternative way of investigating the spectral functions of the electron-boson interaction which in many cases is more productive and simpler than tunneling.

2. POINT-CONTACT SPECTROSCOPY IN THE NORMAL STATE

First, we consider the point-contact spectroscopy (PCS) of electron-boson interaction (EBI) in the normal state [17]. Many superconductors can be driven into the normal state by magnetic field, and the PCS theory is more fundamental and transparent in this case. As a model for 3D constrictions, we consider a circular channel of length L and diameter d connecting two bulk metallic half-spaces (Fig. 5.2). If $L \ll d$, we have an orifice in nontransparent infinitely thin partition, while for the opposite case ($L \gg d$) a one-dimensional channel appears. Since contact sizes L, d are much less than the mean free path of conduction electrons $\mathbf{e}^-$ the latter move through the constriction ballistically on applying the voltage bias eV. The resistance of such a ballistic contact was first derived by Sharvin [18] and is equal to

$$R_{Sh} = R_q \left(\frac{16}{d^2 k_F^2} \right); \quad R_q = \frac{\pi \hbar}{e^2} \simeq 12.9\,\mathbf{k\Omega}$$

Suppose that there are some seldom elastic scatterings off static defects or impurities ($\star$) or inelastic processes with emission phonons and other quasiparticles shown in Fig. 5.2 as a wavy line with momenta q. Let us call this regime quasi-ballistic. The nonequilibrium distribution function in the momentum space consists, like in the ballistic regime, of two spherical segments which are displaced in energy by eV and which have the form of two semispheres at the central cross-section of the constriction [19]. At low temperatures we can neglect the smearing of the Fermi edge. Then, the energy conservation sets a sharp threshold $\hbar\omega \leq eV$ on the quasiparticles which can be emitted by the inelastic transition of conduction electron with momenta p to p'. This backscattering processes create opposite current which depends strongly on the bias because of the energy dependence of the quasiparticle density of states and matrix element of EBI. This current is the basis of PCS. To be complete, we must note that in some cases the elastic scattering can also depend on the energy like, for instance, the scattering off the paramagnetic impurities (in Fig. 5.2 these are shown with $\nearrow$) due to the Kondo effect or to two-level systems.

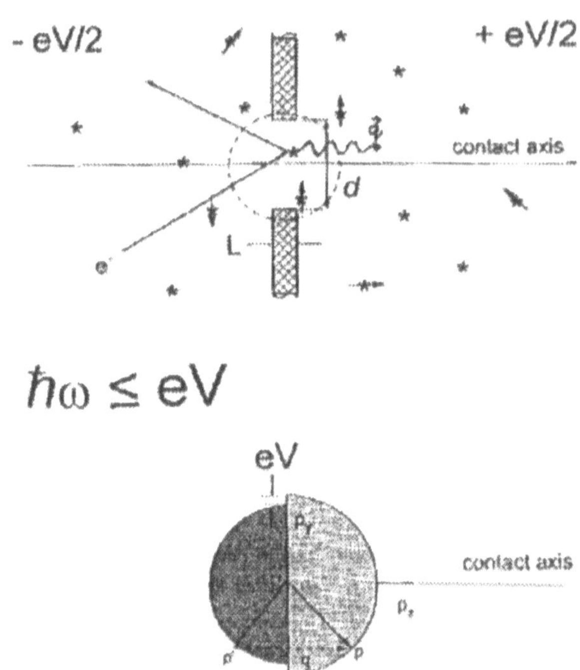

Figure 5.2 Schematic illustration of quasi-ballistic electron flow. Two metallic half spaces are connected by a circular orifice of diameter d in a partition of thickness L. Voltage bias eV is applied between them. Most of the electrons flow ballistically through an orifice but some of them experience rare collisions off static (denoted by stars) and dynamic (denoted by wavy lines) scatterers. Small arrows superimposed on stars denote the impurities with magnetic moments. The dashed circle approximately denotes the area where the nonlinear information to IVC comes from. In the lower part, the nonequilibrium distribution function in the momentum space at the central cross section is drawn. Inelastic backscattering from $\mathbf{p}$ to $\mathbf{p}'$ with emission of quasiparticle (phonon) with momentum $\mathbf{q}$ is shown. These processes have a sharp threshold $\hbar\omega \leq eV$ at low temperatures.

In the quasi-ballistic regime ($l > \max\{d, L\}$) the differential resistance of the point contact is equal to $R(eV) = R_{Sh} + \Delta R(eV)$, where $\Delta R(eV)/R_{Sh} = (8/6\pi)[d/l_i(eV)]$.

For the second derivative of the current-voltage characteristic, which is usually called the point-contact spectrum, we obtain at $T \simeq 0$ a simple relation [19]:

$$\frac{d \ln R(eV)}{d(eV)} = \frac{8}{3}\frac{ed}{\hbar v_F} g_{PC}(eV) \tag{5.1}$$

where $g_{PC}(eV)$ is the transport EBI function which in many cases appears to be the electron-phonon-interaction (EPI) spectral function similar to the Eliashberg EPI function [19]. The point-contact EPI function can be expressed as $g_{PC}(\omega) = \alpha^2_{PC}(\omega)F(\omega)$, where $\alpha^2_{PC}(\omega)$ is the averaged electron-phonon matrix element with kinematic restrictions imposed by the point contact geometry and $F(\omega)$ is the phonon density of states. Thus, we see that PCS allows direct recording of the transport version of EBI (EPI) spectral function which is responsible for the formation of Cooper pairs in superconductors. We show here only one example for tin although many pure metals, alloys and compounds have already been studied by means of PCS [16, 17, 20]-[23]. In Fig. 5.3 (upper panel) the directly recorded PC spectrum of tin is shown by a solid line which is compared with the EPI spectral function (dots) reconstructed from the Eliashberg integral equations by means of the Rowell-McMillan procedure from the superconducting tunneling experiment. Both curve are smeared by the PCS resolution which is shown by the horizontal segment on the graph and which is equal to

$$\delta(eV) = \sqrt{(5.44k_FT)^2 + (1.72eV_1)^2}.$$

The PCS resolution depends on the temperature T and modulation voltage V_1 which is used for recording the second derivative of current-voltage characteristic with a lock-in technique.

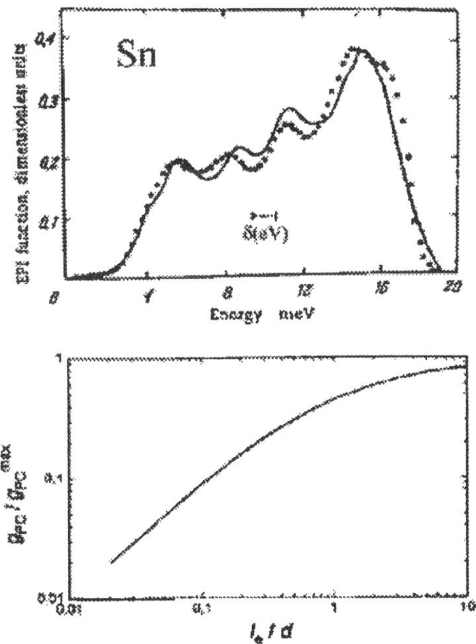

Figure 5.3 (upper panel): Comparison of electron-phonon-interaction spectral functions determined by the superconducting tunneling reconstruction technique [32] (dotted curve) and directly recorded by the point-contact spectroscopy [26] (solid curve). The small horizontal bar denotes the resolution for given parameters. (lower panel): Interpolation curve for spectral intensity between the ballistic limit ($l_e \gg d$) and diffusive regime ($l_e \ll d$) [24].

For symmetrical heterocontacts (point contact with dissimilar electrodes), the PC spectrum at $T \simeq 0$ takes the form:

$$\frac{d \ln R(eV)}{d(eV)} = \frac{4}{3}\frac{ed}{\hbar}\left\{\frac{g_{PC}^{(1)}(eV)}{v_F^{(1)}} + \frac{g_{PC}^{(2)}(eV)}{v_F^{(2)}}\right\} \tag{5.2}$$

where the superscript (i) refers to particular metal. This formula is used for a S-c-N contact.

The heavy elastic scattering around the contact is not necessarily detrimental to PCS as long as the energy relaxation length remains much larger than the contact size:

$$d \ll \sqrt{l_i l_e}.$$

In the case of diffusive movement of charge carriers through the contact ($l_e \ll d$), the contact size d in formula (5.1) should be replaced by the elastic mean free path l_e. The interpolation curve of the spectrum intensity is shown on the lower panel of Fig. 5.3 [24]. It can be used for estimation of l_e inside the contact by measuring the intensity of the PC spectrum. This is important since during making a contact one could introduced a lot of defects into the contact region.

3. EXCESS CURRENT: DEPENDENCE ON PURITY

If one or two of the electrodes of a point contact become superconducting, the excess current emerges, which is superimposed on top of the normal state current. The mechanism is called the Andreev transformation and is as follows: a quasiparticle coming through the constriction has a certain probability to find another electron to form a Cooper pair in the superconducting electrode. This process leads to reflection of quasiparticles which have opposite signs of charge and velocity in the normal metal (in case of S-c-N junction) but the same excitation energy [25]. For biases larger than gap energy this current is constant if strong-coupling, inelastic and nonequilibrium processes are disregarded. Below we consider all of them, but now let us focus on the undisturbed excess current which spreads not too far from the gap region. In Fig. 5.4 (upper panel) the IVC of S-c-S tin junction is plotted [26]. The normal-state zero-bias resistance is compensated by a bridge circuit, hence the deflection of the real IVC in the normal state from the Ohmic behaviour is shown by curve 2. Curve 1 presents the superconducting state of the same contact ($H = 0$). The critical (Josephson) current is clearly visible at $V = 0$, and a small dip is seen at $eV = 2\Delta$. This is because the small barrier appears at the interface between the electrodes due to introduced inhomogeneities while making the contact. The difference between curves 1 and 2 presents the excess current for biases greater than 2Δ which is seen to be approximately constant up to the energies near 5 meV. We shall discuss the nonlinearities of excess current later on and now concentrate on the dependence of $I_{exc,0}$ on purity of the metal. Here we have to consider the purity inside the constriction region which could be quite different from that of the bulk. As a criterion of purity, we may take the maximum intensity of PC spectra which depends on the elastic mean free path averaged over the contact region, as shown by the curve in Fig. 5.3 (lower panel). The dependence of $I_{exc,0}$ in Δ/eR_0 units is shown in the lower panel of Fig. 5.4 as a function of $g_{PC}^{\max}$, where $g_{PC}^{\max}$ is the value of the PC spectrum at the bias around

$$I_{exc}(V) = I_{exc,0} + \delta I_{exc}(V) \tag{5.3}$$

15 meV (see Fig. 5.3 (upper panel)). $I_{exc,0}$ starts approximately from 1.47 for dirty contacts (*i.e.* for small $g_{PC}^{\max}$), as was predicted theoretically by Artemenko, Volkov and Zaitsev (AVZ) [7], and rises with $g_{PC}^{\max}$

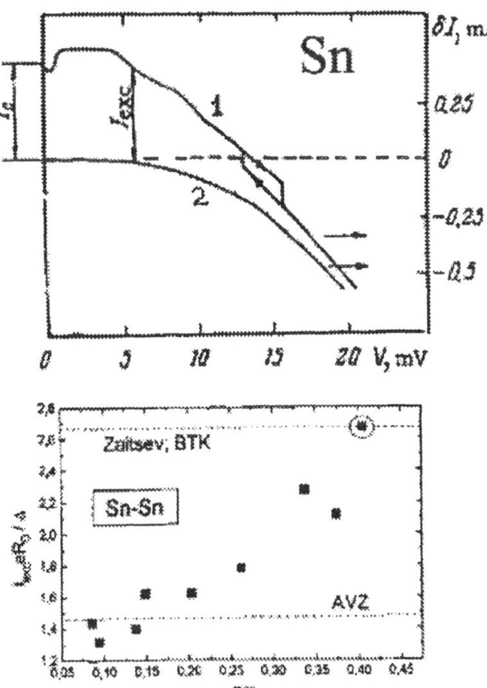

Figure 5.4 The parallel study of S-c-S contacts in superconducting (H=0) and normal (H=0.9 kOe) states of tin [26]. (Upper panel): the excess current as a difference between curves 1 and 2 is plotted versus the voltage bias. Contact resistance R_0=2.55 Ω. (Lower panel): Dependence of excess current in units of Δ/eR_0 on the value of spectral EPI function (determined in the normal state) at the bias of ≃15 meV. Two horizontal straight lines determine the dirty and clean limits according to Refs. [7, 8].

up to the value predicted by Zaitsev and BTK [8, 6] for clean contact[1]. In Fig. 5.4 (lower panel) we show this ultimate value at $g_{PC}^{\max} = 0.41$ taken from Ref. [27], as a dot in a circle. From the graph $g_{PC}/g_{PC}^{\max}(l_e/d)$ of Fig. 5.3 (lower panel) we find that $l_e/d \simeq 0.5$ is quite sufficient to transform a clean S-c-S junction into a dirty one. On the other hand, the BTK theory dealing with *clean* junctions gives the AVZ-dirty value for $I_{exc,0}$ at the barrier parameter $Z \simeq 0.5$ and, what is the most strange, the BTK-fitting leads to approximately the same Z for large diversity of materials, including high-T_c, heavy fermions and conventional superconductors. One might suspect that application of the clean BTK procedure to dirty junctions gives the same $Z \simeq 0.5$ value irrespective of the kind of material.

The parallel study of excess current and intensity of PC spectra allows one to conclude whether the impurities are homogeneously distributed or they are located as a thin barrier at the interface between the electrodes. Indeed, the relative intensity of PC spectra does not saturate for $l_e/d \ll 1$ being proportional to l_e/d, while the excess current does. The same is true when we consider the Z-parameter, instead of the excess current, while fitting the IVC by the BTK theory.

4. ELASTIC CONTRIBUTION TO EXCESS CURRENT

Further on we consider S-c-N junctions which are simpler to interpret due to the absence of the Josephson effect. To be more definite, we restrict ourselves to the electron-phonon interaction. According to Refs. [28, 29], one can write the IVC in the superconducting state as

$$I(V) = V/R_0 - \delta I_{ph}^N(V) + I_{exc}(V) \tag{5.4}$$

where $\delta I_{ph}^N(V)$ is the backscattering current in the normal state whose second derivative is given by Eq. (5.2) and $I_{exc}(V)$ is equal to Eq. (5.3). The order of magnitude of $\delta I_{ph}^N(V)$ is $I(V) \times (d/l_{in})$. The voltage dependent part of $\delta I_{exc}(V)$ can be decomposed in elastic and inelastic parts:

$$\delta I_{exc}(V) = \delta I_{exc}^{el}(V) + \delta I_{exc}^{in}(V)$$

Correspondingly, the order of magnitude of the $\delta I_{exc}^{el}(V)$ and $\delta I_{exc}^{in}(V)$ terms amounts to

$$I_{exc,0} \times (\Delta/\hbar\omega_{ph}) \quad \text{and} \quad I_{exc,0} \times (d/l_{in}) \quad \text{where} \quad I_{exc,0} \simeq \Delta/eR_0. \tag{5.5}$$

[1] For a S-c-N contact the excess current is equal to half of the theoretical values shown in Fig. 5.4 (lower panel).

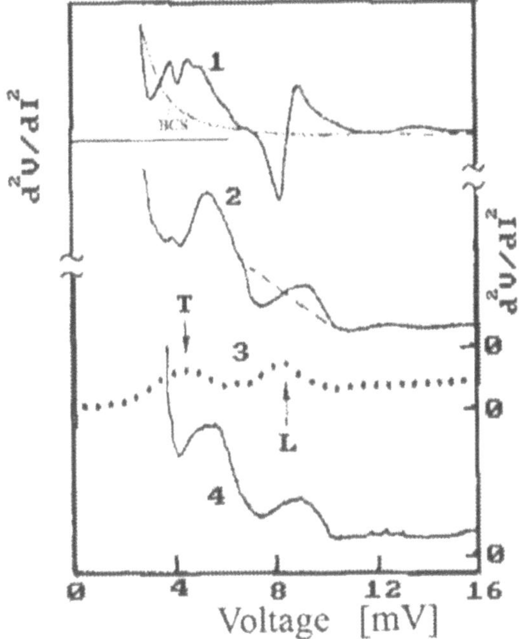

Figure 5.5 Phonon features on the PC spectra of S-c-N contacts for the strong-coupling superconductor. Curves 2 and 4 correspond to the clean Pb-Ru contact and the dirty Pb-Os one, respectively. They are compared with theoretically predicted curve 1 which in turn should be compared with the weak-coupling BCS limit shown with the dashed curve. Dotted curve 3 is the EPI PC spectra of junction 2 with the same ordinate scale for both cases. The curves are shifted vertically for clarity. The letters T and L with arrows mark the positions of transverse and longitudinal phonon peaks in lead, respectively.

Let us consider the elastic component which is essential for strong coupling superconductors when Δ is not much less than ω_{ph}. Fig. 5.5 shows the experimental second derivatives of IVC for lead in the superconducting state of Pb-Ru and Pb-Os point contacts (curves 2 and 4, respectively). Normal metals Ru and Os were chosen since their phonon density of states can be neglected in the energy range where Pb has the highest intensity [30]. The normal state EPI PC spectrum for contact 2 is shown as dotted curve 3 on the same ordinate scale like that for the superconducting state (curve 2). It is clearly seen that in the superconducting state the phonon peaks are shifted to the higher energies roughly by $\Delta(\mathrm{Pb})\simeq 1.3$ mV. Moreover, the intensity of the peaks is larger than that in the normal state. Importantly, the phonon peaks for noticeably dirtier contact (curve 4) are not much smaller, as could be expected from the measurements of EPI-PC-spectrum intensity in the normal state (not shown). These properties qualitatively contradict the theoretical predictions [28, 29] if only the inelastic processes are taken into account.

Let us compare quantitatively the spectra for a clean contact (curve 2) with the predictions of Ref. [31] which takes into account the strong-coupling effect of the superconducting energy gap dependence on energy. For the ballistic S-c-N junction with a strong-coupling superconductor the theory gives the expression of the first derivative of IVC:

$$\left\{\frac{dI}{dV}(eV)\right\}_{S-c-N} = \frac{1}{R_0}\left\{\frac{\Delta(eV)}{eV + \left[(eV)^2 - \Delta^2(eV)\right]^{1/2}}\right\}$$

for the total differential conductance at $T = 0$. The derivative of this curve is displayed as curve 1 in Fig. 5.5 where we used the strong-coupling gap versus energy dependence for lead determined by tunneling spectroscopy [32]. The smooth weak-coupling BCS background (corresponding to $\Delta(eV) = const$) is shown as a dashed line superimposed on curve 1. The experimental characteristic (curve 2) agrees reasonably well with theoretical predictions (curve 1) both in shape and in amplitude. The latter can be estimated by the deflection from the smooth background. The total order of magnitude of the elastic correction to the conductance amounts to $[\Delta(eV)/\hbar\omega_{ph}]^2$ for $\Delta(eV) \ll \hbar\omega_{ph}$, just as in the tunneling spectroscopy of superconductor. Since the elastic term of excess current $\delta I^{el}_{exc}(V)$ does not depend on the contact diameter d (Eq. 5.5), it can be higher than the inelastic contribution $\delta I^{in}_{exc}(V)$. This situation occurs just for lead. Below we consider another case which holds for weak-coupling superconducting tin, where the inelastic term prevails. We conclude this section with the notion that the Eliashberg EPI spectral function can be extracted from the superconducting point-contact characteristic in the same way as in tunneling spectroscopy provided that one should be sure that the inelastic terms are negligible.

5. INELASTIC PROCESSES IN EXCESS CURRENT

For a weak-coupling superconductor ($\Delta(eV) \ll \hbar\omega_{ph}$) the elastic contribution to the excess current is negligible. Consider the Sn-Cu S-c-N point contact which is shown schematically as an inset in Fig. 5.6 [33]. First of all, in the normal state Sn and Cu have approximately the same electronic parameters (v_F) and overlapped energy regions of the phonon density of states. Their PC EPI spectra are shown as curves 5 and 6 for Sn and Cu, respectively. In a symmetrical heterocontact their spectral functions enter almost equally according to the formula (5.2), g_{PC}(Sn-Cu)$\simeq$ (1/2) [g_{PC}(Sn) + g_{PC}(Cu)] which is shown as dotted curve 4. Accordingly, the measured PC spectrum in the normal state of the heterocontact looks like curve 3 which (with a small smooth background subtracted) leads to solid curve 4. The coincidence between dotted and solid curves 4 not only in shape but also in amplitude is fairly well. In the superconducting state of the same contact the excess current ap-

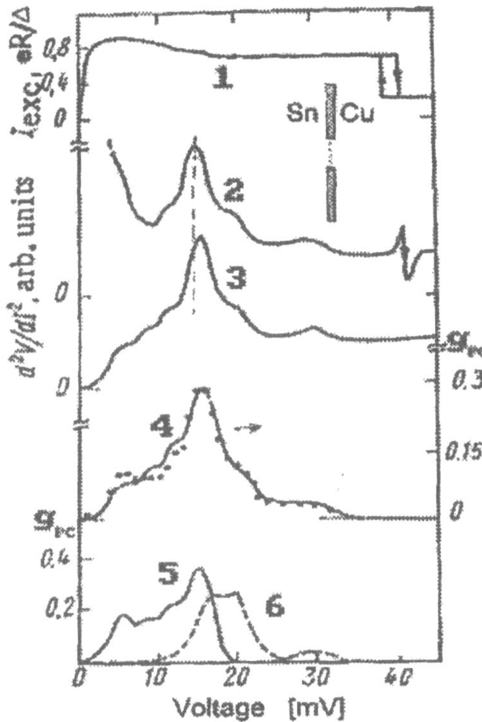

Figure 5.6 Inelastic phonon peculiarities for weak coupling superconductor in Sn-Cu S-c-N junction (see inset). The upper curve shows the dependence of excess current on voltage bias for contact with R_0 = 8.8 Ω with d = 10 nm. The second derivative of IVC in superconducting state is shown as curve 2. Curve 3 gives the EPI spectrum for the same contact in the normal state. It results in the EPI spectral function shown by solid line in curves 4 which should be compared with the calculated one plotted as a dotted curve. Curves 5 and 6 display the PC EPI spectra of Sn and Cu homocontacts, respectively. The curves are displaced vertically for convenience. Note the different zeros along the ordinate axis for each of them.

pears, which is shown by curve 1 in Fig. 5.6 as the difference between the total current and the Ohmic term V/R_0 in superconducting and normal states, respectively. Evidently, the excess current is not constant. We disregard the large hysteresis-like drop at high voltage (larger than the phonon energy band) due to heating and concentrate our attention on the smooth decrease along the phonon energy range. The second derivative (PC spectrum) of IVC in the superconducting state is shown as curve 2. Apart from the smooth background which steeply rises when the voltage approaches the energy gap (Δ(Sn)$\simeq$ 0.6 meV) from above, one sees the curve which almost reproduces the features seen in the normal-state spectrum (curve 3). We thus conclude that the main nonlinearities in the superconducting state come from the backscattering current the same as in the normal state. Yet, there are small peculiarities such as the small shift of the maxima to the lower energy (shown by the vertical dashed line in Fig. 5.6) and a slight broadening of peaks. These peculiarities are predicted by theory [28]. The unexpected shift to the lower energy by the superconducting energy gap is due to electron-hole relaxation as a result of the Andreev reflection from the N-S boundary. The inelastic processes with emission of phonons occur when an electron quasiparticle with energy of the order of eV relaxes to the hole state appearing due to Andreev reflection of another electron with energy of the order of Δ. For an infinitely narrow peak in the EPI function the theory leads to the maximum shifted down by Δ with the width of the order of Δ as well. Hence, in the clean S-c-N contact based on a weak-coupling superconductor the main contribution to formula (5.4) comes from the normal term $\delta I^N_{ph}(V)$ with slight modifications due to $\delta I^{in}_{exc}(V)$ term.

6. NONEQUILIBRIUM PHENOMENA

As soon as the diameter of the contact becomes comparable with the superconducting coherence length ξ_0, the nonequilibrium phenomena may occur which lead to suppression of the gap by i) nonequilibrium populations of normal electrons and phonons with energies greater than 2Δ, ii) entering the magnetic flux vortices generated by the current through the contact, or iii) simply by heating of the contact region.

In Fig. 5.7 the PC spectra of Ta-Cu contact are shown both in the normal (curve 1) and superconducting (curve 2) states [34]. Since the Fermi velocity of Ta is noticeably less and the strength of EPI in Ta is essentially higher than for Cu, only the EPI spectrum of Ta is seen in the experiment: $g_{PC}^{(Ta-Cu)} \simeq \frac{1}{2} g_{PC}^{(Ta)}$ (see Eq. (5.2). Although the relatively high resistance of contact ($R_0 = 80\ \Omega$) implies a small size ($d \simeq 78$ Å), the ultimate (at $eV > \hbar\omega_{ph}$) electron-phonon mean free path is also small ($l_{in} \sim 120$ Å) and the reabsorption of nonequilibrium phonons may occur. Also the superconducting coherence length in Ta ($\xi_0 = 90$ nm) is smaller than in Sn ($\xi_0 = 270$ nm), especially for the dirty Ta-contact region. This is more favorable for nonequilibrium transitions in the superconducting state and assists, by decreasing H_{c1}, the current-generated vortices to enter the contact area. All this makes the EPI spectrum in the superconducting state of Ta quite different from what is observed in Sn. There is no shift of the phonon peaks by Δ to lower biases. Instead, the lower is the contact resistance (the larger is the size), the more fixed are the transverse (T) and longitudinal (L) phonon peaks at energies of 11.3 and 18 meV with very small scattering of ± 0.1 meV. The peak of T-phonon (see curve 2 in Fig. 5.7) grows narrower instead of broadening and with increase of the contact size all the peaks become sharpened (Fig. 5.8).

This contrasts to the behaviour of phonon peaks in clean Ta point contacts in the normal state where one observes the spreading of energy positions in the limit of ± 1 meV probably due to the random anisotropy of the microcrystal orientation with respect to the contact axis. The explanation is as following. The nonequilibrium phonons slowly diffuse

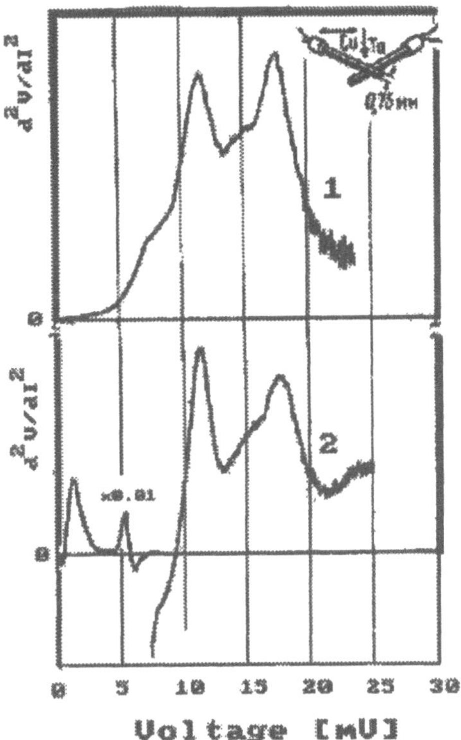

Figure 5.7 Nonequilibrium phenomena in PC spectrum of Ta-contact in the superconducting state. The Ta-Cu contact is made by the "edge-edge" geometry as shown in the inset. It has R_0 =80 Ω and d=78 Å. In the normal state (H=3 kOe) the spectrum is directly recorded as curve 1. Its shape practically coincides with the EPI spectral function calculated from the superconducting tunneling [35]. In the superconducting state (curve 2) anomalous features are seen. The transverse phonon peak at 11.5 meV sharpens, and a part of the spectrum becomes negative. There appears a strong nonlinearity at ≃6 meV which looks like a peak on the differential resistance curve and a sharp decrease of excess current on the IVC.

from the dirty contact region effectively averaging their direction in the momentum space. The phonon peaks at reproducible fixed positions appear due to the slow phonon group velocity (corresponding to the maximum of the averaged phonon density of state) which are accumulated in the contact region and thus effectively depresses the superconducting energy gap and excess current. In addition to the common energy-gap feature at $eV \simeq 0.6$ meV, a new peculiarity appears at V_1 (at about 5÷7 meV in Fig. 5.7), which has strong intensity (note that the ordinate of this fragment has a factor of 0.01!), and whose voltage satisfies the relation $V_1^2/R_0 = const.$ The latter means that either the power input by the current or the magnetic field generated at the contact achieve the threshold value. The threshold value increases with temperature and external magnetic field which is just opposite to the behaviour expected for destruction of superconductivity by simple heating or critical current conditions. We infer that at larger biases the superconductor turns into the nonequilibrium state and the increase in threshold value is explained by enhanced relaxation of nonequilibrium quasiparticles which demand a stronger injection to fulfill the threshold conditions. In this

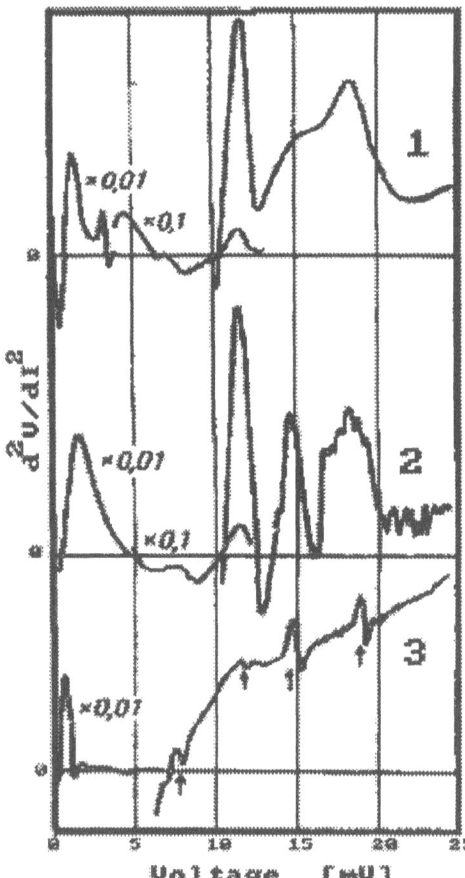

Figure 5.8 The evolution of non-equilibrium peculiarities with decreasing contact resistance (increasing size). Curve 1: contact Ta-Cu, R_0=26.5 Ω, d=13.6 nm. Curves 2 and 3: contacts Ta-Au with R_0=19 Ω (d=16 nm) and R_0=0.76 Ω (d=80.3 nm), respectively.

state the phonon features remain at the proper energies and are seen as a sharp singularity on IVC and its derivatives (Fig. 5.8), whilst in the normal state only the smooth background without any spectral features is observed. Although there is no theory explaining quantitatively these phenomena yet, the experimental extension of the superconducting PCS to the materials with a short inelastic mean free path and coherence length seems to be very encouraging in view of studies on exotic superconductors, such as high-T_c, heavy fermions and organic compounds.

7. CONCLUDING REMARKS

We have briefly described the state of the art of point-contact spectroscopy in the superconducting state. It allows one to penetrate into the mechanism of electron-quasiparticle interaction which may mediate the Cooper pairing. Contrary to tunneling spectroscopy, the point-contact spectroscopy needs not any barrier which inevitably interrupts the homogeneity of the material studied and often hampers investigation of many complicated compounds. At the expense of this, PCS results in principle to the highly nonequilibrium state which in case of superconductors leads to many complications. We hope that with development of new technique of producing well controlled constrictions of nanoscale size and elaborating a new theoretical approach the PCS will conquer more and more supporters and will become in the future as important as the tunneling spectroscopy.

Acknowledgement

The author wishes to acknowledge the financial support of the International Science Foundation (Soros Foundation).

References

[1] I. K. Yanson, Non-linear effects in electrical conductivity of point contacts and electron-phonon interaction in normal metals, Sov. Phys. JETP **39**, 506 (1974).

[2] K. S. Ralls, R. A. Buhrman and R. C. Tiberio, Fabrication of thin-film metal nanobridges, Appl. Phys. Lett. **55**, 2459-2461 (1989).

[3] P. N. Chubov, I. K. Yanson and A. I. Akimenko, Electron-phonon interaction in aluminium microcontacts, Sov. J. Low Temp. Phys. **8**, 32-39 (1982).

[4] A. G. M. Jansen, F.M. Mueller and P. Wyder, Direct measurement of electron-phonon coupling $\alpha^2F(\omega)$ using point contacts: Noble metals, Phys. Rev. B **16**, 1325-1328 (1977).

[5] C. J. Muller, J. M. Ruitenbeek and L. J. de Jongh, Experimental observation of the transition from weak link to tunnel junction, Physica C **191**, 485 (1992).

[6] G. E. Blonder, M. Tinkham and T. M. Klapwijk, Transition from metallic to tunneling regimes in superconducting microconstrictions: Excess current, charge imbalance, and supercurrent conversion, Phys. Rev. B **25**, 4515-4532 (1982).

[7] S. N. Artemenko, A. F. Volkov and A. V. Zaitsev, On the excess current in microbridges S-c-S and S-c-N, Solid State Commun. **30**, 771-773 (1979).

[8] A. V. Zaitsev, Sov. Phys. JETP **51**, 111; **52**, 1018 (1980).
[9] R. C. Dynes, *et al.*, Phys. Rev. Lett. **41**, 1509 (1978).
[10] A. Plecenik, M. Grajcar, S. Benacka, P. Seidel and A. Pfuch, Phys. Rev. B **49**, 10016 (1994).
[11] W. L. McMillan and J. M. Rowell, Tunneling and strong-coupling superconductivity, in: *Superconductivity* (ed. by R. Parks, Vol.**1**., Marcel Dekker, New York, 1969) pp. 561-614.
[12] Ch. Renner and Ø. Fisher, Vacuum tunneling spectroscopy and asymmetric density of states of $Bi_2Sr_2CaCu_2O_{8+\delta}$, Phys. Rev. B **51**, 9208-9218 (1995).
[13] T. Ekino, H. Fujii, M. Kosugi, Y. Zenitani and J. Akimitsu, Tunneling spectroscopy of the superconducting energy gap in $R\mathrm{Ni_2B_2C}$ (R=Y and Lu), Phys. Rev. B **53**, 5640-5649 (1996).
[14] M. Jourdan, M. Huth and H. Adrian, Superconductivity mediated by spin fluctuations in the heavy-fermion compound UPd_2Al_3, Nature **398**, 47-49 (1999).
[15] I. K. Yanson *et al.*, Point-contact spectroscopy of high-temperature superconductor $La_{1.8}Sr_{0.2}CuO_4$, Sov. J. Low Temp. Phys. **12**, 315-316 (1987).
[16] I. K. Yanson, Point-contact spectroscopy of superconducting $R\mathrm{Ni_2B_2C}$ (R=rare earth, Y), in: M. Ausloos and S. Kruchinin, eds., *Symmetry and Pairing in Superconductors*, (Kluwer Academic Publ., Dordrecht, 1999) pp. 271-285.
[17] I. K. Yanson, Point-contact spectroscopy of metals, Physica Scripta, **T23**, 88-94 (1988).
[18] Yu. V. Sharvin, Sov. Phys. JETP **21**, 655 (1965).
[19] I. O. Kulik, A. N. Omel'yanchuk and R. I. Shekhter, Electrical conductivity of point microcontacts and spectroscopy of phonons and impurities in normal metals, Sov. J. Low Temp. Phys. **3**, 740 (1977).
[20] A. G. M. Jansen, A. P. van Gelder and P. Wyder, Point-contact spectroscopy in metals, J. Phys. C: Solid State Phys. **13**, 6073-6118 (1980).
[21] I. K. Yanson, Point-contact spectroscopy of the electron-phonon interaction in pure metals (review), Sov. J. Low Temp. Phys. **9**, 343-360 (1983).
[22] I. K. Yanson and O. I. Shklyarevskii, Point-contact spectroscopy of metallic alloys and compounds (review), Sov. J. Low Temp. Phys. **12**, 509-528 (1986).
[23] A. Duif, A. G. M. Jansen and P. Wyder, J. Phys.: Condense Matter **1**, 3157 (1988).
[24] I. O. Kulik and I. K. Yanson, Microcontact phonon spectroscopy in the dirty limit, Sov. J. Low Temp. Phys. **4**, 596-602 (1978).

[25] A. F. Andreev, Sov. Phys. JETP **19**, 1228 (1964).
[26] A. V. Khotkevich and I. K. Yanson, A parallel study of the energy dependence of the excess current in the superconducting state and the electron-phonon interaction function in the normal state for point microcontacts, Sov. J. Low Temp. Phys. **7**, 354-359 (1981).
[27] I. K. Yanson and A. V. Khotkevich, *Atlas of Point-Contact Spectra of Electron-Phonon Interaction in Metals* (in Russian), Naukova Dumka, Kiev, 1986, 144 pages; [Engl. transl.: Kluwer Academic Publishers, 1995, 151 pages].
[28] V. A. Khlus, Non-linear current-voltage characteristics of the S-c-N point contact, Sov. J. Low Temp. Phys. **9**, 510 (1983).
[29] V. A. Khlus and A. N. Omel'yanchuk, Electron-phonon interaction in superconducting contacts, Sov. J. Low Temp. Phys. **9**, 189 (1983).
[30] A. V. Khotkevich, V. V. Khotkevich, I. K. Yanson and G. B. Kamarchuk, Elastic spectroscopy of electron-phonon interaction in point-contacts of superconductor - normal metal, Sov. J. low Temp. Phys. **16**, 693 (1990).
[31] A. N. Omel'yanchuk, S. I. Beloborod'ko and I. O. Kulik, Point-contact spectroscopy of superconductors with strong electron-phonon interaction, Sov. J. Low Temp. Phys. **14**, 630 (1988).
[32] J. M. Rowell, W. L. McMillan and R. C. Dynes, The electron-phonon interaction in superconducting metals and alloys II, *Bell Laboratories Preprint*, 1969, 145 pages.
[33] I. K. Yanson, G. V. Kamarchuk and A. V. Khotkevich, Electron-phonon interaction-induced nonlinearities of current-voltage characteristics of superconductor - normal metal point contacts, Sov. J. Low Temp. Phys. **10**, 220 (1984).
[34] I. K. Yanson, V. V. Fisun, N. L. Bobrov and L. F. Rybal'chenko, Reabsorption of nonequilibrium phonons at superconducting point contacts, JETP Lett. **45**, 543-352 (1987).
[35] E. L. Wolf, *Principles of Electron Tunneling Spectroscopy*, (Oxford University Press, New York, 1985).

Chapter 6

ATOMIC STRUCTURE, QUANTIZED ELECTRICAL AND THERMAL CONDUCTANCE OF NANOWIRES

S. Ciraci
Department of Physics, Bilkent University
Bilkent 06533, Ankara, Turkey

Abstract Nanowires and monoatomic chains having diameter in the range of Fermi wavelength have shown novel atomic, electronic, and thermal transport properties. Quantization due to the transversal confinement and resulting finite level spacings of electronic and phononic energy states underlie these observed unusual properties. This paper discusses various stretching mechanisms and our predictions on the formation of single monoatomic chain and bundle of chains at the neck of nanowire, and presents a comprehensive analysis of the stepwise variation of force, electrical and thermal conductance.

1. INTRODUCTION

The ballistic conductance of the atomic size point contact, that was realized first in 1987 showed sudden changes, which were attributed to the quantization of conductance [1]. At that time we pointed out that stepwise behavior of electrical conductance is related with the the discontinuous variation of the contact area [2, 3]. Recently, metallic nanowires of better quality and reproducibility and having dimension in the range of Fermi wave length, λ_F, have been produced by STM [4, 5, 6] and also by mechanical break junction [7]. The two terminal electronic conductance, G of the wire showed a stepwise variation with the stretch. The sudden jumps of conductance in the course of stretch have been taken as the manifestation of quantized conductance at room temperature and thus have created much popularity. However, recent experiments [8] have provided evidence that the observed stepwise behavior of conductance

I. O. Kulik and R. Ellialtioğlu (eds.),
Quantum Mesoscopic Phenomena and Mesoscopic Devices in Microelectronics, 79–94.

is actually related to sudden falls of tensile force, F_z and hence abrupt variation of the cross section of the wire, in agreement with our earlier work [2].

The density of states of a perfect metal nanowire can be expressed as $D(\epsilon) \sim \sum_i \theta(\epsilon - \epsilon_i)(\epsilon - \epsilon_i)^{-1/2}$; it has peaks when $\epsilon = \epsilon_i$. The occupancy of the states ϵ_i becomes strongly dependent on the size and the geometry of the nanostructure. Earlier, Ciraci and Batra [9] found that the surface energy, work function, etc of a 2D electron system show discrete changes upon the change of size. They explained this event in terms of an extensive analysis of the effects occurring when an empty band gets occupied. Recently, similar effects have been found for clusters [10] and 1D jellium wires [11]. Based on jellium approximation, it has been shown that the energy position of transversally confined electronic states relative to the Fermi energy, E_F, is related to the diameter of the uniform wire. Accordingly, metal wires having certain diameters shall be energetically favorable. The narrowest diameter of the nanowire prior to the break is only a few angstroms; it has the length scale λ_F, where discontinuous (discrete) nature of the metal dominates over its continuum description. Since $\Delta\epsilon$ is in the range of ~1 eV in this length scale, the peaks of $D(\epsilon)$ of the connective neck and hence the transversal quantization of states become easily resolved even at room temperature. Furthermore, any change in the atomic structure or in the diameter induces significant changes in the level spacing and in the occupancy of the states, that, in turn, may lead to detectable changes in the related properties. Therefore, the ballistic electron transport through nanowire is closely related to its atomic structure and diameter at the narrowest part of the connective neck [12, 13, 14].

In view of the interesting physics underlying the ballistic electron transport through a nanowire between two electrodes, we conjectured [14] earlier that similar effects should occur also in the heat conduction *via phonons*. Similar to a constriction in electrical conduction, we expect that a nanowire (or in general a nanoobject) has a discrete vibrational frequency spectrum with finite spacings, $\Delta\Omega_i = \Omega_{i+1} - \Omega_i$, determined by its atomic configuration and size. Furthermore, when a nanowire is placed between two reservoirs, collective vibrational modes (phonons) which involve the coherent vibrations of the atoms in the nanowire and reservoirs can be developed, leading to the process of ballistic thermal conduction. The thermal conductance, $\mathcal{K}$, depends on the available number of channels and hence it is expected to increase in discrete values each time a new ballistic channel participates in the heat transfer. Each vibrational mode contributes to the heat transfer, but its contribution is weighted by Planck's distribution $n(\Omega, T)$.

In this paper, the structural transformations, force and conductance variations of a nanowire in the course stretching are discussed. In particular, our earlier predictions on the mechanisms of stretching, formation of bundle of atomic chains and single atomic chain at the neck are reviewed [12, 13, 14]. New results on the energetics and structure of uniform monoatomic chains are summarized. The correlation between the conductance variations and the structural transformations is analyzed by using the results of the calculation of conductance as a function of stretch. Similar effects have been briefly discussed for ferromagnetic [15] and layered materials [16]. Finally, the phononic heat transfer through a uniform, as well as finite nanowire and the possible stepwise behavior of thermal conductance have been discussed [17, 18].

2. ATOMIC STRUCTURE OF STRETCHING METALLIC NANOWIRES

Stretching of metallic wires are studied by using classical molecular dynamics (MD) method. The wires have two ends connected by a neck. Last three atomic layers at both ends are robust, and are assumed to be connected to the external agent (or grips) which applies the tensile stress. Atoms in the following three layers adjacent to the robust ones and those in the six layers of the neck are identified as dynamic atoms and are fully relaxed during the MD-steps. These layers are formed initially from either Cu(001) or Cu(111) atomic planes, and the layers at the neck incorporate 12-13 Cu atoms. The interatomic interactions are treated by the EA model potential [19]. The pulling-off (stretching) is realized by displacing the robust end layers solidly along the axis of the wire ($z-$direction) in increments of Δz while keeping the other end fixed. The total stretch, s after m increments is $s = m\Delta z$. Between two consecutive increments of Δz, the dynamic atoms of the wire are relaxed to find their next "steady-state" condition specified by the fact that the system is allowed to evolve dynamically until no significant variation in physical quantities (temperature, force, *etc.*) are observed beyond natural fluctuations. The relaxation process is realized in k_r steps, each step lasting Δt seconds. In each time step, Newton's equation is solved to find new positions of atoms, while temperature is controlled by Hoover drag. The lower limit of Δt is determined from the mobility of atoms in the solid. In the atomistic simulations Δz, Δt and temperature T are crucial parameters. The increment, Δz by itself is important; the break of the junction can be favored by relatively larger Δz. The stretching velocity is defined as $v_s = \Delta z / k_r \Delta t$. Even though the experimental velocities cannot be achieved, we nevertheless examined how the yielding mechanism depends on the velocities by carrying out the MD simulations with different v_s (or Δt) within acceptable limits. The overall features of the results were similar. There are, however, some features that depend on Δt. Further details of the calculations and results can be obtained from Ref. [14].

The simulation of stretching metal nanowires reveals a number of interesting features which are important not only for better understanding of electrical conductance, but also for novel atomic structure occurring under high strain. Here some of these features are outlined:

1. We found that the deformation (or elongation) treated in this study generally occurs in two different and consecutive stages that repeat while the nanowire is stretched. In the first stage, the stored strain energy and average tensile force increase with increasing s, while the layer structure persists. The fluctuations can occur possibly due to displacement and relocation of atoms within the same layer or *atom exchange* between adjacent layers. The variation of $F_z(m)$ in this stage is approximately linear, but it deviates from linearity as the number of atoms in the neck is reduced. This stage was identified as *quasi elastic*. The second stage that follows each quasi elastic stage is the *yielding* stage. A wire can yield by different mechanisms depending on its diameter. The motion of the dislocation and/or the slips on the glide planes are generally responsible for the yielding if the wire maintains an ordered (crystalline) structure and has relatively larger cross section. The type of the ordered structure and its orientation relative to the $z-$axis are expected to affect the yielding. On the other hand, MD simulations predict different mechanical instabilities leading to yielding if the cross section of wires is smaller as considered in the present paper. These are *order-disorder transformation* [5, 14] and *single atom process*. Initially, once the elongation of the nanowire becomes approximately equal to the interlayer separation at the end of the quasi elastic stage, the structure becomes disordered; but after a few increments of stretch, it is recovered with the formation of a new layer. In this yielding stage, $|F_z|$ decreases abruptly; the cross section of narrowest neck layers, A_{neck}, in particular the cross section of the layer formed at the end of yielding stage are abruptly reduced by a few atoms. The force variation and atomic structure of a nanowire calculated by MD method is illustrated in Fig. 6.1. When the neck becomes very narrow (having 3-4 atoms) the yielding is realized, however, by a single atom jumping from one of the adjacent layers to interlayer space. At low temperature, the yielding takes place in relatively shorter time-interval within one or two stretch steps. Owing to the limited mobility of atoms at low temperature, the character of quasi elastic stage changes if the neck is long enough. In this situation, each

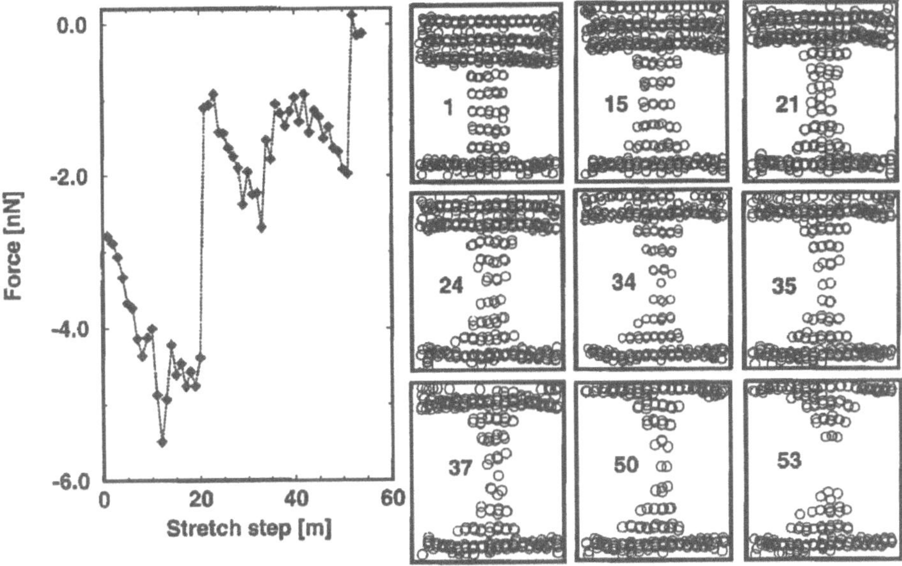

Figure 6.1 The variation of tensile force F_z (in nanoNewtons) with the strain or elongation along the z-axis of the nanowire having Cu(001) structure. The side views show the atomic positions at relevant stretch values m. The MD simulations are performed T=300 K and v_s=0.004 m/s.

layer sequentially ejects one atom to the adjacent interlayer space that widens with the applied strain. At certain circumstances, atoms at the neck form a **pentagon** that becomes staggered in different layers. The interlayer atoms make a chain passing through the center of pentagon rings. In this new phase, the elastic and yielding stages are intermixed and elongation which is more than one interlayer distance can be accommodated. As A_{neck} is further decreased the pentagons are transformed to the **triangles**. In the initial stage of pulling-off the *single atom process*, in which individual atoms also migrate from central layers towards the end layers, can give rise to relatively smaller and also less discontinues changes in the cross section. The tendency to minimize the surface area and hence to reduce the strain energy of the system is the main driving force for this type of necking. While quasi elastic and yielding stages are distinguishable initially, the force variation becomes more complex and more dependent on the migration of atoms for relatively thinner necks. Atomic migrations may have important implications such as dips in the variation of conductance with s.

2. Our simulation studies on the metal wires, that have an initial diameter of $\sim \lambda_F$ at the neck, but their diameters gradually increase as one goes away from the neck region, have indicated structural instabilities; those wires had further necking even in the absence of any tensile strain. This puzzling result is in comply with the recent experiment by using mechanical break junction on gold neck at room temperature that broke by itself [20].

3. In the initial steps of stretch, the layered structure and ordered $2D$ atomic arrangement within the layers are maintained. Upon increased uniaxial strain (or increased m), the layers become wider and rougher due to the atoms departing from the atomic plane, the $2D$ lattice is distorted, and interatomic distances start to deviate from the bulk equilibrium value. In the elastic stage, one can still distinguish layer structure and some kind of order in the atomic arrangement within a layer. In particular, the hollow-site registry between layers are maintained if the layers contain enough number of atoms. The ordered atomic arrangement, however, does not occur any more towards the end of stretching process at low temperature.

4. The quasi elastic and following yielding stages are reminiscent of the stick-slip motion. In the course of stretch, elastic and yielding stages repeat; the surface of the nanowire roughens and deviates strongly from circular symmetry. The narrowest part of the neck, is usually only a layer thick, and is connected to the horn-like ends. In this circumstances the contribution of the tunneling becomes significant. Our results of atomic simulations point to the fact that neither adiabatic evolution of discrete electronic states, nor circular symmetry induced degeneracy can occur in the neck. Consequently any quantized sharp structure shall be smeared out by channel mixing and tunneling [3].

5. If the interlayer interaction is reduced as a result of extensive strain, the $2D$ square-like lattice is transformed into hexagonal structure. At prolonged stretch, just before the break, A_{neck} is reduced to include only $2-3$ atoms. In this case, the hollow-site registry may change to the top-site registry. This leads to the formation of **a bundle of atomic chains (rope)** or of **a single atomic chain** (see Fig. 6.2). We consider this situation a dramatic change in the atomic structure of the wire that may have important implications. Our predictions regarding the formation of a rope or atomic chain at the neck in 1996 [12, 13, 14] have been confirmed later experimentally: Using an ultra high-vacuum electron microscope, Ohnishi *et al.* [21] observed the "strands" and single chain of gold atoms suspended between two gold electrodes. They concluded interatomic distances (as large as 3.5 to 5 Å) much larger than the bulk value. Based on *ab-initio* full-potential, all-electron calculations that uses a liner-muffin-tin-orbital basis set Örmeci and Ciraci [22] studied the stability of the uniform gold chain for different type of

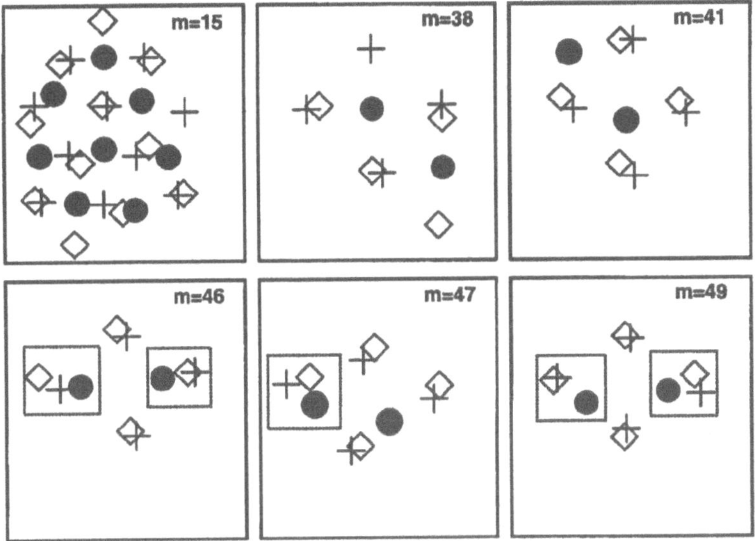

Figure 6.2 The top view of three layers at the neck showing atomic positions and their relative registry at different levels of stretch. $m = 15$ occurs before the first yielding stage. The atomic positions in the second, third and fourth layers are indicated by $+$, $\bullet$ and $\diamond$, respectively. $m = 38$ and $m = 41$ occur after the second yielding stage. $m = 46$, 47 and $m = 49$ show the formation of bundle structure. In the panels $m = 38 - 49$ the positions of atoms in the third, fourth and fifth (central) layers are indicated by $+$, $\bullet$ and $\diamond$, respectively. Atomic chain in a bundle are highlighted by dashed boxes.

structures. They found that the **zig zag chain** of gold atoms with the interatomic distance, d=2.55 Å, and the apex angle α ~140° has the minimum energy.

6. Structural transformations of the nanowire depends on the velocity of stretching, v_s and temperature. For slow or low temperature stretching the nanowire can experience more quasi elastic-yielding stages and hence more elongation. Consequently, at room temperature the necking is faster and sharper.

3. ELECTRON TRANSPORT IN NANOWIRES

The structure of nanowires discussed in the previous section reveals that the length of the uniform region at the narrowest part is smaller than λ_F. This increases the contribution of tunneling and hence smoothes out the sharp step structure that occurs when a conduction channel is opened (or closed). An extreme case is, of course, the quantum Sharvin's conductance [3, 23], $G_S = (2e^2/h)(\pi R_c/\lambda_F)^2$, with contact radius R_c, where the step structure is almost smeared out and plateaus disappeared since the length of the constriction vanishes. [3] Furthermore, the potential $V(r)$ in the wire is nonuniform and strongly roughened due to the irregular atomic arrangement as illustrated in Fig. 6.1. Under these circumstances, the channels mix significantly [3], the opening of the channels are delayed and G attains values lower than the integer multiples of quantum conductance [3, 24, 25]. Then the resulting $G(s)$ curve would trace smoothed structure if A_{neck} were changed continuously with s [26].

Clearly, the understanding of the features in the experimental $G(s)$ curve requires the calculations of the conductance using an SCF potential corresponding to the detailed atomic structure of the wire. A rigorous SCF calculations in $3D$ for a realistic wire is too tedious, however. Here, a method [14] will be presented, which is intermediate between the simple model calculations and rigorous 3D SCF calculations, whereby the atomic structure of the nanowire is taken into account. We start by generating a realistic potential from the linear combination of atomic potentials according to the atomic structure calculated by the present MD simulations. First, we calculate the ionic potential of the free Cu atom, $V_p(r)$ from the generalized norm conserving pseudopotential given by Hamann [27]. We next Fourier transform the atomic potential in order to take into account the electron-electron screening in the neck using Thomas-Fermi approximation.

$$V_p(K) = \frac{4\pi}{8\pi^3 K} \int_0^\infty r V_p(r) sin(Kr) dr, \tag{6.1}$$

where K is $|\mathbf{K}|$. In order to calculate the potential of the wire corresponding to a given stretch s, its charge density at $\mathbf{r}$ is calculated from the linear combination of atomic charge densities, $q_a(r)$,

$$q_s(\mathbf{r}) = \sum_i q_a[\mathbf{r} - \mathbf{R}_i(s)], \tag{6.2}$$

where $\mathbf{R}_i(s)$ are the atomic position vectors obtained from the MD simulations performed at a given s. The index i runs over all atoms of the wire. The potential is given by

$$V_s(\mathbf{r}) = V_{xc}(\mathbf{r}, q_s) + \sum_i \int d\mathbf{k} \frac{V_p(\mathbf{K})}{\varepsilon(\mathbf{K}, q_s)} e^{i\mathbf{k}.[\mathbf{r}-\mathbf{R}_i(s)]}, \tag{6.3}$$

where V_{xc} is the Ceperley-Adler exchange correlation potential [28] and $\varepsilon(\mathbf{K}, q_s)$ is the Fourier component of the static dielectric function. To this end the potential is transformed to finite wall cylindrical potential through radial averaging and then least square fit. In order to make computations of transversally confined states feasible, the walls of $V_s(\mathbf{r})$

are taken infinite. The effect of this change is the upward shift of the energy which is corrected by increasing the radius of the cylindrical infinite wall potential, $\rho_s(z)$ at any z so that the eigenenergy of the ground state coincides with that of finite wall potential. Eventually, potential of the wire reads:

$$V_s(\rho, z) = \begin{cases} \phi_s(z) & \rho \leq \rho_s(z) \\ \infty & \rho \geq \rho_s(z) \end{cases} \tag{6.4}$$

where $\rho = (x^2 + y^2)^{1/2}$, and $\rho_s(z)$ is the radius of the infinite wall potential. The saddle point potential $\phi_s(z)$ is an essential ingredient that omitted in the free and nearly free electron calculations. We calculate the conductance of the nanowire (that is specified by the stretch s it undergoes) by using the transfer matrix method: We divide $V_s(\rho, z)$ into discrete segments of equal widths, and take their average values in each segment $z_j < z < z_{j+1}$, $\phi_s(z)$ and $\rho_s(z)$. The longitudinal wave function $e^{i\gamma^s_{n_j}(z)}$ and circularly transversal wave function $\varphi^s_{n_j}(\rho; z)$ are solved for each segment. Here the propagation constant:

$$\gamma^s_{n_j}(z) = \frac{2m}{\hbar^2}[E - \phi_s(z) - \epsilon^s_{n_j}(\rho; z)]^{1/2}, \tag{6.5}$$

where E is the energy of the incident electron and $\epsilon^s_{n_j}(\rho; z)$ is the eigenenergy of $\varphi^s_{n_j}(\rho; z)$. If $E > \phi_s(z) + \epsilon^s_{n_j}$, all the $\gamma^s_{n_j}$ are real and propagate in the segment. Imaginary $\gamma^s_{n_j}$ occurring for $\epsilon^s_{n_j} > E - \phi_s(z)$ yield longitudinal waves that decay in the segment, and contribute to the tunneling. In order to construct the current transporting state, we start with an incident plane wave $e^{i\mathbf{k}_l \cdot \mathbf{r}}$ for $z \leq 0$ in the left reservoir, and express it in each segment $z_j < z < z_{j+1}$ by the linear combination of the longitudinal and transversal states and determine the coefficients by multiple boundary matching. The current is calculated by evaluating the current operator corresponding to an incident plane wave and summing it over the Fermi surface. Then, the conductance is obtained within linear response theory. In Fig. 6.3. we show the $G(m)$ curve calculated for the stretching nanowire described in Fig. 6.1. We note the conductance changes abruptly between plateau like regions. Many features of $G(m)$ can be understood by comparing it with the force variation curve, $F_z(m)$. Whenever the cross section decreases suddenly following a yielding stage the conductance falls also suddenly. This is in agreement with the recent experiments and also confirms our arguments [2] put forward earlier to explain the stepwise behavior of G. It is also noted that $G(m)$ becomes strongly dependent on the atomic configuration when the neck becomes extremely thin (having one or two atoms) shortly before the

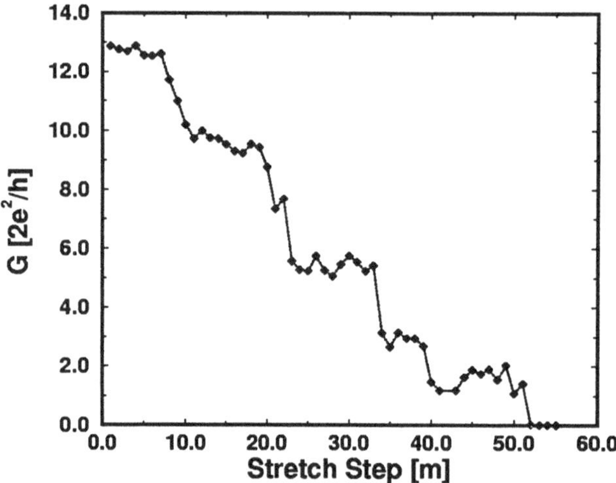

Figure 6.3 Variation of conductance calculated for the stretching nanowire described in Fig. 6.1.

break. While one atom exchange between layers gives rise to fluctuations of G when the cross section is large, it may lead to significant dips for relatively thinner necks. Sometimes $G(m)$ increases with increasing s when the atoms change from hollow-site to the top-site registry. Recent studies demonstrated that not only atomic structure, but also the bonding sites to the electrodes are crucial for the measured conductance [29].

The above situation can be different for the ballistic conductance of ferromagnetic nanowires and layered materials. The conductance distribution obtained from an ensemble of stretching ferromagnetic nanowires or point contact do not exhibit peaks near the integer multiples of $2e^2/h$. This observation has been interpreted as the absence of conductance quantization [30]. Our theoretical analysis suggests that the tunneling through the *closely spaced* states near E_F which originate from crystal field and spin split $3d$ states of Ni prevents the plateaus of conductance from forming in the course of stretch [15]. The conductance through a quantum point contact created by a sharp and hard metal tip on the graphite surface displays features which are rather different from those of metals. The recent study based on the *ab-initio* calculations of electronic energy structure shows that the the metalicity of graphite increases under the uniaxial compressive stress perpendicular to the basal plane [16]. Consequently, the conductance depends inversely on the strain in the tunneling regime. At high compressive stress, the sequential puncturing of the layers characterizes the variation of conductance [16].

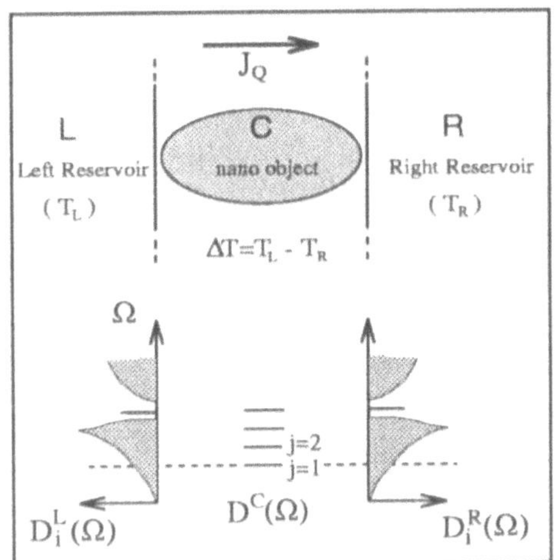

Figure 6.4 A schematic description of the model, for which the quantum heat transfer and related thermal conductance are studied. A nanoobject (either a uniform wire or a finite wire) is coupled to two heat reservoirs and the corresponding density of frequencies. Resonant mode is described by dotted lines.

4. QUANTIZED THERMAL CONDUCTANCE

The model system [17, 18] to study the effect of discrete phonon spectrum on the *phononic* thermal conductance through a nanowire (or nanoobject) is schematically described in Fig. 6.4. Heat flows from the left (L) to the right (R) through a nanowire (C). Here L, R can be sections of nanowire itself or heat reservoirs. The temperature of L and R, T_L and T_R, respectively are kept constant, so that the steady state of the transport is considered.

For this system we investigate the thermal conduction by using a model Hamiltonian approach. To find the modes which are responsible for the ballistic transport of heat, we specify first a zero order vibrational Hamiltonian which characterizes the possible modes in the three regions, $(L + C + R)$,

$$H_o = \sum_{i,q_L} \hbar\Omega_{iq_L} a^+_{iq_L} a_{iq_L} + \sum_{i,q_R} \hbar\Omega_{iq_R} a^+_{iq_R} a_{iq_R} + \sum_{i,q_C} \hbar\Omega_{iq_C} a^+_{iq_C} a_{iq_C} \quad (6.6)$$

Here q_X is the momentum quantum numbers for vibrational mode Ω_{iq_X} belonging to the branch i in one of the three regions ($X = L, R, C$), and a_{iq_X} ($a^+_{iq_X}$) is the corresponding annihilation (creation) operator. The heat transfer via phonons is realized when L,C, and R are coupled and $\Delta T = T_L - T_R > 0$. The coupling can be described by an interaction Hamiltonian,

$$H_{int} = \sum_{ij} T_{ij,q_Lq_C}(a^+_{jq_C} a_{iq_L} + a_{jq_C} a^+_{iq_L}) + \sum_{jl} T_{jl,q_Cq_R}(a^+_{lq_R} a_{jq_C} + a_{lq_R} a^+_{jq_C})$$
$$+ \sum_{ijt} U_{ijt,q_Lq_Cq_R}(a_{iq_L} a^+_{jq_C} a^+_{tq_R} + \cdots) \quad (6.7)$$

in terms of the harmonic ($\mathcal{T}$) and anharmonic (U) coupling coefficients. This is similar to the transfer Hamiltonian approach [31]. Here we consider only the thermal conduction mediated by phonons. Electronic, as well as the electron-phonon interactions are not included in this study. The anharmonic couplings describe the phonon-phonon interactions which are the mechanism responsible for several dissipative processes, such as energy loss to local modes of the nanowire and thermal resistance due to Umklapp process. Depending on the strength of the anharmonic coupling, the continuing energy transfer to the localized modes may result in *energy discharge* to the heat transporting states, or *desorption* or migration of an atom. If the former event, *i.e.* the excitation of a local phonon, followed by deexcitation occurs periodically, it can be manifested in the time variation of the conductance, $\mathcal{K}(t)$. In what follows we apply the above approach to calculate the thermal conductance of a uniform dielectric wire and also of a finite atomic chain by considering only the harmonic couplings.

First we start with a uniform wire and consider three sections on it. Owing to the uniformity of the sections (L, C, R), the phonon modes are the normal modes of the system, hence $q = q_L = q_C = q_R$, and also $\mathcal{T}_{ij,qq} = \mathcal{T}_{ii,qq}$. Furthermore, $\mathcal{T}_{ii,qq} = \hbar\Omega_{iq}$. If the effect of mode decay is considered, the rate of the heat transfer from L to R in this 1D system is given by $J_Q^{LR} = \sum_{iq} \hbar\Omega_{iq}\Gamma_{iq}^L$ by using the decay rate expression [32] for a given i and q,

$$\Gamma_{iq}^L = \Gamma_{iq}^C \frac{V_{qq}^{LC} V_{qq}^{CL}}{(\hbar\Gamma_{iq}^C)^2}. \tag{6.8}$$

with

$$\Gamma_{iq}^C = (2\pi/\hbar) V_{qq}^{CR} V_{qq}^{RC} D_i^R(\Omega_{iq}),$$
$$V_{qq}^{LC} = \langle n_L - 1, n_C + 1, n_R | H_{int} | n_L, n_C, n_R \rangle, \text{ and}$$
$$V_{qq}^{CR} = \langle n_L - 1, n_C, n_R + 1 | H_{int} | n_L - 1, n_C + 1, n_R \rangle.$$

By symmetry $V_{qq}^{BA} = V_{qq}^{AB}$ and for $(X = L, C, R)$ the Planck's distribution $n_X(\Omega_{iq}, T_X) = [\exp(\hbar\Omega_{iq}/k_B T_X) - 1]^{-1}$ is given in terms of the frequency of the mode and temperature of the section, T_X. Hence $V_{qq}^{LC} = \mathcal{T}_{ii,qq}\sqrt{n_L(n_C+1)}$ and $V_{qq}^{CR} = \mathcal{T}_{ii,qq}\sqrt{(n_C+1)(n_R+1)}$ are expressed in terms of Planck's distribution functions. The net rate of heat transfer through the chain is $J_Q = J_Q^{LR} - J_Q^{RL}$. Since n_L and $n_R \ll 1$ at low temperature, J_Q can be expressed in the following form

$$J_Q \simeq \frac{\hbar}{2\pi} \sum_i \int_0^\infty [n_L - n_R]\, \Omega\, d\Omega \tag{6.9}$$

Here index i of the summation sign specifies the branches for n_L and n_R. Note that Eq. (6.9) yields exactly the same expression of current one obtains from the phenomenological approach [33] by assuming the transmission coefficients, $t_i(\Omega) = 1$. Therefore, the present approach justifies the Landauer type expression of heat current by deriving it from the model Hamiltonian specified by Eq. (6.7). Now, we evaluate Eq. (6.9) for $\Delta T \to 0$. We take T_L and T_R small so that the upper limit of the integral is justified. Defining $T = (T_L - T_R)/2$, the following expression for the ballistic thermal conductance, $\mathcal{K} = J_Q/\Delta T$, is obtained [18]

$$\mathcal{K} = \sum_i \frac{\pi^2 k_B^2}{3h} T \tag{6.10}$$

According to this interesting result the ballistic conductance of each branch i of the uniform and harmonic atomic chain is limited by the value $\mathcal{K}_o = \pi^2 k_B^2 T/3h$. It is independent of any material property, and is linearly dependent on T. The total thermal conductance becomes $\mathcal{K} = \mathcal{N}\mathcal{K}_o$, where $\mathcal{N}$ is the total number of branches. For an ideal 1D atomic chain $\mathcal{N} = 3$, if the transverse vibrations are allowed. The result agrees with the well-known expression derived earlier for dielectric bridges [34, 35]. For a homogeneous wire that has a larger cross section, more subbands can now be present due to the transverse confinement of the vibrational motions in the direction perpendicular to the propagation. In this case, the ballistic thermal conductance of a branch with $\Omega_{i,min} \neq 0$ is smaller than $\mathcal{K}_o$ due to the finite frequency of a subband through the occupancy of Planck's distribution. One expects that each subband in the wire having $\Omega_{i,min} \neq 0$ leads to a jump in $\mathcal{K}$ as in the electrical conduction. However, the amplitude of steps decreases exponentially with increasing $\Omega_{i,min}$ at given low temperature T. Furthermore, an important difference between the electrical and thermal conductance is that all the modes of a given branch i contribute to $\mathcal{K}$, whereas the quantum ballistic electrical conduction G in the previous section is governed by the Fermi-Dirac distribution.

The quantum heat transfer through a uniform and harmonic atomic wire is an idealized situation. A more realistic system consists of a finite atomic wire, C coupled to reservoirs L and R. Here the finite atomic wire has discrete vibrational frequency spectrum, while those of reservoirs are continuous. In Fig. 6.4 the corresponding densities of states, $D^X(\Omega)$, are schematically described. It is also noted that cross sections change abruptly where the atomic wire is connected to the reservoirs. As a result, the transmission coefficient, $t_j(\Omega) < 1$ because of reflections. Also, the modes of wire differ from those of the reservoirs; C itself is not uniform in the sense that the atoms near the boundary experience different force constants. For such a wire one can distinguishes three types of modes: i) Modes localized strongly in the chain. ii) Modes localized strongly in the reservoirs. iii) Resonant modes which are the modes of the chain mixed with those of the reservoirs. Heat is transported through the resonant modes. Similar concept was used for the transport of electrical current through a nanoobject or an atom between two metal electrodes [29, 36]. For a given system the local density of states of a resonant state can be calculated numerically [18]. We now consider the ballistic heat transfer through resonances. Starting from the rate expression given above the ballistic thermal conductance for a single resonant mode iq, is given by,

$$\kappa_{iq} \sim k_B v_{iq} x^2 e^x (e^x - 1)^{-2} \tag{6.11}$$

with $x = \hbar\Omega_{iq}/k_B T$. For large x, $\kappa_{iq} \sim v_{iq} k_B x^2 e^{-x}$. On the other hand $\kappa_{iq} \sim v_{iq} k_B$ if x is small. In view of the analysis in Ref. [29], the thermal conductance of a finite atomic chain due to all resonance modes can be obtained from the sum over these modes, iq_C,

$$\mathcal{K} \sim \sum_{iq_C} v_{iq_C} k_B \int_0^{\infty} d\Omega\, \rho_{iq_C}(\Omega) \left(\frac{\hbar\Omega}{k_B T}\right)^2 \exp\left(\frac{\hbar\Omega}{k_B T}\right) \left[\exp\left(\frac{\hbar\Omega}{k_B T}\right) - 1\right]^{-2}$$

Here ρ_{iq_C} is the local density of states for the resonant modes derived from the mode iq_C and v_{iq_C} is average velocity associated with the same mode. This density of states modifies the ideal ballistic transmission as $t_{iq}(\Omega)$ does to J_Q in Ref. [35]. If ρ_{iq_C} is replaced by $D_i^C(\Omega)$ corresponding to a branch of uniform atomic chain, the thermal conductance $\mathcal{K}$ in Eq. (6.11) yields the universal value $\mathcal{K}_o$. The thermal conductance is determined by phonons with energies up to thermal energy ($\hbar\Omega \sim k_B T$) and each ρ_{iq_C} would give rise to a jump at $\mathcal{K}(T)$ curve at low temperature if this effect were not thermally smeared out [18].

5. CONCLUSIONS

This paper shows that the elongation and yielding mechanisms of a nanowire under uniaxial tensile force depend on its size, atomic structure and temperature. For the nanowire considered in this paper the yielding can take place either by order-disorder structural transformation or by single atomic process. At certain range of stretch pentagon or triangle of atoms can form. Shortly before the break the structure is transformed to form bundle of atomic chains or single atomic chain at the neck. The most important aspect of the nanowires that underlies their unusual properties is that the spacings of the energy levels of transversally confined electronic and phononic states are wide. Even if the potential of the wire is not uniform the level spacing is reflected to the related physical properties, especially when the cross section of the neck includes 1-2 atoms. In this respect the structure of the experimental $G(s)$ curve reflects the quantization of electronic states in the neck. At low temperature under ballistic transport, the thermal conductance contributed by each phonon branch of a uniform and harmonic chain cannot exceed the well-known value which depends linearly on temperature but is material independent. For finite atomic chain resonance modes are found which should also lead to a stepwise behavior in $\mathcal{K}(T)$.

Acknowledgements

I would like to thank H. Mehrez, Dr A. Buldum, Ç. Kılıç, , Prof I. P. Batra and Prof C. Y. Fong, who coauthored several papers reviewed in this work.

References

[1] J. K. Gimzewski and R. Möller, Phys. Rev. B **41**, 2763 (1987).

[2] S. Ciraci and E. Tekman, Phys. Rev. B **40**, 11969 (1989); S. Ciraci, Tip-surface interactions, in *Scanning Tunneling Microscopy and Related Methods*, edited by R. J. Bohm, N. Garcia and H. Rohrer (Kluwer, Dordrecht, 1990) Vol. 184, p. 113.

[3] E. Tekman and S. Ciraci, Phys. Rev. B **43**, 7145 (1991).

[4] N. Agraït, J. G. Rodrigo and S. Vieria, Phys. Rev. B **47**, 12345 (1993).

[5] J. I. Pascual, J. I. Mèndez, J. Gòmez-Herrero, A. M. Barò, N. Garcia, U. Landman, W. D. Luedke, E. N. Bogachek and H. P. Cheng, Science **267**, 1793 (1995).

[6] L. Olesen, E. Laegsgaard, I. Stensgaard, F. Besenbacher, J. Schiøtz, P. Stoltze, K. W. Jacobsen and J. K. Nørskov, Phys. Rev. Lett. **72**, 2251 (1994).

[7] J. M. Krans, C. J. Müller, I. K. Yanson, T. C. M. Gowarent, R. Hesper and J. M. Van Ruitenbeek, Phys. Rev. B **50**, 14721 (1994).

[8] N. Agraït, G. Rubio and S. Vieira, Phys. Rev. Lett. **74**, 3995 (1995); G. Rubio, N. Agraït, S. Vieira, ibid **76**, 2302 (1996).

[9] S. Ciraci and I. P. Batra, Phys. Rev. B **33**, 4294 (1986); I. P. Batra, S. Ciraci, G. P. Srivastava, J. S. Nelson and C. Y. Fong, Phys. Rev. B **34**, 8246 (1986).

[10] W. Ekardt, Phys. Rev B **29**, 1558 (1984).

[11] C. A. Stafford, D. Baeriswyl and J. Burki, Phys. Rev. Lett. **79**, 2863 (1997).

[12] H. Mehrez, M.S Thesis (August 1996, Bilkent University).

[13] H. Mehrez and S. Ciraci, *Optical Spectroscopy of Low Dimensional Semiconductors* edited by G. Abstreiter, A. Aydinli and J.-P. Leburton (Kluwer, New York, 1997) Vol. 344, p.213.
[14] H. Mehrez and S. Ciraci, Phys. Rev. B **56**, 12632 (1997).
[15] H. Mehrez and S. Ciraci, Phys. Rev. B **58**, 9674 (1998).
[16] Ç. Kılıç, H. Mehrez and S. Ciraci, Phys. Rev. B **58**, 7872 (1998).
[17] A. Buldum, Ph. D. Thesis (August 1998, Bilkent University).
[18] A. Buldum, S. Ciraci and C. Y. Fong, J. Phys.: Condens. Matter **12**, xxxx (2000).
[19] M. S. Daw and M. I. Baskes, Phys. Rev. B **29**, 6443 (1984); S. M. Foiles, M. I. Baskes and M. S. Daw, Phys. Rev. B **33**, 7983 (1986); M. S. Daw, Phys. Rev. B **39**, 7441 (1989).
[20] C. J. Muller, J. M. Krans, T. N. Todorov and M. A. Reed, Phys. Rev B **53**, 1022 (1996).
[21] H. Ohnishi, Y. Kondo and K. Takayanagi, Nature **395**, 780 (1988).
[22] A. Örmeci and S. Ciraci, (poster paper in this Conference and to be published).
[23] Yu. V. Sharvin, Zh. Eksp. Teor. Fiz. **48**, 984 (1965) [Sov. Phys.-JETP-**21**, 655 (1965)].
[24] E. Tekman and S. Ciraci, Phys. Rev. B **40**, 8559 (1989).
[25] I. Kander, Y. Imry and U. Sivan, Phys. Rev. B **41**, 12941 (1990).
[26] Ç. Kılıç, H.Mehrez and S. Ciraci and I. P. Batra, J. Electron Spectroscopy and Related Phenomena, 98-99, 335 (1999).
[27] D. R. Hamann, Phys. Rev. B **40**, 2980 (1989).
[28] D. M. Ceperley and B. J. Alder, Phys. Rev. Lett. **45**, 556 (1980).
[29] H. Mehrez, S. Ciraci, A. Buldum and I. P. Batra Phys. Rev. B **55**, R1981 (1997).
[30] K. Hansen, E. Laegsgaard, I. Stensgaard, and F. Besenbacher, Phys. Rev. B **56**, 2208 (1997).
[31] J. Bardeen, Phys. Rev, Lett. **6**, 57 (1961).
[32] G. A. Voth, J. Chem. Phys. **88**, 5547 (1988); G. A. Voth, *ibid* **87**, 5272 (1987); G. A. Voth and R. A. Marcus, *ibid* **84**, 2254 (1986).
[33] D. C. Langreth and E. Abrahams, Phys. Rev. B **24**, 2978 (1981).
[34] D. E. Angelescu, M. C. Cross and M. L. Roukes, Superlattice Microstruct. **23**, 673 (1998).
[35] L. G. C. Rego and G. Kirczenow, Phys. Rev. Lett. **81**, 232 (1998).
[36] V. Kalmeyer and R. B. Laughlin, Phys. Rev. B **35**, 9805 (1987).

Chapter 7

MAGNETOTRANSPORT AND MAGNETOCOHESION IN NANOWIRES

E. N. Bogachek, A. G. Scherbakov and U. Landman
School of Physics, Georgia Institute of Technology
Atlanta, Georgia 30332, USA

Abstract The behavior of the cohesive force and electronic conductance in nanowires, modeled by soft-wall confining potentials, under the influence of a magnetic field are studied. The appearance of force oscillations as a function of the magnetic field and their correlation with the corresponding characteristics of the electronic conductance are demonstrated. For materials with a strong Fermi surface anisotropy (*e.g.* bismuth), it is predicted that when the crystallographic axis associated with the largest diagonal element of the effective mass tensor is aligned along the direction of the wire the cohesive force increases dramatically (order of magnitude) compared to the case when that axis is perpendicular to the wire direction.

Formation of interfacial wires of atomic dimensions (nanowires) via elongation of contacts had been predicted through early molecular dynamics (MD) simulations [1], and observed experimentally [2] using tip-based microscopy, pin-disk, and mechanical break junction techniques. Furthermore, it had also been found that the elongation mechanism of the wire involves an oscillatory variation of the pulling force [1], and that such oscillations are correlated with step-wise variations of the conductance through the wire [3-5]. The energetics and mechanical response in nanowires as well as the electrical transport through them and the nature of the oscillatory behavior of the elongation force and the conductance, originating from structural rearrangements in the wire in conjunction with modifications of the electronic structure, have been analyzed using MD simulations and electronic structure calculations [1] as well as through the use of a jellium model [4-6] and free-electron treatments [7-10].

I. O. Kulik and R. Ellialtioğlu (eds.),
Quantum Mesoscopic Phenomena and Mesoscopic Devices in Microelectronics, 95–102.

Because of the nature of the preparation methods currently used to generate such three-dimensional (3D) nanowires it is often difficult to obtain reproducibly well-defined structures of the wires (unlike the case of lithographically fabricated two-dimensional wires, controlled via voltage-gates). Consequently, it is particularly desirable in investigations of such wires to explore the dependence of the wire's properties (mechanical and/or electrical) on external fields (magnetic [11-15], finite bias voltage [16-17], electromagnetic radiation [18], and thermal gradients [19-20]), which may be varied in a controllable manner.

In this paper we study the cohesive force and electric conductance of metallic nanowires under the influence of a magnetic field. Using a free-electron model for a wire modeled by a soft-wall confining potential we demonstrate the effect of a magnetic field on the oscillatory cohesion force and its correlation with the magnetic-field induced variations of the electric conductance through the wire. Furthermore, we investigate effects of the anisotropy of the Fermi surface on the cohesive the force of the nanowire and show that for a bismuth wire the force oscillations are highest when the crystallographic axis corresponding to the largest diagonal element of the electron mass tensor is parallel to the axis of the wire.

The mechanical properties of nanowires may be characterized by the cohesive force F

$$F = -\left(\frac{\partial \Omega}{\partial L}\right)_V , \tag{7.1}$$

where L is the length of the wire, and Ω is the grand-canonical potential. We model the nanowire (see inset in Fig. 7.1a) by a soft-wall harmonic confining potential of the form [17]

$$U_z(x, y) = \frac{\mu}{R^2(z)} \left(x^2 + y^2\right) , \tag{7.2}$$

where μ is the chemical potential and the effective radius of the wire, $R(z)$, is given by (a parabolic well)

$$R(z) = R_0 + (R_{max} - R_0)\frac{(2z)^2}{L^2} , \quad -\frac{L}{2} \leq z \leq \frac{L}{2} , \tag{7.3}$$

with $R_{max} \equiv R(\pm L/2)$, the radius at the ends of constricted section, kept constant. We assume also that the volume of the constricted region, V, remains constant during elongation of the wire in accordance with the results of molecular dynamics simulations [1]. In this case all the geometrical parameters depend only on the length of the wire, L. Considering here wires with adiabatically slow variation of $R(z)$, we can divide the wire into thin cylindrical slices with radii $R(z)$, and the electronic energy levels in each slice may be written as

$$E_{\nu p_z s} = \varepsilon_\nu + \frac{p_z^2}{2m_{||}} + sg^*\beta H , \tag{7.4}$$

where ν denotes the discrete quantum numbers due to transverse confinement, $s = \pm 1/2$, $m_{||}$ is the electronic mass corresponding to the motion along the axis of the cylindrical wire, g^* is the effective g-factor, $\beta = e\hbar/2m_0c$ is the Bohr magneton, and m_0 is the free electron mass. Thermodynamic potential of a wire with a variable shape is given by

$$\Omega_{wire} = \frac{1}{L}\int_{-L/2}^{L/2} \Omega_{cyl} dz , \tag{7.5}$$

with the thermodynamic potential in each cylindrical slice given by

$$\Omega_{cyl} = -k_B T \sum_{\nu p_z s} \ln\left[1 + \exp\left(\frac{\mu - E_{\nu p_z s}}{k_B T}\right)\right] . \tag{7.6}$$

For calculation of the conductance of the nanowire we use the Landauer formalism. The model which we use allows analytic solution of the Schrödinger equation in a longitudinal (directed along the axis of the nanowire) magnetic field. In addition, it is particularly suitable for investigations of the effects of anisotropy of the electronic Fermi surface and may be applied to a study of bismuth nanowires. Consider the case of an ellipsoidal Fermi surface with $m_i (i = 1, 2, 3)$ diagonal elements of the effective mass tensor when one of the crystallographic axes (axis Z) is parallel to the axis of the wire. The transverse energy levels of the electron are expressed in terms of the frequencies $\omega_{1,2}$:

$$\omega_{1,2}^2 = \frac{2\mu}{m_{1,2} R^2(z)} \tag{7.7}$$

and have a form (compare with Ref. [20]).

$$\varepsilon_{n_1,n_2} = \hbar\omega_+ \left(n_1 + \frac{1}{2}\right) + \hbar\omega_- \left(n_2 + \frac{1}{2}\right) , \tag{7.8}$$

with

$$\omega_\pm = \frac{1}{2}\left[\sqrt{\omega_c^2 + (\omega_1 + \omega_2)^2} \pm \sqrt{\omega_c^2 + (\omega_1 - \omega_2)^2}\right] . \tag{7.9}$$

Here, n_1 and n_2 are non-negative integers and $\omega_c = eH/cm_\perp$ is the cyclotron frequency, and $m_\perp = \sqrt{m_1 m_2}$ (m_1 and m_2 are diagonal elements of the mass tensor corresponding to the axes perpendicular to the axis of the wire). In the following we discuss the results of our calculations for bismuth nanowires with different orientations of the crystallographic axes with respect to the axis of the wire. In our calculations we use the following values for the diagonal elements of the mass tensor [21]: m_0, $0.02m_0$, and $0.006m_0$. We use 0.012 eV for the Fermi energy. Effective g-factor for bismuth is of the order of $m_0/m_\perp$ [22].

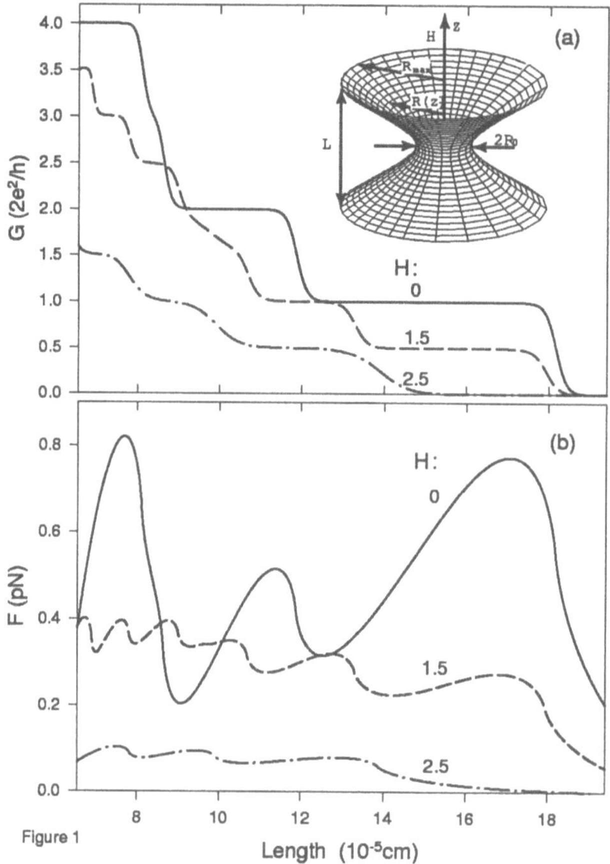

Figure 7.1 (a)Conductance (G, in units of $2e^2/h$) and (b) Cohesive force (F, in pN) of a bismuth wire with a variable shape at zero temperature, plotted vs the length of the wire, L (units of 10^{-5} cm). The different curves correspond to the marked values of the applied longitudinal (*i.e.* directed along the axis of the wire) magnetic field (H, in Teslas). The Fermi energy for bismuth was taken to be 0.012 eV. The diagonal element of the mass tensor corresponding to the direction along the axis of the wire is $m_{||} = m_0$, and the effective g-factor used in our calculations is 50. As the wire elongated its profile was readjusted (see Eq. (7.3)) such that its volume ($V = 2.13 \times 10^{-14}$ cm^3) and the radius of its maximal cross-section ($R_{max} = 1.02 \times 10^{-5}$ cm) remained constant. The geometry of the wire is shown in an inset (notations are in the text).

In Fig. 7.1 we show the conductance (a) and the cohesion force (b) as a function of the length of a bismuth nanowire for different values of an applied magnetic field. The conductance of the wire decreases with the length as the radius of the narrowest part of the wires decreases, thus reducing the number of the conducting electronic channels in the bottleneck of the constriction. The cohesive force oscillates as a function of length of the wire in correspondence with the changes in the conductance. Note that for stronger applied magnetic fields the values for the conductance, as well as the amplitude of the cohesive force, are smaller since a smaller number of conducting channels exists in the narrowest part of the constriction. The main effect of the magnetic field on the force is to modify the oscillatory pattern (which exists already in zero-field), including the appearance of new peaks. This effect, which is correlated with variations in the wire's conductance (compare panels (a) and (b)), originates from magnetic-field-induced removal of both orbital and spin degeneracies and shifts of the electronic transverse energy levels in the nanowire.

In Fig. 7.2 we show the conductance (a) and the cohesion force (b) for wires of different length, oriented such that their largest (m_0) diagonal element of the mass tensor corresponds to the axis of the wire. All wires in Fig. 7.2 have the same volume and the radius at the end of the constriction R_{max}. The behavior of the conductance as a function of the magnetic field demonstrates a "magnetic switch" effect discussed in [13]. The pattern of variation of the cohesion force correlates with the step pattern of the conductance. From comparison of different curves in Fig. 7.2 we observe that wires with a larger number of conductance channels exhibit a higher sensitivity to the magnetic field. Note that the appearance of conductance steps of magnitude e^2/h rather than $2e^2/h$ due to removal of the degeneracy of the electron energy levels by the Zeeman spin splitting.

The calculations of the cohesion force and of the conductance presented above pertain to bismuth wires oriented such that their largest diagonal element of the mass tensor coincides with the axis of the wire ($m_{||} = m_0$). In this case the amplitude of the cohesion force is maximal. Indeed, it may be shown easily that $F \sim m_{||}^{1/2}/(m_{\perp}^{3/2} R_0^3)$. Since for the wire with a fixed number of conducting channels $R_0 \sim 1/m_{\perp}^{3/2}$, one obtains $F \sim \sqrt{m_{||}}$. This estimate demonstrates enhancement (order of magnitude) of the amplitude of the cohesion force oscillations in the wires where $m_{||}$ is maximal (m_0) compared to the wires where $m_{||}$ is minimal ($0.006 m_0$). For bismuth nanowires with a fixed transverse dimension the influence of the crystallographic axis orientation will have

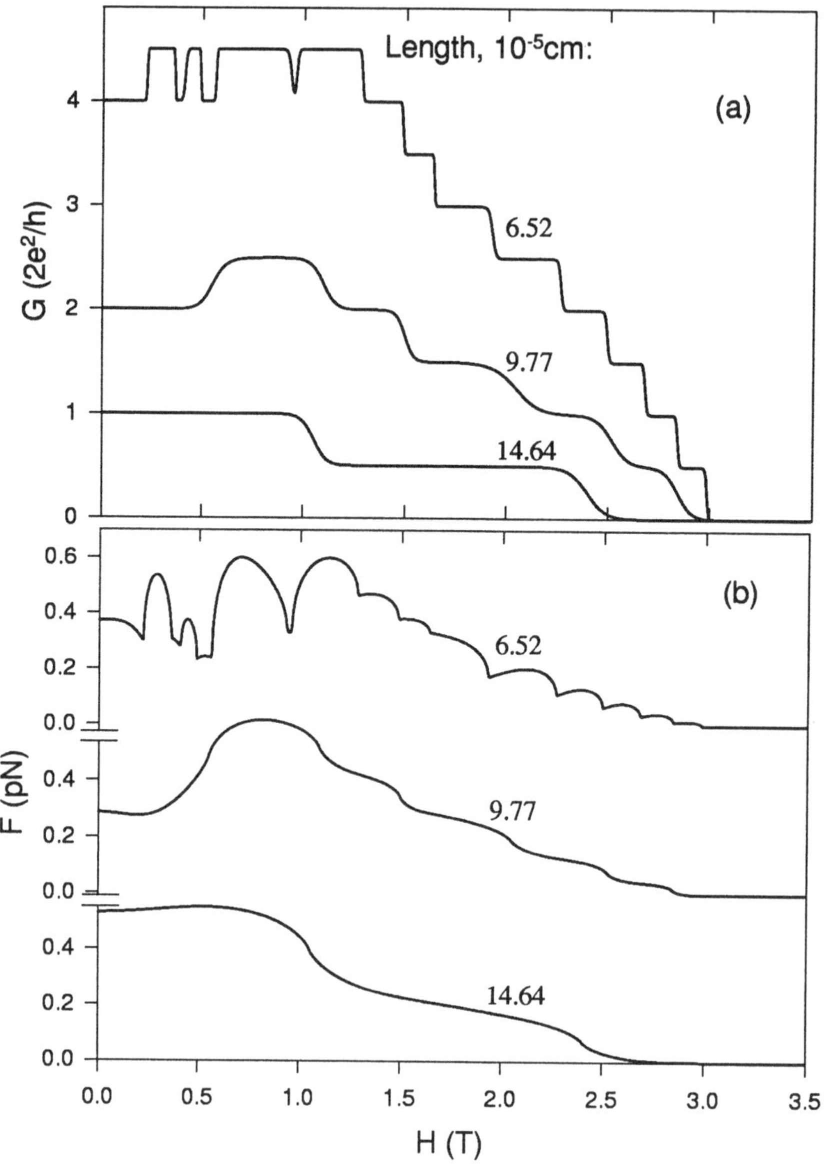

Figure 7.2 (a) Conductance (G, in units of $2e^2/h$) and (b) Cohesive force (F, in pN) of a sample bismuth wire described in Fig. 7.1 plotted vs the magnitude of the applied longitudinal magnetic field. The different curves correspond to the marked values of the length of the wire.

even more profound effect on the electronic transport and cohesion force. Indeed, for a fixed R_0 we find that the amplitude of the force oscillations will be 3 orders of magnitude bigger for wires with their crystallographic axis corresponding to the biggest diagonal element of the mass tensor oriented along the axis of the wire, compared to wires where such crystallographic axis is perpendicular to the axis of the wire. The crystallographic axes orientation will also affect the electronic transport through the nanowire since $m_\perp$, as well as the strong anisotropy, will define the number of conducting channels in the narrowest part of the nanowire. We also note that wires with smaller $m_\perp$ are more sensitive to the magnetic field and are much better candidates for the demonstration of the influence of the magnetic field on the electronic quantum transport (wires with such crystallographic axis orientation demonstrate "magnetic blockade" of the quantum electronic transport [13] at rather modest values of magnetic field).

The model analysis which we performed demonstrates that the elongation process in nanowires may be influenced by an external parameter such as a magnetic field. We studied magneto-mechanical properties in nanowires modeled via soft-wall confining potentials. Oscillations of the cohesive force occur when the quantized conductance of the wire changes from one conductance plateau to another. It is shown that the amplitude of the force oscillations depends on the orientation of the crystallographic axes with respect to the axis of the nanowire. For special orientations of the crystallographic axes the force oscillations might be substantially enhanced. For bismuth nanowires with approximately the same number of conducting channels, the amplitude of the cohesion force is about an order of magnitude larger if the largest diagonal element of the mass tensor lies along the axis of the wire, compared to wires where that crystallographic axis is perpendicular to the symmetry axis of the wire. Experimental investigations of cohesive force fluctuations are of importance because they can probe aspects pertaining to the electronic nature (total or partial) of the cohesive force arising during an elongation process of nanowires. Such effects may be observed more easily in semimetallic nanowires in magnetic fields with fluxes of the order of the flux quantum; *e.g.*, for a bismuth nanowires with a few transverse electronic modes (that is Bi wires with radii of ~20-50 nm) this corresponds to a magnetic field of several Teslas.

Acknowledgements

This study is supported by the U. S. Department of Energy, Grant No. FG05-86ER45234. Calculations performed at the Georgia Tech Center for Computational Materials Science.

References

[1] U. Landman, W. D. Luedtke, N. A. Burnhan, and R. J. Colton, Science **248**, 454 (1990).
[2] See reviews in "Nanowires", edited by P. A. Serena and N. Garcia (Kluwer, Dordrecht, 1997).
[3] E. N. Bogachek, A. M. Zagoskin, and I. O. Kulik, Fiz. Nizk. Temp. **16**, 1404 (1990) [Sov. J. Low Temp. Phys. **16**, 796 (1990)]
[4] C. Yannouleas and U. Landman, J. Phys. Chem. B **101**, 5780 (1997).
[5] C. Yannouleas, E. N. Bogachek, and U. Landman, Phys. Rev. B **57**, 4872 (1998).
[6] N. Zabala, M. J. Puska, and R. M. Nieminen, Phys. Rev. B **59**, 12652 (1999).
[7] C. A. Stafford, D. Baeriswyl, and J. Burki, Phys. Rev. Lett. **79**, 2863 (1997).
[8] J. M. van Ruitenbeek, M. H. Devoret, D. Esteve, and C. Urbina, Phys. Rev. B **56**, 12566 (1997).
[9] S. Blom, H. Olin, J. L. Costa-Kramer, N. Garcia, M. Jonson, P. A. Serena, and R. I. Shekhter, Phys. Rev. B **57**, 8830 (1998).
[10] A. M. Zagoskin, Phys. Rev. B **58**, 15827 (1998).
[11] E. N. Bogachek, M. Jonson, R. I. Shekhter, and T. Swahn, Phys. Rev. B **47**, 16635 (1993); **50**, 18341 (1994).
[12] A. G. Scherbakov, E. N. Bogachek, and U. Landman, Phys. Rev. B **53**, 4054 (1996).
[13] E. N. Bogachek, A. G. Scherbakov, and U. Landman, Phys. Rev. B **53**, R13246 (1996).
[14] A. G. Scherbakov, E. N. Bogachek, and U. Landman, Phys. Rev. B **57**, 6654 (1998).
[15] J. I. Pascual, J. A. Torres, and J. J. Saenz, Phys. Rev. B **55**, R16029 (1997).
[16] E. N. Bogachek, A. G. Scherbakov, and U. Landman, Phys. Rev. B **56**, 14917 (1997).
[17] S. Blom, L. Y. Gorelik, M. Jonson, R. I. Shekhter, A. G. Scherbakov, E. N. Bogachek, and U. Landman, Phys. Rev. B **58**, 16305 (1998).
[18] E. N. Bogachek, A. G. Scherbakov, and U. Landman, Phys. Rev. B **54**, R11094 (1996)
[19] E. N. Bogachek, A. G. Scherbakov, and U. Landman, Solid State Communic. **108**, 851 (1998); Phys. Rev B **60**, 11678 (1999).
[20] B. Schuh, J. Phys. A **18**, 803 (1985).
[21] J.E.Aubrey and R.G.Chambers, J.Phys.Chem.Solids **3**, 128 (1957).
[22] D. Shoenberg, Magnetic oscillations in metals (Cambridge University Press, Cambridge 1984), chapter 9.

II

QOULOMB BLOCKADE AND THE KONDO PROBLEM

Chapter 8

MESOSCOPIC FLUCTUATIONS OF CO–TUNNELING AND KONDO EFFECT IN QUANTUM DOTS

L. I. Glazman
Theoretical Physics Institute, University of Minnesota
Minneapolis, MN 55455, USA

Abstract Conductance of a quantum dot in the Coulomb blockade regime is discussed. At moderately low temperatures, the thermally-assisted transport of electrons through the dot gives way to elastic co-tunneling [1]. The amplitude of the elastic co-tunneling is sensitive to the specific structure of electron wave functions in a given sample. The conductance exhibits strong mesoscopic fluctuations [2]. Statistical properties of these fluctuations are reviewed in this lecture. At lower temperatures, conductance through a dot is altered by the Kondo effect, if the number of electrons on the dot is odd. We describe the universal features of the Kondo conductance [3, 4], and briefly discuss the manifestations of this effect in the limiting cases of weak and strong [5] dot-lead coupling.

1. INTRODUCTION

The effect of charging on the electron transport was conjectured in order to interpret the early data [6, 7] on the in-plane hopping conductivity of granular films. There, the activation energy extracted from the temperature dependence of conductivity was associated with the charging of individual grains composing the film. A clear demonstration of the role of charging, however, appeared later in the experiments of Giaever and Zeller [8] and Lambe and Jaklevic [9] on "vertical" tunneling through a layer of grains.

In Ref. [8] tunneling through a layer of Sn particles coated by a thin insulator and sandwiched between two aluminium electrodes was studied. The authors [8] measured the *I-V* characteristics of the junction both with and without magnetic field applied. That allowed them to

I. O. Kulik and R. Ellialtioğlu (eds.),
Quantum Mesoscopic Phenomena and Mesoscopic Devices in Microelectronics, 105–128.

exclude superconductivity as a source of nonlinearity, and to associate the observed finite-bias offset in the I-V characteristics with the effect of charging. They have realized, that at low temperatures electrons must tunnel through a grain one by one. A single additional electron localized on the grain in the course of tunneling, raises the potential of the grain. This prevents other electrons from hopping onto the grain before that electron have finished its path between the leads, and the grain is discharged. Such a "blockade" is effective only at low temperatures, when the characteristic energy $E_C = e^2/2C$ of a grain carrying one extra electron exceeds temperature, $E_C \gg T$ (hereinafter C is the total capacitance of the grain, and temperature T is measured in energy units).

A comprehensive theory of the effect of charging on tunneling was developed in papers of Shekhter [10] and Kulik and Shekhter [11]. There the events of tunneling to and from the grains were considered as sequential processes. In addition, all the electron-electron interaction was replaced solely by the charging energy. In the course of electron tunneling through the grain, its charge is varied by $\pm e$. In Ref. [11] rate equations controlling the probabilities of various discrete values of the grain charge were derived. These equations formed the basis for calculation of non-linear I-V characteristics. In the case of tunneling through a single grain, the I-V characteristic has a step-like structure with alternating intervals of low and high differential conductance, see Fig. 8.1a. Tunneling through many grains in parallel smears out the structure in the I-V characteristic. However, the average charging energy yields a characteristic offset of the high-voltage linear parts of the characteristic, see Fig. 8.1b.

Tunneling through a single grain may be studied, for instance, by means of the STM technique [12, 13]. At higher biases, $V \gg e/C_{1,2}$, many states differing by the value of charge become available for electron tunneling. This leads to a progressive smearing of the higher steps in the I-V curve, even in the absence of the electron heating effect. On the basis of the rate equations for a strongly asymmetric setup, $R_2 \gg R_1$ in the inset in Fig. 8.1, one can expect that about $C_2R_2/(C_1 + C_2)R_1$ steps are resolved [14].

The investigation of single electron tunneling through a metallic grain in a two-terminal device inevitably involves applying a relatively high bias $eV \sim e^2/C$ to the device, and therefore may easily result in a significant departure of the system from the thermal equilibrium. A device with a third terminal, gate, is free from that drawback. The gate is coupled to the grain only capacitively, and it allows to vary the electrostatic potential on the grain without perturbing its equilibrium state.

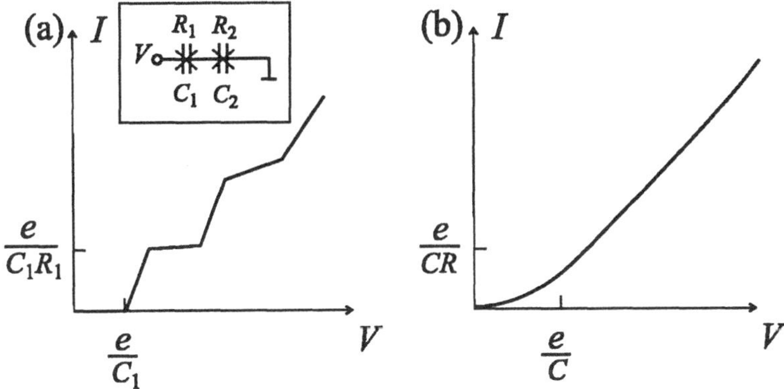

Figure 8.1 Sketches of the $I-V$ characteristics for tunneling (a) through a single grain at $C_1 \ll C_2$ and $R_1 \ll R_2$, and (b) through a layer of grains. In the inset: the equivalent circuit for a single grain attached to the leads by two tunnel junctions.

By tuning the gate voltage of such a three-terminal device, commonly referred to as ***single electron transistor*** [15], one controls the ***Coulomb blockade*** of tunneling, without varying the (small) bias applied to the leads. Experimental implementation of such a device was reported first by Fulton and Dolan [16]. In this experiment, a periodic modulation of the device resistance with the variation of the gate potential was demonstrated.

At some special gate voltages the Coulomb blockade is lifted. Indeed, the two states of the grain with the charges $e\hat{N} = en$ and $e\hat{N} = e(n+1)$, where n is integer, have the same electrostatic energy

$$H_C = E_C\left(\hat{N} - \mathcal{N}\right)^2, \tag{8.1}$$

if the dimensionless gate voltage $\mathcal{N} = C_g V_g/e$ is half-integer, *i.e.*, Coulomb blockade is lifted at $\mathcal{N} = \mathcal{N}^*$ with $\mathcal{N}^* \equiv n + \frac{1}{2}$. Here $E_C = e^2/2C$ is the charging energy, C and C_g are the total capacitance of the grain and its capacitance to the gate, respectively.

Electron transport through the grain is activationless at the degeneracy points $\mathcal{N} = \mathcal{N}^*$. For the gate voltages around these points, and at temperatures $T \ll E_C$, the rate equations [11] yield [17] for the linear conductance

$$G(\mathcal{N}, T) = \frac{1}{2R_\infty}\frac{E_C(\mathcal{N} - \mathcal{N}^*)/T}{\sinh[E_C(\mathcal{N} - \mathcal{N}^*)/T]}. \tag{8.2}$$

Here $R_\infty = (G_1 + G_2)/G_1G_2$ is the high-temperature ($T \gg E_C$) resistance of the device, G_1 and G_2 being the conductances of the two tunnel junctions connecting the grain to the leads, see Fig. 8.2. According to Eq. (8.2), all the *Coulomb blockade peaks* are characterised by a single value of the low-temperature conductance, $G(\mathcal{N}^*) = 1/2R_\infty$. In the *Coulomb blockade valleys* ($\mathcal{N} \neq \mathcal{N}^*$), conductance falls off exponentially with the decreasing temperature (Fig. 8.3); again, all the valleys behave exactly the same way.

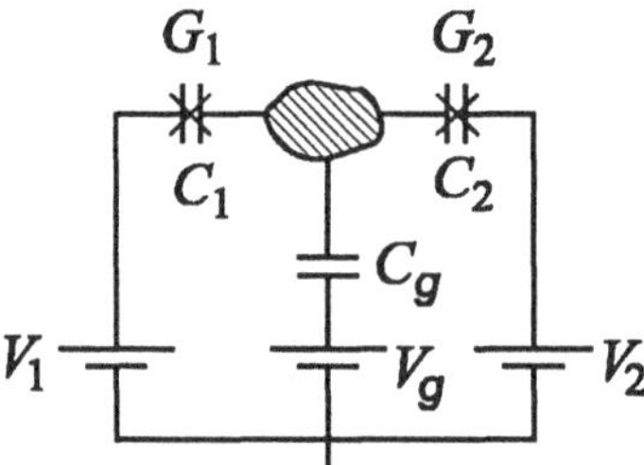

Figure 8.2 An equivalent circuit for a single-electron transistor. The total capacitance of the grain in this circuit is represented by the sum of the gate and junctions capacitances, $C = C_g + C_1 + C_2$. A bias can be applied by disbalancing the voltages V_1 and V_2.

The derivation of Eq. (8.2) disregards entirely the effects of coherent propagation of electrons across the grain, and effects of spectrum discreteness for electrons within the grain. This sets limitations for the application of the simple rate equations theory [17]. The coherence would result in weak-localization corrections to the conductance through a grain, even if one dispenses with the spectrum discreteness. Such corrections are small in the case of multi-channel junctions, no matter how big is the coherence length L_ϕ (the number of channels in a junction can be estimated as its area in units λ_F^2 set by the electron Fermi wavelength). The discreteness of electron spectrum destroys the strict periodicity of the $G(\mathcal{N})$ dependence. However, this discreteness is not important as long as the temperature exceeds the spacing δE between the quasiparticle energy levels in the dot. These two conditions are met usually in experiments with lithographically prepared metallic islands separated from leads by oxide tunnel barriers. Relatively large area of the barriers results in a large number of channels in the tunnel junctions; the Fermi wavelength λ_F in a metal is much smaller than the linear size of a sub-micron grain, so the condition $\delta E \ll T$ is met even for the lowest attainable temperatures. An example of the $G(\mathcal{N}, T)$ behavior well described by Eq. (8.2) can be found in the data of Ref. [18] for the high-resistance devices (sample 1 on Fig. 3 of Ref. [18]). Deviations from the rate equation theory occur at smaller resistance of the junctions ($R_{1,2} \sim h/e^2$), when the treatment of the grain charge as a classical discrete quantity becomes inadequate (data for sample 3 in Ref. [18]).

Single-electron tunneling through metallic nanoparticles and quantum dots formed in semiconductor nanostructures, however, demonstrates

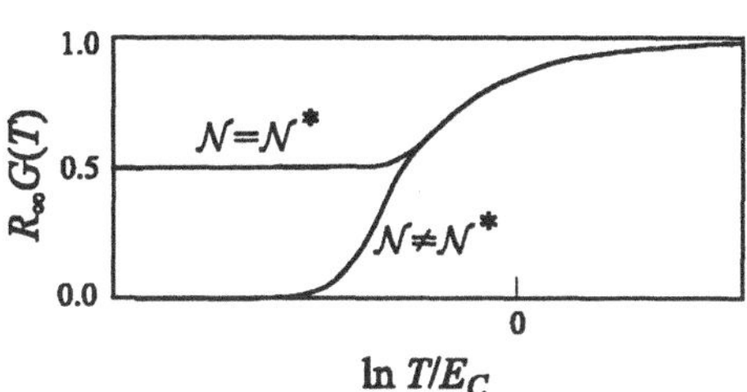

Figure 8.3 Sketches of the temperature dependence of the linear conductance $G(\mathcal{N}, T)$ in the Coulomb blockade peaks $\mathcal{N} = \mathcal{N}^*$ and valleys ($\mathcal{N} \neq \mathcal{N}^*$). Application of the rate equations to a model characterised by continuous electron spectrum in the grain, yields a periodic dependence $G(\mathcal{N})$; all Coulomb blockade peaks have the same height.

significant deviations from the above-described behavior, even in the experiments with high-resistance junctions. The reason is that the level spacing for these objects is considerably larger, and the condition $\delta E \gtrsim T$ is within the reach of an experiment. In addition, in semiconductor devices the number of channels in tunnel junctions connecting a quantum dot to the leads is small and controllable. Mesoscopic fluctuations of the valley conductance become stronger for smaller number of channels. We start the discussion of single-electron tunneling through quantum dots from the estimates of the relevant energy scales.

2. ENERGY SCALES INVOLVED IN THE SINGLE-ELECTRON TUNNELING EFFECTS

As it was discussed above, the Coulomb blockade of tunneling through a conducting grain becomes effective at temperatures $T \ll E_C$. In the Coulomb blockade regime, the linear conductance is finite only at the gate voltages allowing for degeneracy of the ground state with regard to addition of a single electron. The degeneracy condition, in a general form, depends not only on the charging energy, but also on the energy of spatial quantization of an electron confined to the dot. In a disordered grain, or irregularly-shaped quantum dot, we expect no geometrical symmetries, and thus assume the single-particle electron spectrum to be non-degenerate. The average level spacing in such a spectrum is defined by the density of states ν_d in the material, and by the volume L^d of the grain:

$$\delta E \simeq 1/\nu_d L^d. \tag{8.3}$$

Here L is the characteristic linear size of the grain, and d is the dimensionality of the system ($d = 2$ for a quantum dot formed in a two-dimensional electron gas at the interface of a semiconductor heterostructure, and $d = 3$ for a metallic nanoparticle; $d = 1$ for a one-dimensional conductor, like a segment of carbon nanotube). For a mesoscopic ($\lambda_F \ll L$) conductor, the level spacing is small compared to the charging energy. Indeed, using the estimate $E_C \sim e^2/\kappa L$, we find:

$$\frac{\delta E}{E_C} \sim \frac{\kappa \hbar v_F}{e^2}\left(\frac{\lambda_F}{L}\right)^{d-1} \sim \frac{1}{r_s}\left(\frac{\lambda_F}{L}\right)^{d-1}, \tag{8.4}$$

where v_F is the Fermi velocity, κ is the dielectric constant, and r_s is the conventional gas parameter characterising the electron-electron interaction in a non-ideal Fermi gas. Except an exotic case of an extremely weak interaction, having a large number of electrons in a quantum dot or grain, $N_e \sim (L/\lambda_F)^d \gg 1$, guarantees the smallness of the ratio $\delta E/E_C$ in dimensions $d = 2$ and 3. For the smallest quantum dots formed in a GaAs heterostructure, $\delta E/E_C \sim 0.1$; for a metallic nanoparticle with $L \simeq 5$nm this ratio is about 0.01, see, *e.g.*, Refs. [19, 20]. We should mention also that in the $d = 1$ case, the estimate (8.4) is of little help, as even a weak electron-electron interaction results in a formation of a Luttinger liquid [21], significantly affecting the nomenclature of the elementary excitations. In the case of a single-mode finite-length Luttinger liquid, the elementary excitations can be viewed as $1d$ plasmon waves in a confined geometry, so the corresponding level spacing is $\delta E \sim E_C$.

Tunneling between the grain and leads results in broadening of the discrete levels. The characteristic width of the levels Γ can be related to the level spacing δE and to the high-temperature conductance of the junctions connecting grain to the leads. Indeed, consider a grain attached to one lead, and suddenly biased by voltage V, see Fig. 8.4.

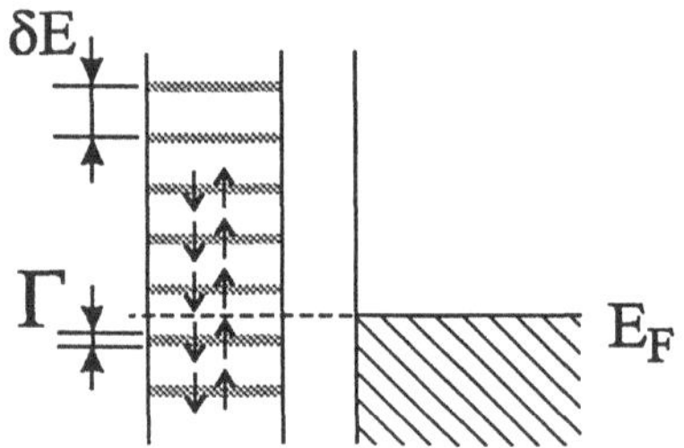

Figure 8.4 A sudden shift V of the potential on a dot results in a transient current I, while the quasi-discrete levels risen above the Fermi level depopulate. Consideration of this current allows one to relate the junction conductance to the levels spacing δE and width Γ.

Such a bias will force $\sim eV/\delta E$ occupied levels in the grain to cross the Fermi level. An electron escapes from an occupied level over time $\tau_{\rm esc} \sim \hbar/\Gamma$. In the absence of Coulomb blockade, the escapes from different levels are uncorrelated with each other, and the current of electrons leaving the grain right after the voltage pulse, is $I \sim (e/\tau_{\rm esc})(eV/\delta E) \sim (e^2/\hbar)(\Gamma/\delta E)V$. This estimate allows one to relate the ratio $\Gamma/\delta E$ to the junction conductance defined as $G = I/V$. Proper accounting for the numerical factors yields the following relation between the partial width

Γ_i of the level and the linear conductance G_i of the junction number i:

$$\langle \Gamma_i \rangle = \frac{\hbar G_i}{2e^2} \delta E. \tag{8.5}$$

Note that for each particular level n in the dot, widths Γ_i reflect the magnitudes of the electron eigenfunctions $|\varphi_n(\vec{r}_i)|^2$ in the vicinities of the respective junctions; $\langle \ldots \rangle$ represents averaging over many discrete eigenstates in the grain. Well-developed Coulomb blockade occurs at small junction conductances, $G_i \ll e^2/h$. Therefore, in the main part of this lecture we will assume the following hierarchy of the energy scales:

$$\Gamma \ll \delta E \ll E_C, \tag{8.6}$$

where $\Gamma = \sum_i \Gamma_i$. We start with the description of a theory of an isolated quantum dot ($\Gamma = 0$).

3. THE CONSTANT INTERACTION MODEL AND ITS JUSTIFICATION

A simple description of the confined electron system in a dot is possible if a number of conditions is met: (*i*) the electron-electron interaction within the dot should be not too strong, $r_s \lesssim 1$, so that the Fermi liquid model is applicable; (*ii*) there should be no degeneracies in the spectrum of the confined quasiparticles with energies near the Fermi level (this is satisfied, in general, for a chaotic electron motion); (*iii*) the conductance within the dot should correspond to the metallic regime (*i.e.*, the electron mean free path must exceed considerably the Fermi wave length).

The dimensionless conductance g of a dot is well-defined, if due to the disorder or irregular shape of the dot the motion of an electron within the dot is chaotic. In a dot of a linear size L the dimensionless conductance g is related to the Thouless energy [22] E_T and level spacing [see Eq. (8.3)] as $g \sim E_T/\delta E$. Here the Thouless energy is

$$E_T = \frac{\hbar D}{L^2}, \quad \text{or} \quad E_T \simeq \frac{\hbar v_F}{L}, \tag{8.7}$$

for the cases of elastic mean free path l shorter and longer than L, respectively; D is the diffusion constant in a disordered dot. If $L \gg l$ in a $2d$ dot, then its dimensionless conductance is $g \sim l/\lambda_F$. For a "ballistic" quantum dot $\hbar/E_T$ equals the time it takes for an electron to traverse the dot; conductance is obviously large in this case, $g \sim L/\lambda_F$.

If the conditions (*i*)–(*iii*) are met, then the Random Matrix Theory Hamiltonian [23] of non-interacting quasiparticles $\hat{H}_F$ is a good starting point for describing the dot,

$$\hat{H}_F = \sum_{\alpha,\gamma} \mathcal{H}_{\alpha\gamma} \psi^\dagger_{\alpha\sigma} \psi_{\gamma\sigma}. \tag{8.8}$$

Elements $\mathcal{H}_{\alpha\gamma}$ of the Hermitian matrix in Eq. (8.8) belong to a Gaussian ensemble of random real (GOE) or complex (GUE) variables. The matrix elements do not depend on spin, and therefore each eigenvalue ξ_n of the single-particle Hamiltonian represents a spin-degenerate orbital level. The spacings $|\xi_{n+1}-\xi_n|$ obey the Wigner-Dyson distribution [24]; the average value of $|\xi_{n+1}-\xi_n|$ is δE.

The two-particle interaction has the form

$$\hat{H}_{int} = \frac{1}{2}\sum \mathcal{H}_{\alpha\beta\gamma\delta}\psi^{\dagger}_{\alpha,\sigma_1}\psi^{\dagger}_{\beta,\sigma_2}\psi_{\gamma,\sigma_2}\psi_{\delta,\sigma_1}, \tag{8.9}$$

with the matrix elements $\mathcal{H}_{\alpha\beta\gamma\delta}$ depending on the electron-electron interaction potential $V(\vec{r}_1-\vec{r}_2)$ and on the chosen basis of orbital states ϕ_α,

$$\mathcal{H}_{\alpha\beta\gamma\delta} = \int d\vec{r}_1 d\vec{r}_2 V(\vec{r}_1-\vec{r}_2)\phi_\alpha(\vec{r}_1)\phi_\beta(\vec{r}_2)\phi^*_\gamma(\vec{r}_2)\phi^*_\delta(\vec{r}_1). \tag{8.10}$$

It turns out [25]-[29] that the majority of these matrix elements are small if $g \gg 1$, and only relatively few "most diagonal" elements remain finite in the limit $g \to \infty$. In other words, the Hamiltonian (8.10) can be separated in two pieces:

$$\hat{H}_{int} = \hat{H}^{(0)}_{int} + \hat{H}^{(1)}_{int}. \tag{8.11}$$

The first term here is universal: it does not depend on the geometry of the dot, or on the realization of the disorder. The second term consists of several parts, each proportional to some power of $1/g$; so $\hat{H}^{(1)}_{int}$ is small and sample-specific.

The form of the universal part of the interaction Hamiltonian can be established from the symmetry requirement [30]. Indeed, in view of the invariance of the distribution function of the Random Matrix Hamiltonian with respect to an arbitrary orthogonal (for GOE) or unitary (in the case of GUE) transformation, the term $\hat{H}^{(0)}_{int}$ should consist of operators invariant under such transformations. There are three such operators, bilinear in $\psi_{\alpha,\sigma}$ and $\psi^{\dagger}_{\alpha,\sigma}$:

$$\hat{N} = \sum_{\alpha,\sigma}\hat{\psi}^{\dagger}_{\alpha,\sigma}\hat{\psi}_{\alpha,\sigma}; \quad \hat{\vec{S}} = \frac{1}{2}\sum_{\alpha}\hat{\psi}^{\dagger}_{\alpha,\sigma_1}\vec{\sigma}_{\sigma_1\sigma_2}\hat{\psi}_{\alpha,\sigma_2}; \quad \hat{T} = \sum_{\alpha}\hat{\psi}_{\alpha,\uparrow}\hat{\psi}_{\alpha,\downarrow}. \tag{8.12}$$

Operator $\hat{N}$ is the total number of electrons, $\hat{\vec{S}}$ is the total spin of the dot; the "superconducting pair" operator $\hat{T}$ is needed for the description of the Cooper instability in grains of superconducting materials. Because of the pair-wise nature of interaction between the particles (8.9), the universal Hamiltonian is quadratic in terms of operators $\hat{T}$, $\hat{\vec{S}}$, and $\hat{N}$:

$$\hat{H}_{int}^{(0)} = E_C\left(\hat{N} - \mathcal{N}\right)^2 + J_S\left(\hat{\vec{S}}\right)^2 + J_c\hat{T}^\dagger\hat{T}. \qquad (8.13)$$

Note that the first term in Eq. (8.13) is just the charging energy (8.1). The second term represents the exchange energy of a dot. As long as the dot is isolated from the leads, both its charge and spin are conserving quantities, because the corresponding operators commute with the non-interacting part (8.8) of the Hamiltonian. On the contrary, operator $\hat{T}$ does not commute with $\hat{H}_F$, so an isolated dot in a ground state cannot have a non-zero average superconducting order parameter.

The exchange integral J_S is small in the case of weak electron-electron interaction: $J_S \sim r_s\delta E$ (there is an additional $\ln(1/r_s)$ factor in this estimate in the case of a $2d$ dot). Smallness of the ratio $J_S/\delta E$ guarantees the absence of a macroscopic (proportional to the volume of the dot) spin in the ground state, in accordance with the well-known Stoner criterion for the itinerant magnetism [31]. If all the orbital levels would be equidistant, then the spin of an even-electron state should be zero, while an odd-electron state should have spin 1/2. However, the level spacings $|\xi_{n+1} - \xi_n|$ are random. If the spacing between some orbital levels is accidentally small, the dot may acquire a spin [32] exceeding 1/2. In particular, a dot with an even number of electrons may have spin 1 in the ground state [33]. At small r_s, higher spins may occur only if several levels come very close to each other, which is quite a rare event [34].

The third term in the Hamiltonian (8.13) is renormalized to zero in the case of repulsive interaction, *i.e.*, $J_c/\delta E \to 0$ in a large ($\delta E \to 0$) dot. For an attractive interaction, $\mathcal{H}_{\alpha\alpha\gamma\gamma} < 0$, the renormalization enhances Δ_c, signalling the superconducting instability. In a large grain with δE significantly smaller than the superconducting gap Δ in the bulk material, the introduced energy constant $J_c \to -\delta E$. The variation of J_c with the decreasing size of the grain (*i.e.*, with the increasing ratio $\delta E/\Delta$) was studied in Refs. [35, 36, 37].

We will not discuss superconducting grains here, and set $J_c = 0$ in the remainder of this paper. If one disregards also the rare configurations in which two orbital levels come very close to each other, then the higher spin states can be also thrown away. The result of these simplifications is the so-called Constant Interaction model,

$$\hat{H}_{\rm CI} = E_C\left(\hat{N} - \mathcal{N}\right)^2 + \sum_{n\sigma}\xi_n a_{n\sigma}^\dagger a_{n\sigma}, \quad \hat{N} = \sum_{n\sigma} a_{n\sigma}^\dagger a_{n\sigma}, \qquad (8.14)$$

which is widely used in the analysis of the experimental data. Note that the single-particle level spacings $|\xi_{n+1} - \xi_n| \sim \delta E$ here are small compared to the charging energy, and each state is spin-degenerate. The eigenfunctions $\varphi_n(\vec{r})$ are random, and, in the leading order in $1/g$, do not correlate with each other and satisfy the Porter-Thomas distribution [24].

The non-universal correction to the Hamiltonian (8.13) can be split in two terms, $\hat{H}^{(1)}_{int} = \hat{H}^{(1/\sqrt{g})}_{int} + \hat{H}^{(1/g)}_{int}$. The leading term here, $\hat{H}^{(1/\sqrt{g})}_{int}$, comes from the joint effect of the interaction and of the electron confinement in dot [38], this correction is proportional to $1/\sqrt{g}$. The corresponding term in $\hat{H}^{(1)}_{int}$ depends also on the gate voltage $\mathcal{N}$. In the simplest case of a distant gate, this term can be cast into the form [30]

$$\hat{H}^{(1/\sqrt{g})} = (\hat{N} - \mathcal{N}) \sum_{\alpha,\gamma} \mathcal{X}_{\alpha\gamma} \psi^\dagger_\alpha \psi_\gamma. \tag{8.15}$$

The average value of the random matrix elements $\mathcal{X}_{\alpha\beta}$ is zero, while their variance is given by

$$\langle \mathcal{X}^2_{\alpha\gamma} \rangle = b_1 \frac{(\delta E)^2}{g}. \tag{8.16}$$

Here $b_1 \sim 1$ is a numerical coefficient which depends on the details of the potential confining the electrons [30, 38]. As one can see, the correction (8.15) causes some shifts of the single-particle energy levels described by Hamiltonian (8.14) of the Constant Interaction model. This shifts however, at large conductance of the dot ($g \gg 1$) are small compared with the level spacing δE, which gives additional support for the Constant Interaction model.

Next in the magnitude mesoscopic term in the Hamiltonian $\hat{H}^{(1)}_{int}$ has the standard four-fermion structure,

$$\hat{H}^{(1/g)} = \frac{1}{2} \sum_{\alpha\beta\gamma\delta} \mathcal{H}^{(1/g)}_{\alpha\beta\gamma\delta} \psi^\dagger_{\alpha,\sigma_1} \psi^\dagger_{\beta,\sigma_2} \psi_{\gamma,\sigma_2} \psi_{\delta,\sigma_1}. \tag{8.17}$$

Here the matrix elements $\mathcal{H}^{(1/g)}_{\alpha\beta\gamma\delta}$ have non-zero average values,

$$\langle H^{(1/g)}_{\alpha\beta\gamma\delta} \rangle = b_2 \frac{\delta E}{g} \left(\delta_{\alpha\delta} \delta_{\beta\gamma} + \delta_{\alpha\gamma} \delta_{\beta\delta} \right). \tag{8.18}$$

The amplitude of the mesoscopic fluctuations $\delta H_{\alpha\beta\gamma\delta}$ of these matrix elements is of the order of their average [25]-[29],

$$\left\langle \left[\delta H^{(1/g)}_{\alpha\beta\gamma\delta} \right]^2 \right\rangle = b_3 \left(\frac{\delta E}{g} \right)^2. \tag{8.19}$$

Here the numerical constants $b_2 \sim 1$ and $b_3 \sim 1$ depend on the details of the dynamics of electron motion within the dot [25]. This part of the Hamiltonian determines the inelastic electron relaxation within the dot.

Discussing electron transport through a quantum dot in the following sections, we will concentrate on the "conventional" case, described by the Constant Interaction model (8.14).

4. ACTIVATIONLESS TRANSPORT THROUGH A BLOCKADED QUANTUM DOT

According to the rate equations theory [11, 17], at low temperatures, $T \ll E_C$, conduction through the dot in the Coulomb blockade valleys is exponentially suppressed. This suppression occurs because the process of electron transport through the dot involves a real state in which the charge of the dot differs by the one-electron charge e from the thermodynamically most probable value. The thermodynamic probability of such a fluctuation is $\sim \exp[-E_C|\mathcal{N}-\mathcal{N}^*|/T]$, which explains the conductance suppression, see Eq. (8.2). Going beyond the lowest-order perturbation theory in conductances G_1 and G_2 allows one to consider processes in which the quantum states of the dot carrying a "wrong" charge participate only as virtual states in the tunneling process. Such higher-order contribution to the tunneling conductance was envisioned in the insightful paper of Giaever and Zeller [8]. The first quantitative theory of this effect, however, was developed much later [1, 39]. The leading contributions to the activationless transport, according to Refs. [1, 39], are provided by the processes of inelastic and elastic ***co-tunneling***.

Unlike the sequential tunneling, in the co-tunneling mechanism, the events of electron tunneling from one of the leads into the dot, and tunneling out from the dot into the other lead occur as a single quantum process. There exists also a complementary process of co-tunneling, in which the order of the tunneling events is reversed. The state of the dot with the "wrong" charge (one extra electron or hole) appears only as an intermediate (virtual) state in such processes.

4.1 Inelastic Co-Tunneling

In the inelastic co-tunneling mechanism, the electron which enters the dot occupies one of the empty levels of spatial quantization, and the electron that leaves the dot, vacates one of the *other* levels of spatial quantization, see Fig. 8.5. Therefore, the process of inelastic co-tunneling results in the transfer of charge e between the leads and by a simultaneous creation of an electron-hole pair in the dot.

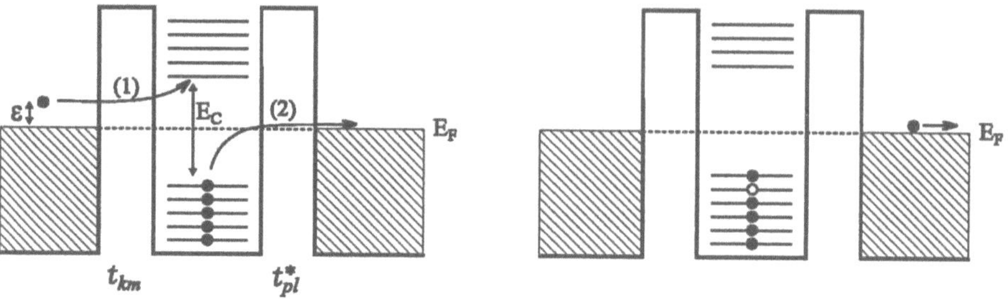

Figure 8.5 Inelastic co-tunneling. (*a*): The virtual state is formed after step 1 of the process is accomplished, and the dot acquires an extra electron. In step 2, some other electron tunnels out. (The reverted sequence of processes also contributes to the tunneling amplitude.) (*b*): In the final state, a charge e is transferred from the left lead to the right one, and an electron-hole excitation is left in the quantum dot.

The amplitude of the co-tunneling process may be calculated by means of perturbation theory in the tunnel Hamiltonian $\hat{H}_T$ describing the electron transport between the dot and leads:

$$\hat{H}_T = \sum_{k,n} \left(t_{kn} a_k^\dagger a_n + t_{kn}^* a_n^\dagger a_k \right) + \sum_{p,n} \left(t_{pn} a_p^\dagger a_n + t_{pn}^* a_n^\dagger a_p \right). \quad (8.20)$$

Here $a_k^\dagger$ and $a_p^\dagger$ are the creation operators describing electrons in the leads, and t_{kn}, t_{pn} are the tunneling matrix elements connecting state n in the dot with the states k and p in the two opposite leads. The typical values of these matrix elements can be related to the junctions conductances,

$$\langle |t_{kn}|^2 \rangle = \frac{\langle \Gamma_1 \rangle}{2\pi\nu_1} = \frac{\hbar G_1}{2\pi e^2 \nu_1} \delta E, \quad \langle |t_{kp}|^2 \rangle = \frac{\langle \Gamma_2 \rangle}{2\pi\nu_2} = \frac{\hbar G_2}{2\pi e^2 \nu_2} \delta E, \quad (8.21)$$

where $\nu_{1,2}$ are the densities of states in the leads.

For simplicity, we will illustrate here the estimate of the inelastic co-tunneling conductance deep in the Coulomb blockade valley, *i.e.*, at almost integer $\mathcal{N}$. Suppose the energy of the incoming electron ε, measured from the Fermi level, is confined by condition:

$$\varepsilon \ll E_C. \quad (8.22)$$

Then the energy deficit of the virtual state involved in the co-tunneling process is close to E_C. The amplitude A_{in} of the inelastic transition is

$$A_{\text{in}} = \frac{2t_{km}^* t_{pl}}{E_C}. \quad (8.23)$$

In such a transition, the initial state has an extra electron in the single-particle state k, and the final state has an extra electron in the opposite lead (state p) and an electron-hole pair in the dot (discrete state m is occupied, and state l is empty). The number of final states available at given initial energy ε can be estimated from the phase space argument, familiar from the calculation of the lifetime of a quasiparticle in the Fermi liquid [40]. If the initial electron energy $\varepsilon \gg \delta E$, then the number of final states available is $\sim \varepsilon^2/\delta E$. Using this estimate and Eqs. (8.21), (8.23), we can find now, up to a numerical factor, the conductance for the inelastic process:

$$G_{\text{in}} = \frac{4\pi^3}{3}|A_{\text{in}}|^2 \left(\frac{T}{\delta E}\right)^2 = \frac{4\pi}{3}\frac{G_1 G_2}{e^2/\hbar}\left(\frac{T}{E_C}\right)^2 . \tag{8.24}$$

(The numerical constants here are fixed so that the final result coincides with the result of rigorous calculation [1]). A comparison of G_{in} with the result of the rate equations theory (8.2) shows that the inelastic co-tunneling takes over the thermally-activated hopping at moderately low temperatures

$$T \lesssim T_{\text{in}} \equiv E_C \left[\ln\left(\frac{e^2/\hbar}{G_1 + G_2}\right)\right]^{-1} . \tag{8.25}$$

The smallest energy of the electron-hole pair is of the order of δE. At temperatures below that threshold the contribution of the inelastic co-tunneling mechanism to the conductance becomes exponentially small. It turns out, however, that even at much higher temperature this mechanism becomes less effective than the elastic co-tunneling.

4.2 Elastic Co-Tunneling. Mesoscopic Fluctuations Of The Conductance In The Coulomb Blockade Valleys

In the process of elastic co-tunneling, an electron tunnels in and out of the dot to/from the same level of spatial quantization. Therefore, no electron-hole pairs are created in such a process: after the charge is transferred between the leads, the quantum dot remains in its initial state.

Estimating the elastic co-tunneling contribution to the conductance we will consider the edge of a Coulomb-blockade valley,

$$\frac{\delta E}{E_C} \ll \mathcal{N} - \mathcal{N}^* \ll 1/2. \tag{8.26}$$

This will allow us to account only for the hole-like contributions to the tunneling amplitude, see Fig. 8.6. (The general case is qualitatively similar to this simplified one.) Each of these contributions A_n utilizes

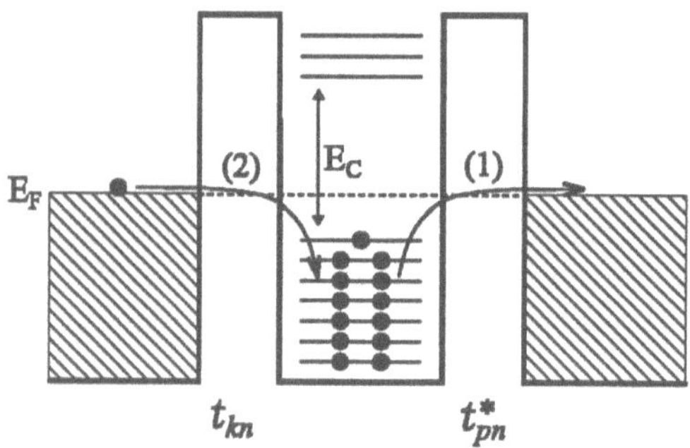

Figure 8.6 Elastic co-tunneling. Under the condition (8.26), the hole-like processes dominate: first an electron is extracted from the dot (1), and then the resulting hole is filled by an electron from the other lead (2). The special role of higher-order tunneling through the top-most filled level in the case of an odd number of electrons on the dot will be discussed in Section 5.

one of the discrete levels as a virtual state for tunneling:

$$A_n = \frac{t_{kn}t^*_{pn}}{2E_C(\mathcal{N}-\mathcal{N}^*)+|\xi_n|}, \quad \xi_n < 0. \tag{8.27}$$

To derive the energy denominators here, we have utilized the Constant Interaction model (8.14); the quasiparticle energies are measured from the Fermi level. The total tunneling amplitude is the sum of the partial amplitudes A_n. The elastic co-tunneling conductance can be calculated then as

$$G_{\rm el} = \frac{e^2}{\pi\hbar}\left|\sum_n A_n\right|^2 = \frac{e^2}{\pi\hbar}\sum_{lm} A_l A^*_m \; ; \quad \xi_l, \xi_m < 0. \tag{8.28}$$

If the junctions to the dot are point contacts, then $G_{\rm el}$ exhibits strong mesoscopic fluctuations. Indeed, tunneling matrix elements entering Eq. (8.27) depend on the values of the electron wave functions at the points of contacts, $t_{kn} \propto \varphi_n(\vec{r}_1)$, $t_{pn} \propto \varphi_n(\vec{r}_2)$. As it was discussed in Section 3, the electron eigenfunctions in the dot $\varphi_n(\vec{r}_i)$ are random and uncorrelated. Therefore, the partial amplitudes

$$A_n = \frac{\hbar}{e^2}\sqrt{G_1 G_2}\frac{\delta E}{2E_C(\mathcal{N}-\mathcal{N}^*)+|\xi_n|}\frac{\varphi_n(\vec{r}_1)\varphi_n(\vec{r}_2)}{\langle|\varphi_n|^2\rangle} \tag{8.29}$$

are random and uncorrelated as well, and $\langle A_n\rangle = 0$.

Now it is clear that only diagonal terms ($l = m$) in the double-sum of Eq. (8.28) contribute to the average conductance $\langle G_{\rm el}\rangle$. The sum of diagonal terms $\langle|A_n|^2\rangle$ is converging, and the characteristic number of terms contributing to it, $\sim E_C|\mathcal{N}-\mathcal{N}^*|/\delta E$, is large under the conditions (8.26). The resulting average contribution to the conductance is [1]

$$\langle G_{\rm el}\rangle = \frac{\hbar}{4\pi e^2}G_1 G_2\frac{\delta E}{E_C}\left(\frac{1}{\mathcal{N}-\mathcal{N}^*}+\frac{1}{\mathcal{N}^*-\mathcal{N}+1}\right). \tag{8.30}$$

(Here both the hole-like and electron-like contributions are taken into account.)

Comparing Eq. (8.30) with Eq. (8.24), we see that near the bottom of the valleys the elastic co-tunneling dominates the electron transport already at temperatures

$$T \lesssim T_{\rm el} \equiv \sqrt{E_C \delta E}, \tag{8.31}$$

which may exceed significantly the level spacing.

All the off-diagonal terms ($l \neq m$) in Eq. (8.28) contribute to the mesoscopic fluctuations of the conductance. Given the random Gaussian behavior of the amplitudes (8.29), it is obvious that the variance of the conductance $\langle \delta G_{\rm el}^2 \rangle$ is of the order of $\langle G_{\rm el} \rangle^2$. The exact relation [2] between these quantities

$$\langle \delta G_{\rm el}^2 \rangle = \frac{2}{\beta} \langle G_{\rm el} \rangle^2 \tag{8.32}$$

depends on the existence of the time reversal symmetry in the system. In zero magnetic field such symmetry is present, and the wave functions in the dot are described by the orthogonal statistical ensemble (GOE), $\beta = 1$. With the increase of the magnetic field, all states contributing to the elastic co-tunneling eventually cross over into the unitary ensemble (GUE), and $\beta = 2$.

The structure of the electron wave functions in the dot, and the sample-specific value of $\delta G_{\rm el}$ is affected by a relatively weak magnetic field B. An increase of the field from B to $B + B_{\rm corr}$ results in scrambling of the wave functions; the corresponding random values $\delta G_{\rm el}(B)$ and $\delta G_{\rm el}(B + B_{\rm corr})$ become uncorrelated at a sufficiently large $B_{\rm corr}$. The characteristic field increment $B_{\rm corr}$ needed for suppression of correlations, can be rigorously defined after the correlation function $\langle \delta G_{\rm el}(B_1) \delta G_{\rm el}(B_2) \rangle$ is calculated [2]. Here we will give only an estimate of $B_{\rm corr}$ for a disordered quantum dot, appealing to the semi-classical picture of electron motion.

The characteristic time τ the dot spends in the virtual state during the process of elastic co-tunneling is determined by the energy deficit of such a state, $\tau \sim \hbar/[E_C|\mathcal{N} - \mathcal{N}^*|]$. During this time, the trajectory of the tunneling electron would cover an area $\sim D\tau$, if there would be no boundaries of the dot (here D is the electron diffusion constant in the dot). Because of the finite linear size of the dot L, the trajectory winds inside it, approximately $\eta = D\tau/L^2$ times. The winding direction is random, therefore the effective area $S_{\rm eff}$ under the electron trajectory is determined by the fluctuation of η; the typical value of this area is $S_{\rm eff} \sim L^2\sqrt{\eta}$. Introducing here Thouless energy (8.7), we find $S_{\rm eff} \sim L^2 (E_T/E_C)^{1/2} |\mathcal{N} - \mathcal{N}^*|^{-1/2}$. The electron wave functions which are important for the elastic co-tunneling vary substantially if the magnetic flux through the typical trajectory is increased by one quantum, $B_{\rm corr}S_{\rm eff} \sim \Phi_0$. This relation yields:

$$B_{\rm corr}(\mathcal{N}) \sim \frac{\Phi_0}{L^2}\sqrt{\frac{E_C|\mathcal{N}-\mathcal{N}^*|}{E_T}}. \tag{8.33}$$

Note that the correlation field depends on the gate voltage. It is increasing while the gate voltage is approaching the bottom of the valley. This can be understood as the result of the decrease of the time τ allowed for the virtual localization of an electron in the dot. The shorter this time, the stronger field should be applied in order to affect the phases of the partial tunneling amplitudes (8.29).

The fact that the number of discrete single-particle levels participating in the co-tunneling process is large, affects also the correlation of the conductance fluctuations in different valleys. Indeed, it is necessary to shift the gate voltage $\mathcal{N}$ by $\sim E_C/\delta E$ in order to replace all the discrete eigenstates participating (as virtual states) in the elastic co-tunneling process. So one would expect[42] that the conductance fluctuations are correlated over about $E_C/\delta E$ valleys.

Equations (8.30) and (8.33) explain, why conductance $G(\mathcal{N})$ and correlation magnetic field $B_{\rm corr}(\mathcal{N})$ were observed [41] to vary with opposite phases when the gate voltage was varied. Later, the calculated [2] correlation function $\langle \delta G_{\rm el}(B_1)\delta G_{\rm el}(B_2)\rangle$ was used to extract the values of the correlation field from the data, and a quantitative agreement with Eq. (8.33) was found [43]. At the same time, the correlation of conductance fluctuations in different valleys was found to fall off substantially faster than the theoretical prediction.

5. KONDO CONDUCTANCE OF A BLOCKADED QUANTUM DOT

Among the $E_C|\mathcal{N}-\mathcal{N}^*|/\delta E$ virtual states participating in the elastic co-tunneling through a blockaded dot, the top-most occupied discrete level plays a special role. If the number of electrons in the dot is odd, this level is filled by a single electron and therefore is spin-degenerate. The amplitude of an electron transfer through such a level, calculated in the fourth order in tunnel matrix elements t_{kn}, t_{pn} logarithmically diverges at low temperatures. This divergence signals the Kondo singularity [44] in the transmission amplitude.

If the junctions conductances are small, $G_{1,2} \ll e^2/\hbar$, then the other levels of the dot, which are doubly-filled or empty, are unimportant in the discussion of Kondo effect. The model of the dot attached to two leads then can be truncated to the Anderson single-level impurity model with the electron continuum consisting of two bands:

$$\hat{H}_A = \sum_{q,\sigma} \xi_q (a^\dagger_{1q\sigma} a_{1q\sigma} + a^\dagger_{2q\sigma} a_{2q\sigma}) + \sum_\sigma \varepsilon_0 a^\dagger_{0\sigma} a_{0\sigma} + U \hat{n}_\uparrow \hat{n}_\downarrow \qquad (8.34)$$
$$+ \sum_{q,\sigma} (t_1 a^\dagger_{1q\sigma} + t_2 a^\dagger_{2q\sigma}) a_{0\sigma} + a^\dagger_{0\sigma} (t_1 a_{1q\sigma} + t_2 a_{2q\sigma}), \; \hat{n}_\sigma = a^\dagger_{0\sigma} a_{0\sigma} \, .$$

Here $a^\dagger_{1q\sigma}$, $a^\dagger_{2q\sigma}$, and $a^\dagger_{0\sigma}$ are the electron creation operators in the leads 1 and 2, and on the upper occupied discrete level in the dot, respectively; ξ_q and $\varepsilon_0 = 2E_C(\mathcal{N}^* - \mathcal{N})$ are the corresponding energies in the electron continuum and on the dot; $U \sim E_C$. For brevity, the tunneling matrix elements t_1 and t_2 connecting the discrete state in the dot with the states in the leads are taken here to be q-independent; for the same reason, we consider here the "depth" of the localized state to be confined by the conditions: $\delta E \ll -\varepsilon_0 \ll E_C$. At first sight, the Anderson model with two bands (8.34) may be associated with a two-channel Kondo model [46]. However, it is easy to show that in fact such a two-channel model is degenerate, and can be reduced to the conventional single-channel one. Indeed, by a unitary transformation

$$\left\{ \begin{array}{c} \alpha_{q\sigma} \\ \beta_{q\sigma} \end{array} \right\} = u a_{1q\sigma} \pm v a_{2q\sigma} \text{ with } \left\{ \begin{array}{c} u \\ v \end{array} \right\} = \frac{1}{\sqrt{|t_1|^2 + |t_2|^2}} \left\{ \begin{array}{c} t_1 \\ t_2 \end{array} \right\}, \qquad (8.35)$$

Hamiltonian (8.34) can be converted [4] to the conventional one-band Anderson impurity model [45]. The localized state and band "α" form the usual Anderson impurity model, which is characterized by three parameters: U, ε_0, and $\Gamma = \Gamma_1 + \Gamma_2$. Band "$\beta$" is entirely decoupled from the impurity.

The unitary transformation (8.35) establishes the relation between the tunneling conductance G_K associated with the spin-degenerate level ("Kondo conductance") in a quantum dot, and the known results for the Kondo-impurity contribution to the resistivity in a bulk conductor,

$$G_K = \frac{e^2}{\pi\hbar} \frac{4\Gamma_1\Gamma_2}{(\Gamma_1 + \Gamma_2)^2} f\left(\frac{T}{T_K}\right). \qquad (8.36)$$

Here T_K is the characteristic temperature of the problem (Kondo temperature), and $f(x)$ is some universal function (it has been evaluated numerically and plotted, *e.g.*, in Ref. [48]). A remarkable property of scattering on a Kondo impurity, is that the corresponding cross-section approaches the unitary limit at low energies, $f(0) = 1$. The low-temperature correction to the unitary limit is proportional to T^2 and

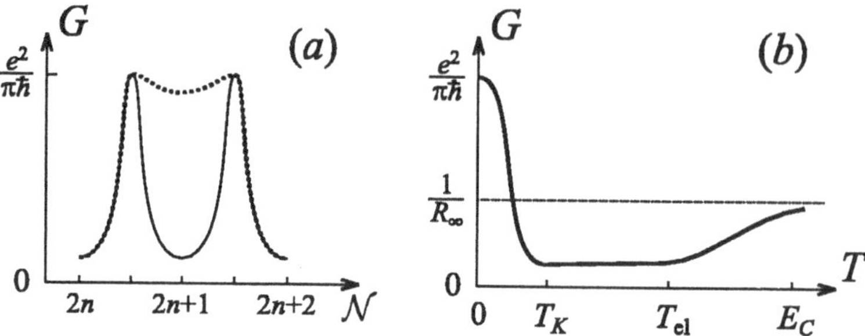

Figure 8.7 (*a*) Linear conductance at $T \gg T_K$ and $T \lesssim T_K$. Due to the Kondo effect, function $G(\mathcal{N})$ develops plateaus in place of the "odd" valleys of the Coulomb blockade. (*b*) Sketch of the temperature dependence of the linear conductance, $G(T)$, in the "odd" valley. Conductance decreases with the decrease of temperature from $T \sim E_C$ down to $T \sim T_{el}$. At very low temperatures, $T \lesssim T_K$, conductance grows again, reaching the unitary limit $G = e^2/\pi\hbar$ in the case of equal partial widths Γ_1 and Γ_2 of the singly-occupied level, which gives rise to the Kondo effect.

described by Nozières' Fermi-liquid theory [49]. The conventional [40] representation of the low-temperature expansion of $f(x)$,

$$f\left(\frac{T}{T_K}\right) = 1 - \frac{\pi^2 T^2}{3T_K^2}, \qquad T \ll T_K, \tag{8.37}$$

yields one of the ways of defining the Kondo temperature. Being expressed [47] in terms of the parameters of the Anderson impurity model, T_K is given, up to some unimportant pre-exponential factor, by the following relation:

$$T_K \simeq (U\Gamma)^{1/2} \exp\{\pi\varepsilon_0(\varepsilon_0 + U)/2\Gamma U\}. \tag{8.38}$$

Equations (8.36) and (8.37) tell us that upon sufficiently deep cooling, the gate voltage dependence of the conductance through a quantum dot should exhibit a drastic change. Instead of the "odd" valleys, which correspond to the intervals of gate voltage

$$|\mathcal{N} - (2n+1)| < \frac{1}{2}, \quad n = \text{integer}, \tag{8.39}$$

plateaus in the function $G(\mathcal{N})$ develop, see Fig. 8.7*a*. In other words, the temperature dependence of the conductance in the "odd" valleys should be very different from the one in "even" valleys. The conductance decreases monotonously with the decrease of temperature down to $T \lesssim T_{el}$, and then saturates at the value of G_{el} in the even valleys. On the contrary, if the gate voltage is tuned to one of the intervals (8.39), the

$G(T)$ dependence is non-monotonous. After the initial drop occurs with lowering the temperature in the interval $T_{\rm el} \gtrsim T \gtrsim E_C$, the conductance starts to increase again at $T \lesssim T_K$, see Fig. 8.7*b*. Its $T = 0$ saturation value depends on the ratio of the partial level widths Γ_1 and Γ_2 of a particular discrete level; these widths are random quantities described by Porter-Thomas distribution, see Sections 2 and 3.

The experimental search for a tunable Kondo effect brought positive results [19] only recently. In retrospect it is clear, why such experiments were hard to perform. Let us express the Kondo temperature (8.38) in terms of the quantum dot parameters. Replacing the partial widths by their average values, we find:

$$T_K \simeq \delta E \sqrt{\frac{\hbar(G_1 + G_2)}{e^2}\frac{\delta E}{E_C}} \exp\left[-\frac{2\pi e^2}{\hbar(G_1 + G_2)}\frac{E_C}{\delta E}(\mathcal{N} - \mathcal{N}^*)\right]. \quad (8.40)$$

The negative exponent in the above formula contains a product of two large parameters, $E_C/\delta E$ and $e^2/h(G_1 + G_2)$, leading to a strong suppression of T_K. For a quantum dot device, it is also hard to see the conventional signature of the Kondo effect, which is the logarithmic temperature dependence of the conductance in the perturbative regime [50], at temperatures significantly exceeding T_K. The proper expansion of the function $f(T/T_K)$ in Eq. (8.36) yields the temperature-dependent correction

$$\Delta G_K \sim G_{\rm el}\frac{\hbar(G_1 + G_2)}{e^2}\left(\frac{\delta E}{E_C}\right)^2 \ln\left(\frac{E_C}{T}\right). \quad (8.41)$$

As one can see from Eq. (8.41), the Kondo correction to the conductance remains particularly small compared to $G_{\rm el}$ everywhere in the temperature region $T \gtrsim T_K$.

To bring T_K within the reach of a modern low-temperature experiment, one may try smaller quantum dots in order to decrease $E_C/\delta E$; this route obviously has technological limitations. Another, complementary option is to increase the junction conductances, so that $G_{1,2}$ come close to $2e^2/h$, which is the maximal conductance of a single-mode quantum point contact. Junctions in the experiment [19] were tuned to $G \simeq (0.3-0.5)e^2/\pi\hbar$. A clear evidence for the Kondo effect was found at the gate voltages away from the very bottom of the odd-number valley, where $\mathcal{N}^* - \mathcal{N}$ is relatively small. Only in this domain of gate voltages the anomalous increase of conductance $G(T)$ with lowering the temperature T was clearly observed. (The unitary limit and saturation of G, indicating that $T \ll T_K$, were not reached even there.) At $\mathcal{N} = 2n + 1$, where the Kondo temperature is the smallest, the anomalous temperature dependence of the conductance was hardly seen.

To increase the Kondo temperature, it is useful to make the junctions conductances larger. However, if $G_{1,2}$ come close to $e^2/\pi\hbar$, the discreteness of the number of electrons on the dot is almost completely washed out. This raises the question about the nature of the Kondo effect in the absence of charge quantization. The way charge quantization is destroyed at $G_{1,2} \sim e^2/\pi\hbar$ depends on the detailed properties of the junctions. For the experimentally relevant case of a dot connected to leads by single-mode quantum point contacts, this problem was addressed in Ref. [51]. Spin quantization and Kondo effect for an almost open quantum dot were considered in Ref. [5]. Here we will reproduce the result for T_K only in the case of a strongly asymmetric setup, assuming one of the junctions is in the weak tunneling regime, $G_1 \ll e^2/\pi\hbar$, and the other one is in the regime of weak reflection, $G_2 = (e^2/\pi\hbar)(1-|r_2|^2)$ with $|r_2|^2 \ll 1$. Kondo effect develops in the "odd" valleys (8.39) at

$$T_K \simeq \delta E \sqrt{\frac{\delta E}{T_0(\mathcal{N})}} \exp\left\{-\frac{T_0(\mathcal{N})}{\delta E}\right\}, \quad \text{with} \qquad (8.42)$$
$$T_0(\mathcal{N}) = (4e^{\mathbf{C}}/\pi)E_C|r_2|^2\cos^2\pi\mathcal{N}.$$

Here $\mathbf{C} = 0.5772...$ is the Euler constant. Note, that in the case of weak reflection the exponent in the Kondo temperature is $\sim (E_C/\delta E)|r_2|^2$, and the large ratio $E_C/\delta E$ can be partially compensated by the smallness of $|r|^2$. So, the spin of a quantum dot may remain quantized even if charge quantization is destroyed, and the average charge $e\langle\hat{N}\rangle$ is not integer. This spin-charge separation is possible because charge and spin excitations of the dot are controlled by two very different energies: E_C and δE, respectively. Kondo effect is distinguishable on the background of the elastic co-tunneling, as long as $|r_2|^2 \gtrsim \delta E/E_C$, and, correspondingly, $T_K \ll \delta E$. Note, that similarly to the case of weak tunneling, the Kondo temperature (8.42), and the conductance at $T \ll T_K$ exhibits strong mesoscopic fluctuations which can be described with the help of the Random Matrix Theory.

6. CONCLUSION

We have discussed various regimes of electron transport through a blockaded dot weakly coupled to the leads, see Eq. (8.6). Transport is controlled by the sequential tunneling of electrons above temperature $T_{\rm in}$, see Eq. (8.25). In the temperature interval $T_{\rm el} \leq T \leq T_{\rm in}$ the main contribution to the conductance comes from the inelastic co-tunneling. Everywhere above the temperature $T_{\rm el}$, see Eq. (8.31), the mesoscopic fluctuations of the conductance δG are small compared to the average conductance $\langle G\rangle$. Below that temperature, the electrons passing through

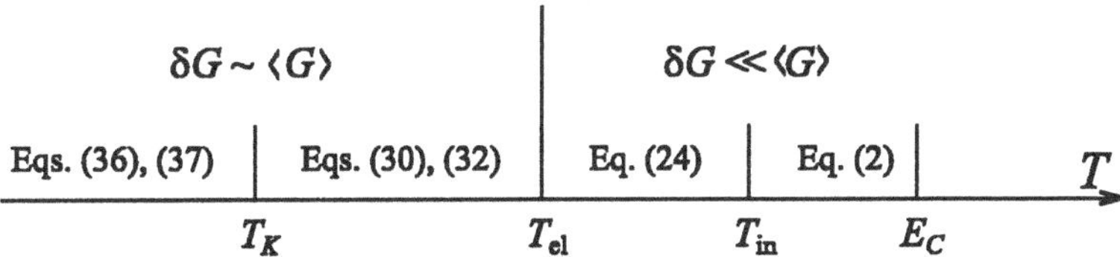

Figure 8.8 Evolution of the linear conductance mechanisms with temperature. The region where the inelastic co-tunneling dominates the electron transport, is confined by the temperatures $T_{\rm in}$, see Eq. (8.25), and $T_{\rm el}$, see Eq. (8.31). At temperatures below $T_{\rm el}$, mesoscopic conductance fluctuations δG are of the order of the average value of the conductance.The Kondo temperature T_K is given by Eq. (8.38).

the dot preserve the coherence; at $T \leq T_{\rm el}$ mesoscopic fluctuations are strong, $\delta G \sim \langle G \rangle$. If the dot carries a spin (*i.e.*, if the number of electrons on the dot is odd), then Kondo effect should develop at sufficiently low temperatures. This results in a non-monotonous temperature dependence of the conductance in the "odd" valleys of the Coulomb blockade, see Fig. 8.7. The evolution of the conductance mechanisms with temperature is summarized in Fig. 8.8.

In the weak tunneling regime, see Eq. (8.6), we are able to apply the perturbation theory in the junction conductances. Such an approach cannot be used in a consideration of an interesting and important for experiments regime of strong dot-lead coupling ($G_i \sim e^2/\pi\hbar$). To treat this regime, one needs to develop methods allowing to simultaneously account for the Coulomb interaction (8.1) and tunneling (8.20) in a non-perturbative manner. Despite this difficulty, there is quite complete understanding by now of the most relevant case of a quantum dot connected to the leads by single-channel junctions. The special limit of the partially-open dot case, described by the conditions $G_1 \ll e^2/\pi\hbar$, and $e^2/\pi\hbar - G_2 \ll e^2/\pi\hbar$, allows for a straightforward generalization of the scheme of Fig. 8.8. In that limit, no room is left for the thermally-activated transport: $T_{\rm in} \sim E_C$. The inelastic co-tunneling mechanism controls the electron transport at temperatures $T_{\rm el} \lesssim T \lesssim E_C$; proper results for the conductance $G(T)$ can be found in Ref. [52]. At $T_K \lesssim T \lesssim T_{\rm el}$, elastic co-tunneling is the leading mechanism of conduction, $\delta G \sim \langle G \rangle$; the detailed results for the conductance and its mesoscopic fluctuations can be found in Ref. [53]. At lower temperatures, the Kondo effect may develop. The condition for the Kondo effect to be distinguishable from the mesoscopic conductance fluctuations, is $T_K \lesssim \delta E$. The Kondo effect is washed out if the conductance G_2 comes too close to $e^2/\pi\hbar$.

Acknowledgements

The author is grateful to A. Kaminski for the useful comments and significant help in preparation of the manuscript. This work was supported by NSF Grant DMR-9731756.

References

[1] D. V. Averin and Yu. N. Nazarov, Phys. Rev. Lett. **65**, 2446 (1990).
[2] I. L. Aleiner and L. I. Glazman, Phys. Rev. Lett. **77**, 2057 (1996).
[3] T. K. Ng and P. A. Lee, Phys. Rev. Lett. **61**, 1768 (1988).
[4] L. I. Glazman and M. E. Raikh, JETP Lett. **47**, 452 (1988).
[5] L. I. Glazman, F. W. J. Hekking, and A. I. Larkin, Phys. Rev. Lett. **83**, 1830 (1999).
[6] C. J. Gorter, Physica **17**, 777 (1951).
[7] C. A. Neugebauer and M. B. Webb, J. Appl. Phys. **33**, 74 (1962).
[8] I. Giaever and H. R. Zeller, Phys. Rev. Lett. **20**, 1504 (1968); Phys. Rev. **181**, 789 (1969).
[9] J. Lambe and R. C. Jaklevic, Phys. Rev. Lett. **22**, 1371 (1961).
[10] R. I. Shekhter, Sov. Phys. JETP **36**, 747 (1973).
[11] I. O. Kulik and R. I. Shekhter, Sov. Phys. JETP **41**, 308 (1975).
[12] R. Wilkins, E. Ben-Jacob, and R. C. Jaklevic, Phys. Rev. Lett. **63**, 801 (1989).
[13] K. A. McGreer, J.-C. Wan, N. Anand, and A. M. Goldman, Phys. Rev. B **39**, 12260 (1989).
[14] J.-C. Wan, K. A. McGreer, L. I. Glazman, A. M. Goldman, R.I. Shekter, Phys. Rev B **43**, R9381 (1991).
[15] K. K. Likharev and T. Claeson, "Single Electronics", Scientific American, June 1992, pp. 80-85.
[16] T. A. Fulton and G. J. Dolan, Phys. Rev. Lett. **59**, 109 (1987).
[17] L. I. Glazman and R. I. Shekhter, J. Phys. Condens. Matter. **1**, 5811 (1989).
[18] P. Joyez, V. Bouchiat, D. Esteve, C. Urbina, M. H. Devoret, Phys. Rev. Lett. **79**, 1349 (1997).
[19] D. Goldhaber-Gordon, H. Shtrikman, D. Mahalu, D. Adusch-Madger, U. Meirav, M. A. Kastner, Nature (London) **391**, 156 (1998); D. Goldhaber-Gordon, J. Gores, M. A. Kastner, H. Shtrikman, D. Mahalu, U. Meirav, Phys. Rev. Lett. **81**, 5225 (1998); S. M. Cronenwett, T. H. Oosterkamp, and L. P. Kouwenhoven, Science **281**, 540 (1998); J. Schmid, Physica B **256-258**, 182 (1998).
[20] D. Davidovič and M. Tinkham, preprint cond-mat/9905043; M. Tinkham, this issue.

[21] M. P. A. Fisher and L. I. Glazman, in: *Mesoscopic Electron Transport*, ed. by L. L. Sohn, L. P. Kouwenhoven, and G. Schön, (Kluwer Press, Netherlands, 1997), p. 331.
[22] D. J. Thouless, Phys. Rev. Lett. **39**, 1167 (1977).
[23] For a review, see: C. W. J. Beenakker, Rev. Mod. Phys. **69**, 731 (1997).
[24] M. L. Mehta, *Random Matrices* (Academic, New York, 1991).
[25] B. L. Altshuler, Y. Gefen, A. Kamenev, L. S. Levitov, Phys. Rev. Lett. **78**, 2803 (1997).
[26] O. Agam, N. S. Wingreen, B.L. Altshuler, D. C. Ralph, and M. Tinkham, Phys. Rev. Lett. **78**, 1956 (1997).
[27] Ya. M. Blanter, Phys. Rev. B **54**, 12807 (1996).
[28] Ya. M. Blanter and A. D. Mirlin, Phys. Rev. E **55**, 6514 (1997).
[29] I.L. Aleiner and L. I. Glazman, Phys. Rev. B **57**, 9608 (1998).
[30] I. L. Aleiner, P. W. Brouwer, and L. I. Glazman, Quantum Effects in Coulomb Blockade, manuscript in preparation.
[31] J. M. Ziman, *Principles of the Theory of Solids*, (Cambridge University Press, Cambridge, 1972), p.339.
[32] I. Kurlyand, I. L. Aleiner, and B. L. Altshuler, cond-mat/0004205.
[33] H. U. Baranger, D. Ullmo, and L. I. Glazman, preprint cond-mat/9907151.
[34] P. W. Brouwer, Y. Oreg, and B. I. Halperin, preprint cond-mat/9907148.
[35] K. A. Matveev and A. I. Larkin, Phys. Rev. Lett. **78**, 3749 (1997).
[36] S. D. Berger and B. I. Halperin, Phys. Rev. B, **58**, 5213 (1997).
[37] A. Mastellone, G. Falci, and R. Fazio, Phys. Rev. Lett., **80**, 4542 (1998).
[38] Ya. M. Blanter, A. D. Mirlin, and B. A. Muzykantskii, Phys. Rev. Lett. **78**, 2449 (1997).
[39] D. V. Averin and A. A. Odintsov, Phys. Lett. A **140**, 251 (1989).
[40] A. A. Abrikosov, *Fundamentals of the Theory of Metals*, (North-Holland, Amsterdam, 1988) 620 p.
[41] J. A. Folk, S. R. Patel, S. F. Godijn, A. G. Huibers, S. M. Cronenwett, C. M. Marcus, K. Campman, A. C. Gossard *et al.*, Phys. Rev. Lett. **76**, 1699 (1996).
[42] A. Kaminski, I. L. Aleiner and L. I. Glazman, Phys. Rev. Lett. **81**, 685 (1998).
[43] S. M. Cronenwett, S. R. Patel, C. M. Marcus, K. Campman, A. C. Gossard, Phys. Rev. Lett. **79**, 2312 (1997).
[44] J. Kondo, Prog. Theor. Phys. **32**, 37 (1964).
[45] P. W. Anderson, Phys. Rev. **124**, 41 (1961).

[46] Ph. Nozières and A. Blandin, J. de Physique **41**, 193 (1980).
[47] F. D. M. Haldane, Phys. Rev. Lett. **40**, 416 (1979).
[48] T. A. Costi, A. C. Hewson, and V. Zlatić, J. Phys. Condens. Matter **6**, 2519 (1994).
[49] P. Nozières, J. Low Temp. Phys. **17**, 31, 1974.
[50] J. Appelbaum, Phys. Rev. Lett. **17**, 91 (1966); P. W. Anderson, Phys. Rev. Lett. **17**, 95 (1966); J. M. Rowell, in *Tunneling Phenomena in Solids*, edited by E. Burstein and S. Lundquist (Plenum, New York, 1969), p. 385.
[51] K. A. Matveev, Phys. Rev. B **51**, 1743 (1995).
[52] A. Furusaki and K. A. Matveev, Phys. Rev. B **52**, 16676 (1995)
[53] I. L. Aleiner and L. I. Glazman, Phys. Rev. B **57**, 9608 (1998).

Chapter 9

QUANTUM SMEARING OF COULOMB BLOCKADE

K. A. Matveev
Duke University, Durham, NC 27708-0305, USA

Abstract The simplest system in which the phenomenon of Coulomb blockade can be observed consists of a small metallic grain connected to a large conductor via a tunnel junction and capacitively coupled to a gate electrode. At temperatures much lower than the charging energy of the grain the system prefers the values of the grain charge which are integer in the units of electron charge. A review of the effects of small but finite tunneling strength on the accuracy of charge quantization is presented. It turns out that this problem can be mapped onto the well known Kondo problem of electron scattering by a magnetic impurity in a metal. By exploring this analogy one can find the complete solution of the problem of grain charge in the case of weak tunneling.

1. INTRODUCTION

The phenomenon of Coulomb blockade was discovered experimentally many years ago [1]. In its simplest form it can be discussed in a system consisting of a small metal grain or a quantum dot connected to a large metal lead, Fig. 9.1. If an electron tunnels from the lead into the grain, the latter acquires charge e of one electron. Clearly, this charge now generates an electric field around the grain. The energy of this field can be found from classical electrostatics as $E_C = e^2/2C$, where C is the capacitance of the grain. If the temperature is very low, and the grain is very small, one can achieve the regime when $T \ll E_C$. In this case vast majority of electrons in the lead cannot tunnel into the grain, as their typical energy T is not sufficient to charge the grain. Hence the tunneling through the barrier is *blocked.*

In order to observe this phenomenon one can apply voltage V_g to a gate as shown in Fig. 9.1. The effect of the gate is that the positive charges

I. O. Kulik and R. Ellialtioğlu (eds.),
Quantum Mesoscopic Phenomena and Mesoscopic Devices in Microelectronics, 129–143.

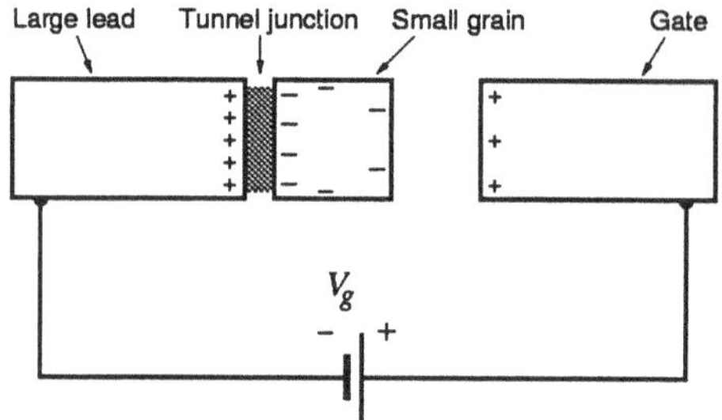

Figure 9.1 A small metallic grain is coupled to a lead electrode via a tunnel junction. The electrostatic energy of the system can be tuned by applying voltage V_g to the gate electrode.

on it attract the negatively charged electrons and lower the charging gap for tunneling. Mathematically this can be described with the help of the following expression for the charging energy of the system in the presence of the gate:

$$U(n) = E_C(n - N)^2. \tag{9.1}$$

Here n is the number of extra electrons in the grain, and N is a dimensionless parameter proportional to the gate voltage V_g. One can easily see from Eq. (9.1) that as the gate voltage N is changed, the number of particles in the grain changes as well. Indeed, at fixed N the minimum of the charging energy (9.1) is achieved when n is the integer nearest to N. Thus, the dependence of the grain charge on the gate voltage shows staircase behavior shown by solid line in Fig. 9.2. This behavior, which is commonly referred to as *the Coulomb staircase*, was convincingly demonstrated by experiments [2].

It is important to note, however, that in the preceding discussion we essentially assumed that the tunneling between the grain and the lead is negligible. In this limit one expects that the number of electrons n in the grain is an integral of motion and should be quantized. On the other hand, even a weak tunneling will destroy the conservation of the grain charge. Consequently, the actual dependence of n on the gate voltage should deviate from the simple Coulomb staircase behavior.

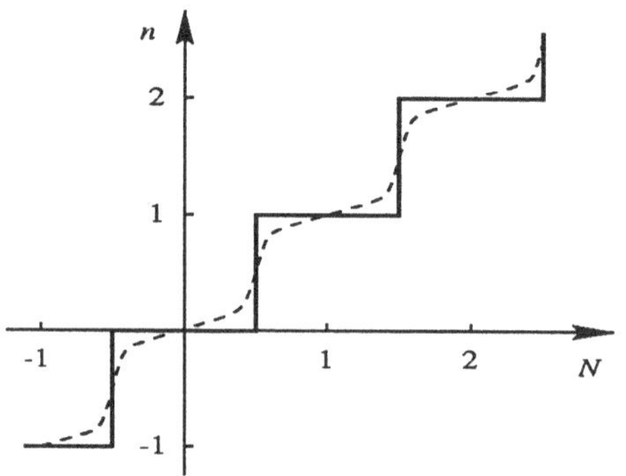

Figure 9.2 The dependence of the number of extra electrons n in the grain on the gate voltage parameter N. Solid line shows the classical behavior found by minimizing the charging energy (9.1) with respect to n. The dashed line shows schematically the dependence of the expectation value of n in the ground state of the system for the case of weak but non-vanishing tunneling between the grain and the lead.

To discuss the charge of the grain at non-vanishing tunnel coupling, one needs to define the measurable quantity describing the grain charge. The most natural such quantity is the expectation value $\langle n \rangle$ of the number of particles operator in the ground state of the system. This quantity should be measurable in the experiment similar to [2] at zero temperature and weak coupling between the grain and the lead.

In the presence of tunneling the electrons are not confined to the grain. Instead, even in the ground state the electrons tunnel from the grain to the lead. Of course, when this happens the energy of the system increases. This means that the new quantum state is just a virtual state, and the electron will subsequently return to the grain. Similarly, the electrons in the lead experience virtual tunneling into the grain. These processes of virtual tunneling destroy the quantization of charge, resulting in *quantum smearing* of the Coulomb staircase, see the dashed line in Fig. 9.2.

In Sec. 2. we use the perturbation theory approach to find the corrections to the Coulomb staircase caused by weak tunneling between the grain and the lead. We will see that the perturbative result diverges logarithmically at half-integer values of N. These divergences indicate that the complete description of the charge dependence on the gate voltage can only be obtained by accounting properly for the high-order terms of the perturbation theory. This can be accomplished by mapping the Coulomb blockade problem onto a Kondo model, as described in Sec. 3.

2. PERTURBATION THEORY

2.1 Energy Scales

To discuss the quantum smearing of the Coulomb staircase quantitatively, we need to establish the relations between the relevant energy scales. In our system they are:

- Fermi energy E_F. It can be estimated as $E_F = \hbar^2 k_F^2/2m$, where k_F is the typical Fermi wavevector of the electrons, and m is their effective mass.

- Charging energy E_C. It is defined as $e^2/2C$ and can be estimated as e^2/L, where L is the size of the grain.

- Temperature T.

- Quantum level spacing in the grain δ. Since the number of conduction electrons in the grain $N_e \sim (k_F L)^d$, we can estimate $\delta \sim E_F/(k_F L)^d$, where the dimension of the space d is 3 for metallic grains and 2 for quantum dots.

In the following discussion we will make a number of assumptions regarding these energy scales:

- $E_C \ll E_F$. Comparing the above estimates of E_C and E_F, we find

$$\frac{E_C}{E_F} \sim \frac{e^2}{\hbar v_F} \frac{1}{k_F L},$$

where $v_F = \hbar k_F / m$ is the electron Fermi velocity. In almost all experiments the interaction parameter $e^2/\hbar v_F$ is of order unity. We can also assume that the number of conduction electrons in the grain is large, $N_e \gg 1$, and therefore the parameter $k_F L \sim N_e^{1/d} \gg 1$. This justifies the original assumption that the charging energy is small compared to the Fermi energy.

- $T \ll E_C$. In fact, we will study the ground state properties of the system, *i.e.*, we will assume $T = 0$. Since the subject of this discussion is *quantum* smearing of Coulomb blockade, we ignore the additional smearing due to the thermal fluctuations of the grain charge.

- $\delta \ll E_C$. Using the above estimate of δ, we find

$$\frac{\delta}{E_C} \sim \left(\frac{e^2}{\hbar v_F}\right)^{-1} \frac{1}{(k_F L)^{d-1}} \ll 1.$$

Below we will simply assume $\delta = 0$. It is worth mentioning that a number of interesting phenomena in quantum dots are finite-size effects, which disappear in the limit $\delta \to 0$. An example of such a phenomenon is the recently observed [3] Kondo effect in single electron transistors. On the other hand, the Kondo physics of the quantum charge fluctuations discussed in Sec. 3. is most prominent in the limit of infinitesimal δ.

2.2 Tunneling Hamiltonian

A convenient mathematical description of tunneling between the grain and the lead is given by the following Hamiltonian:

$$H = \sum_k \epsilon_k a_k^\dagger a_k + \sum_p \epsilon_p a_p^\dagger a_p + \sum_{kp} t(a_k^\dagger a_p + a_p^\dagger a_k) + E_C(n - N)^2. \quad (9.2)$$

Here the first two terms describe the electrons occupying states k and p in the lead and grain, respectively; $a_{k(p)}$ are electron annihilation operators, and $\epsilon_{k(p)}$ denote electron energies measured from the Fermi level. The

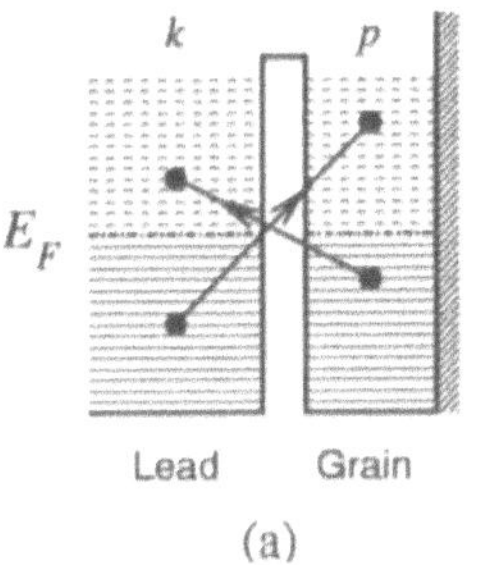

(a)

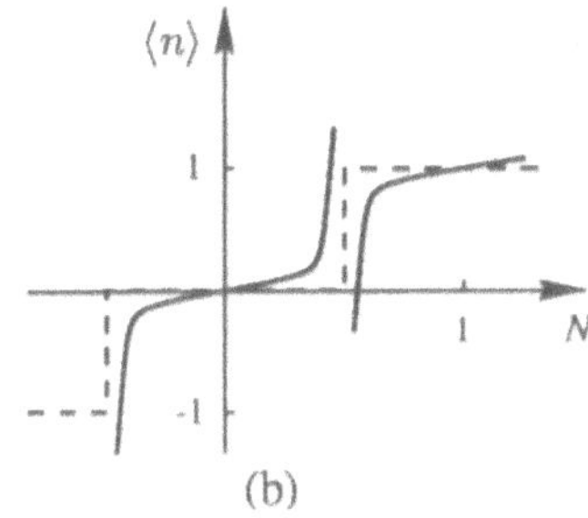

(b)

Figure 9.3 (a) The energy levels in the system under consideration. The levels below and above the Fermi level E_F are denoted by solid and dashed lines, respectively. Arrows indicate the two simplest tunneling processes smearing the quantization of the grain charge. (b) Schematic view of the perturbation theory result (9.6) for the expectation value $\langle n \rangle$ of the number of extra electrons in the grain. Dashed line shows the unperturbed behavior at $g = 0$.

third term describes the tunneling trough the barrier separating the grain from the lead. The strength of tunneling is determined by the value of the matrix element t. Our notations are illustrated in Fig. 9.3*a*.

The last term in (9.2) represents the charging energy (9.1). It should be noted that the number of extra particles in the grain is now an operator:

$$n = \sum_p \left[a_p^\dagger a_p - \theta(-\epsilon_p) \right] . \tag{9.3}$$

The step function $\theta(-\epsilon_p)$ coincides with the occupation numbers of electrons in a reference ground state, so that the operator (9.3) accounts only for the additional electrons brought from the lead.

2.3 Second-Order Perturbation Theory

Let us assume that the dimensionless gate voltage N is between $-\frac{1}{2}$ and $\frac{1}{2}$, *i.e.*, we concentrate on only one of the plateaus of the Coulomb staircase, Fig. 9.3*b*. In this case in the absence of tunneling the ground state of the system has $n = 0$. Let us now include tunneling perturbatively. One can identify two basic first order processes caused by the tunneling term in the Hamiltonian (9.2):

1. An electron from a filled state k in the lead tunnels into an empty state p in the grain; $n = 0 \to n = 1$. The amplitude of such a process is

$$A_{kp} = \frac{t}{\epsilon_k - \epsilon_p - [U(1) - U(0)]}, \tag{9.4}$$

and the energies satisfy the relation $\epsilon_k < 0 < \epsilon_p$.

2. An electron from a filled state p in the grain tunnels into an empty state k in the lead; $n = 0 \to n = -1$. The amplitude of this process is

$$A_{pk} = \frac{t}{\epsilon_p - \epsilon_k - [U(-1) - U(0)]}. \tag{9.5}$$

Here $\epsilon_p < 0 < \epsilon_k$.

Since the processes of the first type increase the charge n by 1, and the processes of the second type decrease it by 1, the average charge of the grain is

$$\langle n \rangle = \sum_{kp} \left[|A_{kp}|^2 \theta(-\epsilon_k)\theta(\epsilon_p) - |A_{pk}|^2 \theta(-\epsilon_p)\theta(\epsilon_k) \right].$$

One can now substitute the above expressions for the amplitudes A_{kp} and A_{pk} and convert the sums over the states k and p into integrals with respect to ϵ_k and ϵ_p. The resulting expression for the average number of particles in the ground state is

$$\langle n \rangle = g \ln \frac{\frac{1}{2} + N}{\frac{1}{2} - N}, \tag{9.6}$$

Ref. [5]. Here we have introduced a dimensionless parameter $g = \nu_l \nu_g t^2$, where ν_l and ν_g are the densities of states in the lead and the grain. Physically, g is the conductance of the tunneling barrier in units of $2\pi e^2/\hbar$. The dependence (9.6) is shown schematically by solid line in Fig. 9.3*b*.

As expected, we found the average charge $\langle n \rangle$ to be no longer quantized. In agreement with the qualitative behavior of Fig. 9.2, the plateaus now have a finite slope proportional to the conductance of the barrier. It is important to notice that the perturbative result (9.6) diverges logarithmically at $N = \pm\frac{1}{2}$, *i.e.*, at the positions of the steps. Since the physical charge cannot diverge, it is clear that the divergence is an artifact of the second-order perturbation theory. Thus, to get the correct shape of the steps one has to account for the higher-order terms of the perturbation theory.

3. THE SHAPE OF THE STEPS OF COULOMB STAIRCASE

The divergent behavior of $\langle n \rangle$ near the steps is not limited to the second-order perturbation theory (9.6) in tunneling matrix element t. In fact, one can show that all the terms of the perturbation theory diverge at $N = \frac{1}{2}$. Consequently, in order to describe the shape of the steps in Fig. 9.2 one has to take into account an infinite number of terms of the perturbation theory. Instead of trying to sum up the whole series of the perturbation theory, we will show that the problem of the shape of the steps can be mapped onto a certain Kondo model. The advantage of this approach is that the Kondo problem has been studied extensively, and the techniques developed and the results obtained in those studies can be directly applied to the charge smearing problem.

3.1 Higher-Order Terms Of The Perturbation Theory

First, we need to understand the origin of the divergent behavior of the average charge (9.6). Let us consider the vicinity of a step at gate voltage $N = \frac{1}{2}$. At this value of N the charging energy (9.1) is degenerate: $U(0) = U(1)$. As a result, the energy of the virtual state created by tunneling of an electron from a state k in the lead to a state p in the grain vanishes at $\epsilon_k \to -0$ and $\epsilon_p \to +0$. In this limit the tunneling amplitude (9.4) diverges, which gives rise to the divergent behavior of the charge (9.6). Therefore the divergence of the perturbation theory for $\langle n \rangle$ originates from the virtual tunneling events with small increment of the energy, $\Delta E \ll E_C$, which is only possible at half-integer values of the gate voltage N where the charging energy (9.1) is degenerate. Clearly, similar divergences will show up in higher-order terms of the perturbation theory.

In order to describe the shape of the steps, we will collect the sub-series of the most divergent terms of the perturbation theory. Since the divergences originate at small energies $\Delta E \ll E_C$, at N near $\frac{1}{2}$ we can neglect all states of the system with charge $n \neq 0, 1$. We should simultaneously reduce the bandwidth of the model from $D_0 \sim E_F$ to $D \sim E_C$. Indeed tunneling into states with energies exceeding D would then result in energies of the virtual states ΔE greater than E_C.

To illustrate the new model, let us discuss the fourth-order correction to the ground-state energy E_0 of the system. Suppose N is slightly less than $\frac{1}{2}$. Then the unperturbed ground state has charge $n = 0$. At the first step the charge should change to $n = 1$, *i.e.*, an electron from a filled state k_1 in the lead should tunnel to an empty state p_1 in the grain. At the next step an electron should tunnel from the grain to the lead, $n = 1 \to 0$, because otherwise we get a "forbidden" high energy state with $n = 2$. For example, the electron at p_1 can tunnel to an empty state k_2 in the lead. At the third step the electron can tunnel from k_2 to a state p_2 in the grain, changing the charge n from 0 to 1. Finally, at the fourth step an electron tunnels from the state p_2 in the grain to the original state k_1. The system returns to the initial state, and the corresponding contribution to the ground-state energy of our Coulomb blockade system is

$$\delta E_{\mathrm{CB}} = t\,\frac{1}{\epsilon_{k_1} - (\epsilon_{p_1} + \Delta U)}\,t\,\frac{1}{\epsilon_{k_1} - \epsilon_{k_2}}\,t\,\frac{1}{\epsilon_{k_1} - (\epsilon_{p_2} + \Delta U)}\,t, \tag{9.7}$$

where $\Delta U = U(1) - U(0)$. Note that the charge n of the grain alternates at each step: $n = 0 \to 1 \to 0 \to 1 \to 0$. This property will obviously occur for all terms in all orders of the perturbation theory in t.

3.2 Kondo Problem

To sum up the subseries of the most divergent terms (such as (9.7)) of the perturbation theory for the ground state energy of a grain coupled to a lead, it will be helpful to use a number of results obtained in connection to the well-known Kondo problem [4]. The latter refers to the problem of scattering of electrons in a metal off an impurity with a spin S. The interaction of electrons with such an impurity can be described by a Hamiltonian

$$H = \sum_{k\alpha} \epsilon_k a^\dagger_{k\alpha} a_{k\alpha} + \sum_{kp\alpha\beta} [J_z \sigma^z_{\alpha\beta} S^z + J_\perp(\sigma^x_{\alpha\beta} S^x + \sigma^y_{\alpha\beta} S^y)] a^\dagger_{k\alpha} a_{p\beta} - hS^z. \tag{9.8}$$

Here α and β are the projections of the electron spin on the z axis, $\sigma^i_{\alpha\beta}$ are Pauli matrices, S^i are the operators of the impurity spin, and h describes Zeeman energy of the impurity spin in the external magnetic field. The Hamiltonian (9.8) assumes anisotropic exchange described by the constants J_z and $J_\perp$. The more usual isotropic case corresponds to $J_z = J_\perp = J$.

Let us consider the special case of the Hamiltonian (9.8) when $J_z = 0$. It is then convenient to rewrite the exchange term as

$$J_\perp(\sigma^x_{\alpha\beta} S^x + \sigma^y_{\alpha\beta} S^y) = \frac{1}{2} J_\perp(\sigma^+_{\alpha\beta} S^- + \sigma^-_{\alpha\beta} S^+), \tag{9.9}$$

where $S^\pm = S^x \pm iS^y$ and $\sigma^\pm = \sigma^x \pm i\sigma^y$. Substituting the Pauli matrices explicitly, one can easily show that

$$\frac{1}{2}\sigma^+ = \begin{pmatrix} 0 & 1 \\ 0 & 0 \end{pmatrix}, \quad \frac{1}{2}\sigma^- = \begin{pmatrix} 0 & 0 \\ 1 & 0 \end{pmatrix}, \tag{9.10}$$

in agreement with the familiar spin-raising and spin-lowering meaning of matrices σ^+ and σ^-. Substituting these expressions for σ^+ and σ^- into the exchange term (9.9), we can re-write the Hamiltonian (9.8) for $J_z = 0$ as

$$H = \sum_{k\alpha} \epsilon_k a^\dagger_{k\alpha} a_{k\alpha} + \sum_{kp} J_\perp (S^+ a^\dagger_{k\downarrow} a_{p\uparrow} + S^- a^\dagger_{p\uparrow} a_{k\downarrow}) - hS^z. \tag{9.11}$$

It is interesting to note a special feature of the exchange term if the Hamiltonian (9.11). For a spin-$\frac{1}{2}$ impurity the spin S can take only two values, $\uparrow$ and $\downarrow$. The exchange term in (9.11) flips the impurity spin between these two positions and simultaneously scatters an electron in the conduction band while flipping its spin as well.

It is instructive to calculate a 4-th order in $J_\perp$ contribution to the ground state energy, similar to the calculation for the Coulomb blockade problem in Sec. 3.1. Consider the following sequence of scattering events: $(k_1\downarrow, S{=}\uparrow) \to (p_1\uparrow, S{=}\downarrow) \to (k_2\downarrow, S{=}\uparrow) \to (p_2\uparrow, S{=}\downarrow) \to (k_1\downarrow, S{=}\uparrow)$. The matrix element associated with each scattering event is $J_\perp$, so the contribution to the energy of the anisotropic Kondo system is

$$\delta E_{\mathrm{AK}} = J_\perp \frac{1}{\epsilon_{k_1} - (\epsilon_{p_1} + h)} J_\perp \frac{1}{\epsilon_{k_1} - \epsilon_{k_2}} J_\perp \frac{1}{\epsilon_{k_1} - (\epsilon_{p_2} + h)} J_\perp . \quad (9.12)$$

In the denominators we have accounted for the fact that the energy of a spin-down impurity is increased by the Zeeman energy h.

3.3 Mapping Onto A Kondo Model

It is interesting to compare the corrections to the ground state energy for a grain coupled to a lead (9.7) and for the anisotropic Kondo problem (9.12). One can see that the expressions coincide if $t = J_\perp$ and $\Delta U = h$. Although equations (9.7) and (9.12) represent only single terms of the perturbation theory expansions for the ground state energy, one can demonstrate that the two series will coincide in all orders.

In fact, one can establish a formal mapping of the Coulomb blockade problem onto the anisotropic Kondo model (9.11). To do this, one can assign fictitious electron and impurity "spins" as shown in Fig. 9.4. By assigning different "spins" to the electrons on the two sides of the tunneling barrier we ensure that each tunneling event flips the "spin" of the tunneling electron. Since the charge of the grain changes at any such event, the "impurity spin" flips simultaneously. Therefore, all the tunneling events are properly described by the exchange terms of the Kondo Hamiltonian (9.11). A slight deviation of the gate voltage N from $\frac{1}{2}$ caused small splitting of the unperturbed ground state energy $\Delta U = U(1) - U(0)$. It is natural to interpret ΔU as Zeeman splitting h of the fictitious impurity spin in the Hamiltonian (9.11). This completes the mapping of the problem of grain charge smearing onto anisotropic Kondo model.

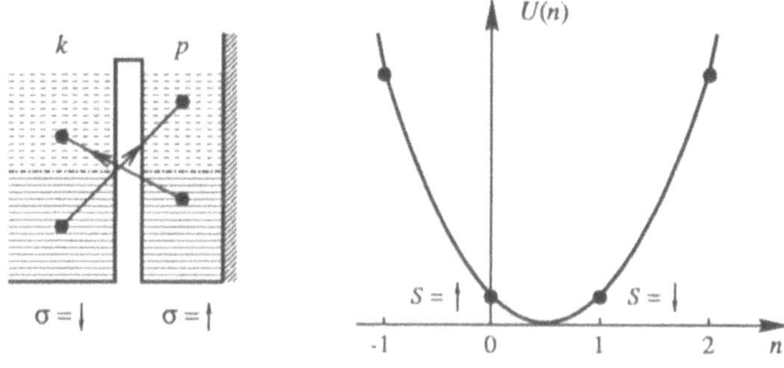

Figure 9.4 The fictitious spins ↑ and ↓ are assigned to the electrons in the grain and in the lead, respectively. The "impurity spin" indicates which of the two charge states the system is in: $n = 0$ and $n = 1$ states are assigned fictitious impurity spins ↑ and ↓, respectively. (We assume that N is near $\frac{1}{2}$.) The other charge states, $n = -1, -2, \ldots$ and $n = 2, 3, \ldots$, have a substantially higher energy, and are therefore neglected.

3.4 Average Charge

Let us see how the mapping described in Sec. 3.3 enables us to evaluate the expectation value $\langle n \rangle$ of the grain charge. Since the values of n are limited to 0 and 1, we can write the charging energy as

$$U(n) = U(0) + n\Delta U. \tag{9.13}$$

This results enables us to find the average charge of the grain $\langle n \rangle$ by differentiating the ground state energy of the system E_{CB} with respect to the energy difference ΔU,

$$\langle n \rangle = \left\langle \frac{\partial H_{\text{CB}}}{\partial \Delta U} \right\rangle = \frac{\partial E_{\text{CB}}}{\partial \Delta U}. \tag{9.14}$$

The mapping of Sec. 3.3 ensures that the corrections to the ground state energies E_{CB} and E_{AK} due to tunneling (exchange) coincide. Since the unperturbed ground state energies are $E^0_{\text{CB}} = 0$ and $E^0_{\text{AK}} = -h/2$ (for the impurity spin ↓), we have

$$E_{\text{CB}}(h) = \frac{1}{2}h + E_{\text{AK}}(h). \tag{9.15}$$

Here $E_{\text{CB}}(h)$ denotes the ground state energy of the Coulomb blockade problem at $\Delta U = h$. If we now differentiate both parts of the identity (9.15) with respect to h, then using (9.14) and a similar relation $\partial E_{\text{AK}}/\partial h = -\langle S^z \rangle$, we find

$$\langle n \rangle = \frac{1}{2} - \langle S^z \rangle. \tag{9.16}$$

Thus the expectation value of the grain charge is directly related with the average spin (or the magnetic moment) of the Kondo impurity.

We can now use the relation (9.16) to find the shape of the charge steps. To begin with, it is interesting to check simple limiting cases. In the absence of tunneling, $t = J_\perp = 0$, we have $\langle S^z \rangle = \frac{1}{2}$, and consequently $\langle n \rangle = 0$, as expected. Let us now allow for some weak tunneling and find out the grain charge at the symmetry point $N = \frac{1}{2}$. This corresponds to the zero magnetic field case of the Kondo problem. It is well known [4] that in the ground state and at $h = 0$ the spin of the Kondo impurity is completely screened, i.e., $\langle S^z \rangle = 0$. It then follows from (9.16) that at $N = \frac{1}{2}$ the average grain charge $\langle n \rangle = \frac{1}{2}$. Although this may appear to be a natural consequence of the symmetry of the problem, see Fig. 9.2, it is identical to a rather non-trivial statement of perfect screening of a Kondo spin by band electrons.

The exact shape of the steps of Coulomb staircase can be found by using the exact results for the magnetic moment of a Kondo impurity as a function of the magnetic field [6, 7]. The exact formula [8, 9] for the $\langle n \rangle$ is rather complicated. We will only mention that the slope of the Coulomb staircase at the center of the step is given by

$$\left. \frac{d\langle n \rangle}{dN} \right|_{N=\frac{1}{2}} \sim \frac{1}{g^{1/4}} \exp\left(\frac{\pi}{4\sqrt{g}} \right). \tag{9.17}$$

Note that the dimensionless conductance is the small parameter of the problem, $g \ll 1$. Thus the slopes of the steps are exponentially large. This can be easily understood from the analogy with the Kondo problem. Since the $\langle n \rangle$ is proportional to the magnetic moment of the Kondo impurity, and $\Delta U = 2E_C(1/2 - N)$ is the analog of the external magnetic field, the slope (9.17) has to be proportional to the magnetic susceptibility of the Kondo impurity. The latter is known to be inversely proportional to the Kondo temperature T_K, which is exponentially small for weak exchange coupling.

To summarize the rules of the mapping we discussed in this section, we list the physical parameters of the Coulomb blockade problem and their analogs in the Kondo problem in Table 9.1.

3.5 Effect of Physical Spins and Transverse Channels

In the above discussion we have neglected the spins of electrons completely. In an experiment the physical spins of electrons will affect the smearing of the steps of Coulomb staircase (Fig. 9.2), unless a strong magnetic field is applied. To account for the possibility of electron spin

Table 9.1 Rules of the mapping to the Kondo model

Coulomb blockade	Anisotropic Kondo problem
Charging energy E_C	Bandwidth D
Tunneling matrix element t	Exchange constant $J_\perp$
Gate voltage, $\Delta U = E_C(1-2N)$	Magnetic field h
Average charge $\langle n\rangle$	Average spin, $\frac{1}{2} - \langle S^z\rangle$
Differential capacitance $\delta C \propto \frac{d\langle n\rangle}{dN}$	Magnetic susceptibility $\chi \propto \frac{d\langle S^z\rangle}{dh}$

having two directions, $\tau = \uparrow$ and $\downarrow$, one should notice that τ does not change when electron tunnels through the barrier. Therefore, one can think of two species of electrons which can tunnel through the barrier independently. On the other hand, the Coulomb interaction is insensitive to the spin, *i.e.*, the electrons with different τ interact with each other via the charging energy (9.1).

In the second order perturbation theory the presence of spins is not important. One can easily see that the result (9.6) for the average charge doubles, because twice as many electrons can now tunnel through the barrier as in the spinless case. On the other hand, the conductance of the barrier G is now twice as large, so if we define g in terms of the conductance of the barrier, $g = G\hbar/2\pi e^2$, the result (9.6) will remain correct.

To describe the shape of the steps in the presence of electron spins, one has to modify the Hamiltonian (9.11) to account for the two species of electrons:

$$H = \sum_{k\alpha\tau} \epsilon_k a^\dagger_{k\alpha\tau} a_{k\alpha\tau} + \sum_{kp\tau} J_\perp (S^+ a^\dagger_{k\downarrow\tau} a_{p\uparrow\tau} + S^- a^\dagger_{p\uparrow\tau} a_{k\downarrow\tau}) - hS^z. \quad (9.18)$$

Since the index τ takes two values, $\uparrow$ and $\downarrow$, this Hamiltonian describes a 2-channel spin-$\frac{1}{2}$ anisotropic Kondo model. Fortunately, multichannel Kondo problem is also exactly solvable [10, 11]. Therefore, one can use these exact results to describe the shape of the steps of Coulomb staircase in the presence of electron spins. The exact expression for the shape of the step is rather complicated [8]. The most interesting effect of the spins is that in the 2-channel Kondo problem the susceptibility is logarithmically divergent, $\chi \sim T_K^{-1} \ln(T_K/h)$, and, therefore, the slopes of the steps in Fig. 9.2 diverge logarithmically as well:

$$\left.\frac{d\langle n\rangle}{dN}\right|_{N\to\frac{1}{2}} \sim \frac{1}{\sqrt{g}} \exp\left(\frac{\pi}{\sqrt{8g}}\right) \ln \frac{\exp(-\pi/\sqrt{8g})}{|N-1/2|}. \quad (9.19)$$

This result is not easy to test experimentally, because in an experiment with small g the step (9.19) is exponentially narrow and is difficult to resolve. On the other hand, the non-analytic behavior of $\langle n\rangle$ at $N \to \frac{1}{2}$ should not be sensitive to the exact value of g. One can show [12] that even in the strong coupling regime $G \to e^2/\pi\hbar$ the logarithmic singularity (9.19) should be present. This statement was supported by recent experiment [13].

Another feature of a tunneling junction which affects the quantum smearing of the Coulomb staircase is its geometry. The model we have discussed so far assumed that the coupling of any state in the lead to any state in the grain is described by a single tunneling constant t. This is an adequate model for the case of a quantum dot (such as in Ref. [13]) coupled to a lead by a quantum point contact in the weak tunneling regime, where the dimensions of the contact are comparable to the Fermi wavelength of the electrons λ_F. An opposite situation is realized in the experiments with metallic grains, such as Ref. [2]. There the area of the junction A is usually much greater than λ_F^2. Such junctions can be modeled by a Hamiltonian with very large number of channels $M \sim A/\lambda_F^2$. Similar to the case of spins, the perturbative result (9.6) remains valid if g is defined as the conductance of the barrier in units of $2\pi e^2/\hbar$.

To find the shape of the steps one can treat the Hamiltonian (9.18) by methods of perturbative renormalization group. The large number of channels ensures that the exchange constants remain small, so that the complete solution of the problem can be obtained. The result for the shape of the steps is remarkably simple:

$$\langle n\rangle = \frac{g \ln \frac{1}{1/2-N}}{1 + 2g \ln \frac{1}{1/2-N}}. \qquad (9.20)$$

Here the gate voltage N is assumed to approach $1/2$ from below. The limit $g \to 0$ at fixed N coincides with the perturbation theory result (9.6) near the step, as expected. On the other hand, the limit $N \to \frac{1}{2}$ at fixed g gives $\langle n\rangle = \frac{1}{2}$, so that the smeared dependence of $\langle n\rangle$ on N is continuous, Fig. 9.2.

4. SUMMARY

We discussed the quantum smearing of the charge in a grain coupled to a lead via a tunneling junction, Fig. 9.1. Schematically the results are shown in Fig. 9.2. In the presence of weak tunneling the charge is no longer quantized, so that the plateaus of the Coulomb staircase acquire finite slope. The behavior of the average charge for small tunneling conductance $g = G\hbar/e^2 \ll 1$ is given by the expression (9.6) obtained within the framework of the second-order perturbation theory. This result is illustrated in Fig. 9.3(b). The divergence of the result (9.6) at the positions of the steps of Coulomb staircase indicates that a more elaborate approach is needed to describe the expectation value of the grain charge in that region.

The shape of the steps can be obtained by mapping of the Coulomb blockade problem onto an anisotropic multichannel Kondo model. The common feature of all the relevant cases is that in the center of the step the grain charge is smeared completely, *i.e.*, at $N \to \frac{1}{2}$ we have $\langle n \rangle \to \frac{1}{2}$. This means that the smearing of the steps makes $\langle n \rangle$ a continuous function of the gate voltage, Fig. 9.2. The exact shape of the steps depends on the presence of electron spins and on the geometry of the tunneling junction. In the case of point contact geometry relevant for the experiments with semiconductor quantum dots, the slope of the steps is exponentially large, Eqs. (9.17) and (9.19). In the language of the Kondo problem this is identical to the statement that the susceptibility of a Kondo impurity is inversely proportional to an exponentially small Kondo temperature, $\chi \sim 1/T_K$. It is also interesting that the presence of electron spins results in a logarithmically divergent slope (9.19) of the Coulomb staircase at the centers of the steps. This divergence becomes stronger (9.20) in the case of a wide junction relevant for the experiments with metallic grains. The technical details of the calculations can be found in Ref. [8].

Acknowledgements

This work was supported by A.P. Sloan Foundation and by NSF Grant DMR-9974435.

References

[1] H. R. Zeller and I. Giaver, Phys. Rev. **181**, 789 (1969).

[2] P. Lafarge, P. Joyez, D. Esteve, C. Urbina, and M. H. Devoret, Nature **365**, 422 (1993).

[3] D. Goldhaber-Gordon, H. Shtrikman, D. Mahalu, D. Abusch-Magder, U. Meirav, M.A. Kastner, Nature **391**, 156 (1998).

[4] A good discussion of the Kondo effect can be found, *e.g.*, in Abrikosov A. A., *Fundamentals of the theory of metals* (Elsevier, Amsterdam, 1988).

[5] L. I. Glazman and K. A. Matveev, Sov. Phys. JETP **71**, 1031 (1990).

[6] A. M. Tsvelick and P. B. Wiegmann, Adv. Phys., **32**, 453 (1983).

[7] N. Andrei, K. Furuya, and J. H. Lowenstein, Rev. Mod. Phys., **55**, 331 (1983).

[8] K. A. Matveev, Sov. Phys. JETP **72**, 892 (1991).

[9] Strictly speaking, the exact results [6, 7] refer to the case of isotropic exchange, $J_z = J_\perp$. To apply those results to the anisotropic case $J_z = 0$, one has to start with perturbative renormalization of the exchange constants. It is well known that the model becomes isotropic within the range of applicability of the perturbative techniques. After that one can use the exact solutions to arrive at the exact step shape.

[10] N. Andrei and C. Destri, Phy. Rev. Lett. **52**, 364 (1984).

[11] A. M. Tsvelick and P. B. Wiegmann, Z. Phys. B **54**, 201 (1984).

[12] K. A. Matveev, Phys. Rev. B **51**, 1743 (1995).

[13] D. Berman, N. B. Zhitenev, R. C. Ashoori, and M. Shayegan, Phys. Rev. Lett. **82**, 161 (1999).

Chapter 10

COULOMB BLOCKADE IN SINGLE TUNNEL JUNCTION CONNECTED TO NANOWIRE AND CARBON NANOTUBE

J. Haruyama, I. Takesue, Y. Sato and K. Hijioka

Aoyama Gkuin University, Dept. Electrical Engineering and Electronics

6-16-1 Chitosedai, Setagaya, Tokyo 157-8572 Japan

Abstract It is well known that Coulomb blockade (CB) in single tunnel junction (STJ) system strongly depends on its external electromagnetic environment (EME), so called phase correlation theory. Tunneling electron can transfer its charging energy E_c to the EME and charges on junction surface can be isolated from external phase fluctuation, only when the impedance of EME is larger than resistance quantum (R_Q=h/e^2 ~ 25.8 kΩ). It is quite interesting condition because of pure requirement from quantum mechanics. Here, some mesoscopic phenomena can also yield high impedance. In this work, we connect STJ directly to Ni-nanowire and multi-walled Carbon nanotube (MWNT). It is, for the first time, confirmed that mutual Coulomb interaction (MCI) in the Ni-wire and weak localization (WL) in the MWNT can play the role of high impedance EME of CB. It is also found that the CB is very sensitive to phase fluctuation of EME.

1. INTRODUCTION

Single electron tunneling (SET) is one of the luminescent mesoscopic phenomena studied in the last decade [35]. It provides us macroscopic observation of quantum mechanics (*i.e.* duality of wave "tunneling effect" and particle "single electron charge") and chances to develop ultimate microelectronic devices.Correlation of SET with mesoscopic phenomena in the EME has recently attracted much attention. For instance, phase coherence of electron wave was never destroyed even in a quantum dot inserted into an AB ring [5]. Such a study for an artificial atom has been carried out in detail using many body effects including

I. O. Kulik and R. Ellialtioğlu (eds.),
Quantum Mesoscopic Phenomena and Mesoscopic Devices in Microelectronics, 145–160.

spin interactions (*i.e.*, SET spectroscopy) [20]. These systems are basically multi-tunnel junction (MTJ) systems. In contrast, EME plays the key role for SET in STJ systems because the tunnel junction is directly connected to voltage (current) source by leads, unlike MTJ systems [4]. Thus, its CB may be more sensitive to mesoscopic phenomena in the EME.

Here, EME plays the following two key roles for CB in STJ system [4]. (*i*) Tunneling electron transfers its E_c to the EME, leading to CB. (*ii*) Phase fluctuation in EME is combined with zero-point fluctuation, smearing out CB. For both roles, it is the most important that the impedance of EME (R_{ext}) is larger than R_Q. Otherwise, tunneling electron can not dissipate the E_c in the EME, and phase fluctuation in the EME causes the charges on junction surface go through zero-point fluctuation. They smear out CB. It is the requirement of quantum mechanics (*e.g.*, uncertainty between the phase of tunneling electron and the charge on junction surface) and as it is well known as phase correlation (PC) theory in STJ system, although one may not be familiar with SET in MTJ system.

Here, there are some mechanisms for high impedance in solid states. Although electron-phonon scattering will be dominant at high temperature T, it decreases at low T. At low T, some of mesoscopic phenomena can also yield high impedance, *e.g.* repulsive MCI, WL, Anderson localization, metal-insulator transition. Can they play the roles of high impedance EME for CB? In this work, we will connect (*i*) disordered Ni-nanowire and (*ii*) MWNT directly to the STJ. One dimensional (1D) MCI and 2D WL are observable in *i* and *ii*, respectively. In both cases, we confirm that CB is observable and CB is sensitive to phase fluctuation in the EME. In these cases, where does E_c dissipate? It is very questionable since, for instance, WL is basically an elastic process. In addition, what unique phenomena are caused and how the results are different from prediction of PC theory attract our attention due to the relation between SET and mesoscopic phenomena.

2. SAMPLE STRUCTURES

We employ novel material to fabricate STJ system, so called nanoporous Alumina film template (NAT). Fig. 10.1*a* shows schematic cross section of the samples. Since NAT is self-organized grown by anodizing Al substrate in some solutions, it exhibits nice uniformity, controllability, and reproducibility of nano-structure parameters (*e.g.* ϕ, L_{space}, d) [5]. One can deposit electrochemically varieties of materials into the nanopores (*e.g.*, metals, semiconductors, organic materials). Because the pore has automatically STJ at the bottom, it results in STJ system directly connected to nano-materials deposited into the pores. In this work, Ni [6] and MWNT [7] were deposited. Fig. 10.1*b* indicates that the Ni-wire is actually straight without any branches. Fig. 10.1*c* indicates actually the presence of the tunnel barrier layer and amorphous like structure of Ni-wire implying high disorder. Fig. 10.1*d* indicates the presence of multi-shells (near 26 layers) along the inner wall of the pore, implying MWNT. The homogeneity at the surface of MWNT is bad reflecting bad homogeneous of the pore inner wall.

In order to measure electrical characteristics, Ag and Au contact layers were deposited on top of the Ni-wires and MWNTs ($\sim 10^{5-6}$) respectively. Here it is very important to note that no tunnel barrier layer was confirmed at these interfaces and the insides of Ni-wire and MWNT themselves by high resolution TEM. These systems are undoubtedly array of STJs directly connected to Ni-nanowires and MWNTs.

3. CB IN AN ARRAY OF STJ/NI-NANOWIRE (AL/AL_2O_3/NI-NANOWIRE)

3.1 Identification of CB by junction array model

Fig. 10.2*a* shows temperature dependence of typical $G - V^{1/2}$ curves. It exhibits clear G_0 anomaly. We reported that 1st derivative of the $G - V$ curves in the three samples with different ϕs could be fit by Nazarov's theory, including Fig. 10.2*a* [6]. Nazarov extended PC theory to the system in which MCI interaction exists in the EME [10]. Thus, the fitting implies that our $G-V$ curves originate from CB with MCI in the EME. However, the feature of MCI and its correlation with CB were not revealed in it. Here, we try to reveal them focusing on temperature dependence of G.

The inset of Fig. 10.2*a* shows temperature dependence of G_0. It indicates a linear $G_0 - T$ relation below 4 K. This G_0 anomaly with the linear $G_0 - T$ relation can be interpreted as CB from the comparison with Zeller's [9] and Cleland's [10] reports. Cleland reported on the linear normalized $R_0 - T^{-1}$ relation and its saturation at very low temperature in a STJ system. They attributed this to CB. He also turned out qualitatively PC theory. On the other hand, Zeller *et al.* reported the observation of the G_0 anomaly and the linear $G_0 - T$ relation in Sn nano-particle array. Both are in good agreement with our result. Based on his junction array model, our linear $G_0 - T$ relation can be also interpreted as CB as follows.

Tunnel probability Γ for one STJ is given by the following equation in the orthodox theory of CB when tunnel transition probability is simply

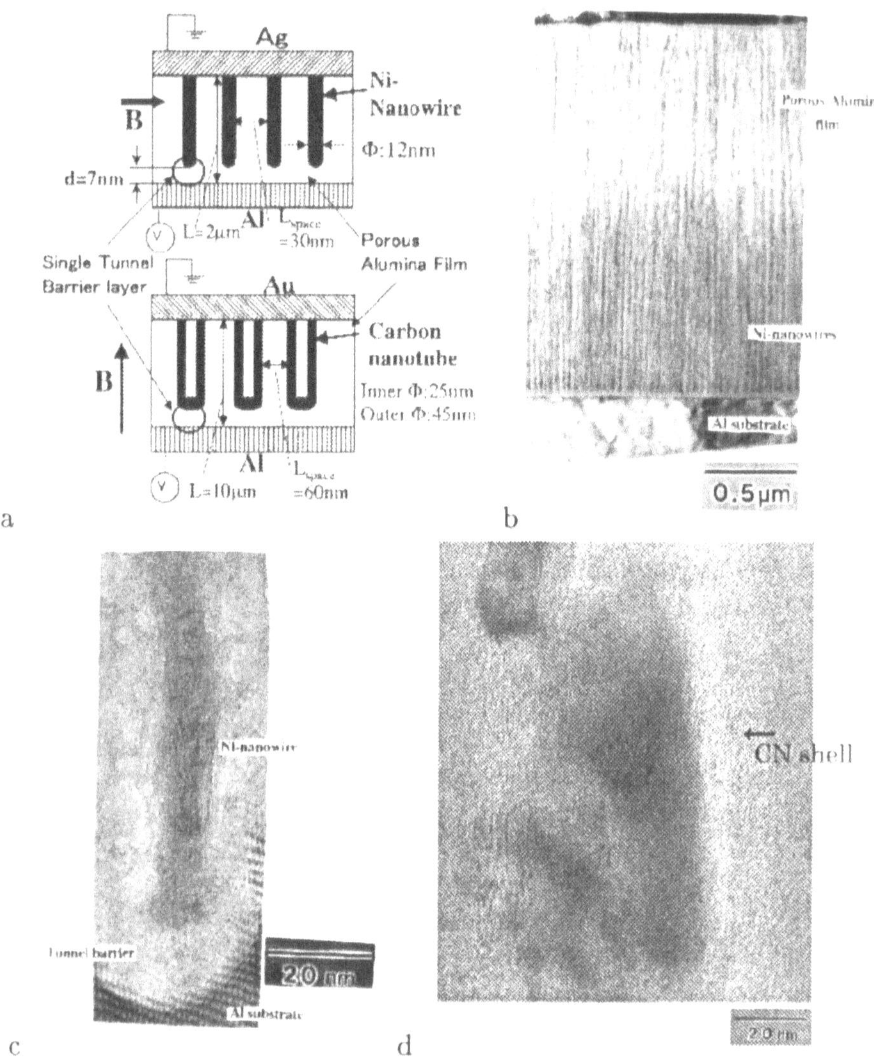

Figure 10.1 (*a*) Schematic cross section of sample structures. Nano-porous Alumina film template (NAT) consists of an array of nano-sized diameter pores, so called honeycomb structure on the top view. In this work, Ni and Carbon was electrochemically deposited into the nano-pores, resulting in STJ arrays of Al/Al_2O_3/Ni-nanowire and Al/Al_2O_3/carbon nanotube (CN). (*b*) Cross sectional TEM image of the sample whole part (*c*) High resolution cross sectional TEM of one Ni-wire around the bottom part (*d*) High resolution cross sectional TEM of one CN, indicating the multi-walled structure with about 26 shells.

assumed to be $(e^2R_t)^{-1}$ [35].

$$\Gamma(V,T) = \frac{1}{e^2R_t}\int_{-\infty}^{\infty} dE\, f(E)[1-f(E+\delta E)]P$$

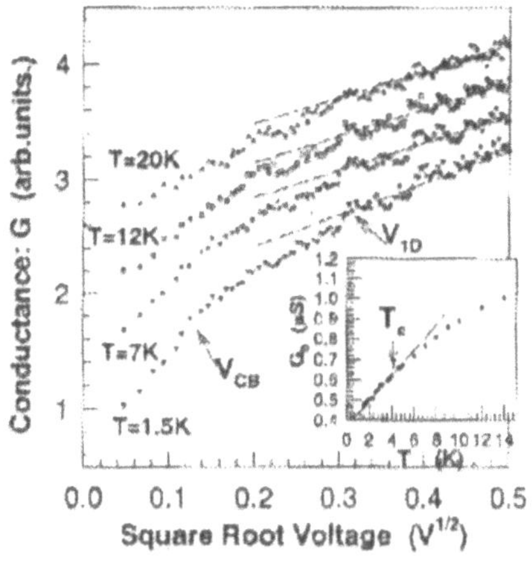

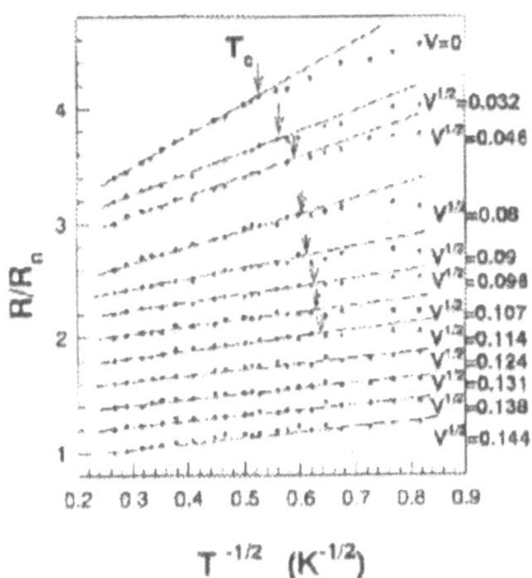

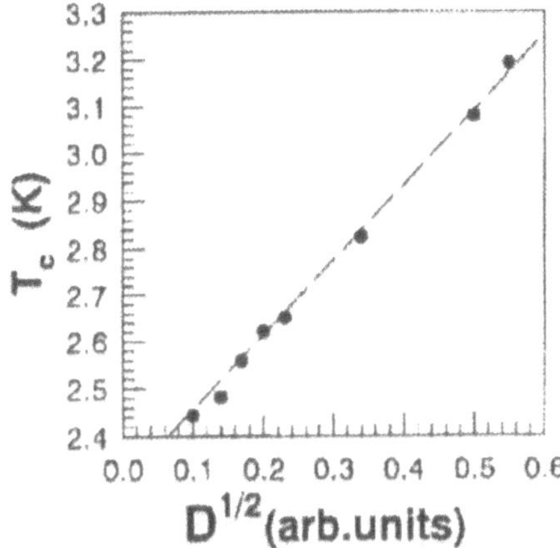

Figure 10.2 (*a*) Temperature dependence of typical G–$V^{1/2}$ characteristics in STJ arrays of Al/ Al_2O_3/Ni-nanowire; Inset: temperature dependence of G_0, (*b*) Temperature dependence of normalized resistance at each voltage point of (*a*), (*c*) Dependence of transition temperature T_c on diffusion constant D, exhibiting mostly linear relation.

$$= \frac{V - e/2C_j}{eR_t} P\left(1 - \exp\left[-(\delta E/kT)\right]\right)^{-1} \qquad (10.1)$$

where δE is the resultant energy change of system by SET event(*i.e.*, $\delta E = eV - e^2/2C_j$) and C_j and R_t are the junction capacitance and the tunnel resistance of STJ, respectively. P, which is discussed in section 4.1, is the probability density for tunneling electron to emit the energy to EME in PC theory. Here we assumed P as a constant. Eq. (10.0) can be rewritten by placing V=0 and taking into account the charging energy$e^2/2C_j (= E_c) \gg kT$ as follows.

$$\Gamma(0, T) = \frac{P}{2C_j R_t} \exp[-(E_c/kT)] \qquad (10.2)$$

This equation gives the T dependence of zero-bias SET probability caused by occasional tunneling of a thermally excited electron in one STJ. More detailed calculation for tunnel transition probability leads to the T dependence of G_0 in the orthodox theory [35]. Here, the contribution of tunnel junction array with the distributed parameters can be introduced by integrating E_c in Eq. (10.2) as follows:

$$\Gamma_{total}(0,T) = \frac{P}{2C_jR_{t(total)}} \int_0^\infty \exp[-(E_c/kT)]dE_c = \frac{Pk}{2C_jR_{t(total)}}T \tag{10.3}$$

where $R_t(total)$ is the total R_t. It implies actually the linear $G_0 - T$ relation that is qualitatively consistent with our observation.

Here, since $R_{t(total)} \sim 10\ \Omega$ can be estimated from the R_t of $10^6\ \Omega$ and the number of junctions of 10^5 [6] and $C_j \sim 10^{-17}$ F is also estimated from [6], the coefficient of T in Eq. (10.3) can be estimated to be on the order of 10^{-8}, assuming P to be 0.1. This value is in nice quantitative agreement with the slope value of the observed linear G_0-T (of the order of 10^{-8} /K). Therefore, one can conclude that the linear G_0-T relation can be qualitatively and quantitatively the strong evidence for the CB in the junction array. For better quantitative agreement, contribution of P and actual distribution function of the junction parameters have to be introduced into Eq. (10.0).

When the G_0 anomaly in Fig. 10.2*a* is due to CB, our system at least has to satisfy the following four necessary conditions [6]. 1) $R_t \gg R_Q$, 2) $E_c \gg kT$, 3) $R_{ext} \gg R_Q$ 4) Half width of distributed tunnel junction parameters <25%. It was confirmed that our STJ system satisfied all of these conditions [6], although there remained one problem about the parasitic capacitance. Here, the 3rd condition is the core importance only in STJ systems and is what we want to discuss here. To confirm it, we measured the resistance R_{Ni} of one Ni-wire by STM and revealed that the R_{Ni} was near 120 kΩ ($> R_Q$) [6]. The Ni-wire is also directly connected to the STJ. Therefore, one can conclude that the Ni-wire plays automatically the role of high impedance EME in our system. The R_{Ni} of 120 kΩ is, however, three orders of magnitude higher than that of bulk Ni. What is the origin of this?

3.2 1D-MCI as high impedance EME of CB

Fig. 10.2*b* shows temperature dependence of the normalized resistance at each voltage point. Data fitting by using the following Eq. (10.4), which is Altshuler's formula for one dimensional MCI in disordered conductors [11], gave the best fit to the linear part, compared with the fitting by the other functions.

$$\delta R(T)/R_n = (\rho e^2/8\hbar A)(4 + 3\lambda/2)(D\hbar/T)^{1/2} \qquad (10.4)$$

where R_n, ρ, A, and λ are the resistance at the highest T, the resistivity (of order 10^{-7}), the cross sectional area (order of 10^{-16}), and the effective constant for MCI (of order 10), respectively. D is also given by $v_F^2 \tau_{e-e}/d$, where v_F, τ_{e-e}, and d are the Fermi velocity, the relaxation time for MCI, and the sample dimension, respectively [11]. In the data fitting, D was used as a free parameter only, fixing R_n at one end. The T dependence can be classified into the following two regions, **1. High voltage region:** lower 4 features. The data can be well fit by Eq. (10.4) in all the Ts measured. **2. Low voltage region:** upper 8 features. In the T region above T_c, it can be fit by Eq. (10.4). In contrast, deviations are observed in the T region below T_c.

The data fit in the first voltage region qualitatively indicates that only Altshuler's 1D MCI in disordered Ni-wire is the dominant G mechanism. Here, we estimated and compared three characteristics lengths as follows; *i.e.*, thermal diffusion length $l_T = (\hbar D/kT)^{1/2} \sim 10^{-8}$ < localization length $\xi_{loc} = (2\hbar/e^2)A\sigma \sim 10^{-7}$ < sample length of 2μm. They support the WL regime of the sample. This l_T is also larger than ϕ, supporting the presence of 1D structure. Consequently, these results imply that the origin of high R_{Ni} is in the 1D MCI. However, note that any CB related characteristics are not observed in this first voltage region in spite of this high impedance.

In the secopnd voltage region, the T dependence was classified to two regimes with respect to T_c, **1. High T region ($T > T_c$) 2. Low T region ($T < T_c$).** In the low T region, $G-T$ features exhibit mostly linear relation as well as the linear $G_0 - T$ regime. This indicates directly that the second T region is the CB T regime. This CB T regime emerges only within V_{CB} at T=1.5 K. In the sense, V_{CB} can be a transition voltage from the 1D MCI to the CB voltage regimes. In contrast, the linearity in the high T region ($T > T_c$) indicates the presence of Altshuler's 1D MCI as well as that of the first voltage region. Although there is one large difference that the slope value α of the linear part increases drastically with reducing voltage, it can be explained also by Altshuler's theory [11]. Consequently, it is concluded that the 1D MCI in the Ni-nanowire plays the role of high impedance EME for the CB in the T region below T_c. Here, MCI is basically an elastic process. Since, however, this MCI occurs in disordered conductors, electron-phonon scattering is indispensable as the origin [11]. It can act qualitatively the role of energy transfer for tunneling electron as EME.

3.3 Smearing of CB by phase fluctuation in EME

Concerning the role as the isolator from the external fluctuation, this CB T regime depends strongly on the diffusion constant D of the MCI as follows. T_c shifts to the lower T region with the increase of α (by reducing the voltage) in Fig. 10.2*b*. Fig. 10.2*c* shows T_c *vs.* $D^{1/2}$ relation. D was determined from the parameter fitting to α by using Eq. (10.0). The relation is mostly linear. Since D is assumed to be proportional to τ_{e-e}, this linearity indicates that T_c strongly depends on τ_{e-e} in the Ni-wire. It can be interpreted as follows. Altshuler *et al.* pointed out that multiple quasielastic MCI processes dominated the phase breaking process in 1D and 2D disordered conductors in the absence of other phase breaking mechanisms: so called Nyquist phase breaking [12]. According to Ref. [12], its fluctuation energy E_N can be given by

$$E_N = h\omega_N \sim h[T/D^{1/2}N(E)]^{2/3} \qquad (10.5)$$

Where ω_N and $N(E)$ are the frequency of fluctuating field and the density of state in 1D conductor, respectively. Since V_{CB} is in the 1D MCI voltage regime, Eq. (10.5) is relevant for this discussion. If the Ec is smaller than the E_N, the CB will be smeared out by the fluctuation even if T satisfies the thermal condition of $E_C \gg kT$. Hence, the T which satisfies $E_N = E_C$ can be a critical $T(T_c')$. This T_c' can be given as follows, from $E_N = E_C$ and Eq. (10.5).

$$T_c' \sim (E_c/h)^{3/2}N(E)^{-1}D^{1/2} \qquad (10.6)$$

If it is assumed that both E_c and $N(E)$ are basically constants, this equation indicates the linear $T_c' - D^{1/2}$ relation which explains qualitatively the linear $T_c - D^{1/2}$ relation in Fig. 10.2*c*, suggesting $T_c = T_c'$. In fact, E_c is a constant in the same one sample and $N(E)$ can be assumed to be almost constant because this voltage region (*i.e.*, $V < V_{CB}$) is around 0 V.

This result implies that the CB is sensitive to the τ_{e-e}, D, and E_N in the EME. The MCI can play the role of high impedance isolator as the EME. However, this result implies that the CB is smeared out if too large E_N caused by the high τ_{e-e}^{-1} exists in EME. In such case, the Ni-wire can no longer act as the high impedance isolator because it becomes only the Nyquist noise source. This is also consistent with Cleland's report of CB with Nyquist voltage noise in the external lead line [10].

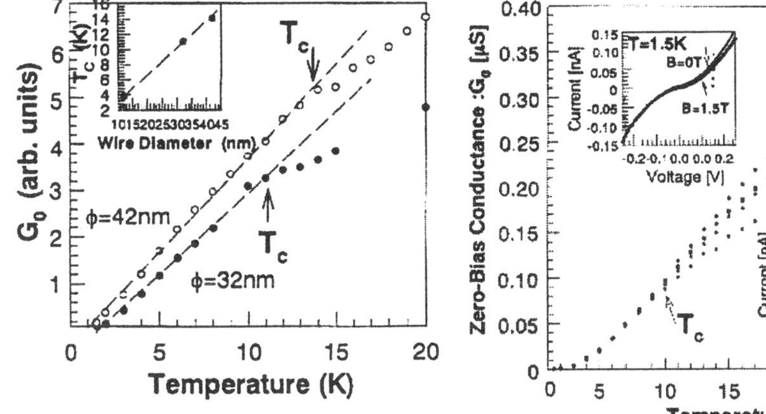

Figure 10.3 Wire diameter dependence of $G_0 - T$ characteristics Inset: Wire diameter dependence of T_c, exhibiting mostly linear relation.

Figure 10.4 Magnetic field dependence of $G_0 - T$ feature; Upper inset: $I - V$ curve at CB T regime; Lower inset: $I-V$ curve at 1D-MCI regime.

3.4 Wire diameter and magnetic field dependencies

In order to reconfirm this discussion, dependence of T_c on ϕ of Ni-nanowire is shown in Fig. 10.3. In both wire features, the linear $G_0 - T$ relation and T_c can be clearly observed. The $T_c - \phi$ relation is mostly linear as shown in the inset, although the sample number is only three. Here, D has the following relation with the electron density n and the cross sectional area of Ni-wire A, *i.e.*, $D \propto 1/n \propto A = \pi(\phi/2)^2$. Thus, T_c can have the following relation with D, when the linear T_c-ϕ relation actually exists, *i.e.*, $T_c \propto \phi = 2(A/\pi)^{1/2} \propto D^{1/2}$. Therefore, the linear $T_c - \phi$ relation observed in Fig. 10.3 supports qualitatively the presence of the linear $T_c - D^{1/2}$ relation.

Fig. 10.4 shows magnetic field dependence of the $G_0 - T$ relation that varies drastically around T_c. G_0 is mostly independent of B below T_c. On the contrary, positive magnet-conductance emerges above T_c. The insets show change of the $I - V$ curves by B applied in each temperature region. The change of $I - V$ curve is quite asymmetric at T=1.5 K ($< T_c$). The current is reduced in the plus voltage region. In contrast, it does not mostly change in the minus voltage region. They leads to mostly independent G_0 on B. On the other hand, the change of $I - V$ is symmetric at T=15 K ($> T_c$). Absolute values of the current are much increased in all the voltage regions, leading to the positive magnet-conductance. Therefore, we can reconfirm at least that the T_c exists actually and the G-T mechanisms changes drastically at the T_c. Although these are quite interesting results and may be understood by spin polarization [6], the origin is not yet revealed.

4. CB IN AN ARRAY OF STJ/MWNT ($AL/AL_2O_3/MWNT$)

In this section, WL as high impedance EME is discussed for the first time. Since WL is the result of phase interference of electron wave in diffusive regime, it has an impedance mechanism different from the others (*e.g.*, electron-phonon scattering) [13]. Can it play the role of high impedance EME? Is the PC theory relevant for WL? Where does energy transfer occur? In order to realize such STJ/WL system, we, for the first time, employed MWNT. Carbon nanotube (CN) has recently attracted much attention because it provides us unique mesoscopic phenomena and chances for fabricating nano-molecular electronic devices [14]. Coulomb oscillation was successfully observed in single walled CN (SWNT) [14]. In contrast, MWNT has been characterized only by interference effects of electron phase by disorder scattering (*e.g.*, 2D WL, AAS effect, UCF) [7] and none reported on SET related phenomena. Here we, for the first time, connect novel MWNT array to STJs.

4.1 Identification of CB by PC theory

The temperature dependence of the typical conductance $G-V$ curve is shown in Fig. 10.5*a*. They exhibit clearly G_0 anomaly. The shape of $G-V$ curves was drastically changed near T=5 K. Fig. 10.5*b* shows temperature dependence of the G_0. It is distinguished to the following 3 temperature regions, 1– Above 10 K: linear G_0-log(T), 2– 5 K-10 K: Saturation region, 3– Below 5 K: linear G_0-T.

At first, we identify the presence of CB in the temperature region below 5 K from the linear G_0-T characteristics and the $G-V$ curve fitting by PC theory. The linear G_0-T relation is also evidence for CB in junction array as well as the case of Section 3. The coefficient of Eq. (10.3) agrees quantitatively with the experimental result also in this case. The coefficient can be estimated to be on the order of 10^{-6}, from $R_{t(total)} \sim 10^{-1}\,\Omega$, $C_j \sim 10^{-17}$ F, and $P \sim 10^{-1}$. This value is in good agreement with the slope value of the linear part (order of 10^{-6} /K) in the inset of Fig. 10.5*b*.

Here, in order to clarify the correlation of this CB with EME and identify junction parameters, we calculated numerically $G-V$ curve, which was normalized by R_t and the number of junctions, using the

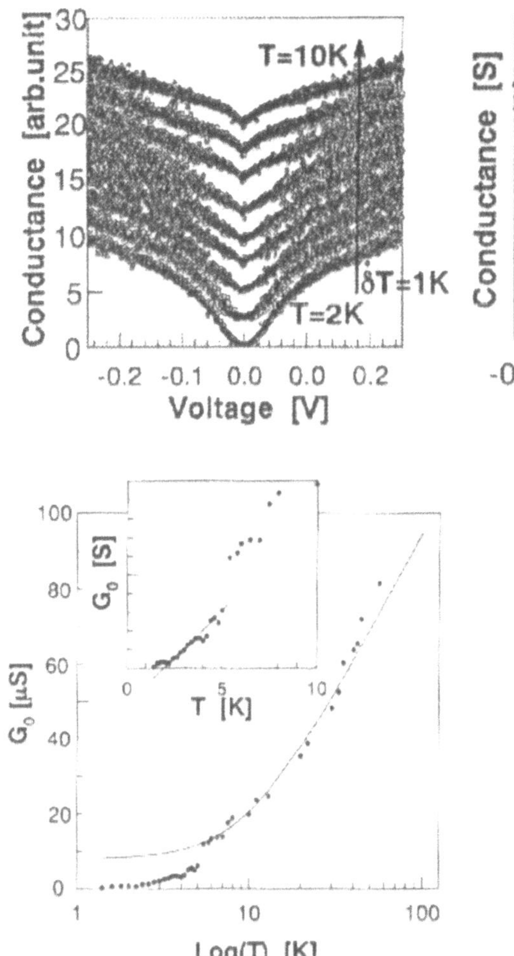

Figure 10.5 (*a*) Temperature dependence of the typical $G-V$ curve in STJ arrays of Al/Al_2O_3/MWNT. (*b*) Temperature dependence of G_0. The solid line is the result calculated by 2D WL formula of MWNT. It is in nice agreement with the data, except for CB region below T=5 K. Inset: linear $G_0 - T$ relation at the temperature below 5 K. (*c*) $G-V$ curve at T=2 K. Solid line is the result calculated by Phase correlation (PC) theory. It is in excellent agreement with measured $G-V$.

following equations of PC theory of CB [4] and fit the $G-V$ curve in Fig. 10.5*a* by utilizing it. Since we have R_t of 0.3 MΩ ($> R_Q$), fitting parameters are the impedance (R_{ext}) and capacitance (C_{ext}) of EME as $\omega_{RC} = (R_{ext}C_{ext})^{-1}$.

$$I(V) = \frac{1-e^{-\beta eV}}{eR_t}\int_{-\infty}^{\infty} dE \, \frac{E}{1-e^{-\beta E}} \, P(eV-E) \tag{10.7}$$

$$P(E) = \frac{1}{2\pi\hbar}\int_{-\infty}^{\infty} dt \; e^{J(t)+i\frac{E}{\hbar}t} \tag{10.8}$$

$$J(t) = 2\int_{-\infty}^{\infty} \frac{d\omega}{\omega} \, \frac{Re[Z_t(\omega)]}{R_Q} \, \frac{e^{-i\omega t}}{1-e^{-\beta\hbar\omega}} \tag{10.9}$$

where β and $Z_t(\omega) = [i\omega C + Z(\omega)^{-1}]^{-1}$ are $1/kT$ and the total impedance of EME circuit. $J(t)$, $P(E)$, and $I(V)$ are the PC function, the Fourier transform of $J(t)$, and the tunnel current which is basically the same as Eq. (10.0). In the PC theory, the phase on the tunnel junction is introduced as

$$\varphi(t) = \frac{e}{\hbar} \int_{-\infty}^{t} dt' \, U(t') \tag{10.10}$$

where $U(t') = Q(t')/C_j$ is the voltage across the tunnel junction. Since its fluctuation around the mean voltage $(e/\hbar)Vt$ is defined by $\widetilde{\Psi}(t) = \Psi(t) - (e/\hbar)Vt$, $J(t)$ is defined by $J(t) =< [\widetilde{\Psi}(t) - \widetilde{\Psi}(0)]\widetilde{\Psi}(t) >$ as an indication of time evolution of phase fluctuation. Here, if EME is described by a set of harmonic oscillator with energy quantum $\hbar\omega$, the $J(t)$ results in Eq. (10.11). Tunnel current (probability) is obtained by Eq. (10.7) by perturbatively treating tunneling Hamiltonian using Fermi golden rule and $P(E)$. Thus, $P(E)$ is interpreted as probability density for tunneling electron to transfer the energy E to EME thorough the $J(t)$. This PC theory implies the two key roles of EME mentioned in the introduction, especially by Eq. (10.11) which compares the EME impedance with R_Q. Cleland reconfirmed it experimentally [10], and some works reported on good agreement of this PC theory with $I-V$ or $G-V$ features observed in the weak tunneling case (*i.e.*, the case of $R_t \gg R_Q$) [17], recently even in the strong tunneling case [18].

As shown by the solid line in Fig. 10.5*c*, the fitting between the experimental result and theory is in excellent agreement also in our case (*i.e.*, R_t of 0.3 $M\Omega > R_Q$). From the best fitting, R_{ext} is given as 0.45 $M\Omega$. This R_{ext} corresponds to resistance of CN (R_{CN}) in our system because the CN is directly connected to the STJ. The R_{CN} is in good agreement with the past report of MWNT [7]. In addition, it should be noted that the R_{CN} is actually larger than R_Q. Therefore, we conclude that our CB is consistent with the PC theory and MWNT, which connected directly to STJ, can play the role of high impedance EME.

4.2 WL as high impedance EME

Here, the origin of this high impedance of MWNT is the key point for our argument. It can easily be understood from the result of WL by curve fitting as in Fig. 10.5*b*. Solid line in Fig. 10.5*b* shows the $G_0 - T$ characteristics in good agreement with the following 2D WL formula of MWNT [7], except for the CB temperature region.

$$G(t) = G(0) + \frac{e^2}{2\pi^2\hbar} \frac{n\pi d}{L} \ln \left[1 + \left(\frac{T}{T_c(B, \tau_s)} \right)^p \right] \tag{10.11}$$

where n, d, L, and τ_s are the number of shells, diameter of inner shell of CN, length of CN, and relaxation time of spin flip scattering, respectively. The contribution of number of CNs and R_t were taken into the term of $G(0)$.

The best fitting gives n=18, p=2.1, and T_c=10 K. Here, the most different feature from the past report of MWNT is the higher T_c of 10 K. T_c means the critical temperature between the linear G-log(T) and saturation regions. At the temperature above T_c, inelastic scattering by electron-phonon governs the dephasing process. In contrast, at the temperature below T_c spin flip scattering dominates the dephasing. Consequently total scattering time τ_ϕ is given by $\tau_\phi^{-1} = \tau_{in}^{-1} + 2\tau_s^{-1}$. Thus, high T_c indicates the presence of strong spin flip scattering in our system. In addition, the fitting line in the spin scattering regime in Fig. 10.5*b* is very flat, although actual data plot could not be measured owing to CB. These are interpreted as contribution of presence of magnetic impurity [7]. In fact, we could find out Co, which we used as a reactor, with a little volume in some of MWNTs by high resolution TEM. Detailed will be reported in [7].

Consequently, we conclude that 2D WL effect exists also in our MWNT with strong spin flip scattering. The high impedance of MWNT originates from this WL effect. The R_{CN} estimated from the normalized $G-V$ curve fitting by PC theory was 0.45 $M\Omega$. The $\xi_{loc} \sim 1$ μm can be estimated from this R_{CN}. Since it is shorter than CN length of 10 μm, it is consistent with WL regime of our MWNT. Furthermore, this result implies also that tunnel effect at STJ never destroys electron phase coherence in the MWNT, at least except for CB temperature region.

Based on these confirmations (*i.e.*, CB following PC theory and WL in the MWNT), we can conclude that WL in the MWNT plays the role of high impedance EME for the CB. Here, the origin of PC theory is in the phase fluctuation of tunneling electron. Since WL is also the result of interference of electron phase, it can act the role of high impedance isolator. In contrast, how can WL play the role of energy dissipation because interference of electron phase itself is elastic process, particularly in spin flip scattering regime in which the CB emerges in Fig. 10.5*b*. Although one possibility may be the dephasing process by magnetic impurity scattering, it is not sure here. We can find out answer in this work. At lower temperature, small energy transfer (*i.e.*, Nyquist noise) and large energy transfer (electron-electron scattering) become the main contributor to dephasing [19]. The saturation of dephasing time also is recently attracting attention based on quantum fluctuation [20]. Since, however, our measurement was done down to 1.5 K, we neglected those contribution here.

4.3 Magnetic field dependence: AAS effect in CB regime

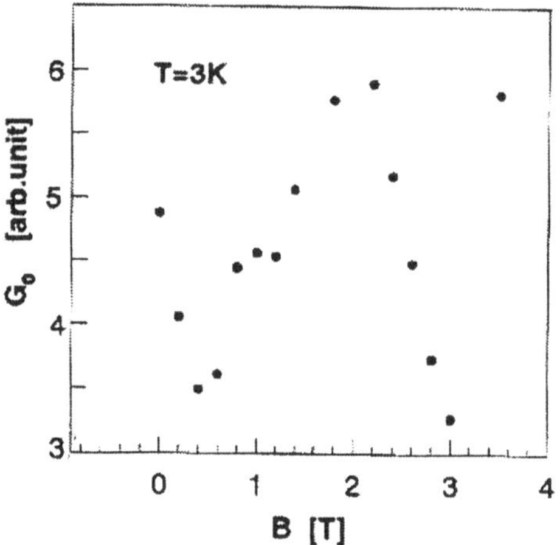

Figure 10.6 Magnetic field dependence of G_0 in the CB temperature regime. B was applied parallel to the axis of CN as shown in Fig. 10.1*a*.

Concerning the role of high impedance isolator, we can find out also an interesting phenomenon. Fig. 10.6 shows magnetic field dependence of G_0 in the CB temperature regime. Oscillatory behavior is clearly observed, although it is only one period and measured points are rough. Such behavior in MWNT has been understood as AAS effect in graphite cylinder [21]. In general, AAS effect exhibits periodical G oscillation by magnetic flux applied to inner area (πr^2, where r is the radius) of AAS ring, with oscillation period $\Delta B = (h/2e)/(\pi r^2)$. We can determine simply the effective r from the B at the 1st oscillation peak. Since B is 2.2 T, r is determined to be 17.3 nm. This value of r is in excellent agreement with the mean r of 17.5 nm of our MWNT. Therefore, we conclude that this oscillatory $G-B$ behavior originates from AAS effect. We measured many CNs. Thus, this result indicates very high uniformity of the radius r. Of course, for exact confirmation, the effects of the length L of CN, misalignment angle θ to the tube axis, the contribution of CB, and number of MWNTs on the magnitude of the oscillation have to be taken into consideration.

Here, it is surprising that this AAS effect is observable even in the CB temperature regime. The CB is yielded by high impedance EME ($R_{CN} > R_Q$), which is caused by WL, of the MWNT at B=0. Since the R_{CN} at the bottom of G-oscillation (*i.e.*, B=0.4 T) is larger than that at B=0, the CB should never be influenced by this increase of

R_{CN}, because one still has $R_{CN} > R_Q$, and hence the AAS oscillation should be isolated from CB. Nevertheless, CB can be modulated and AAS effect can be observed as in Fig. 10.3. This may imply that the CB can be smeared by phase fluctuation (or dephasing) of EME even under the high impedance EME, in the case of WL depending EME by applying B. This is the characteristics which is quite different from the PC theory. This may lead to the control of CB by phasing-dephasing effect of WL in the EME, and to novel quantum computer devices.

5. CONCLUSION

We reported, for the first time, that 1D-MCI and WL could play the role of high impedance EME for CB in STJ system. Energy transfer for the PC theory can be caused by electron-phonon scattering in the case of 1D-MCI in disordered conductor. In contrast, it is still questionable where it occurs in the case of WL. In both cases, it was also found that the CB was very sensitive to phase fluctuations (or dephasing) in the EME.

Acknowledgements

The authors thank D. Averin, B.L. Altshuler, M. Buttiker, X. Wang, and M. Ueda for very useful discussion and suggestion, and also University of Toronto group for sample preparing, and MST Japan for nice TEM images. This work was financially supported by the NSERC, MST, and the scientific research project both on the basic study B and on the priority area of Japanese Ministry of Education, Science, Sport, and Culture. J.H. also sincerely thanks Yuki, Wakana, and Rika for continuous encouragement.

References

[1] D. V. Averin and K. K. Likharev, in *Mesoscopic Phenomena in Solids*, eds. B. L. Altshuler *et al.*, (North-Holland, 1991) p. 173; *Single Charge Tunneling*, eds. H. Grabert and M. H. Devoret, (Plenum Press, New York and London, 1991).

[2] A. Yacoby, M. Heiblum, D. Mahalu and H. Shtrikman, Phys. Rev. Lett. **74**, 4047 (1995).

[3] S. Tarucha, D. G. Austing and T. Honda *et al.*, Phys. Rev. Lett. **77**, 3613 (1996).

[4] G.-L. Ingold and Yu. V. Nazarov, in the 2nd article of Ref. 1, p.21; M. H. Devoret, D. Esteve, H. Grabert, *et al.*, Phys. Rev. Lett. **64**, 1824 (1990).

[5] D. Routkevitch, A. Tager, J. Haruyama, *et al.*, IEEE-ED **43**, 1646 (1996); A. Tager, D. Routkevitch, J. Haruyama, *et al.*, *Future Trends*

in Microelectronics, eds. S. Luryi, J.M. Xu and A. Zalavsky, NATO ASI Series E-**323**, 171 (1996)

[6] J. Haruyama, D. Davydov, M. Moskovits and J. M. Xu, *et al.*, Solid-state Electronics, the proc. NPE'97, **42**, 1257 (1998); D. Davydov, J. Haruyama, M. Moskovits and J. M. Xu *et al.*, Phys. Rev. B **57**, 13550 (1998); J. Haruyama, K. Hijioka, Y. Sato *et al.*, Phys. Rev. B and Appl. Phys. Lett., submitted.

[7] J. Haruyama, I. Takesue, K. Ohta *et al.*, Nature, submitted; J. Li, M. Moskovits *et al.*, Chem. Mater. **10**, 1963 (1998).

[8] Yu. V. Nazarov, Sov. Phys. JETP. **68**, 561 (1989); in the 2nd article of Ref. 1, p.99.

[9] H. R. Zeller and I. Giaever, Phys. Rev. **181**, 789 (1969).

[10] A. N. Cleland, J. M. Schmidt and J. Clarke, Phys. Rev. Lett. **64**, 1565 (1990).

[11] B. L. Altshuler and A. G. Aronov, in *Electron-electron Interactions in Disordered Systems*, eds. A. L. Efros and M. Pollak, (North-Holland, 1985); B. L. Altshuler and A. G. Aronov, Solid State Comm. **30**, 115 (1979).

[12] B. L. Altshuler, A. G. Aronov and D. E. Khmelnitsky, J. Phys. C **15**, 7367 (1982).

[13] Y. Imry, in *Introduction to Mesoscopic Physics*, (Oxford University Press, 1997); P. W. Anderson, Phys. Rev. **109**, 1492 (1958); E. Abrahams, P. W. Anderson, *et al.*, Phys. Rev. Lett. **42**, 673 (1979); S. Kobayashi, Surf. Sci. Rep., **16**(1) (1992).

[14] *e.g.*, many mesoscopic phenomena in SWNT such as large mean free path across one SWNT, Luttinger liquid like behavior, chirality dependent electrical features, many body effect with MCI.

[15] S. J. Tans, M. H. Devoret, C. Dekker *et al.*, Nature **386**, 474 (1997), and **394**, 761 (1998).

[16] L. Langer, V. Bayot, *et al.*, Phys. Rev. Lett. **76**, 479 (1996); T.W. Ebbesen, *et al.*, Nature **382**, 54 (1996).

[17] P. Delsing, K. K. Likharev, *et al.*, Phys. Rev. Lett. **63**, 1180 (1989).

[18] X. H. Wang and K. A. Chao, Phys. Rev. B **59**, 13094 (1999).

[19] A. G. Huibers and C. M. Marcus, Phys. Rev. Lett. **81**, 200 (1998).

[20] Y. Imry, Abstract of this NATO conference, p.16 (1999).

[21] A. Bachtold, C. Strunk, *et al.*, Nature **397**, 673 (1999).

Chapter 11

TRANSPORT THROUGH QUANTUM DOTS AND THE KONDO PROBLEM

J. König
Department of Physics, Indiana University
Bloomington, Indiana 47405, USA

T. Pohjola
Institut für Theoretische Festkörperphysik, Universität Karlsruhe
76128 Karlsruhe, Germany
and
Materials Physics Laboratory, HUT
02150 Espoo, Finland

H. Schoeller
Institut für Nanotechnologie, Forschungszentrum Karlsruh
76021 Karlsruhe, Germany

G. Schön
Institut für Theoretische Festkörperphysik, Universität Karlsruhe
76128 Karlsruhe, Germany

Abstract Transport through quantum dots with large level spacing and charging energy is considered. At low temperature and strong coupling to the leads, quantum fluctuations and the Kondo effect become important. They show up, e.g., as zero-bias anomalies in the current-voltage characteristics. We use a recently developed diagrammatic technique as well as a new real-time renormalization-group approach to describe charge and spin fluctuations. Both approaches cover the linear as well as the nonlinear response regime. The spin fluctuations give rise to a Kondo-assisted enhancement of the current through the dot as seen in experiments.

I. O. Kulik and R. Ellialtioğlu (eds.),
Quantum Mesoscopic Phenomena and Mesoscopic Devices in Microelectronics, 161–167.

1. INTRODUCTION

Transport through single levels in small quantum dots is strongly affected by Coulomb interaction between the dot electrons. At temperatures lower than the charging energy Coulomb blockade phenomena arise. Due to the coupling to leads the dot states have a finite life time and both the charge and the spin of the dot fluctuate.

In recent experiments [1, 2, 3] on quantum dots with both strong Coulomb interaction and strong coupling to the leads Kondo-like physics has been observed as an enhancement of the conductance with lowering the temperature or bias voltage. This enhancement is interpreted as a reminiscence of the fact that at exactly zero temperature and bias voltage the transmittivity reaches the unitary limit [4]. The combination of both charging effects *and* quantum fluctuations is necessary for Kondo-like behavior.

We describe this interplay of charging effects and quantum fluctuations in terms of a recently developed diagrammatic technique as well as a new renormalization-group (RG) method. Both formulations cover the linear and non-linear response regime. We find zero-bias anomalies in the current-voltage characteristics and discuss their dependence on temperature, a gate voltage and an external magnetic field.

2. MODEL HAMILTONIAN

The spacing of the dot levels in the experiments [1, 2, 3] is larger or at least of the order of the charging energy. We, therefore, concentrate on the extreme limit of a quantum dot containing only one spin-degenerate level. This is equivalent to the single-impurity Anderson model, $H = H_0 + H_T$,

$$H_0 = \sum_{k\sigma r} \epsilon_{k\sigma r} a^\dagger_{k\sigma r} a_{k\sigma r} + \sum_\sigma \epsilon_\sigma c^\dagger_\sigma c_\sigma + U n_\uparrow n_\downarrow \tag{11.1}$$

$$H_T = \sum_{k\sigma r} \left(T^r_k a^\dagger_{k\sigma r} c_\sigma + h.c. \right) . \tag{11.2}$$

The first term describes noninteracting electrons in state k with spin σ in the left or right lead, $r = L/R$, and the second one the dot electrons characterized by spin $\sigma =\uparrow, \downarrow$. The position of the dot levels can be tuned by a gate voltage. The spin degeneracy may be lifted by a finite Zeeman energy due to an external magnetic field B. The charging energy U is accounted for in the third term, and the coupling of the dot states to the leads is modeled by the standard tunnel term. The spin σ is conserved during a tunnel process. The tunnel coupling H_T leads to a finite lifetime τ of the dot states and, therefore, to an intrinsic level broadening $\Gamma = \hbar/\tau$. The latter is related to the tunnel rates $2\pi\gamma_r^+(\omega) = \Gamma_r f_r(\omega)$ into and $2\pi\gamma_r^-(\omega) = \Gamma_r(1-f_r(\omega))$ out of the dot through barrier r, obtained from simple golden-rule arguments, by $\Gamma = \sum_r \Gamma_r$ and $\Gamma_r = (2\pi/\hbar)\sum_k |T_k^r|^2 \delta(\omega - \epsilon_{k\sigma r})$. Here, $f_r(\omega)$ is the Fermi distribution of reservoir r.

3. RELATION TO THE KONDO MODEL

There are three regimes for the Anderson model (Kondo, mixed-valence, and empty-orbital regime) distinguished by the position of the dot level ϵ relative to the Fermi level of the leads.

For a low-lying level ϵ (Kondo regime) the quantum dot is always filled with one electron, either with up or down spin, and the Anderson model can be mapped onto the Kondo model. Spin flip occurs via a virtual empty dot. At low temperature, the spin of the dot electron is screened by the reservoir electrons, and the Kondo effect shows up as an increase of the transmittivity through the dot, expressed by a spectral density with a sharp resonance near the Fermi level of the leads. As a result the current is increased. This can also be seen from Friedel's sum rule which yields Langreth's formula [4]

$$G|_{T,V=0} = 2\frac{e^2}{h}\frac{4\Gamma_L\Gamma_R}{\Gamma^2}\sin^2(\pi\langle n_\sigma\rangle) \tag{11.3}$$

for the conductance $G = \partial I/\partial V$ at zero temperature and bias voltage. The conductance reaches the unitary limit in the Kondo regime since there $\langle n_\sigma\rangle = 1/2$. To achieve the value of the unitary limit, the conductance increases with decreasing temperature and bias voltage.

By varying the gate voltage the level position can be shifted near the Fermi level of the leads (mixed-valence regime). The dot can be either empty or singly occupied. While in the Kondo regime spin fluctuations are important, in the mixed-valence regime charge fluctuations dominate. Finally for the level being well above the Fermi levels (empty-orbital regime) the dot is always empty.

4. REAL-TIME TRANSPORT THEORY

In general, a quantum-statistical expectation value of an operator A at time t can be written as

$$\langle A(t)\rangle = \mathrm{tr}\left[\rho_0\, T_K \exp\left(-i\int_K dt' H_T(t')_I\right) A(t)_I\right] \tag{11.4}$$

where the index I denotes the interaction picture with respect to H_0 and T_K is the time-ordering operator on the Keldysh contour going from

$-\infty$ to t and then back to $-\infty$. We expand the exponential in powers of H_T and perform the trace of each term of the expansion. Since the Hamiltonian H_0 is bilinear in the lead electron operators, Wick's theorem holds for the lead operators. They are contracted in pairs representing equilibrium distribution functions $\gamma_r^\pm(\omega)$. For the dot electrons Wick's theorem does not hold since the Coulomb interaction is expressed by a quartic term. The remaining dot electron operators have to be treated explicitly.

The tunneling current $I = I_L = -I_R$ is given by

$$I(t) = -ie \sum_{k\sigma} \left\{ T_k^L \langle (a_{k\sigma L}^\dagger c_\sigma)(t) \rangle \; - \; H.c. \right\} \tag{11.5}$$

or, using the spectral density $A_\sigma(\omega)$ of the dot level

$$I = e \frac{\Gamma_L \Gamma_R}{\Gamma} \sum_\sigma \int_{-\infty}^{\infty} d\omega \, A_\sigma(\omega) [f_R(\omega) - f_L(\omega)] \, . \tag{11.6}$$

In the following we concentrate on the limit of strong Coulomb repulsion such that double occupancy is forbidden. In order to describe the Kondo regime, we need a nonperturbative treatment of the tunneling, where quantum fluctuations yield finite life-time broadening and renormalization effects of the dot levels.

5. RESONANT-TUNNELING APPROXIMATION: ZERO-BIAS ANOMALIES AND MAGNETIC-FIELD DEPENDENCE

In Refs. [5, 6] we developed a diagrammatic technique to evaluate Eq. (11.5). Within the "resonant-tunneling approximation" we take into account non-diagonal matrix elements of the total density matrix with a nondiagonality of at most one electron-hole pair excitation in the leads.

For a low-lying level Kondo resonances arise at the Fermi levels of the leads, and at zero bias a peak in the conductance evolves (zero-bias maximum; see Fig. 11.1).

An external magnetic field shifts the Kondo resonances in the spectral density $A_\sigma(\omega)$ by the level splitting $\Delta\epsilon = \epsilon_\uparrow - \epsilon_\downarrow$. For this reason, the Kondo-assisted tunneling, which is suppressed at low transport voltage, sets in if transport voltage and level splitting are equal. Therefore, the conductance peak at zero bias splits up into two peaks separated by twice the level splitting (see r.h.s of Fig. 11.1) in agreement with experiments.

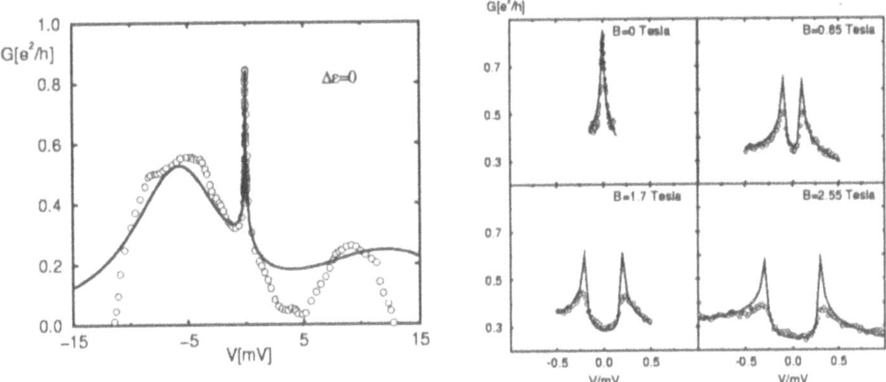

Figure 11.1 LHS: The differential conductance vs. bias voltage for $T = 4.3$ μeV, $\epsilon_\sigma(B=0) = -5.2$ meV, $\Gamma = 3.4$ meV, $a_c = 0.33$, and $U = 30$ meV. The circles are experimental data from Ref. [3] in which Kondo-assisted tunneling via a single charge trap of a point contact tunnel barrier has been measured. RHS: The same but with a finite magnetic field. The zero-bias anomaly splits due to a finite magnetic field.

6. TWO-LEVEL QUANTUM DOT

The resonant-tunneling approximation can be extended to more complex systems [6]. In Fig. 11.2 we show an example for a quantum dot with two levels (each level is spin degenerate). We expect Kondo resonances in the spectral density at the Fermi level as well as at energies shifted by the level splitting. This leads to a triple-peak structure near zero-bias for low-lying levels.

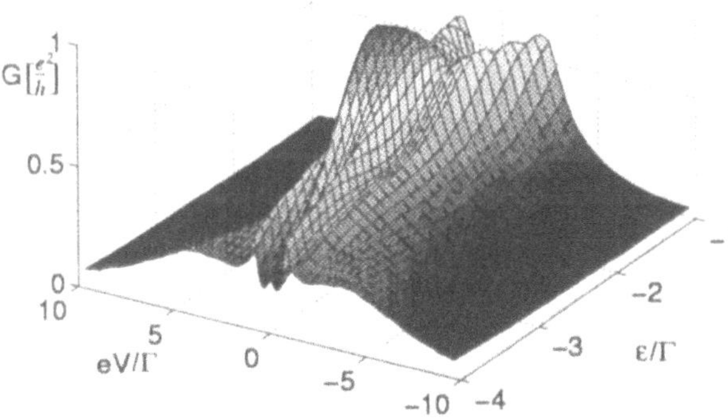

Figure 11.2 The differential conductance as a function of the bias voltage and the level position ϵ of the lower level. The level splitting is fixed to $\Delta\epsilon = 0.8\Gamma$, and the temperature is $k_B T = 0.04\Gamma$.

7. REAL-TIME RENORMALIZATION-GROUP APPROACH

For a more quantitative analysis we formulate an RG approach working in Liouville space with $L = [H, \cdot] = L_0 + L_T$. We continuously integrate out all contractions with time difference t_c from 0 to ∞. The result can be interpreted as a renormalization of L_0, L_T or as a generation of double (and higher) vertex terms (see Ref. [7] for details), symbolically written as

$$\frac{dL_0}{dt_c} = -i\gamma(t_c)e^{iL_0 t_c}Ge^{-iL_0 t_c}G \tag{11.7}$$

$$\frac{dG}{dt_c} = -\gamma(t_c)\int_0^{t_c} dt[G(t), G]_+ G(t - t_c) \tag{11.8}$$

where we have neglected double and higher vertex generation, and G represents the tunnel coupling vertex. As an essential difference to conventional poor man scaling and operator product expansion methods (besides the fact that we work in Liouville space), we have not expanded the propagation $\exp(\pm iL_0 t_c)$ of the dot in t_c. This is why we can send the final cutoff t_c to infinity, *i.e.*, we can integrate out all time scales. Furthermore, the renormalized Liouville operator of the dot is no longer represented as the commutator with a renormalized Hamiltonian $L_0 \neq [H_0, \cdot]$. This non-Hamiltonian dynamics describes the physics of dissipation and finite life times.

We find that the mixed-valence and empty-orbital regime, $\epsilon > -0.4\Gamma$, are sufficiently well described by propagator and single vertex renormalization. At $T = V = 0$, Friedel sum rule is satisfied with $2-3\%$ precision. Fig. 11.3 shows the temperature dependence of the linear conductance for various values of the dot level in comparison to experimental data.

For a low-lying level we find a logarithmic increase of conductance with decreasing temperature and finally a saturation. We note that in this region Kondo-like physics is not important but nevertheless we find pronounced zero-bias maxima for the nonlinear conductance, see Fig. 11.4. On the other hand, for a level close to the Fermi levels of the reservoirs, the linear conductance shows a local maximum as function of temperature. Here, the renormalized level is above the Fermi levels of the reservoirs, and therefore the conductance increases with increasing temperature due to thermal occupation of the local level. In this region one obtains zero-bias minima for the nonlinear conductance.

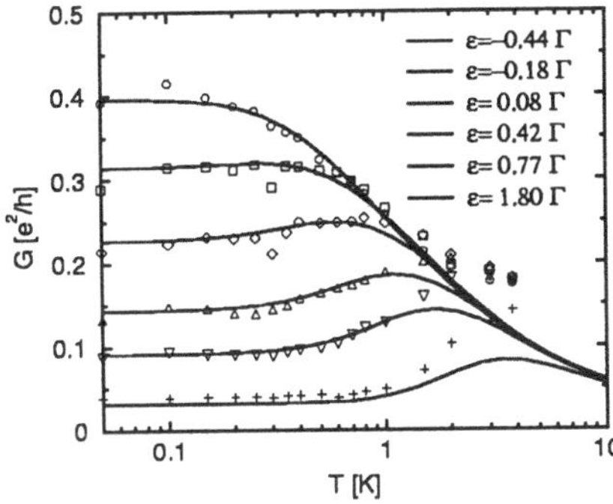

Figure 11.3 Linear conductance as function of temperature in comparison with experiment [1]. Chosen parameters are $U = 6.2\Gamma$, $8\Gamma_L\Gamma_R/\Gamma^2 = 0.6$ and $\Gamma = 3423$ mK.

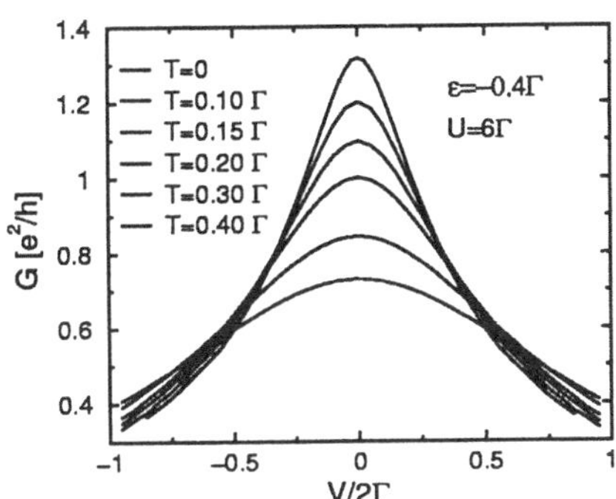

Figure 11.4 Nonlinear conductance as function of bias voltage.

References

[1] D. Goldhaber-Gordon *et al.*, Nature (London), **391**, 156 (1998); D. Goldhaber-Gordon *et al.*, Phys. Rev. Lett. **81**, 5225 (1998).

[2] S. M. Cronenwett *et al.*, Science **281**, 540 (1998); J. Schmid *et al.*, Physica B**256-258**, 182 (1998); F. Simmel *et al.*, Phys. Rev. Lett. **83**, 804 (1999).

[3] D. C. Ralph and R. A. Buhrman, Phys. Rev. Lett. **72**, 3401 (1994).

[4] D. C. Langreth, Phys. Rev. **150**, 516 (1966); T. K. Ng and P. A. Lee, Phys. Rev. Lett. **61**, 1768 (1988).

[5] H. Schoeller, in *'Mesoscopic Electron Transport'*, eds. L. L. Sohn *et al.* (Kluwer 1997), p. 291; J. König, *Quantum Fluctuations in the Single-Electron Transistor*, Shaker, Aachen (1999).

[6] H. Schoeller and G. Schön, Phys. Rev. B **50**, 18436 (1994); J. König *et al.* Europhys. Lett. **31**, 31 (1995); Phys. Rev. Lett. **76**, 1715 (1996); J. König *et al.*, Phys. Rev. B **54**, 16820 (1996); T. Pohjola *et al.*, Europhys. Lett. **40**, 189 (1997).

[7] J. König and H. Schoeller, Phys. Rev. Lett. **81**, 3511 (1998); H. Schoeller and J. König, cond-mat/9908404.

Chapter 12

COULOMB BLOCKADE IN QUANTUM DOTS WITH OVERLAPPING RESONANCES:

Towards an Explanation of the Phase Behaviour in the Mesoscopic Double-Slit Experiment

P. G. Silvestrov[1,2]
[1] *Budker Institute of Nuclear Physics*
630090 Novosibirsk, Russia

Y. Imry[2]
[2] *Weizmann Institute of Science*
Rehovot 76100, Israel

Abstract Coulomb blockade (CB) in a quantum dot (QD) with one anomalously broad level is considered. In this case many consecutive pronounced CB peaks correspond to occupation of one and the same broad level. Between the peaks the electron jumps from this level to one of the narrow levels and the transmission through the dot at the next resonance essentially repeats that at the previous one. This offers a natural explanation to the recently observed behavior of the transmission phase in an interferometer with a QD. Single particle resonances of very different width are natural if the dot is not fully chaotic. This idea is illustrated by the numerical simulations for a non-integrable QD whose classical dynamics is intermediate between integrable and chaotic. Possible manifestations for the Kondo experiments in the QD are discussed.

1. INTRODUCTION. THE DOUBLE-SLIT EXPERIMENT

Much progress has recently been achieved in the fabrication and experimental investigation of ultrasmall few-electron devices - such as Quantum dots [1]. A useful theoretical tool for the investigation of chaotic

I. O. Kulik and R. Ellialtioğlu (eds.),
Quantum Mesoscopic Phenomena and Mesoscopic Devices in Microelectronics, 169–182.

QD-s is the Random Matrix theory (see reviews [2, 3]). Nevertheless, many experimentally observed features of these artificial multi-electron systems still have not found a reasonable theoretical explanation.

A challenging problem which has resisted adequate theoretical interpretation arises from the experiments [4] which determine the phase of the wave transmitted through the QD[1]. The goal of this talk will be to present a mechanism which may lead to a satisfactory explanation of these results. The QD in the experiment was imbedded into one arm of the Aharonov–Bohm (AB) interferometer, which allowed to measure not only the magnitude of the transmission, but also the phase acquired by the electrons traversing the dot. The total current through the interferometer is [4]

$$J_{total} \sim \left| t_{QD} + e^{i\phi_{AB}} t_{sl} \right|^2 , \tag{12.1}$$

where t_{QD} and t_{sl} are the transmission amplitudes through the QD and the second reference arm of the interferometer. The AB phase is proportional to the magnetic flux Φ threading the interferometer $\phi_{AB} = 2\pi\Phi/\Phi_0$. The *complex* amplitude t_{QD} (as well as the number of electrons in the dot) is slowly changed by the plunger gate voltage V_g and does not depend on the weak magnetic field. Parametrising the amplitudes as $t = |t| \exp(i\theta)$ the interference term in Eq. (12.1) becomes, in obvious notation:

$$|t_{QD}|\,|t_{sl}| \cos\left(\phi_{AB} + \theta_{sl} - \theta_{QD}\right) \ . \tag{12.2}$$

Thus by independently changing V_g and ϕ_{AB} one may find both $|t_{QD}|$ and the variation of the phase $\Delta\theta_{QD}$ as a function of the gate voltage.

In the experiments [4, 5] the coherent component of the electron transport through the QD was measured directly for the first time. Later experiments allowed also an investigation of the dephasing rate of the electron state in the QD [6]. Here we discuss the unexpected and not understood yet observed behavior of the phase θ_{QD} of the transmitted electron.

This unusual behavior of the phase is illustrated in Fig. 12.1 (an approximate drawing, very close to the real experimental figure of ref. [4]). As one can see from the figure, in accordance with the Breit-Wigner picture, the measured phase increased by π around each CB peak. Absolutely unexpected, was a fast jump of the phase by $-\pi$ between the resonances just at the minimum of the transmitted current. Such a behavior is in evident contradiction with the theoretical expectation for

[1] The two terminal setup of the early experiment [5] does not allow to measure the phase of the transmission amplitude through a QD.

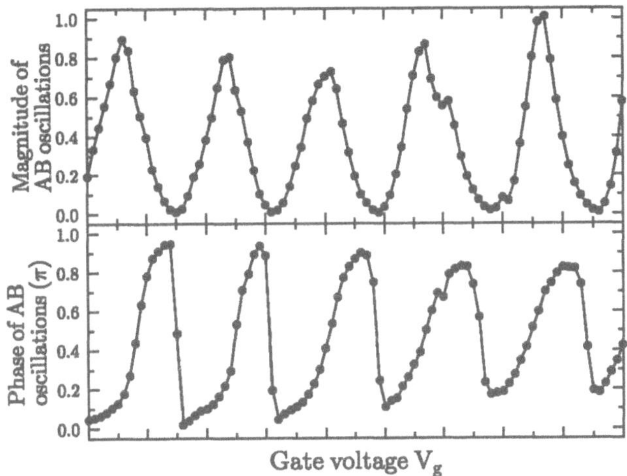

Figure 12.1 The magnitude of AB oscillations $\sim |t_{QD}|$ (arbtrary units) and the variation of the phase $\Delta\theta_{QD}$ (in units of π) as a function of the gate voltage.

transmission via the consecutive levels in 1-dimensional quantum well. The phase shift between the resonances in this case simply accounts for the number of nodes of the (almost real) wave function inside the dot, which is changed by one in $1d$ at the consecutive resonances.

A number of theoretical papers attempted to find an explanation of the $-\pi$ jumps. It is relatively easy to find a mechanism which leads to the fast drops of phase in some part of the CB valleys. It is more complicated to explain why for the long series of peaks the increase of the phase by π at the resonance will be accompanied by the $-\pi$ drop at the valley. In two–dimensional QD the phase drops associated with the nodes of the transmission amplitude arise already within the single–particle picture. However, in order to have a sequence of such events one should consider a QD of a very special form [7, 8]. The model of Ref. [9] also was not designed to explain the series of drops. For electrons with spin, the phase drop occurs between the resonances corresponding to adding of the first and second electron to the same level, but again this effect is not easy to generalize for a series of peaks. The mechanism of Refs. [10, 11] makes nontrivial assumptions on the geometry of the QD and the way it changes under the change of plunger gate voltage. An interesting generic mechanism, which indeed may lead to the sequence of $\sim U_{CB}/\Delta$ drops between the resonances was suggested in Ref. [12]. However, the predicted phase behavior differs from what has been seen experimentally.

In this paper we propose a mechanism according to which the transmission at many CB peaks proceeds through one and the same level in the QD. This means that the phases at the wings of different resonances should coincide and therefore the increase by π at the resonance must be compensated. This compensation occurs via narrow jumps between the resonances and is accompanied by a fast rearrangement of the electrons in the dot.

The experiment [4] was clearly done in the CB regime. However, in order to be able to measure the AB oscillations in the CB valleys the dot was significantly "opened". As a result, the widths of the resonances turn out to be rather large, being only few times smaller than the charging energy (see Fig. 12.1). Also the widths and heights of all observed resonances are quite similar. These features of the experimental results, which have not attracted so wide an attention as the phase jumps, also naturally follow from our mechanism.

It is generally believed that the CB is observed only if the widths of resonances are small compared to the single-particle level spacing in the dot Δ. This condition assumes that couplings of all levels to the leads are of the same order of magnitude. However, as we will show, even for nonintegrable ballistic QD-s the widths of the resonances may vary by orders of magnitude. In this case it does not make sense to compare the width of few broad resonances with the level spacing, determined by the majority of narrow, practically decoupled, levels. Our goal in this paper is two-fold. First, in the following section we will consider a simple numerical example which shows that coexistence of narrow and broad resonances is quite natural in ballistic QD-s with relatively strong coupling to the leads. Second, we will consider how the CB and filling of the levels in the dot are changed in the presence of a dominant single broad resonance.

A useful theoretical model for the description of charging effects in QD-s is the tunneling Hamiltonian (see *e.g.* [13])

$$H = \sum_i \varepsilon_i a_i^+ a_i + \frac{U_{CB}}{2}\left(\sum_i a_i^+ a_i - N_0\right)^2 + \sum_k \varepsilon_k^L b_k^{L+} b_k^L + \sum_{k,j}\left(t_j^L a_j^+ b_k^L + h.c.\right) + L \leftrightarrow R\,. \tag{12.3}$$

Here $a(a^+)$ and $b(b^+)$ are the annihilation(creation) operators for electron in the dot and in the lead, L and R stand for the "left" and "right" lead and summation over spin orientations is included everywhere.

Refined theoretical treatment of the charging effect within the Hamiltonian (12.3) includes modern theoretical tools such as bosonisation and mapping onto the Kondo problem [14]. For our purposes, it will be possible to simplify further Eq. (12.3). We will consider the case of only one level N in the dot significantly coupled to the leads (only $t_N^{L,R} \neq 0$). If in addition the width of this level is larger than the single-particle level spacing Δ, a very nontrivial regime of CB may be described by means of simple second order of perturbation theory estimates. Surprisingly this simple limit of CB in the QD with broad level have not been considered yet.

2. SEMI-CHAOTIC QUANTUM DOTS

In this section we will show by explicit numerical example how the anomalously broad levels may appear even in sufficiently large and irregular QD.

A simple example of a system for which the widths differ drastically is the integrable QD [10, 11]. Integrable QD-s having only few $(2 \div 4)$ electrons have been produced recently [15]. However, for large numbers of electrons $N_e \sim 100 \div 1000$ it will be much more difficult to produce an integrable QD[2]. Nevertheless, at least in classical mechanics, one finds a considerable gap left between integrable and fully chaotic systems. Even in the nonintegrable dot there may coexist two kinds of trajectories - quasi-periodic and chaotic. In this case, in 2-dimensions any trajectory (even the chaotic one!) does not cover all the phase space formally allowed by energy conservation. Consequently, the corresponding wave functions do not cover all the area of the QD. If such a regime is or may be realized in QD-s, it easily explains, why the widths of the resonances may vary by orders of magnitude. Moreover, many other features of such a QD may differ strongly from those of the chaotic QD (CB intervals, energy level statistics [21]). To *illustrate* this idea we have performed numerical simulations for a model QD of the size l coupled smoothly to the leads with the potential

$$V = -4x^2 (1 - x/l)^2 + \left(y + 0.2x^2/l\right)^2 \left[1 + 2(2x/l - 1)^2\right] . \qquad (12.4)$$

The dot is strongly nonintegrable, but, similarly to the experimental geometry [4], sufficiently symmetric. For numerical simulations we considered the QD on the lattice having a lattice spacing equal to unity and $l = 40$. The kinetic term is given by the standard nearest-neighbor hopping Hamiltonian. With the hopping matrix element $\tau \approx 40$ our dot has

[2] It is not clear how close to integrable was the QD used in the experiment [4]. However, the QD containing ~ 200 electrons was ~ 50 times smaller than the nominal elastic mean free path. Thus, disorder should not be essential for the dynamics of the electrons.

about 100 electrons (~ 200 with spin). The calculations were performed for the right lead closed by a hard wall at $x = l$ (see Ref. [22] for some more details). Only one mode may propagate along the left lead within the energy interval $5.8 < \varepsilon < 10.3$ (the energy band is $V < \varepsilon < V + 8\tau$). The phase of the electron in the lead was not fixed and we were able to find a solution of the Schrödinger equation at any energy. The probability to find an electron in the dot as a function of energy for fixed in- and out-going current in the lead is shown in Fig. 12.2. Above-barrier resonances with very different widths are seen. Moreover, the width of the resonance #111 turns out to be few times larger than the level spacing. The origin of this hierarchy of widths becomes clear from Fig. 12.3, where we have plotted $|\psi|^2$ in the QD at these resonances. The quantized version of different variants of classical motion may be found on this figure. The most narrow level #113 corresponds to short transverse periodic orbit. Other broader levels, such as #109, may be considered as the projections of the invariant tori corresponding to quasi–periodic classical motion. This classical trajectory reaches the line $V(x, y) = \varepsilon$ only at few points. The presence of different types of (quasi)periodic motion is natural for the nonintegrable dot (see the wave functions #113, #110, #111 and less pronounced #115). The candidates for chaotic classical motion (*e.g.* #114) also correspond to relatively broad resonances. Even in this case only a part of QD is covered by the trajectory.

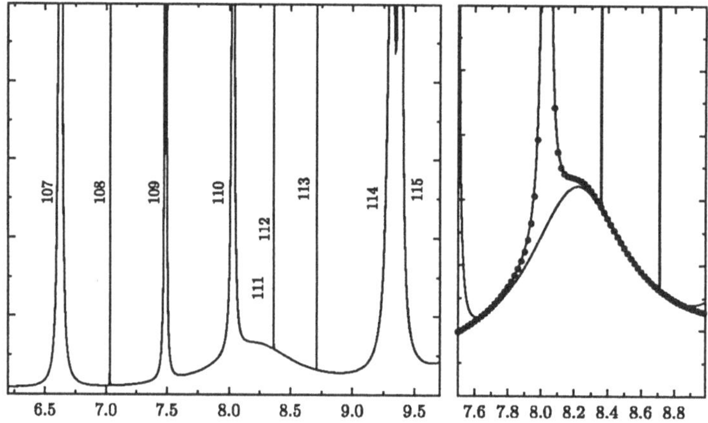

Figure 12.2 The energy dependence of the probability to find the electron in the QD for fixed incoming (outgoing) current. The numbers of the corresponding levels in the dot are shown. Right panel: A part of figure close to the resonance #111 together with the fit by two Breit-Wigner peaks (circles) and single broad peak with $\Gamma = 0.7$.

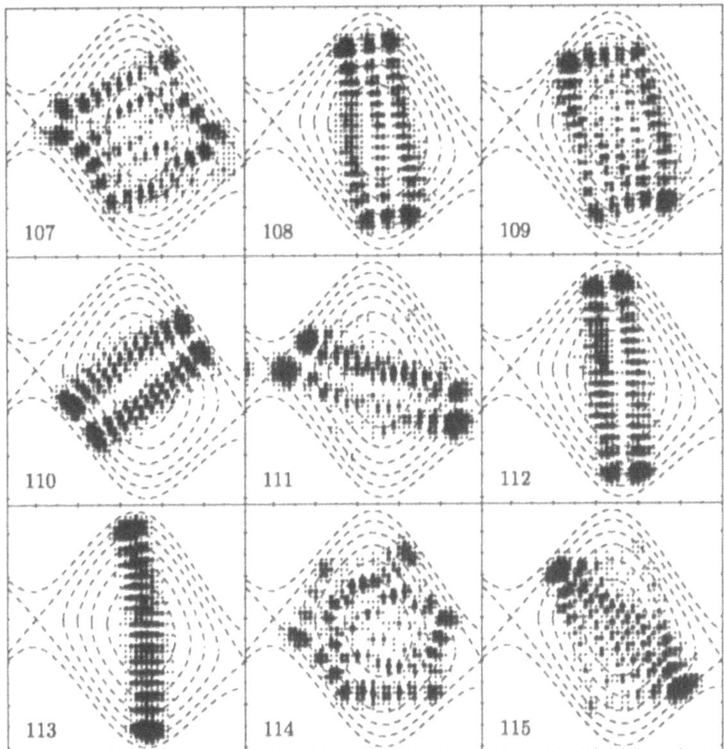

Figure 12.3 The density (~dark area) of electrons in the dot for the nine levels shown in Fig. 12.1. Dashed lines are the lines of constant potential $V = 20, 0, -20, -40, -60$ and -80. The single-channel wire is attached from the left. About 95% of the norm of the wave function in the dot is shown.

Most interesting is the level #111, which is well coupled to the leads. The corresponding classical trajectory covers some invariant tori in the dot starting from the left contact. The Breit-Wigner width of this resonance is $\Gamma \approx 0.7$ (same units as in Fig. 12.2), which is sufficiently larger than Δ.

Essentially the same pictures with "regular" and "chaotic" wave functions may be found in Ref. [21], where also the potential of the QD was treated self-consistently. Other mechanisms which may lead to the "stability" of the broad level will be considered in the next section.

3. COULOMB BLOCKADE FOR A SINGLE BROAD LEVEL

With our numerical *example* we have investigated only the single particle properties of the ballistic nonintegrable QD. Now let us turn to the many–particle effects. First of all, the parameters of the dot are slowly changed by the change of the plunger gate voltage V_g. We describe this effect by letting the levels in the dot flow as

$$\varepsilon_i = \varepsilon_i(V_g = 0) - V_g \ , \quad i = 1, 2, 3... \ . \tag{12.5}$$

The occupied levels are now those with $i \leq 0$. The energies of the electrons in the wire are given by

$$\varepsilon(k) = k^2/2m - E_F. \tag{12.6}$$

Here $k = n\pi/L$ with large integer n and L is the length of the wire. As it was mentioned in the introduction, we suppose that only one (N-th) level in the dot is coupled strongly to the leads. It is enough to consider coupling with only one lead [18]. Let the corresponding matrix element be t_N and

$$\Gamma_N = 2\pi |t_N|^2 dn/d\varepsilon \gg \Delta \ . \tag{12.7}$$

In our example (after the doubling of number of levels due to spin) the values were $\Gamma_N/\Delta \sim 5$. The charging energy is assumed even larger, $U_{CB} \gg \Gamma$. The widths of other levels are much smaller than Δ and may be neglected. Our aim is to show, that transmission of a current at about Γ/Δ consecutive CB peaks will proceed through one and the same level ε_N in the dot.

Let us start with spinless electrons. As long as we assume $t_i = 0$ for all $i \neq N$, the spectrum of the tunneling Hamiltonian (12.3) consists of completely decoupled branches corresponding to different occupation numbers of narrow levels. Let at first only the levels with $i \leq 0$ be occupied. It is easy to find the second order correction to the total energy of the wire and QD, when the N-th level lies far above the Fermi energy ($\varepsilon_N(V_g) \gg \Gamma$)

$$\Delta E_{tot}^{(0)} = \int_0^{k_F} \frac{|t_N|^2}{\varepsilon(k) - \varepsilon_N} \frac{L}{\pi} dk = \frac{-\Gamma}{2\pi} \ln\left(\frac{4E_F}{\varepsilon_N(V_g)}\right) . \tag{12.8}$$

The levels in the wire are lowered due to the repulsion from the unoccupied level ε_N. We will not include into E_{tot} the trivial constant which arose due to occupation of the decoupled levels with $i \leq 0$ and unperturbed electron energies in the lead. Generalization of Eq. (12.8) for the case of negative ε_N, $\varepsilon_N \ll -\Gamma$ (the broad level being below the Fermi energy) is straightforward (note that the level ε_N is occupied, *not the level* ε_1 *as one might expect*):

$$E_{tot}^{(0)} = \varepsilon_N - \frac{\Gamma}{2\pi} \ln\left(\frac{4E_F}{|\varepsilon_N|}\right) . \tag{12.9}$$

Here the first term, $\varepsilon_N < 0$, accounts for the energy loss due to replacement of one electron from lead to the dot. The correction due to the second order of perturbation theory now includes both lowering of levels with $\varepsilon(k) < \varepsilon_N$ and rising of those with $\varepsilon(k) > \varepsilon_N$. The perturbative treatment fails for $|\varepsilon(k) - \varepsilon_N| \leq \Gamma$, but the corresponding shifts of levels below and above ε_N evidently compensate each other, which is equivalent to taking the principal value of the integral in Eq. (12.8).

Finally, the exact solution for a single state interacting with a continuum is also known (see *e.g.* [16]). A precise treatment of this Breit-Wigner-type situation, along the lines of Ref. [17], yields for $E_F \gg \varepsilon_N, \Gamma$:

$$E_{tot}^{(0)} = \frac{-\Gamma}{4\pi}\left[\ln\left(\frac{16E_F^2}{\varepsilon_N^2 + \Gamma^2/4}\right) + 2\right] + \frac{\varepsilon_N}{\pi}\cot^{-1}\frac{2\varepsilon_N}{\Gamma}, \tag{12.10}$$

which coincides with Eqs. (12.8,12.9) for $|\varepsilon_N| \gg \Gamma$.

Let us now consider the branch for which the level ε_1 is occupied. The energy of this electron is $\varepsilon_1(V_g)$. However, adding one more electron via the hopping t_N *now costs* $\varepsilon_N(V_g) + U_{CB}$. The ensuing reduction of the downward shift of the level $E_{tot}^{(1)}$ is of crucial importance. The analog of Eq. (12.8) for $\varepsilon_N + U_{CB} \gg \Gamma$ now reads

$$E_{tot}^{(1)} = \varepsilon_1 - \frac{\Gamma}{2\pi}\ln\left(\frac{4E_F}{\varepsilon_N + U_{CB}}\right) . \tag{12.11}$$

The initial energy of the electron gas in the leads as well as the contribution from ε_i with $i \leq 0$ may be included as an additive constant in Eqs. (12.9-12.11).

Within $-U_{CB} < \varepsilon_{1,N} < 0$, both Eqs. (12.9) and (12.11) are valid. Since $\varepsilon_N > \varepsilon_1$ one expects $E_{tot}^{(1)}$ to be the true ground state in this region. However, because in our case $\Gamma \gg \Delta$ the second term in Eqs. (12.9, 12.11) may invalidate this expectation. For small V_g one has $E_{tot}^{(0)} < E_{tot}^{(1)}$ and the Eqs. (12.8-12.10) describe the true ground state of the system. However, the two functions $E_{tot}^{(0)}(V_g)$ and $E_{tot}^{(1)}(V_g)$ cross at

$$\varepsilon_N(V_g) = -U_{CB}/[\exp\{2\pi(\varepsilon_N - \varepsilon_1)/\Gamma\} + 1] \tag{12.12}$$

and the ground state jumps onto the branch $E_{tot}^{(1)}$. The current-transmitting *virtual* state N now again has a positive energy. Thus, the phase of this transmission has returned to what it was before the process of filling of state N and the subsequent sharp jump into the state where level 1 is filled. It is the latter jump which provides the sharp drop by π of the phase of the transmission amplitude, following the increase by π through the broad resonance. Thus, many ($\sim (\Gamma/\Delta)\ln(U_{CB}/\Gamma)$)

consecutive resonances are due to the transition via one and the same level N.

Taking into account the spin of the electron leads to even more spectacular scenario for charging of the QD with a single broad level. We are not able to discuss in detail the role of spin in this short note. Very briefly, the corrections to the energy in the case of the broad level being above the Fermi energy (12.8,12.11) now simply acquire a factor 2. However, after the broad level crosses the Fermi energy a larger energy gain is achieved (see Eq. (12.9)) if it is occupied by two electrons with opposite spins. Since only one electron may be added to the QD from the lead within one charging event, the second electron is taken to the broad level from one of the narrow levels in the dot. Nevertheless, in this case also the many charging events proceed via one and the same broad resonance, each accompanied by the increase of phase by π which is compensated by the $-\pi$ jump in the valley.

The crossing of energy levels $E_{tot}^{(0)}$ (12.9) and $E_{tot}^{(1)}$ (12.11) becomes avoided if one introduces the small individual width Γ_1 of the narrow level $|1\rangle$.

In the simplified model Eq. (12.5) all levels in the dot are changed in the same way by the plunger voltage. Taking into account the different sensitivities of longitudinal and transverse modes to the plunger [10, 11] may allow to keep our broad level ε_N even longer within the relevant strip of energy. This may provide an explanation of even longer sequences of resonances accompanied by the $-\pi$ jumps.

Generalization of our approach for $N < 0$ (still $|\varepsilon_N| < \Gamma$, the broad level immersed into the sea of occupied narrow levels) is straightforward.

Also in a more refined approach, adding new electrons into the QD should cause a slow change of the self-consistent potential $V(x, y)$. The total energy of the dot and the wire will be lowered in the presence of strongly coupled levels. This may cause the potential of the QD to automatically adjust to allow such levels, which will support our explanation of the experiment of Ref. [4].

Even though we have shown in the Section 3 that the broad levels are natural for the strongly coupled ballistic QD-s, one may wonder, why it happens that in the experiment such a level was found just close enough to the Fermi energy. We see two possible explanations. First, the self-consistent shape of the QD may indeed be automatically adjusted in order to have such a level. Second, the parameters of the QD might have been tuned in the course of preparation of the experiment while opening up the dot.

4. KONDO EFFECT

Recent experiments [23-26] observed the increase of the conductance G of an appropriate QD at low temperature in the CB valleys corresponding to the odd number of electrons, which has stimulated the renewed interest in the Kondo effect in QD-s [18-20].

The correction to G is due to the spin-flip interaction of the unpaired electron in the dot with the leads. For example, at the upper wing of the first charging resonance

$$G_K = G_0 \left[1 + \frac{const \times \Gamma}{E_F - \varepsilon} \ln \left(\frac{U_{CB}}{T} \right) \right], \quad \Gamma \ll E_F - \varepsilon \ll U_{CB}. \qquad (12.13)$$

Here ε is the energy of single particle state in the dot and the unperturbed conductance at the wings of resonance is $G_0 \sim \Gamma_L \Gamma_R / (E_F - \varepsilon)^2$. However, already in the first experiments, effects which do not find an explanation within the straightforward Kondo mechanism were observed [25, 26].

The correction associated with the spin flip blows up at the Kondo temperature $T_K = \sqrt{U_{CB}\Gamma} \exp\{\pi(\varepsilon - E_F)(\varepsilon + U_{CB} - E_F)/2\Gamma U_{CB}\}$ [27]. The natural way to observe the Kondo effect is to fabricate the devices with large T_K. T_K is large in very small (up to ~ 50 electrons in real experiments) and sufficiently open QD-s.

In this section we will consider how the Kondo effect behaves in the case of overlapping resonances. In fact we do not need now to have $\Gamma \gg \Delta$. The interesting new effects appear already if one has two levels ε_1 and ε_2 one broad with $\Gamma \sim \varepsilon_2 - \varepsilon_1$ and another narrow.

As we have seen in the Section 2, the strongly coupled levels in small ballistic QD-s may be associated with the special trajectories covering only the part of area of the QD. In this case it is natural to disregard the approximation of uniform Coulomb interaction (12.3) and to describe the interaction of electrons at different levels by different matrix elements U_{ij}. In particular let us consider the case when the repulsion between the two electrons with opposite spin at the "localized" strongly coupled level U_{11} exceeds all other Coulomb matrix elements. (This model allows to overcome the tendency to occupy the broad level by two electrons at once, considered in the previous section.) The resulting Kondo corrections for the conductance are illustrated in fig. 12.4.

In Fig. 12.4*a*. we have shown schematically the conductance as a function of gate voltage following the occupation of two narrow levels (the usual Kondo effect). The first two similar peaks correspond to occupation of level ε_1, the other two peaks correspond to occupation of the second level ε_2, having slightly different width. At low temperature $T \sim T_K$ the conductance grows up at the odd valleys 1 and 3.

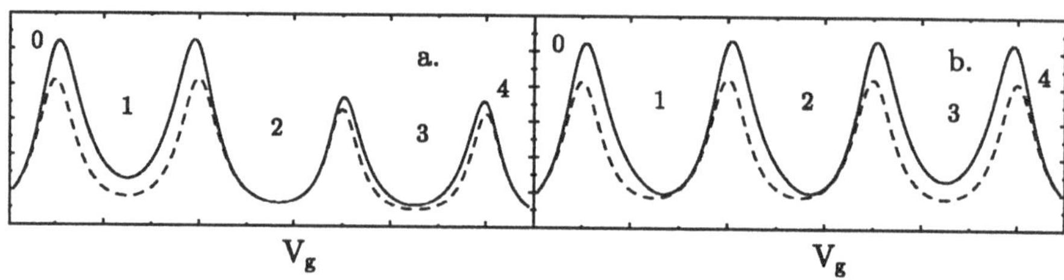

Figure 12.4 The gate voltage dependence of the conductance (schematic) for: (*a*) two narrow resonances ($\Gamma \ll \Delta$), (*b*) one broad and one narrow resonance. Dashed line $T \gg T_K$ (but $\Gamma \gg T$), solid line $T \sim T_K$. The valley number coincides with the number of electrons added to the dot.

Quite different is the Kondo effect in the case of the broad peak shown in Fig. 12.4*b*. At the left peak the broad level crosses the Fermi energy for the first time. Since $\Gamma \sim \varepsilon_2 - \varepsilon_1$ no matter which energy is larger, the broad level is occupied at this resonance. At low $T \sim T_K$ the conductance increases at the right wing of the resonance (1 electron in the QD) and does not change at the left wing (no electrons in the QD). Something of interest happens at the middle of the valley 1. Here, in accordance with the theory developed in the previous section, the electron in the dot jumps from the broad to the narrow level and becomes "invisible" for the exchange with the leads. Thus though formally the spin of the QD equals $s = 1/2$, there is no Kondo effect at the left wing of the second resonance[3]. The second peak is associated again with the population of the broad level by one electron. As for the valley 1, one finds the Kondo effect only in the left half of the second valley. In the middle of this valley the electron configuration is rearranged again and the second electron jumps to the narrow level. After that the occupied narrow level does not play any role in the transmission. The usual Kondo effect takes place in the valley 3 in Fig. 12.4*b*.

In this section we proposed a mechanism which may lead to Kondo effect in even valleys in the QD. Remarkably, something very close to this prediction was observed in the recent experiment in Stuttgart [25]. The authors of Ref. [25] have observed in a few samples the increase of the zero-bias differential conductance for odd number of electrons in the dot (such as the valley 3 on our Fig. 12.4*b*). In addition, in two samples they have found the Kondo–type peak in the preceding valley, which starts at zero bias and is then (with increase of V_g) shifted smoothly to the (small) non-zero bias voltages. In this paper we consider

[3]Some increase of the conductance due to spin exchange still may take place here, but the correction to G is proportional to the small width of narrow level

only the zero bias conductance, but already the zero bias part of this observation of Ref. [25] agrees well with what we have shown on the valley 2 of Fig. 12.4b.[4]. The work on the explanation of the finite bias part (including nonequilibrium effects) is in progress.

5. CONCLUSIONS

To conclude, motivated by the measurements of the transmission phase in the QD [4, 5] we investigated the charging effects in the QD with a single anomalously broad level. Contrary to the common expectation the pronounced CB takes place here even for $\Gamma \gg \Delta$. In this case upon increasing V_g, it is energetically favorable to first populate in the dot the level strongly coupled to the leads. At a somewhat larger V_g a sharp jump occurs to a state where the "next in line" narrow level becomes populated. This jump accounts for the sharp decrease by $\sim \pi$ of the transmission phase observed in the experiment. The absence of significant fluctuations of the strength of resonances seen in the experiment [4] and their large widths are clear within our mechanism. The current transmission through such QD resembles the behavior of rare earth elements, whose chemical properties are determined not by the electrons with highest energy, but by the "strongly coupled" valence electrons. An explicit numerical example shows how such broad levels may appear in ballistic (non–integrable) QD-s.

The overlapping of single-particle resonances may take place also in the Kondo experiments in QD-s, where in order to increase the Kondo temperature the dot is usually sufficiently opened. New low-temperature effects take place in this case, including the Kondo effect in even valleys. This may provide an explanation of the unexpected results of recent experiments [25, 26].

Acknowledgements

The work of PGS was supported by RFBR, grant 98-02-17905. Work at WIS was supported by the Albert Einstein Minerva Center for Theoretical Physics and by grants from the German-Israeli Foundation (GIF), the Bond fund and the Israel Science Foundation, Jerusalem.

References

[1] L. P. Kouwenhoven *et al., Mesosopic Electron Transport, Proceedings of the NATO ASI,* edited by L. L. Sohn, L. P. Kouwenhoven

[4]The "elementary event" shown on the Fig. 12.4*b*. consists of the sequence of 4 peaks (valleys), but Ref. [25] reports measurements for two valleys only

and G. Schön (Kluwer 1997).
[2] C. W. J. Beenakker, Rev. Mod. Phys. **69**, 731 (1997).
[3] T. Guhr, A. Müller-Groeling and H. A. Weidenmüller, Phys. Rep. **299**, 189 (1998).
[4] E. Schuster *et al.*, Nature, **385**, 417 (1997);
[5] A. Yacoby *et al.*, Phys. Rev. Lett., **74**, 4047 (1995).
[6] E. Buks *et al.*, Physica, **249-251**, 295, 1998; D. Sprinzak and M. Heiblum, unpublished.
[7] P. S. Deo and A. M. Jayannavar, Mod. Phys. Lett. **10**, 787 (1996)
[8] H. Xu and W. Sheng, Phys. Rev. **B 57**, 11903 (1998); C.-M. Ryu and S. Y. Cho, Phys. Rev. **B 58**, 3572 (1998); H.-W. Lee, Phys. Rev. Lett. **82**, 2358 (1999).
[9] Y. Oreg and Y. Gefen, Phys. Rev. **B 55**, 13726 (1997).
[10] G. Hackenbroich, W. D. Heiss and H. A. Weidenmüller, Phys. Rev. Lett. **79**, 127 (1997).
[11] R. Baltin *et al.*, cond-mat/9807286.
[12] R. Baltin and Y. Gefen, cond-mat/9907205.
[13] *Single charge tunneling: Coulomb blockade phenomena in nanostructures, NATO ASI series*, edited by H. Grabert and M. H. Devoret (Plenum press 1992).
[14] K. A. Matveev, Sov. Phys. JETP **72**, 892 (1991); Phys. Rev. **B 51**, 1743 (1995).
[15] S. Tarucha *et al.*, Phys. Rev. Lett. **77**, 3613 (1996).
[16] A. Bohr and B. R. Mottelson, *Nuclear Structure*, Benjamin, New York **1**, 284 (1969).
[17] L. D. Landau and E. M. Lifschitz *Quantum Mechanics*, p. 555, Pergamon, Oxford (1976).
[18] L. I. Glazman and M. E. Raikh, JETP Lett. **47**, 452 (1988)
[19] T. K. Ng and P. A. Lee, Phys. Rev. Lett. **61**, 1768 (1988)
[20] N. S. Wingreen and Y. Meir, Phys. Rev. **B 49**, 11040 (1994).
[21] M. Stopa, Physica **B 251**, 228 (1998).
[22] P. G. Silvestrov and Y. Imry, cond-mat/9903299
[23] D. Goldhaber-Gordon *et al.*, Nature **391**, 156 (1998); Phys. Rev. Lett. **81**, 5225 (1998).
[24] S. M. Cronenwett, T. H. Osterkamp, and L. P. Kouwenhoven, Science **281**, 540 (1998).
[25] J. Schmid *et al.*, Physica B **256-258**, 182 (1998).
[26] F. Simmel *et al.*, Phys. Rev. Lett. **83**, 804 (1999).
[27] F. D. M. Haldane, Phys. Rev. Lett. **40**, 416 (1979).

III
DEPHASING AND SHOT NOISE

Chapter 13

DEPHASING AND SHOT-NOISE IN MESOSCOPIC SYSTEMS

Y. Levinson and Y. Imry
Weizmann Institute of Science, Rehovot 76100, Israel

Abstract The general principles of dephasing of quantum-mechanical interference by coupling to the environment degrees of freedom are summarized. A general expression for the dephasing rate is obtained and used to rederive known results for dirty conductors. The theory of quantum shot noise within the Landauer transport formulation is reviewed and explained. A recent example of dephasing by a current-carrying system is discussed in some detail. The detectability of the zero-temperature and finite frequency noise by coupling to an electromagnetic field is examined. Finally, the problem of the apparently finite dephasing rate that was recently reported at very low temperatures in several systems is considered. We show that the "standard model" of a conductor with static defects can not have such an effect. However, allowing some dynamics of the defects may significantly increase the dephasing rate at low temperatures.

1. INTRODUCTION

Mesoscopic physics [1]-[10] deals with the realm which is in-between the microscopic (atomic and molecular) scale and the macroscopic one. It can give us fundamental information on the crossover between the microscopic (quantum), and the macroscopic regimes. Macroscopic-type electrical measurements can be sensitive to nontrivial quantum phenomena in mesoscopic samples at low temperatures, where the phase coherence of the electrons is preserved over the relevant scales. One of the important insights gained from these studies is that elastic scattering *does not* destroy phase coherence. It takes *inelastic scattering* – changing the quantum state of other degrees of freedom ("the environment"), to do that. This paper is devoted to the detailed consideration of how the phase coherence of the electrons is lost by the coupling to the envi-

I. O. Kulik and R. Ellialtioğlu (eds.),
Quantum Mesoscopic Phenomena and Mesoscopic Devices in Microelectronics, 185–210.

ronment [11, 12, 13]. In particular, dephasing by shot-noise fluctuations is reviewed. In order to do that, the theory of quantum shot-noise is presented.

In Section 2, we discuss the general principles of dephasing of quantum-mechanical interference by coupling to the degrees of freedom which do not directly participate in the interference. We derive a useful expression for the dephasing rate, and apply it to linear transport in disordered conductors. The theory of quantum shot noise is presented qualitatively in Section 3 and explained more formally in Section 4. The recent Weizmann experiment on dephasing by a current-carrying quantum point contact detector is then discussed in Section 5. More recent developments, mainly having to do with dephasing by phase shifting edge states, are presented in Section 6. In Section 7, the detectability of the zero-temperature noise is discussed. In Section 8 the recent issue of the apparent saturation of the dephasing rate as $T \to 0$ is reviewed. While this does not exist in the usual model with static defects, we show that allowing for some motion of the defects may produce this effect.

2. GENERAL PICTURE OF QUANTUM-MECHANICAL DEPHASING

One of the principal insights gained from mesoscopic physics is that it takes *inelastic* scattering[1] off other degrees of freedom (the "environment"), to eliminate phase coherence of the degree of freedom considered. This is characterized by the "phase-breaking" time, τ_ϕ, for that degree of freedom – the conducting electron in a typical mesoscopic case.

Two descriptions have been used for the dephasing process. The first relies on the electron leaving a trace, in the form of an excitation of the environment, which identifies which path it took. The second description is based on the phase uncertainty induced on the electron by the fluctuating environment (what counts physically is the uncertainty of the *relative* phases of the paths). In Ref. [14] these two descriptions were proven to be equivalent, which turned out to be a useful insight.

Some important physical comments:

1. The phase uncertainty remains constant when the interfering wave does not interact with the environment. Thus, if a trace is left by a partial wave on its environment, this trace cannot be wiped out after the interaction is over. The proof of this statement follows from unitarity

[1] Here the term "inelastic" implies just changing the quantum state of the environment. It is irrelevant how much energy is transferred in this process. This includes zero energy transfer – flipping the environment to a degenerate state.

[14]. Interesting subtleties were discussed in Refs. [14, 15]. This is relevant for the dephasing problem treated in the Section 5.

2. If the same environment interacts with the two interfering waves and when both waves emit the same excitation of the medium, each of the partial waves' phases becomes uncertain, but the relative phase is unchanged. A well-known example is that of "coherent inelastic neutron scattering" in crystals (see *e.g.* Ref. [16]). This process is due to the coherent addition of the amplitudes for the neutron exchanging *the same* phonon with *all* scatterers in the crystal.

3. Long-wave excitations (phonons, photons) may not dephase the interference. It might be thought that this is because of their energy being too low to cause dephasing. However, the amount of energy transferred is *not* the issue, but rather whether an excitation of the environment which can identify the path has occurred. If the dephasing time is shorter than the time it takes to produce this excitation, its effectiveness will be lost [17]. Another important issue is that the excitation must influence the *relative* phase of the paths. An equivalent way to state this (see Refs. [13, 14]) is that, as in the Heisenberg microscope, the radiation with wavelength λ can not resolve the two paths if their separation is smaller than λ.

4. Dephasing may occur by coupling to a discrete or a continuous environment. In the latter case, the excitation *may* move away to infinity and the loss of phase can be regarded as, practically speaking, irreversible. However, in special cases it is possible, even in the continuum case, to have a finite probability to reabsorb the created excitation and thus retain coherence. This happens, for example, in a quantum interference model due to Holstein for the Hall effect in insulators.

We now derive a useful general expression for the dephasing rate in terms of known properties of a given system. In impure (realistic) conductors, the issue is the dephasing of electrons performing diffusive motion due to defects, just above the Fermi energy, and interacting strongly with all the other electrons. This electron-electron interaction provides in many cases the dominant dephasing mechanism. It is easy then to obtain the dephasing rate from the strength of the inelastic scattering of the considered electron by the electron sea, *i.e.* using the trace left in the environment [14]. A similar result was first obtained in Ref. [17] by using the effect of the electromagnetic fluctuations due to the electron gas on the considered electron. The equivalence of these two points of view is guaranteed by the fluctuation-dissipation theorem.

The Coulomb interaction of the interfering electron with the rest of the electrons is, in the interaction picture,

$$\hat{V}_I(\mathbf{x},t) = \int \frac{\hat{\rho}_I(\mathbf{r}',t)d^3r'}{|\mathbf{x}-\mathbf{r}'|} \tag{13.1}$$

where $\hat{\rho}_I(r,t) = e\Sigma_i\delta(r-\hat{y}_I^i(t))$. For the brevity of the following expressions, we first consider only the interaction of the electron bath with the right partial wave of the interfering electron, we omit the corresponding subscript, and we begin by assuming (this will be relaxed later, where the averaging over the ground state will be replaced by the appropriate averaging, for example at a finite temperature) that the electron bath is initially in its ground state, $|0\rangle$. Assuming that the "left" partial wave does not interact with the electron bath, the intensity of the interference pattern is reduced by the probability that the bath's state coupled to the "right" partial wave becomes different from $|0\rangle$. Up to second order in the interaction, this probability is

$$P = \frac{1}{\hbar^2}\sum_{|n\rangle\neq|0\rangle}\int_0^{\tau_0} dt\int_0^{\tau_0} dt'\langle 0|V_I(\mathbf{x}(t),t)|n\rangle\langle n|V_I(\mathbf{x}(t'),t')|0\rangle \tag{13.2}$$

For the ground state $\langle 0|\hat{\rho}|0\rangle = 0$, so that the summation in Eq. (13.2) can be extended to include all states. We neglect the changes in the paths $x(t)$ due to the interaction, thus only the phase due to the latter is taken into account. We now express P in terms of the response of the environment. Using equation (13.1) and the convolution theorem, $\int d^3r' f(\mathbf{r}-\mathbf{r}')g(\mathbf{r}') = \frac{1}{(2\pi)^3}\int d^3q f_{\mathbf{q}} g_{\mathbf{q}} e^{-i\mathbf{q}\cdot\mathbf{r}}$ where $f_{\mathbf{q}}$ and $g_{\mathbf{q}}$ are the Fourier transforms of f and g, we write:

$$P = \frac{1}{\hbar^2(2\pi)^6}\int_0^{\tau_0} dt\int_0^{\tau_0} dt'\int d^3q\int d^3q'\frac{4\pi e}{q^2}\frac{4\pi e}{q'^2}\langle\rho_{\mathbf{q}}(t)\rho_{\mathbf{q}'}(t')\rangle e^{i\mathbf{q}\cdot\mathbf{x}(t)-i\mathbf{q}'\cdot\mathbf{x}(t')} \tag{13.3}$$

We assume translational invariance

$$\langle\rho_q\rho_{q'}\rangle = \frac{(2\pi)^3}{Vol}\delta(\mathbf{q}+\mathbf{q}')\langle\rho_{\mathbf{q}}\rho_{-\mathbf{q}}\rangle \tag{13.4}$$

(for a finite system the q's are discrete and one just has $\delta_{qq'}$. Going to the continuum the Kronecker delta is replaced by $\frac{(2\pi)^3}{Vol}$ times the Dirac delta). By performing one q integration and inserting a complete set of intermediate states, we obtain:

$$P=\frac{1}{Vol(2\pi)^3\hbar^2}\sum_{|n\rangle}\int_0^{\tau_0}dt\int_0^{\tau_0}dt'\int d^3q\frac{(4\pi e)^2}{q^4}\langle 0|\rho_I^{\mathbf{q}}(t)|n\rangle\langle n|\rho_I^{\mathbf{q}}(t')|0\rangle e^{i\mathbf{q}\cdot(\mathbf{x}(t)-\mathbf{x}(t'))} \tag{13.5}$$

By transforming into Schrödinger picture operators and inserting a dummy integration variable ω, P can be rewritten in the following form:

$$P = \frac{1}{Vol(2\pi)^3\hbar^2}\sum_{|n\rangle}\int_0^{\tau_0}dt\int_0^{\tau_0}dt'\int d^3q\int d\omega\frac{(4\pi e)^2}{q^4}\left|\langle 0|\rho_S^{\mathbf{q}}|n\rangle\right|^2$$
$$\times\delta(\omega-\omega_{n0})e^{i\mathbf{q}\cdot(\mathbf{x}(t)-\mathbf{x}(t'))-i\omega(t-t')} \tag{13.6}$$

The usefulness of that Eq. (13.6) stems from its relation to the linear response expression for dynamic structure factor $S(\mathbf{q},\omega)$ and the imaginary part of the complex dielectric function $\mathrm{Im}\left(\frac{1}{\epsilon(\mathbf{q},\omega)}\right)$. These quantities are related by the fluctuation-dissipation theorem:

$$\mathrm{Im}\left(\frac{1}{\epsilon(\mathbf{q},\omega)}\right)=\frac{4\pi^2e^2}{Vol\,q^2\hbar}\sum_{|n\rangle}|\langle 0|\rho_S^q|n\rangle|^2\;\delta(\omega-\omega_{n0})=\frac{4\pi^2e^2}{q^2}S(\mathbf{q},\omega) \tag{13.7}$$

Thus, Eq. (13.6) becomes,

$$P=\frac{1}{\hbar(2\pi)^3}\int_0^{\tau_0}dt\int_0^{\tau_0}dt'\int d^3q\int d\omega\frac{4e^2}{q^2}\mathrm{Im}\left(\frac{1}{\epsilon(\mathbf{q},\omega)}\right)e^{i\mathbf{q}\cdot[\mathbf{x}(t)-\mathbf{x}(t')]-i\omega(t-t')} \tag{13.8}$$

We now notice that $e^{i\mathbf{q}\cdot\mathbf{x}(t)}$ is the qth Fourier component of the electron's density at time t. We should average the product of the latter with $e^{-i\mathbf{q}\cdot\mathbf{x}(t')}$ over all appropriate classical paths $x(t)$. This semiclassical averaging can be argued to be equivalent to the equilibrium, $T=0$, one and hence to yield the dynamic correlation function of these density operators, which is the Fourier transform of the well-known dynamic structure factor [18] $S_p(q,\omega)$ of the diffusing particle. Thus:

$$1/\tau_\phi=\int\int d\mathbf{q}d\omega|V_q|^2S_p(q,\omega)S_{env}(-q,-\omega). \tag{13.9}$$

where $S_p(q,\omega)$ is the dynamic structure factor of the diffusing electron (a Lorentzian with width Dq^2) and $S_{env}(-q,-\omega)$ is the same for the environment. Thus, we expressed $1/\tau_\phi$ in terms of the fluctuations of the environment. The FDT can be used to replace the dynamic structure factor by the dissipative part of the response function of the environment in the last expression. This would yield the expression for $1/\tau_\phi$ via the probability to excite the environment. These two expressions are exactly equivalent, which is consistent with the theorem of Ref. [14]. Equation (13.9) obviously implies that the total rate for inelastic scattering is given by the sum over all $(\mathbf{q}, \omega)$ of the rate for transfer of momentum $\hbar\mathbf{q}$ and energy $\hbar\omega$ between the electron and the environment. These elementary rates are added incoherently.

The, by now well known, pioneering results of Ref. [17] for linear transport dephasing in dirty conductors are very easily obtained by straightforwardly evaluating the integrals in Eq. (13.9) in the various dimensions. One finds in 3D:

$$1/\tau_\phi \sim T^{3/2}. \tag{13.10}$$

For 2D and 1D (thin films and wires) the integrations over the appropriate components of $\mathbf{q}$ are replaced by summations and it is found that the remaining integrations are infrared (small q)-divergent. A careful evaluation of the phase *difference* of two paths shows that this divergence is cured by a cutoff whose physical meaning is exactly that a low q excitation can not distinguish paths that are separated in space by less than $1/q$. This is in agreement with the "Heisenberg microscope"-type argument [13, 14] mentioned above. A different way, which yields similar results, to cure the divergence [17] is by introducing an infrared cutoff (of $1/\tau_\phi$) to the ω integration. This is justified by the inability to transfer an energy $\hbar\omega$ within a time shorter than $1/\omega$. The results are:

$$1/\tau_\phi \sim T \quad \text{(in 2D)}. \qquad (1/\tau_\phi) \sim T^{2/3} \quad \text{(in 1D)}. \tag{13.11}$$

These results are in a *quantitative agreement* with experiments.

3. PHYSICAL DERIVATION OF THE SHOT NOISE FROM THE LANDAUER FORMULATION

3.1 Model And Preliminaries

We consider here the simplest Landauer model for single-channel transport between two infinite electron reservoirs whose electrochemical potentials differ by eV. They are ideally connected via a 1D wire of length L, the wire is pure, except its having at $x = 0$ an obstacle whose transmission and reflection amplitudes from the left are t and r and from the right t' and r'. The scattering matrix of the obstacle is unitary. On the wire we have the scattering states coming from the left ($|l\rangle$) and from the right ($|r\rangle$). These states are *assumed* to be populated according to the equilibrium distribution of the left- and right- reservoir respectively. With these assumptions, one easily obtains the two-terminal Landauer formula for the conductance, G:

$$G = \frac{e^2}{\pi\hbar}T, \tag{13.12}$$

where T is the transmission coefficient of the obstacle. We denote the reflection coefficient by $R = 1 - T$.

We shall need the matrix elements of the current operator between the above states. As in Section 4, we normalize our states in a length $\mathcal{L} \leq L$. For $\mathcal{L}$ much larger than all microscopic lengths, the results become independent of $\mathcal{L}$. For states having the same k one finds:

$$|j_{ll}|^2 = |j_{rr}|^2 = (evT/\mathcal{L})^2; \qquad |j_{rl}|^2 = |j_{lr}|^2 = (ev/\mathcal{L})^2 TR, \tag{13.13}$$

where the indices r and l stand for scattering states coming from the right and from the left respectively. These matrix elements will be essentially the same as above for two states having k and k' such that $(k-k')\mathcal{L} \ll 1$, and they can be taken to vanish at the opposite limit. Energywise, the condition for constancy of the matrix elements reads:

$$E(k) - E(k') \ll \hbar v_F/L, \tag{13.14}$$

where v_F is the Fermi velocity. This will follow taking eV and $\hbar\omega$ to satisfy the inequality:

$$eV, \hbar\omega \ll \hbar v_F/L. \tag{13.15}$$

The above results for the matrix elements can be obtained from the expression for the current operator in Section 4. For a simple derivation valid for low frequencies, given by Eq. (13.15), one may note that the current is approximately conserved along the wire. Therefore its matrix elements are given by those of the velocity operator times the charge density $e/\mathcal{L}$.

As is well known, the power spectrum of the current noise, $S_{jj}(\omega)$ is given by the Fourier transform of the $\langle j(0)j(t)\rangle$ correlation function. This can be shown to be given (as in Ref. [18]) by:

$$S_{jj}(\omega) = \hbar \sum_i P_i |\langle f|j|i\rangle|^2 \delta(E_i - E_f - \hbar\omega). \tag{13.16}$$

Here, P_i is the population of the initial state. At zero temperature, only the many-body ground state, $|g\rangle$ appears with $P_g = 1$. In the independent electron, or quasiparticle, approximation one sums in Eq. (13.16) over all possible electron-hole excitations. This is often verbally described as exciting from the single-particle state $|i\rangle$ below the Fermi energy to state $|f\rangle$ above it.

3.2 Equilibrium Zero Temperature Noise

At equilibrium the sum in Eq. (13.16) has, as mentioned above, to be taken over initial states below the Fermi level and final states above it. The δ-function in Eq. (13.16) constrains these two states to differ by an energy $\hbar\omega$. As long as the matrix elements are approximately constant, this will give

$$S_{jj}(\omega) = G\omega/\pi, \tag{13.17}$$

for $\omega < 0$, and $S_{jj}(\omega) = 0$ otherwise. We replaced each sum over states by an integration with the 1D density of states[2] $\mathcal{L}/2\pi\hbar v_F$ and added a factor of 2 due to the spin. The matrix elements will cut-off the above once $\omega \gg v_F/\mathcal{L}$. In real systems, $\mathcal{L}$ is limited by the elastic mean free path l, and this yields the usual noise cutoff of the inverse elastic mean free time. We do not discuss here the observability of this zero point noise, it will be discussed in Section 7.

3.3 Nonequilibrium Shot-Noise

At zero temperature, the only new terms added by the voltage V to Eq. (13.16) are those corresponding to initial $|l\rangle$ states and final $|r\rangle$ states within the interval eV. Using the appropriate matrix elements from Eq. (13.13) in Eq. (13.16), we find under all the above assumptions for the nonequilibrium shot-noise contribution, for $\hbar|\omega| < eV$:

$$S_{jj}(\omega) = \frac{1}{\pi}(eV - \hbar|\omega|)(e^2/\pi\hbar)TR \to eIR/\pi, \tag{13.18}$$

where we took the low-frequency limit for the last expression, and for $\hbar|\omega| > eV, S_{jj}(\omega) = 0$. This is the well-known Khlus-Lesovik [20] result.

An important remark is that even in the case where the $|r\rangle$ states are not occupied, and one might have naively thought that they should play no role, they must be included in the calculation. This is due to the fact that one *must* use a complete set of states, spanning the whole Hilbert space, in order to faithfully represent the current operator, as discussed in Section 4.

---

[2] The scattering states satisfy no boundary condition in the length $\mathcal{L}$. But the density of states is independent of the boundary conditions and is proportional to $\mathcal{L}$ for large $\mathcal{L}$.

3.4 Effects Of Temperature

The straightforward effect of the temperature is to introduce nonzero P_i for initial states that are within k_BT of the ground state in Eq. (13.16). In the independent particle picture, this is handled by the usual Fermi factors. This replaces ω on the RHS of Eq. (13.17) by the usual thermal factor

$$\omega + \frac{2\omega}{\exp(\hbar\omega/k_BT) - 1}. \tag{13.19}$$

The latter becomes $2k_BT$ for $\hbar\omega \ll k_BT$. For the shot noise, the $T = 0$ result holds as long as $eV \gg k_BT$, and the result goes over to the equilibrium one in the opposite limit.

More subtle is the effect on the shot-noise of inelastic scattering, which leads to energy relaxation.[3] Let us assume for simplicity a single type of inelastic scattering which creates both a local temperature and electrochemical potential (as is the case with electron-electron scattering) on its length scale L_ϵ, and assures that the local temperature equals that of the bath (*e.g.* such as with electron-phonon scattering). The distinction between these two types of inelastic scattering is important but will not be discussed here. The relevant case for us here is when $L_\epsilon \ll L_{tot}$, L_{tot} being the total length of the system. For a rough physical understanding, one can use a Landauer-type picture, for a section of the system with $L \sim L_\epsilon$, taking the electrochemical potential and the temperature to be constants over lengths up to L_ϵ and the voltage imposed on this section by the rest of the system given by $V(L_\epsilon) = IG(L_\epsilon) = V_{tot}(L_\epsilon)/L \ll V_{tot}$. Once the temperature is high enough such that $V(L_\epsilon) \ll k_BT$, the shot noise essentially disappears. On the other hand, at lower temperatures, such that $V(L_\epsilon) \gg k_BT$, each L_ϵ-section of the system has its full shot noise. However, for the whole system, one must look at the voltage fluctuations which are additive over the sections. For each L_ϵ-section, the voltage low-frequency shot-noise is smaller by a factor L_ϵ/L than the full $T = 0$ shot noise over the whole system. These fluctuations are uncorrelated among different L_ϵ-sections. Thus the total rms shot-noise on the system is reduced by a factor of $(L_\epsilon/L_{tot})^{1/2}$ for $L_\epsilon \ll L_{tot}$. The shot-noise obviously vanishes in the large L_{tot} limit.

[3] We consider scattering which takes the system towards equilibrium, and not just causes dephasing of the wavefunction. It is believed that since the shot-noise is *not* an interference phenomenon, just losing the coherence of the wavefunction is not the relevant issue.

4. FORMAL DERIVATION OF THE SHOT NOISE FROM SCATTERING STATES FORMULATION

Consider a 1D Schrödinger equation $H\psi = \epsilon\psi$ with a Hamiltonian ($\hbar = 1$) $H = -(1/2m)(d/dx)^2 + U(x)$, where the barrier potential $U(x) \to 0$ at $x \to \pm\infty$. For each $\epsilon > 0$ we define two scattering states, χ_k and ϕ_k, coming from the left and right, respectively, which satisfy the Schrödinger equation and the following boundary conditions

$$x \to -\infty: \quad \chi_k(x) = \frac{1}{\sqrt{\mathcal{L}}}(e^{+ikx} + r_k e^{-ikx}), \quad \phi_k(x) = \frac{1}{\sqrt{\mathcal{L}}} t_k e^{-ikx}, \tag{13.20}$$

$$x \to +\infty: \quad \phi_k(x) = \frac{1}{\sqrt{\mathcal{L}}}(e^{-ikx} + \tilde{r}_k e^{+ikx}), \quad \chi_k(x) = \frac{1}{\sqrt{\mathcal{L}}} t_k e^{+ikx}. \tag{13.21}$$

Here we use k instead of $\epsilon \equiv k^2/2m$, with $k > 0$. $\mathcal{L}$ is the normalization length, on which the final results do not depend, once it is much larger than all microscopic lengths. t_k is the transmission amplitude, r_k and r'_k are the reflection amplitudes from left and from right.

The *time-dependent* field operator can be represented in the following way

$$\Psi(x,t) = \sum_k [a_k \chi_k(x) + b_k \phi_k(x)] e^{-i\epsilon_k t}, \tag{13.22}$$

where a_k and b_k are Fermi or Bose operators of particles in the left reservoir a at $x = -\infty$ and the right reservoir b at $x = +\infty$, respectively.

The *time-dependent* current operator at point x is

$$j(x,t) = -\frac{ie}{2m} \Psi(x,t)^+ \nabla \Psi(x,t) + h.c., \tag{13.23}$$

where e is the particle charge. Introducing the field operator from Eq. (13.22) we have

$$j(x,t) = \sum_{kk'} (a^+_{k'} a_k A_{k'k} + b^+_{k'} b_k B_{k'k} + a^+_{k'} b_k C_{k'k} + b^+_{k'} a_k C^*_{kk'}) e^{-i(\epsilon_k - \epsilon_{k'})t}, \tag{13.24}$$

where A, B, C are bilinear combination of scattering states

$$\begin{aligned} A_{k'k} &= -\frac{ie}{2m}(\chi^*_{k'} \nabla \chi_k - \chi_k \nabla \chi^*_{k'}) = A^*_{kk'}, \\ B_{k'k} &= -\frac{ie}{2m}(\varphi^*_{k'} \nabla \varphi_k - \varphi_k \nabla \varphi^*_{k'}) = B^*_{kk'}, \\ C_{k'k} &= -\frac{ie}{2m}(\chi^*_{k'} \nabla \varphi_k - \varphi_k \nabla \chi^*_{k'}), \end{aligned} \tag{13.25}$$

which depend on x.

To calculate the current one averages the current operator over the states of reservoirs a and b using

$$\langle a_k^+ b_{k'}\rangle = \langle b_k^+ a_{k'}\rangle = 0, \quad \langle a_k^+ a_{k'}\rangle = \delta_{kk'} n_k^a, \quad \langle b_k^+ b_{k'}\rangle = \delta_{kk'} n_k^b. \quad (13.26)$$

The first equations means that particles in different reservoirs do not correlate. In the second equations n^a and n^b are the Fermi or Bose distributions in reservoirs a and b with temperatures $T_{a,b}$ and chemical potentials $\mu_{a,b}$. The result of averaging is

$$\langle j(x)\rangle = \sum_k [n_k^a A_{kk}(x) + n_k^b B_{kk}(x)]. \quad (13.27)$$

The diagonal combinations A_{kk} and B_{kk} and hence the current Eq. (13.27) does not depend on x. Calculating these combinations at infinity, using the asymptotics of the scattering states, one finds $A_{kk} = -(ev_k/\mathcal{L})|t_k|^2 = -B_{kk}$, where $v_k = k/m$. As a result the current through the barrier is

$$\langle j\rangle = -e\sum_k \frac{v_k}{\mathcal{L}}|t_k|^2(n_k^a - n_k^b) = -e\int \frac{d\epsilon}{2\pi}|t(\epsilon)|^2[n^a(\epsilon) - n^b(\epsilon)]. \quad (13.28)$$

The second expression is obtained replacing $\sum_k = \int_0^\infty dk/2\pi$. For a small voltage bias $eV = \mu_a - \mu_b$ and zero temperature bias $T_a = T_b$ in the case of electrons this gives the Landauer formula. We write $n_{\epsilon_k}^{a,b} = n(\epsilon_k - \delta\mu_{a,b})$, where $\delta\mu_{a,b} = \mu_{a,b} - \mu$, and $n(\epsilon) = [e^{(\epsilon-\mu)/T} + 1]^{-1}$. Expanding the Fermi distributions one finds

$$\langle j\rangle = GV, \qquad G = \frac{e^2}{2\pi}\int d\epsilon \left(-\frac{\partial n(\epsilon)}{\partial\epsilon}\right)|t(\epsilon)|^2, \quad (13.29)$$

where G is the Landauer conductance.

In the same way one can calculate the current noise, described by the current correlator.

$$\langle\langle \delta j(1)\delta j(2)\rangle\rangle = \frac{1}{2}\langle j(1)j(2) + j(2)j(1)\rangle - \langle j(1)\rangle\langle j(2)\rangle, \quad (13.30)$$

where the short notation means current operators defined by Eq. (13.23) $j(1) \equiv j(x_1 t_1)$ and $\delta j(1) = j(1) - \langle j(1)\rangle$. Here we displayed the symmetrized version of the current correlator. Comments on this procedure will be given in Section 7. To perform the averages in the current correlator we use

$$
\begin{aligned}
\langle a_k^+ a_{k'} a_{p'}^+ a_p \rangle &= \delta_{kk'} n_k^a \delta_{pp'} n_p^a + \delta_{kp} n_k^a \delta_{k'p'} (1 \mp n_{k'}^a), \\
\langle b_k^+ b_{k'} b_{p'}^+ b_p \rangle &= \delta_{kk'} n_k^b \delta_{pp'} n_p^b + \delta_{kp} n_k^b \delta_{k'p'} (1 \mp n_{k'}^b), \\
\langle a_k^+ a_{k'} b_{p'}^+ b_p \rangle &= \delta_{kk'} n_k^a \delta_{pp'} n_p^b, \\
\langle b_k^+ b_{k'} a_{p'}^+ a_p \rangle &= \delta_{kk'} n_k^b \delta_{pp'} n_p^a, \\
\langle b_k^+ a_{k'} a_{p'}^+ b_p \rangle &= \delta_{kp} n_k^b \delta_{k'p'} (1 \mp n_{k'}^a), \\
\langle a_k^+ b_{k'} b_{p'}^+ a_p \rangle &= \delta_{kp} n_k^a \delta_{k'p'} (1 \mp n_{k'}^b),
\end{aligned}
\tag{13.31}
$$

where the upper sign is for fermions and the lower for bosons. Using these averages one finds

$$
\begin{aligned}
&\langle\langle \delta j(1) \delta j(2) \rangle\rangle = \\
&\frac{1}{2} \sum_{kk'} [n_k^a (1 \mp n_{k'}^a) A_{k'k}(1)^* A_{k'k}(2) + n_k^b (1 \mp n_{k'}^b) B_{k'k}(1)^* B_{k'k}(2) + \\
&\quad + n_k^b (1 \mp n_{k'}^a) C_{k'k}(1)^* C_{k'k}(2) + n_k^a (1 \mp n_{k'}^b) C_{kk'}(1) C_{kk'}(2)^*] \\
&\qquad \exp\{-i(\epsilon_{k'} - \epsilon_k)(t_1 - t_2)\} + c.c.
\end{aligned}
\tag{13.32}
$$

Contrary to the current the current correlator contains combinations A, B, C nondiagonal in k, which depend on x and hence the correlator depends on x_1 and x_2. Only for low frequencies, smaller than the inverse time of flight, one can expect, and we shall show it later, that the correlator will be independent on x_1 and x_2.

To get some insight we consider fermions in a case of a "simple" barrier, when the height of the barrier is of the order of ϵ_F its length d is of the order of the Fermi wave length $2\pi/k_F$. In this case the energy scale for $t_k, r_k, \tilde{r}_k$ is ϵ_F. Same is the scale for A, B, C if these quantities are calculated for $x \leq d$. At $x \gg d$ a new scale appears. To see it we can calculate A, B, C using Eq. (13.21) giving at $x \gg d$

$$
A_{k'k} = -\frac{e}{2m\mathcal{L}} e^{i(\epsilon_{k'} - \epsilon_k)t} t_k t_{k'}^* (k + k') e^{-i(k'-k)x}
\tag{13.33}
$$

$$
\begin{aligned}
B_{k'k} = &-\frac{e}{2m\mathcal{L}} e^{i(\epsilon_{k'} - \epsilon_k)t} \\
&[-(k + k') e^{+i(k'-k)x} + (k + k') r'_k r'^*_{k'} e^{-i(k'-k)x} \\
&+ (k - k') r'_k e^{+i(k'+k)x} - (k - k') r'^*_{k'} e^{-i(k'+k)x}]
\end{aligned}
\tag{13.34}
$$

$$
C_{k'k} = -\frac{e}{2m\mathcal{L}} e^{i(\epsilon_{k'} - \epsilon_k)t} t_{k'}^* \left[(k + k') r'_k e^{-i(k'-k)x} - (k - k') e^{+i(k'+k)x} \right]
\tag{13.35}
$$

As we will see later for a small bias $eV = \mu_a - \mu_b$ and low temperature T the relevant momenta k, k' correspond to energies $\epsilon_k, \epsilon_{k'}$ within the "transport" window between the Fermi distributions in both leads $|\epsilon_k - \epsilon_F| \leq \max(eV, T)$. So for a simple barrier one can put $k = k' = k_F$ everywhere except the exponents (since x and t can be large). As a result the fast oscillating exponents (of Friedel type) $e^{\pm i(k'+k)x}$ disappear. The slow oscillating exponents $e^{\pm i(k'-k)x}$ introduce a new energy scale v_F/x which is smaller than ϵ_F if $x \gg d$. This scale is the inverse time of flight from the barrier to the point $x = L$ (or, more generally, between two points x_1 and x_2) where current correlations are measured.

For fluctuation frequencies $\omega \ll \epsilon_F$ it follows from the time exponents that the relevant $\epsilon_{k'} - \epsilon_k \simeq \omega$ and hence the relevant $k - k' \simeq \omega/v_F$. We can choose L in the interval $d \ll L \ll v_F/\omega$ which means $(k-k')L \ll 1$. As a result we see that if the current correlations at frequencies ω are measured not too far from the barrier, at $x = L \ll v_F/\omega$, the current fluctuations are quasistationary and are the same in all crossections of the PC.

With these assumptions one can replace the slow exponents by unity and find

$$A_{k'k}(1)^* A_{k'k}(2) = B_{k'k}(1)^* B_{k'k}(2) = e^{i(\epsilon_{k'}-\epsilon_k)(t_1-t_2)} \left(\frac{ev_F}{L}\right)^2 |t_F|^4, \tag{13.36}$$

$$C_{k'k}(1)^* C_{k'k}(2) = C_{kk'}(1) C_{kk'}(2)^* = e^{i(\epsilon_{k'}-\epsilon_k)(t_1-t_2)} \left(\frac{ev_F}{L}\right)^2 |t_F|^2 |r_F|^2.$$

Now we obtain the final result for the current correlator

$$\begin{aligned}\langle\langle \delta j(t_1) \delta j(t_2) \rangle\rangle = \frac{1}{2} \left(\frac{ev_F}{L}\right)^2 \sum_{kk'} e^{i(\epsilon_{k'}-\epsilon_k)(t_1-t_2)} \\ \times \Big\{ |t_F|^4 [n_k^a(1-n_{k'}^a) + n_k^b(1-n_{k'}^b)] + \\ |t_F|^2 |r_F|^2 [n_k^a(1-n_{k'}^b) + n_k^b(1-n_{k'}^a)] \Big\} + \text{c.c.}\end{aligned} \tag{13.37}$$

The spectral noise density is defined as follows

$$S(\omega) = \frac{1}{2\pi} \int_{-\infty}^{+\infty} dt \langle\langle \delta j(t) \delta j(0) \rangle\rangle \exp(i\omega t). \tag{13.38}$$

In case of small voltage bias and zero temperature bias we follow the way used to derive the Landauer formula. In the two terms proportional to $|t_F|^4$ we shift the integration as follows: $\epsilon_k - \delta\mu_{a,b} \Rightarrow \epsilon$, $\epsilon_{k'} - \delta\mu_{a,b} \Rightarrow \epsilon'$ and as a result these terms do not depend on the bias eV. The bias-depending terms are the two terms proportional to $|t_F|^2 |r_F|^2$ where the proper integration shift is as follows: $\epsilon_k - \delta\mu_{a,b} \Rightarrow \epsilon$, $\epsilon_{k'} - \delta\mu_{b,a} \Rightarrow \epsilon'$.

The final result for the spectral density of the current fluctuation is

$$S(\omega) = \frac{e^2}{8\pi^2}\left\{2|t_F|^4 F(\omega) + |t_F|^2|r_F|^2[F(\omega - eV) + F(\omega + eV)]\right\} \quad (13.39)$$

where the function $F(\omega)$ is defined as follows

$$F(\omega) = \int d\epsilon \int d\epsilon' n(\epsilon)[1 - n(\epsilon')][\delta(\epsilon - \epsilon' + \omega) + \delta(\epsilon - \epsilon' - \omega)] \quad (13.40)$$
$$= \omega \coth \frac{\omega}{2T} = 2\omega \left[\mathcal{N}(\omega) + \frac{1}{2}\right]$$

where $\mathcal{N}(\omega) = \{\exp(\omega/T) - 1\}^{-1}$ is the Planck distribution.

The equilibrium noise can be obtained from Eq. (13.39) with $V = 0$ giving $S(\omega)|_{V=0} = (e^2/4\pi^2)|t_F|^2 F(\omega)$. For $\omega = 0$ it is in agreement with the Nyquist theorem: $S(0)|_{V=0} = (T/\pi)G$ with the Landauer conductance, Eq. (13.29).

The nonequilibrium shot-noise is

$$S_V(\omega) \equiv S(\omega) - S(\omega)|_{V=0} \quad (13.41)$$
$$= (e^2/8\pi^2)|t_F|^2|r_F|^2[F(\omega - eV) + F(\omega + eV) - 2F(\omega)]$$

For $\omega = 0$, this result was obtained by Lesovik.

The function $F(\omega)$ does not have a cutoff a high frequencies and hence the spectral distribution of the equilibrium noise is valid only for $\omega \ll \epsilon_F$. At $\omega \geq \epsilon_F$ the equilibrium noise has a cutoff due to the fast oscillations of $A_{k'k}$ and $B_{k'k}$ when $|\epsilon_k - \epsilon_{k'}| \geq \epsilon_F$. Note that the function $F(\omega)$ contains the zero-point oscillations. Probably the fluctuations due to the zero-point oscillations are not directly measurable, since their "energy" can not be transferred to the measuring device (compare to black-body radiation which do not contains the contribution of the zero point fluctuations of the radiating current). If following this remark one replaces $F(\omega)$ by $2\omega\mathcal{N}(\omega)$ the frequencies of the equilibrium noise will be limited by temperature. Note also that the zero point oscillations do not contribute to the shot-noise given by Eq. (13.41), which has a cutoff at $\max(eV, T)$, *i.e.* the transport window between the Fermi distribution in the leads.

For bosons the situation is more complicated since the product of two Bose distributions $n^a(1 - n^b)$ with close chemical potentials and temperatures do not ensure a narrow transport window between the reservoirs. As a result the terms containing $|t|^4$ will also contribute to nonequilibrium noise, however one can see that the appearance of the product $|t|^2|r|^2$ has nothing to do with the fermion nature of the particles, and hence the factor $1 - |t|^2$ has nothing to do with the Pauli factor $1-n$. (The factor $1-|t|^2$ has not to be confused with the factor $1 \mp n|t|^2$ which appears for the case when one reservoir is "empty").

The formal derivation of the current noise spectra reveals a very important point of the physical picture. Assume that the right reservoir is empty, $n^b = 0$. Naively thinking one can forget about particles coming from the right, *i.e.* about the scattering states ϕ_k, and neglect in the field operator Eq. (13.22) the sum containing states ϕ_k. Calculating in this way the current one will obtain only the first term in Eq. (13.28), which is correct when $n^b = 0$. However in this way for the current correlator one will obtain only the term in Eq. (13.32) with $n^a(1 \mp n^a)$, which is not correct when $n^b = 0$, since the term with $n^a(1 \mp n^b)$ also contribute. It means that if a reservoir radiates to vacuum through a barrier the physical picture for the noise has to incorporate not only the "outgoing" states, but also the "incoming" states. Otherwise, the representation of the field operator Eq. (13.22) is not complete.

5. DEPHASING BY A CURRENT-CARRYING QUANTUM DETECTOR

The case where the "environment" is *far* from equilibrium is of interest since the fluctuations are not given by their well-known equilibrium values and the fluctuation-dissipation theorem is not applicable. Examples are provided in this and in the next sections.

The quantum point contact, briefly alluded to in Section 1, can serve as a detector [21] sensitive to small changes in parameters, such as the electrostatic field nearby. This sensitivity may be used to detect the presence of an electron in one of the arms of an interferometer, provided the two arms are placed asymmetrically with respect to the QPC. Following discussions by Gurvitz [22], Buks *et al.* [23] performed measurements confirming "which path" detection by the QPC. The AB oscillations were measured in a ring. On one of its arms the transmission was limited by a "quantum dot" where the electron wave would resonate for a relatively long and controllable (to a degree) dwell time τ_d. A QPC was placed near that arm, see Fig. 13.1, and the degree of dephasing due to its detecting if the electron is on the quantum dot, could be inferred from the strength of the AB conductance oscillations. Clearly, a necessary and sufficient condition for strong dephasing is that $\tau_\phi \ll \tau_d$. The results were in good agreement with the theory developed by the authors [23], by Levinson [24] and, independently, by Aleiner *et al.* [26]. The new interesting feature of this nonequilibrium dephasing is that a finite current is flowing in the detector and, with increasing time, each electron transmitted there contributes to the decrease of the overlap of

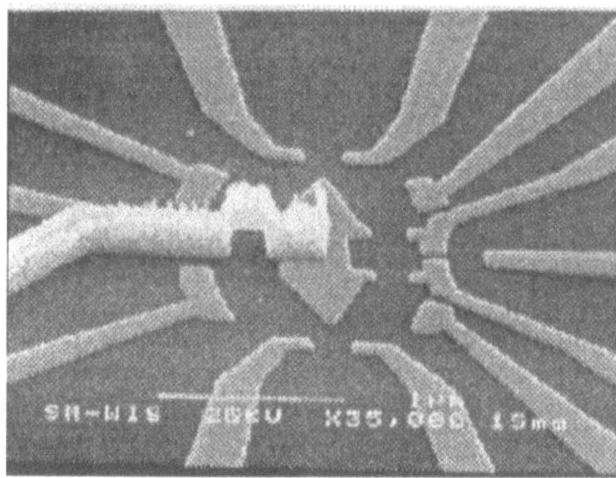

Figure 13.1 A SEM view of the device used by Buks *et al.* [23].

the environment wavefunctions. As discussed in the previous section, the reduction in overlap is conserved when further thermalization of the transferred electrons in the downstream reservoir occurs. The alternative picture is that nonequilibrium (shot-noise) fluctuations of the current in the QPC create a phase uncertainty for the electron in the quantum dot. While the equivalence of these two pictures is guaranteed by the discussion of the previous section (Ref. [14]) it is interesting and non-trivial to see how it emerges in detail, as was demonstrated in Ref. [27]. This is all the more interesting, since the former point of view by which the current fluctuations in the QPC cause dephasing, seems superficially to contradict the idea behind Eq. (13.42). According to the latter, dephasing appears to have to *overcome* the shot-noise fluctuations, which therefore may be thought to *oppose* dephasing.

In the model considered, the QPC is taken for simplicity to be single-channel and symmetric and the temperature T is zero (*i.e.* T is much smaller than the voltage on the QPC). The existence of an electron in the quantum dot is taken to change the transmission coefficient $\mathcal{T} = |t|^2$ from $\mathcal{T}$ to $\mathcal{T} + \Delta\mathcal{T}$ and the conductance by $\frac{e^2}{\pi\hbar}\Delta\mathcal{T}$. The change in phase of t was neglected. It was later considered by Stodolsky [28], and experiments realizing it were conducted by Sprinzak *et al.* [29]. The theory for this case will be discussed in Section 6. τ_ϕ can be physically estimated from the condition that the change in the number of electrons, $\langle N\rangle = (I/e)\tau_\phi$, streaming across the QPC within τ_ϕ, $\langle\Delta N\rangle = \frac{e}{\pi\hbar}V\Delta\mathcal{T}\tau_\phi$ be larger than the rms fluctuations of N during the same time. For the latter one has the quantum shot-noise result [20] according to which the mean-square fluctuation $\langle(\Delta N)^2\rangle$ is given by $(I/e)\tau_\phi(1-\mathcal{T})$ Thus (the numerical factor follows from more detailed calculations):

$$\frac{1}{\tau_\phi} \sim \frac{e}{8\pi\hbar}\frac{(\Delta\mathcal{T})^2 V}{\mathcal{T}(1-\mathcal{T})} \tag{13.42}$$

Several derivations, whose equivalence [27] is guaranteed by the discussion of Section 2 have been given of this result.

The experiments of Ref. [23] agree better than qualitatively with the above picture. For QPC voltages larger than thermal, the visibility of the AB interference contribution to the conductance of the ring decreased roughly linearly in V and the coefficient was in reasonable agreement with the above. The parameter ΔT was directly measured and the dependence on T was qualitatively observed as well.

6. DEPHASING DUE TO EDGE STATES

We start with some general considerations for a nanostructure (NS) coupled capacitively to a quantum dot (QD), for example, as shown in Fig. 13.2. The electron density fluctuations in the NS create electric fields in the QD which are random in time and due to these fields the electron level in the QD, ϵ_0, is a random function in time as well. Its fluctuations are $\delta\epsilon_0(t) = \int d\mathbf{r}W(\mathbf{r})\delta\rho(\mathbf{r},t)$, where $\delta\rho(\mathbf{r},t)$ is the electron density fluctuations and $W(\mathbf{r})$ is a Coulomb interaction kernel. The integration is over the part of the NS close enough to the QD. This "interaction region" is shown in Fig. 13.2. The dephasing rate γ of the QD state is related [24] to the level fluctuation correlator

$$\gamma = \pi K(0), \qquad K(\omega) = (1/2\pi)\int dt\, e^{i\omega t}\langle\delta\epsilon_0(t)\delta\epsilon_0(0)\rangle. \tag{13.43}$$

The electron density fluctuations can be calculated using the scattering state (SS) approach in the same way as current fluctuations. However, in the case of a large magnetic field when the SS's are composed of edge states, one can not consider a usual 1D model, since waves propagating in different directions are located at opposite boundaries of the NS. The SS emitted from terminal α at energy ϵ in Landau level n is $\chi_{\alpha n}(\epsilon,\mathbf{r})$.

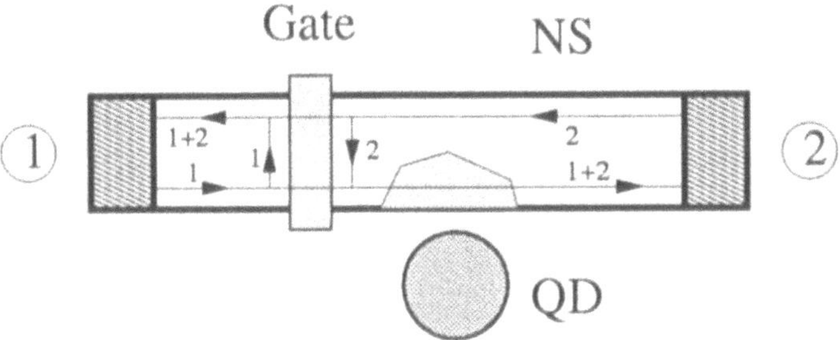

Figure 13.2 A nanostructure with edge states, coupled capacitively to a quantum dot.

One finds [25] the dephasing rate as a sum over *single terminals* and *pairs of terminals,*

$$\gamma = \sum_{\alpha} \gamma^{(\alpha\alpha)} + \sum_{\alpha<\alpha'} \gamma^{(\alpha\alpha')}, \qquad (13.44)$$

$$\gamma^{(\alpha\alpha')} = \pi \int \frac{d\epsilon}{2\pi} \sum_{nn'} f_{\alpha}(\epsilon)[1 - f_{\alpha'}(\epsilon)] |W_{\alpha n,\alpha' n'}(\epsilon)|^2.$$

Here $f_{\alpha}(\epsilon)$ is the Fermi distribution in terminal α and the matrix element contains the SS's: $W_{\alpha n,\alpha' n'}(\epsilon) = \int d\mathbf{r} W(\mathbf{r}) \chi_{\alpha n}(\epsilon, \mathbf{r})^* \chi_{\alpha' n'}(\epsilon, \mathbf{r})$. When the transport window is small one can neglect the energy dependence of the matrix elements. As a result we find $\gamma^{\alpha\alpha} = (1/8\pi)|W_{\alpha,\alpha}|^2 F(0)$ and $\gamma^{(\alpha\alpha')} = (1/4\pi)|W_{\alpha,\alpha'}|^2 F(e|V_{\alpha} - V_{\alpha'}|)$, where V_{α} is the voltage at terminal α and the effective matrix element is $|W_{\alpha,\alpha'}|^2 \equiv \sum_{nn'} |W_{\alpha n,\alpha' n'}|^2$. (The function $F(\omega)$ is defined in Section 4.) Now one can see that a *single terminal* contributes to the dephasing only if SS's emitted from this terminal reach the interaction region, and that this is always an equilibrium contribution. A *pair of terminals* contribute to the dephasing only if SS's emitted from both terminals overlap in the interaction region. When both terminals are at the same voltage, the contribution of this pair to dephasing is an equilibrium one. If one is interested in nonequilibrium dephasing one has to look only at pairs of terminals which are at different voltages, and which send SS's that overlap in the interaction region. Note that $F(\omega)$ contains the "zero point fluctuations", but they do not contribute to the equilibrium dephasing rate, given by $K(\omega)$ at $\omega = 0$, hence for zero temperature there is no equilibrium dephasing.

One can consider $W(\mathbf{r})$ as a small variation of the potential which confines the NS, due to an electron occupying the QD. This allows one to express the matrix elements in terms of the variation of the scattering matrix $S_{\beta m,\alpha n}(\epsilon)$ of the NS as follows:

$$W_{\alpha n,\alpha' n'}(\epsilon) = i \sum_{\beta m} S_{\beta m,\alpha n}(\epsilon)^* \delta S_{\beta m,\alpha' n'}(\epsilon).$$

Consider now the NS+QD shown on Fig. 13.2, which is a schematic representation of the device used in the experiment [29]. The gate describes a point contact (PC). We assume only one Landau level to be relevant and that the edge states at different boundaries do not overlap. The scattering matrix of the NS (in the absence of an electron in the QD) is

$$S = \begin{vmatrix} r & t' \\ t & r' \end{vmatrix} \equiv \begin{vmatrix} \cos\theta e^{i\alpha} & \sin\theta e^{i\beta'} \\ \sin\theta e^{i\beta} & \cos\theta e^{i\alpha'} \end{vmatrix}, \qquad \alpha' - \beta' = \pi - (\alpha - \beta), \tag{13.45}$$

where r and t correspond to reflection and transmission of SS's approaching the gate from left, while r' and t' correspond to SS's approaching the gate from the right. Note that in a magnetic field the scattering matrix is not symmetric, $t \neq t'$.

In case of zero temperature the equilibrium part of the dephasing rate of the QD state vanishes, and the nonequilibrium part is $\gamma = \pi|W|^2|eV_{12}|$, where $|W|^2 = |r^*\delta t' + t^*\delta r'|^2 = |\delta\theta - i(\delta\beta - \delta\alpha)\sin\theta\cos\theta|^2$ and $\delta t'$, $\delta r'$ are the changes of the transmission and the reflection amplitudes due to the electron in the QD. For a symmetric PC $\alpha = \beta$. When in addition the QD is located symmetrically with respect to the PC, $\delta\alpha = \delta\beta$. In this case the matrix element contains only the variation of the absolute value of the transmission amplitude,

$$|W|^2 = (\delta|t|^2)^2/4|t|^2|r|^2, \tag{13.46}$$

(see Eq. (13.42)) The second term in $|W|^2$ which is due to the variation of the phase of the transmission amplitude appears only for a nonsymmetric system. If the QD is far from the gate (which is the case in experiment [29]) this is the only term which contributes, since an edge state can not be reflected by the QD, if the interaction region do not reach the opposite boundary. In this case

$$|W|^2 = |\delta\alpha - \delta\beta|^2|t|^2|r|^2. \tag{13.47}$$

7. THE DETECTABILITY OF $T = 0$ NOISE.

7.1 Generalities, A Simple Model

In this section we consider the simplest model of the detection of the current noise by looking at, for example, the photons it produces by linear coupling to an electromagnetic (EM) field. Obviously, this includes such coupling to any set of harmonic oscillators, a simple case being an ac resonant circuit. Following Ref. [30], we examine the two time-reversed current-current correlation functions (we drop the jj subscripts for brevity):

$$C_+(t) = \langle j(0)j(t)\rangle, \qquad \text{and} \qquad C_-(t) = \langle j(t)j(0)\rangle; \tag{13.48}$$

and their Fourier transforms, $S_+(\omega)$ and $S_-(\omega) = S_+(-\omega)$, respectively. $S_+(\omega)$ is equal to the van Hove-type dynamic structure factor, $S_{jj}(\omega)$, considered in Section 3. Thus, as in Eq. (13.16):

$$S_+(\omega) = S_{jj}(\omega) = \hbar \sum_i P_i |\langle f|j|i\rangle|^2 \delta(E_i - E_f - \hbar\omega). \tag{13.49}$$

It immediately follows from (13.49) that this function satisfies, in equilibrium, the usual detailed balance condition, $S_+(\omega) = S_+(-\omega)e^{-\hbar\omega/k_BT}$, as in Eq. (13.53) for the density-density dynamic structure factor.

A free EM field described by a vector potential $\mathbf{A}(x,t)$ is introduced. $\mathbf{A}$ is expanded in terms of photon creation and annihilation operators. By treating the interaction $\mathbf{j.A}$ of the electronic system with the EM field by lowest-order perturbation theory, it is straightforward to see [32] that $S_+(\omega)$ determines the cross section for energy transfer between the fluctuating currents and the EM field. This means that for $\omega > 0$, $S_+(\omega)$ gives the emission of a photon into the vacuum, zero photon number, state and $S_+(-\omega)$ gives the absorption of a given single photon by the electrons. When the electrons are in equilibrium at $T = 0$, the former vanishes, as it should. Regarding the photons emitted into the vacuum state as a means to detect the current noise of the system, it follows that the zero-point fluctuations of the latter are not detectable in this fashion!

It is customary to take for the power spectrum of the current noise a symmetrized version $S_s(\omega) = [S_+(\omega) + S_-(\omega)]/2$, given by the Fourier transform of $[C_+(t) + C_-(t)]/2$, as in Eq. (13.30). This procedure is fine in the classical limit, but it is *not appropriate* for the quantum fluctuation case, when $k_BT \ll \hbar\omega$. It implies that the zero-point fluctuations are detectable by a simple detector. This is obviously unacceptable, the second law of thermodynamics prohibits the transfer of *anything* from a $T = 0$ system. In the quantum limit, it is the function $S_+(\omega)$, which corresponds to straightforward physical measurements, which is a *proper noise power spectrum*. Obviously, it is also possible to use $S_-(\omega) = S_+(-\omega)$, instead of $S_+(\omega)$, but *not* the symmetrized version $S_s(\omega)$. This problem was considered by Lesovik and Loosen [30]. Our conclusions agree with theirs but our treatment is more straightforward and free from a mathematical convergence difficulty which is properly acknowledged in their paper. We note that the asymmetry of $S_+(\omega)$, which is dictated by the detailed-balance condition, implies that the time correlator, $C_+(t)$, has an imaginary part. This is of course a quantum effect [18], due to the noncommutativity of the current operators at different times. It is related, in turn, to the dissipative part of the response.

We do not claim that the zero-point fluctuations (ZPF) are not detectable, only that they are not seen by a simple *passive* harmonic oscillator detector. It is well-known that the ZPF appear in various other physical effects such as the Lamb shift, the Debye-Waller exponent and the Casimir force. How they influence a linear amplifier is considered for example in Ref. [31]. They may well be detectable by various nonlinear phenomena but *not* by the simple direct measurement considered above. Furthermore, we are going to demonstrate now that augmenting the model by introducing an arbitrary number of photons in the EM field does produce interesting results, but does not make the ZPF directly detectable by emitting photons. The way the system *absorbs* photons from the field can however be regarded as an indirect observation of the ZPF. We emphasize that the statement that $S_+(\omega)$ is the proper noise power spectrum is valid for an arbitrary state (used to perform the averaging in Eq. (13.48)) of the electronic system. This includes nonequilibrium states, for example current-carrying ones, where shot-noise is relevant. The equilibrium assumption is necessary only if one would like to have the detailed balance condition for the correlators.

7.2 Generalization To An Arbitrary State Of The EM Field

When the number of photons in the EM field at frequency ω is $N(\omega)$ the emission and absorption of photons by the system are modified by the well-known "enhanced emission and absorption" factors $N(\omega)+1$ and $N(\omega)$, respectively. This makes the net energy flow, $I(\omega)$, from the electrons to the field, be given by (we continue to take $\omega > 0$)

$$I(\omega) \sim [N(\omega)+1]S_+(\omega) - N(\omega)S_+(-\omega) = S_+(\omega) + N(\omega)[S_+(\omega) - S_+(-\omega)]. \tag{13.50}$$

In the particular case that the electronic system is in equilibrium at a temperature T whose inverse is given by $\beta = 1/(k_B T)$, and denoting the equilibrium $N(\omega)$ at the temperature T by $N_T(\omega) = 1/[exp(\beta\omega) - 1]$, we find:

$$I(\omega) \sim S_+(\omega)\left[1 - \frac{N(\omega)}{N_T(\omega)}\right], \tag{13.51}$$

in agreement with Ref. [30].

If one assumes, further, that the EM field is at equilibrium at temperature T_0, one obtains:

$$I(\omega) \sim S_+(\omega)\left[1 - \frac{N_{T_0}(\omega)}{N_T(\omega)}\right] = S_+(\omega)N_{T_0}(\omega)[\exp(\beta_0\omega) - \exp(\beta\omega)]. \tag{13.52}$$

It is worthwhile to check that the net energy flow is from the hotter to the colder system. Such a flow is a way to detect the fluctuations of the system. One may say that at $T = 0$ the zero point noise does not send anything to the detector. In a sense the zero point noise can be observed by *taking* energy from a detector having a nonzero $N(\omega)$. That means that an energetic detector may work at $T = 0$ by exciting rather than by deexciting the system. It is seen from Eq. (13.16) that the noise spectrum contains information on the excited states. At $T = 0$ the current correlator is proportional to the Fourier transform of the absorption spectrum.

8. DEPHASING WHEN $T \to 0$.

Recently, Mohanty *et al.* [34] have published extensive experimental data indicating that contrary to general theoretical expectations and to Eq. (13.11), the dephasing rate in films and wires does not vanish as $T \to 0$. Serious precautions [35] were taken to eliminate experimental artifacts. It was speculated that such a saturation of the dephasing rate when $T \to 0$, might follow from interactions with the zero point motion of the environment. These speculations have received apparent support from calculations in Ref. [36]. However, the latter were severely criticized in Refs. [37, 38] and were in disagreement with experiments in Ref. [39]. In fact it is clear that since dephasing must be associated with an excitation of the environment, it cannot happen as $T \to 0$. In that limit neither the electron nor the environment has any energy to exchange. Below, we convert this qualitative argument to a proof. While proving unequivocally that zero point motion does not dephase, our proof does show what *further* physical assumptions can in fact produce a finite dephasing rate for $T \to 0$.

We [33, 40] use Eq. (13.9) and apply the very general detailed-balance relationship

$$S(q,\omega) = S(-q,-\omega)e^{-\hbar\omega/k_BT}, \tag{13.53}$$

to either $S_p(q,\omega)$ or $S_{env}(-q,-\omega)$. It is immediately seen that the integrand of Eq. (13.9) is a product of two factors one of which vanishes for $\omega > 0$ and the other for $\omega < 0$, as $T \to 0$. Thus the integral and the dephasing rate vanish in general when $T \to 0$. However, if $S_{env}(-q,-\omega)$

has an approximate delta-function peak at small ω due to an abundance of low-energy excitations, one may get a finite dephasing rate at temperatures higher than the width of that peak. Such near-degeneracies of the ground state are known to exist in disordered, glassy, systems. These follow from the many "mesoscopic" realizations of the disorder configuration. The system slowly fluctuates among these many states and it may in fact not be in full equilibrium. This may cause [41] the commonly observed low-frequency (often "$1/f$") noise [42].

Let us now estimate [40] the inelastic scattering rate [43] from the set of impurities that are rearranging by tunneling at low temperatures at a rate $1/\tau_0$ satisfying:

$$\hbar/\tau_\phi \ll \hbar/\tau_0 \ll k_B T. \tag{13.54}$$

we denote the fraction of the defects that move on these time scales by $p \ll 1$. From the results of Ref. [43], it is seen that the inelastic component of the scattering from those impurities is smaller by $A \cong (k_F d)^2 \ll 1$ than the elastic rate. Thus the inelastic scattering rate from these defects is pA/τ, where τ is the elastic scattering time by all defects. Therefore, to get a low temperature dephasing rate of $10^{-4}/\tau$ [34], we need, for $A \sim 10^{-2}$, a value of $p \sim 10^{-2}$. The impurity density, for impurities whose scattering length is atomic, is of the order of $1/k_F\ell$. This typically corresponds to total impurity concentrations of the order of 10^{-2}. Thus, we have to assume a $\sim$ 100 ppm concentration of *appropriate* low-energy "two-level" defects, to get a low temperature dephasing rate comparable to that of Ref. [34]. This does not sound impossible, however the consistency with observed levels of low-temperature $1/f$ noise must be checked. Experimental studies of the possible correlation between the saturation of τ_ϕ and low-frequency noise, would be very valuable. The above result for the dephasing rate by the mobile impurities can be obtained [40] from the well-known "two-level-system" (TLS) model of disordered systems [44]. The near-constancy of the dephasing rate with temperature is obtained as long as $k_B T > \Omega_{max}$, where Ω_{max} is the maximal tunneling matrix element between the two potential minima for the defect. In the opposite limit, the dephasing rate vanishes linearly with the temperature.

Thus, while the "standard model" of disordered metals (in which the defects are strictly frozen) gives of course an infinite τ_ϕ at $T = 0$, there may be other physical ingredients that can make τ_ϕ finite at very low temperatures (but *not at the $T \to 0$ limit*), *without* contradicting any basic law of physics. The TLS model is a particular example and its requirements may or may not be satisfied in the real samples. But, other models with similar dynamics might exist as well. We reemphasize that this does *not* imply dephasing by zero-point fluctuations, which has been repeatedly, and wrongly, claimed in the literature. The failure of the semiclassical approximation used in these considerations was clarified in Ref. [33].

Acknowledgements

This research was supported by grants from the German-Israel Foundation (GIF), the Israel Science Foundation, Jerusalem and from the Israeli Ministry of Science and the French Ministry of Research and Technology.. The authors thank Y. Aharonov, D. Cohen and A. Stern for collaborations on these problems. I. L. Aleiner, B. L. Altshuler, N. Argaman, C. W. J. Beenakker, M. Berry, E. Buks, Y. Gefen, B. I. Halperin, M. Heiblum, D. E. Khmelnitskii, R. Landauer, Y. Meir, Z. Ovadyahu, M. Schechter, G. Schön, T. D. Schultz, B.Z. Spivak, D. Sprinzak, A. Stern, H. A. Weidenmüller, P. Wölfle and A. Zaikin are thanked for discussions.

References

[1] Y. Imry, *Introduction to Mesoscopic Physics*, Oxford Unversity Press (1997).

[2] R. Landauer, IBM J. Res. Dev. **1**, 223 (1957); R. Landauer, Philosoph. Mag. **21**, 863 (1970).

[3] B. J. van Wees, H. Van Houten, C. W. J. Beenakker, J. G., Williamson, L. P. Kouendhoven, D. van der Marel and C. T. Foxon, Phys. Rev. Lett. **60**, 848 (1988); D. A. Wharam, T. J. Thornton, R. Newbury, M., Pepper, H. Ahmed, J. E. F. Frost, D. G. Husko, D. C. Peacock, D. A. Ritchie, and G. A. C. Jones J. Phys. **C21**, L209 (1988).

[4] Y. Gefen, Y. Imry, and M. Ya Azbel, Phys. Rev. Lett. **52**, 129 (1984).

[5] R. A. Webb, S. Washburn, C. P. Umbach and R. B. Laibowitz, Phys. Rev. Lett. **54**, 2696 (1985).

[6] B. L. Altshuler, A. G. Aronov and B. Z. Spivak, JETP Lett. **33**, 94 (1981).

[7] D. Yu Sharvin, and Yu V. Sharvin, JETP Lett. **34**, 272 (1981).

[8] B. L. Altshuler, JETP Lett. **41**, 649 (1985); P. A. Lee and A. D. Stone, Phys. Rev. Lett. **55**, 1622 (1985). P.A. Lee, A. D. Stone and H. Fukuyama, Phys. Rev. **B35**, 1039 (1986).

[9] M. Büttiker, Y. Imry and R. Landauer, Phys. Lett. **96A**, 365 (1983).

[10] L. P. Levy, G. Dolan, J. Dunsmuir, and H. Bouchiat, Phys. Rev. Lett. **64**, 2074 (1990); V. Chandrasekhar, R. A. Webb, M. J. Brady, M. B. Ketchen, W. J. Gallagher and A. Kleinsasser, Phys. Rev. Lett. **67**, 3578 (1991); D. Mailly, C. Chapelier and A. Benoit, Phys. Rev. Lett. **70**, 2020 (1993).

[11] R. P. Feynman and F. L. Vernon, Ann. Phys. NY **24**, 118 (1963).

[12] A. O. Caldeira and A. J. Leggett Ann. Phys. **149**, 374 (1983) .

[13] R. P. Feynman, R. B. Leighton and M. Sands *The Feyman Lectures on Physics*, Addison Wesley, Reading, MA, Vol. III, pp. 21.14 (1965), contains a beautiful discussion of dephasing.

[14] A. Stern, Y. Aharonov, and Y. Imry, (1990) Phys. Rev. **A40**, 3436 and in G. Kramer, ed. *Quantum Coherence in Mesoscopic Systems*, NATO ASI Series no. 254, Plenum., p. 99 (1991).

[15] G. Hackenbroich, B. Rosenow and H. A. Weidenmüller, Cond-mat/9807317 and to be published.

[16] C. Kittel, *Quantum Theory of Solids*, John Wiley, NY (1963).

[17] B. L. Altshuler, A. G. Aronov, and D. E. Khmelnitskii, J. Phys. **C15**, 7367 (1982).

[18] L. van Hove, Phys. Rev. **95**, 249 (1954).

[19] M. Reznikov, M. Heiblum, H. Shtrikman and D. Mahalu, Phys. Rev. Lett. **75**, 3340 (1995).

[20] V. A. Khlus, JETP **66**, 1243 (1987); G. B. Lesovik, JETP Lett. **49**, 592 (1989); Th. Martin and R. Landauer, Phys. Rev. **B45**, 1742 (1992); M. Büttiker, Phys. Rev. **B46**, 12485 (1992).

[21] M. Field, C. G. Smith, M. Pepper, D. A. Ritchie, J. E. F. Frost, G. A. Jones and D. G. Hasko, Phys. Rev. Lett. **70**, 1311 (1993).

[22] S. A. Gurvitz, Phys. Rev. **B56**, 15215 (1997) and Quant-ph/9697029 (1997)

[23] E. Buks, R. Schuster, M. Heiblum, D. Mahalu and V. Umansky, Nature **391**, 871 (1998).

[24] Y. Levinson, Europhys. Lett. **39**, 299 (1997).

[25] Y. Levinson, Phys. Rev. **B61**, 4748 (2000).

[26] I. L. Aleiner, N.S. Wingreen and Y. Meir, Phys. Rev. Lett. **79**, 3740 (1997).

[27] Y. Imry, Physica Scripta, in press.

[28] L. Stodolsky, quant-ph/9805081.

[29] D. Sprinzak, E. Buks, M. Heiblum and H. Shtrikman, *Controlled dephasing of electrons via a phase sensitive detector*, cond-mat/9907162, Phys. Rev. Lett., in press (2000).

[30] G. B. Lesovik and R. Loosen, JETP Lett. **65**, 269 (1997); see also: G. B. Lesovik and L. S. Levitov, Phys. Rev. Lett. **72**, 538 (1994).

[31] J. R. Tucker and M. J. Feldman, Revs. Mod. Phys. **57**, 1107 (1985).

[32] G. Baym, *Lectures on Quantum Mechanics*, p. 271-276, Addison-Wesley (1993).

[33] D. Cohen and Y. Imry, Cond-mat/9807038.

[34] P. Mohanty, E. M. Jariwala and R. A. Webb, Phys. Rev. Lett. **77**, 3366 (1997).

[35] P. Mohanty, E. M. Jariwala and R. A. Webb, proceedings of ISQM, Tokyo (1998).

[36] D. S. Golubev and A. D. Zaikin, Phys. Rev. Lett. **81**, 1074 (1998); cond-mat/9712203.

[37] I. L. Aleiner, B. L. Altshuler and M. E. Gershenson, cond-mat/9808078, cond-mat/9808053.

[38] B. L. Altshuler, M. E. Gershenson and I. L. Aleiner, cond-mat/9803125.

[39] Yu. B. Khavin, M. E. Gershenson, A. L. Bogdanov, Sov. Phys. Uspekhi **168**, 200 (1998); Phys. Rev. Lett. **81**, 1066 (1998); Phys. Rev. **B58**, 8009 (1998).

[40] Y. Imry, H. Fukuyama and P. Schwab, submitted for publication to Europhys. Lett.

[41] S. Feng, P. A. Lee and A. D. Stone, Phys. Rev. Lett. **56**, 1970, 2272(E) (1986).

[42] N. O. Birge, B. Golding and W. H. Haemmerle, Phys. Rev. **B42**, 2735 (1990)

[43] Y. Imry, chapter 35 in *Tunneling in Solids*, proceedings of the 1967 Nato Conference, E. Burstein and S. Lundquist, eds., Plenum Press (N.Y.) 1969.

[44] P. W. Anderson, B. I. Halperin and C.Varma, Phil. Mag. **25**, 1 (1972); W. Philips, J. Low Temp. Phys. **7**, 351 (1972).

Chapter 14

CHARGE FLUCTUATIONS AND DEPHASING IN COULOMB COUPLED CONDUCTORS

M. Büttiker
Département de Physique Théorique, Université de Genève,
CH-1211 Genève 4, Switzerland

1. INTRODUCTION

It is the purpose of this work to provide a self-consistent discussion of charge and potential fluctuations in Coulomb coupled mesoscopic conductors and to apply the results to evaluate dephasing rates of Coulomb coupled conductors. Charge and potential fluctuation spectra are important in a number of problems: the theory of dynamical (frequency-dependent) conductance of mesoscopic systems can be developed from a fluctuation theory [1] (the dynamical conductance is then obtained from the fluctuation dissipation theorem and the Kramers-Kronig relations); already in the white noise limit, shot noise outside the ohmic range, is renormalized by charge fluctuations [2]; furthermore, the currents induced into gates capacitively coupled to a conductor, or a tunneling microscope tip (used as a small movable gate) depend on the charge fluctuations of the mesoscopic conductor [3]. At low temperatures, the dephasing rates, which give the time over which a quasi-particle retains phase memory, are determined by potential fluctuations. It is clearly desirable to have a theory of charge fluctuations which works in all these situations.

The theory of fluctuations in mesoscopic conductors, has been predominantly concerned with current fluctuations [4]-[7], and correlations of currents at different terminals of a multiprobe conductor [7, 8, 9]. For an extended review of shot noise in mesoscopic conductors, we refer the reader to Ref. [10]. As the above mentioned rare examples [1, 2, 3] demonstrate, the theory can, however, be extended to treat charge and

I. O. Kulik and R. Ellialtioğlu (eds.),
Quantum Mesoscopic Phenomena and Mesoscopic Devices in Microelectronics, 211–242.

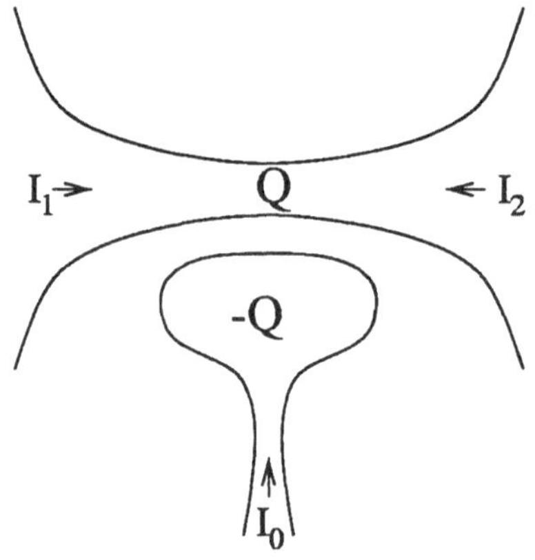

Figure 14.1 Cavity with a charge deficit $-Q$ in the proximity of a quantum point contact with an excess charge Q. The dipolar nature of the charge distribution ensures the conservation of currents I_0, I_1 and I_2 flowing into this structure. After [3].

potential fluctuations. In contrast to fluctuations in the total current at a contact of a mesoscopic sample, a discussion of charge fluctuations is more difficult, since it necessitates a treatment of interactions.

To emphasize the need for a self-consistent treatment of charge fluctuations, consider the conductor shown in Fig. 14.1. A quantum point contact [11, 12] is capacitively coupled to a second mesoscopic conductor which might represent a gate or a small quantum chaotic cavity coupled only via a single lead to an electron reservoir [3]. The currents to the right and left of the quantum point contact (QPC) are I_1 and I_2 and the current that flows from the cavity to its electron reservoir is denoted by I_0. We want to assume that the QPC and the cavity are so close to one another that every electrical field line which emanates from the cavity ends either again on the cavity or on the QPC. There exists then a volume which encloses these two conductors with the property that the net electric flux through the surface of this volume is zero. According to Gauss, the net electric charge inside this volume is thus zero. Hence any charge fluctuation Q_1 on the cavity must be counter-balanced by a charge fluctuation Q_2 on the QPC such that the total charge is preserved, $Q_1 + Q_2 = 0$. Hence the charge on the cavity $Q_1 \equiv -Q$ is totally correlated with the excess charge $Q_2 \equiv Q$ on the QPC. The excess charge is dipolar with a charge accumulation on one of the conductors and a charge depletion on the other conductor. It is the conservation (and complete correlation) of the excess charges on the two conductors which ensures the conservation of currents,

$$I_0(t) + I_1(t) + I_2(t) = 0. \tag{14.1}$$

This conservation holds at any instant of time. A fluctuation theory based on independent particles, cannot describe the correlations in the charge fluctuations which are necessary to establish Eq. (14.1). In an independent particle approach the number of particles in the cavity would be independent of the number of particles on the QPC. Thus an approach which takes the Coulomb interactions into account is necessary to establish such a basic property like current conservation.

Dephasing rates in disordered conductors are mostly discussed in connection with weak localization [13, 14, 15]. Weak localization is a quantum effect which arises from the interference of time-reversed particle trajectories and which survives ensemble averaging. The calculations of the dephasing rate for this effect is performed by first ensemble averaging such that the dynamics is effectively diffusive. This permits a treatment of the Coulomb interactions for a sample that is uniform on the scale of the elastic scattering length: screening is treated with the help of a frequency and wave-vector dependent dielectric constant. The charge-fluctuations are essentially electron-hole pairs which leave the total charge on the conductor invariant.

In contrast, the dephasing rates discussed here, *cannot* be applied to weak localization. The main effect which is investigated comes from carriers which leave or enter the conductor. These carriers thus change the total charge on the conductor. It is clear that for the rates considered here the connections of the sample to the outside world, the properties of its contacts, play an important role. This is again in contrast to dephasing rates determined by electron-hole pair excitations. Both approaches have in common that the charge excitations considered are dipolar: here we place the electron on one conductor and the hole on the other conductor.

The experiments we have in mind are provided in recent works by Buks *et al.* [16] and by Sprinzak *et al.* [17]. In these experiments, the effect of a current carrying QPC on a Coulomb coupled nearby phase-coherent system is investigated. The dephasing rate which is measured is proportional to the voltage applied to the QPC. In addition to the discussion provided in the experimental works, dephasing in Coulomb coupled structures, has found attention in a number of theoretical works [18, 19, 20]. Widely different approaches have been used, but the point of view provided here, which emphasizes the correlations of the charge fluctuations on the two conductors and the need for a self-consistent discussion, seems novel. A brief discussion of the results of our work is presented in Ref. [21] for the geometry of the experiment of Sprinzak *et al.* [17]. Different and closely related aspects of the work presented here are discussed in two conference proceedings [22, 23]. The experiment [17] is also discussed in Ref. [24] but without an attempt to provide a self-consistent discussion of charge fluctuations.

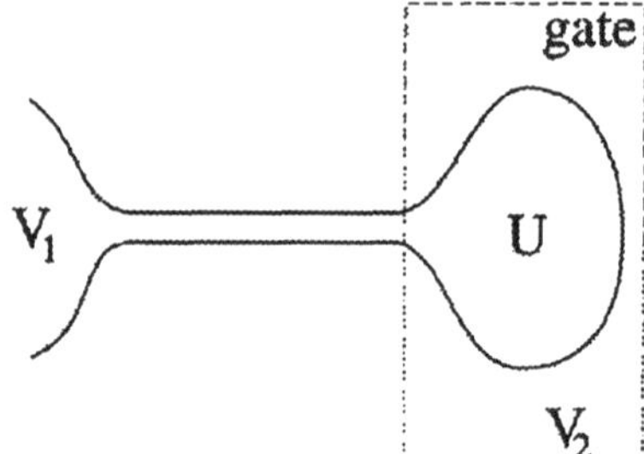

Figure 14.2 Mesoscopic capacitor connected via a single lead to an electron reservoir and capacitively coupled to a gate. V_1 and V_2 are the potentials applied to the contacts, U is the electrostatic potential of the cavity. After Ref. [35].

2. THE MESOSCOPIC CAPACITOR (MACROSCOPIC BACKGATE)

The simplest system for which it is instructive to investigate the fluctuations of charge is a mesoscopic capacitor [1]. Fig. 14.2 shows a cavity capacitively coupled to a backgate and connected via a single lead to an electron reservoir. In a first step we will treat the gate as a macroscopic conductor. We assume that the electrostatic potential on the cavity can be described be a single parameter U. The theory presented below is not limited to this simplifying assumption but can be extended to treat the microscopic landscape [25]. In the presence of an oscillating potential at the reservoir with Fourier amplitude $V_1(\omega)$ or a potential $V_2(\omega)$ at the gate, the electrostatic potential on the cavity oscillates with an amplitude $U(\omega)$. These potential oscillations are connected to the charge oscillations $Q(\omega)$ on the cavity via $Q(\omega) = C(U(\omega) - V_2(\omega))$. Here C is the geometrical capacitance coupling the charge on the cavity to that on the gate. The resulting dynamic conductance, that relates the ac-current through this structure to a small ac-voltage applied between the reservoir and the backgate is [1]

$$G(\omega) = \frac{-i\omega C_\mu}{1 - i\omega R_q C_\mu}. \tag{14.2}$$

In Eq. (14.2) we have retained only the pole in the complex frequency-plane with the smallest imaginary part. The dynamic conductance of the mesoscopic capacitor is, like that of a macroscopic capacitor, determined by an RC-time. But instead of only purely classical quantities, we obtain now expressions that contain quantum corrections due to the phase-coherent electron motion in the cavity. It turns out that the RC-time can be expressed with the help of the Wigner-Smith time-delay matrix [26]

$$\mathcal{N} = \frac{1}{2\pi i}\mathbf{s}^\dagger \frac{d\mathbf{s}}{dE} . \tag{14.3}$$

Here **s** is the scattering matrix which relates the incident current amplitudes in the lead connecting the cavity to the reservoir 1 to the out-going current amplitudes in this lead (see also Appendix A). The sum of the diagonal elements of this matrix determines the density of states [27]

$$N = \mathrm{Tr}\mathcal{N} = \frac{1}{2\pi i}\mathrm{Tr}\left[\mathbf{s}^{\dagger}\frac{d\mathbf{s}}{dE}\right] \tag{14.4}$$

and gives rise to a "quantum capacitance" e^2N which in series with the geometrical capacitance determines the electrochemical capacitance [1]

$$C_{\mu}^{-1} = C^{-1} + (e^2N)^{-1}. \tag{14.5}$$

The resistance which counts is the charge relaxation resistance [1]

$$R_q = \frac{h}{2e^2}\frac{\mathrm{Tr}[\mathcal{N}^{\dagger}\mathcal{N}]}{[\mathrm{Tr}\mathcal{N}]^2}. \tag{14.6}$$

For simplicity we have given these results, Eqs. (14.3-14.6), only in the zero temperature limit. It is instructive to consider a basis in which the scattering matrix is diagonal. Since we have only reflection, all eigenvalues of the scattering matrix are of the form $\exp(i\zeta_n)$ where ζ_n is the phase which a carrier accumulates from the entrance to the cavity through multiple scattering inside the cavity until it finally exits the cavity. Thus the density of states can also be expressed as

$$N = (1/2\pi)\sum_{n}(d\zeta_n/dE) \tag{14.7}$$

and is seen to be proportional to the total Wigner time delay carriers experience in the cavity. The time delay for channel n is [28] $\tau_n = \hbar d\zeta_n/dE$. Similarly we can express the charge relaxation resistance in terms of the energy derivatives of phases and we obtain in the zero-temperature limit,

$$R_q = \frac{h}{2e^2}\frac{\sum_n(d\zeta_n/dE)^2}{[\sum_n d\zeta_n/dE]^2}. \tag{14.8}$$

R_q is thus determined by the sum of the squares of the delay times divided by the square of the sum of the delay times.

We now briefly discuss these results. First, our Eq. (14.5) for the electrochemical capacitance predicts that it is not a purely geometrical quantity but that it depends on the density of states of the cavity. This effect is well known from investigations of the capacitance of the quantized Hall effect. More recent work investigates the mesoscopic capacitance of quantum dots and wires and is often termed *capacitance spectroscopy*. In addition to describing the average behavior, our results can also be used to investigate the fluctuations in the capacitance. Similar to the universal conductance fluctuations there are capacitance fluctuations in mesoscopic samples due to the fluctuation of the density of states. Such effects can be expected to be most pronounced if the contact permits just the transmission of a single channel. Then it is necessary not only to investigate the fluctuations of the mean square fluctuations but the entire distribution function. Such an investigation was carried out by Gopar *et al.* [29] in the single channel limit and by Brouwer and the author [30], and Brouwer *et al.* [31] for chaotic cavities with quantum point contacts which are wide open (many channel limit). Since in experiments [32] the Coulomb energy e^2/C is typically much larger than the level separation Δ these fluctuations are small. A very interesting prediction which follows from the density of states dependence of the electrochemical capacitance is that for a sample of the form of a loop with an Aharonov-Bohm flux through the hole of the loop, the capacitance should exhibit Aharonov-Bohm oscillations [33]. Aharonov-Bohm oscillations in the capacitance of small rings have recently been measured by Deblock *et al.* [34].

Next let us discuss briefly the charge relaxation resistance R_q. An over- view of charge relaxation resistances in mesoscopic systems is presented in Ref. [35]. First we note that the resistance unit is not the resistance quantum h/e^2 but $h/2e^2$. The factor two arises since the cavity is coupled to one reservoir only. Thus only half the energy is dissipated as compared to dc-transport through a two terminal conductor. Second, we note that in the single channel limit, equation (14.8) is *universal* and given just by $h/2e^2$. This is astonishing since if we imagine that a barrier is inserted into the lead connecting the cavity to the reservoir one would expect a charge relaxation resistance that increases as the transparency of the barrier is lowered. Indeed, if there is a barrier with transmission probability $\mathcal{T}$ per channel in the lead connecting the cavity and the reservoir, then in the large channel limit, for $\mathcal{T}M \gg 1$, R_q is [36],

$$R_q = (h/e^2)(1/\mathcal{T}M). \tag{14.9}$$

In the large channel-number limit, Eq. (14.8) is proportional to $1/M$, where M is the number of scattering channels, and proportional to $1/\mathcal{T}$. Thus in the large channel limit Eq. (14.8) behaves as expected.

Let us next consider the fluctuation spectra of the current, charge and potential. Ref. [1] gives a derivation of these spectra using a dynamic fluctuation theory of mesoscopic conductors. The current fluctuation spectra must, however, in any case, be connected to the dynamical conductance via the fluctuation-dissipation theorem. This gives for the current noise spectrum [1],

$$S_{II}(\omega) = 2kT \frac{\omega^2 C_\mu^2 R_q}{1 + \omega^2 R_q^2 C_\mu^2}. \tag{14.10}$$

Since the charge on the cavity is the time-derivative of the current, we find for the fluctuation spectrum of the total charge on the cavity,

$$S_{QQ}(\omega) = 2kT \frac{C_\mu^2 R_q}{1 + \omega^2 R_q^2 C_\mu^2}, \tag{14.11}$$

and since the charge is related via the geometrical capacitance to the potential, $Q(\omega) = CU(\omega)$ we find

$$S_{UU}(\omega) = 2kT \frac{C_\mu^2}{C^2} \frac{R_q}{1 + \omega^2 R_q^2 C_\mu^2}. \tag{14.12}$$

Let us pause here and consider two limiting cases. First, if the charging energy is unimportant, the geometrical capacitance becomes very large, and the electrochemical capacitance is essentially determined by the quantum capacitance $C_\mu \simeq e^2 N$. In this case the spectrum for the voltage fluctuations, Eq. (14.12), tends to zero. There is no screening of the charge pile-up. In the opposite limit, when the geometrical capacitance tends to zero, the quantum capacitance becomes unimportant and $C_\mu \simeq C$. In this case, both the current fluctuation spectrum and the charge fluctuation spectrum become very small. Charging of the cavity is now energetically very expensive, and we can neither drive a current through the structure nor can we pile-up charge. In this limit, the spectrum of the potential fluctuations as a function of the geometrical capacitance reaches its maximum amplitude, whereas its width in frequency becomes increasingly narrow.

The reason that in the small capacitance (charge neutral limit) both the current and the charge fluctuation spectrum tend to zero, whereas the voltage fluctuation spectrum stays finite, is also due to electron-hole pair excitations. If simultaneously an electron and hole enter the cavity, the total charge remains unchanged, there is no net current generated but there are nevertheless potential fluctuations.

Let us next consider the dephasing of a carrier in the mesoscopic cavity due to the potential fluctuations. We follow Ref. [20] and relate the phase ϕ of a carrier to the fluctuations in the potential via, $\hbar d\phi/dt = eU(t)$. Integrating this equation, defines for $t \gg RC$ a dephasing rate

$$\Gamma_\phi = \frac{\langle[\phi(t) - \phi(0)]^2\rangle}{t} = \left(\frac{e^2}{\hbar^2 t}\right) \langle \int_0^t dt' \int_0^t dt'' U(t')U(t'')\rangle \tag{14.13}$$

which is related to the zero-frequency limit of the the voltage fluctuation spectrum via,

$$\Gamma_\phi = (e^2/2\hbar^2)S_{UU}(0). \tag{14.14}$$

Thus up to fundamental constants, the dephasing rate is determined by the white noise limit of the potential fluctuation spectrum. Using Eq. (14.12) for the mesoscopic capacitor considered, we thus obtain,

$$\Gamma_\phi = (e^2/\hbar^2)kT(C_\mu^2/C^2)R_q. \tag{14.15}$$

This result expresses the dephasing rate in terms of *electrical* quantities. The dephasing rate is essentially determined by the charge relaxation resistance R_q and a ratio of capacitances. We have already noticed, that in the free-electron limit, the voltage fluctuation spectrum vanishes, since C_μ/C tends to zero. Consequently, in this limit there is no phase-breaking. In the charge neutral limit, the dephasing rate is determined by the charge relaxation resistance alone, since C_μ^2/C^2 tends to 1. It is this later limit, that describes most often the physical situation which we encounter in mesoscopic conductors. The charging energy $E_c = e^2/2C$ is typically an order of magnitude larger than the level spacing [32]. Thus the dephasing rate is essentially determined by the charge relaxation resistance. Consequently, the knowledge we have gained on charge relaxation resistances can immediately be applied to the dephasing time.

We re-emphasize that the dephasing rate, as given by Eqs. (14.13, 14.15), cannot be compared with a weak localization dephasing time. We considered here only a uniform potential inside the cavity, and a uniform potential does not affect a weak localization loop, since both the backward and forward trajectories experience the same potential. Eqs. (14.13, 14.15) are sample specific results and as has been pointed out before are due to carriers entering or leaving the sample.

We must leave it as an open question, to what extent the dephasing time introduced above is physically meaningful. For small conductors, R_q is not well defined by its average nor its mean square fluctuations, but must be characterized by an entire distribution. Thus for small samples, it is similarly only meaningful to define a distribution of dephasing rates. But if the dephasing rate can only be characterized in terms of a distribution function, it becomes a less useful object. In principle, rather than defining a rate, it is desirable to evaluate the quantity of interest, such as the density of states of the cavity, the conductance which we consider next, directly, without relying on the concept of a dephasing rate.

3. ROLE OF EXTERNAL IMPEDANCE

The fluctuations in a mesoscopic conductor depend not only on the conductor itself, but also on the impedance of the external circuit. The considerations given above apply only for the case of vanishing external impedance. Let us now briefly discuss the case where the external impedance $Z_{ext}(\omega)$ does not vanish. The entire circuit can be described by writing two Langevin equations [1, 7] for the mesoscopic conductor and for the external circuit. For the mesoscopic conductor we have

$$\Delta I(\omega) = G(\omega)V(\omega) + \delta I(\omega). \tag{14.16}$$

Here $G(\omega)$ is determined by Eq. (14.2), $V(\omega)$ is the voltage drop between reservoir and gate, and $\delta I(\omega)$ are the current fluctuations for infinite impedance with a spectrum determined by Eq. (14.10). The fluctuating current in the external circuit is

$$\Delta I(\omega) = -Z_{ext}^{-1}(\omega)V(\omega) + \delta I_{ext}(\omega), \tag{14.17}$$

where $\delta I_{ext}(\omega)$ has a spectrum $S_{ext}(\omega) = 2kT/Z_{ext}(\omega)$. Consider the important case where the external impedance is just a dc-resistance R_{ext}. The resulting current-, charge-, and voltage-fluctuation spectrum can be found by replacing in Eqs. (14.10-14.12), the charge relaxation resistance R_q by $R_q + R_{ext}$. An external resistive impedance leads to an increase in the RC-time. As a consequence the dephasing rate is given by

$$\Gamma_\phi = (e^2/\hbar^2)kT(C_\mu^2/C^2)(R_q + R_{ext}). \tag{14.18}$$

A finite external resistive impedance increases the voltage fluctuations and shortens the dephasing time. The impedance conditions of the external circuit are thus important if we want to make a quantitative comparison with experiment. Below, we now return to investigate charge fluctuations and dephasing rates for the case of zero-impedance external circuits.

4. EQUILIBRIUM DEPHASING IN MULTILEAD SYSTEMS

Consider next a mesoscopic conductor that is connected to two or more reservoirs and capacitively coupled to a backgate. As an example, Fig. 14.3 shows a chaotic cavity which is connected via two point contacts with M_1 and M_2 open channels to two reservoirs and capacitively coupled to a back gate. The equilibrium charge fluctuations in this conductor correspond to a situation in which all the potentials at all contacts are at equilibrium. Thus for the determination of the equilibrium charge fluctuations (and for a zero impedance external circuit),

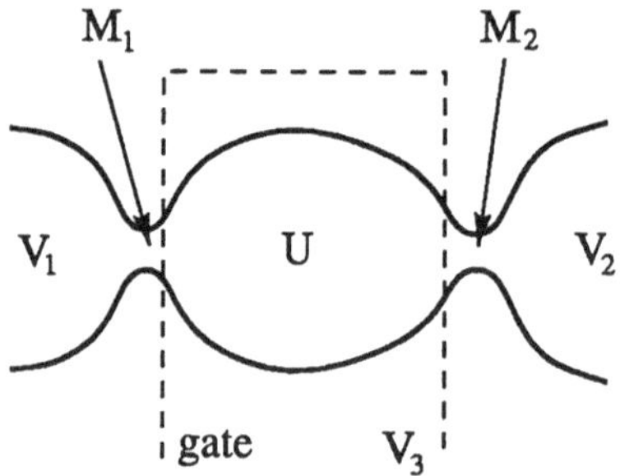

Figure 14.3 Two probe conductor connected to two electron-reservoirs and capacitively coupled to a gate. After Ref. [35].

it is irrelevant whether we consider the leads connected to different reservoirs or consider all the leads of the conductor connected to the same reservoir. But if all the leads are connected to the same reservoir it is evident that the theory presented above for the case of a single lead also applies to a conductor with many leads. All that is necessary is to use the full scattering matrix in the evaluation of the density of states and of the charge relaxation resistance. Therefore, for the discussion of equilibrium charge relaxation resistances we only need to distinguish the dimension of the scattering matrix (number of quantum channels) but not the number of leads.

Following is a summary of what is known on charge relaxation resistances for various mesoscopic systems (see also Ref. [35] for an overview):

For a capacitor with a single channel lead the charge relaxation resistance is universal and given by $R_q = h/2e^2$. For a nearly charge neutral cavity such that $C_\mu = C$, Eq. (14.15) gives

$$\Gamma_\phi = (4\pi^2/h)kT. \tag{14.19}$$

For a chaotic cavity connected via two *perfect single channel* leads to a reservoir the charge relaxation resistance R_q must be characterized by a distribution (which based on Dyson's circular ensemble can be given analytically [3]). Consequently, the dephasing rate is also given by a distribution function $P(\Gamma_\phi/\Gamma_0)$. Here $\Gamma_0 = (4\pi^2/h)kT$. For the orthogonal ensemble the symmetry parameter is $\beta = 1$ and for the unitary ensemble it is $\beta = 2$. For these two symmetry classes, Ref. [3] finds a distribution function which is non-zero only between $\Gamma_\phi/\Gamma_0 = 1/4$ and $\Gamma_\phi/\Gamma_0 = 1/2$ and is given by

$$P(\Gamma_\phi/\Gamma_0) = \begin{cases} 4, & \beta = 1, \\ 30(1 - 2\Gamma_\phi/\Gamma_0)\sqrt{4\Gamma_\phi/\Gamma_0 - 1}, & \beta = 2. \end{cases} \tag{14.20}$$

For a chaotic cavity (see Fig. 14.3) coupled to leads with many channels $M = M_1 + M_2$, R_q becomes well defined and exhibits only small fluctuations around its average [30]. Here N is the total number of channels of all contacts and M_1 and M_2 are the number of channels for a chaotic cavity coupled by contacts with M_1 and M_2 channels to reservoirs. On the ensemble average $R_q = (h/e^2)(M_1 + M_2)^{-1}$. Note that the conductances of the two quantum point contacts add in *parallel*. This is in contrast to the (ensemble averaged) conductance of the cavity which is $G^{-1} = (h/e^2)(M_1^{-1} + M_2^{-1})$ and corresponds to the series addition of the resistances of the contacts. Thus whether or not the effect discussed here contributes to the dephasing rate in a chaotic cavity is easy to test experimentally: Whereas for two very unequal contacts $M_1 \ll M_2$ the resistance is dominated by the smaller contact, (*i.e.* it is determined by M_1), the charge relaxation resistance and thus the dephasing rate discussed here is dominated by the larger contact M_2.

For a perfect ballistic one-channel wire [37] connecting two reservoirs, and capacitively coupled to a gate, the charge relaxation resistance is $R_q = h/4e^2$. For a two dimensional wire subject to a high magnetic field, such that the only extended states at the Fermi energy are edge states, and for a geometry in which a gate couples to one edge state [38], the charge relaxation resistance is [39] $R_q = h/2e^2$. If more than one edge state contributes to transport, R_q depends on the density of states of the different edge channels. Of much interest is the charge relaxation resistance of a quantum point contact and we discuss it now in some detail.

5. CHARGE RELAXATION RESISTANCE OF A QUANTUM POINT CONTACT

We consider a quantum point contact (QPC) formed with the help of gates such that it connects two two-dimensional electron gases [11, 12]. Here we treat these gates as macroscopic. For simplicity, we consider a symmetric contact: We assume that the electrostatic potential is symmetric for electrons approaching the contact from the left or from the right. This implies that the two gates are located symmetrically. We can combine the capacitances of the conduction channel to the two gates and consider a single gate as schematically shown in Fig. 14.1. For a symmetric scattering potential the scattering matrix (in a basis in which the transmission and reflection matrices are diagonal) is for the n-th channel of the form

$$s_n(E) = \begin{pmatrix} -i\sqrt{R_n}\exp(i\phi_n) & \sqrt{T_n}\exp(i\phi_n) \\ \sqrt{T_n}\exp(i\phi_n) & -i\sqrt{R_n}\exp(i\phi_n) \end{pmatrix}, \tag{14.21}$$

where T_n and $R_n = 1 - T_n$ are the transmission and reflection probabilities and ϕ_n is the phase accumulated by a carrier during reflection or transmission at the QPC. For the eigenvalues $exp(i\zeta_{n\pm})$ of this matrix a little algebra shows that the derivatives with respect to energy of the eigenphases $\zeta_{n\pm}$ are given by

$$\frac{d\zeta_{n\pm}}{dE} = \frac{d\phi_n}{dE} \pm \frac{1}{2T_n^{1/2}R_n^{1/2}}\frac{dT_n}{dE}. \tag{14.22}$$

For the density of states, Eq. (14.4), this leads to $N = (1/2\pi)\sum_n(d\zeta_{n+} + d\zeta_{n-}) = (1/\pi)\sum_n d\phi_n/dE$, and for the charge relaxation resistance, Eq. (14.6), we find

$$R_q = \frac{h}{4e^2}\frac{\sum_n\left[(\frac{d\phi_n}{dE})^2 + \frac{1}{4T_nR_n}(\frac{dT_n}{dE})^2\right]}{\left[\sum_n \frac{d\phi_n}{dE}\right]^2}. \tag{14.23}$$

If only a few channels are open the potential has in the center of the conduction channel the form of a saddle [40]:

$$V(x,y) = V_0 + \frac{1}{2}m\omega_y^2 y^2 - \frac{1}{2}m\omega_x^2 x^2, \tag{14.24}$$

where V_0 is the electrostatic potential at the saddle and the curvatures of the potential are parametrized by ω_x and ω_y. The resulting transmission probabilities have the form of Fermi functions $T(E) \equiv f(E) = 1/[e^{\beta(E-\mu)} + 1]$ (with a negative temperature $\beta = -2\pi/\hbar\omega_x$ and $\mu = \hbar\omega_y[n + 1/2] + V_0$). As a function of energy (gate voltage) the conductance rises step-like [11, 12]. The energy derivative of the transmission probability

$$\frac{dT_n}{dE} = -\beta f(1-f) = \frac{2\pi}{\hbar\omega_x}T_n(1-T_n) \tag{14.25}$$

is itself proportional to the transmission probability times the reflection probability. We note that such a relation holds not only for the saddle point model of a QPC but also for instance for the adiabatic model of Glazman *et al.* [41]. As a consequence

$$\frac{1}{4T_nR_n}\left(\frac{dT_n}{dE}\right)^2 = \left(\frac{\pi}{\hbar\omega_x}\right)^2 T_nR_n \tag{14.26}$$

is proportional to T_nR_n. Thus we can re-write the charge relaxation resistance of a saddle QPC as

$$R_q = \frac{h}{4e^2}\frac{\sum_n\left[(\frac{d\phi_n}{dE})^2 + (\frac{\pi}{\hbar\omega_x})^2T_nR_n\right]}{[\sum_n \frac{d\phi_n}{dE}]^2}. \tag{14.27}$$

To obtain the density of states of the n-th eigenchannel, we use the relation between density and phase (action) and $N_n = (1/\pi)d\phi_n/dE$. We evaluate the phase semi-classically. The spatial region of interest for which we have to find the density of states is the region over which the electron density in the contact is not screened completely. We denote this length by λ. The density of states is then found from $N_n = 1/h \int_{-\lambda}^{\lambda} \frac{dp_n}{dE} dx$ where p_n is the classically allowed momentum. A simple calculation gives a density of states [3]

$$N_n(E) = \frac{4}{h\omega_x} asinh \left(\sqrt{\frac{1}{2} \frac{m\omega_x^2}{E - E_n}} \lambda \right), \tag{14.28}$$

for energies E exceeding the channel threshold E_n and gives a density of states

$$N_n(E) = \frac{4}{h\omega_x} acosh \left(\sqrt{\frac{1}{2} \frac{m\omega_x^2}{E_n - E}} \lambda \right), \tag{14.29}$$

for energies in the interval $E_n - (1/2)m\omega_x^2\lambda \leq E < E_n$ below the channel threshold. Electrons with energies less than $E_n - \frac{1}{2}m\omega_x^2\lambda$ are reflected before reaching the region of interest, and thus do not contribute to the DOS. The resulting density of states has a logarithmic singularity at the threshold $E_n = \hbar\omega_y(n + \frac{1}{2}) + V_0$ of the n-th quantum channel. (A fully quantum mechanical calculation gives a density of states which exhibits also a peak at the threshold but which is not singular). The charge relaxation resistance for a saddle QPC is shown in Fig. 14.4 for a set of parameters given in the figure caption. The charge relaxation resistance exhibits a sharp spike at each opening of a quantum channel. Physically this implies that the relaxation of charge, determined by the RC-time is very rapid at the opening of a quantum channel.

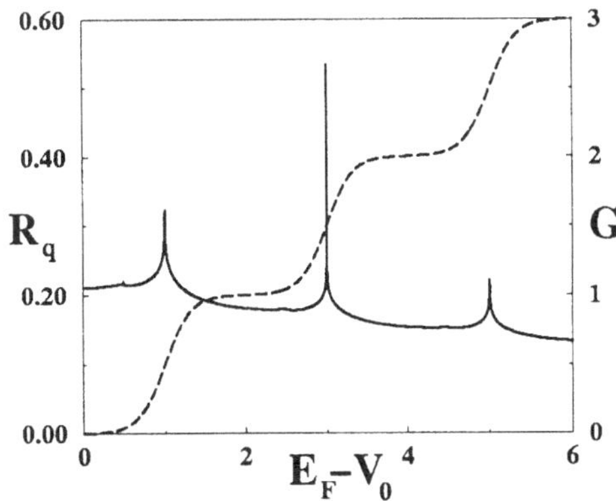

Figure 14.4 Charge relaxation resistance R_q of a saddle QPC in units of $h/4e^2$ for $\omega_y/\omega_x = 2$ and a screening length of $m\omega_x\lambda^2/\hbar = 25$ as a function of $E_F - V_0$ in units of $\hbar\omega_x$ (full line). The broken line shows the conductance of the QPC. After Ref. [42].

6. CHARGE FLUCTUATIONS AND THE SCATTERING MATRIX

We now present an approach which can be used to determine the fluctuations of the charge on Coulomb coupled conductors not only in the equilibrium state of the conductors but also if one or both conductors are in a transport state. First, we are concerned with the fluctuations of the total charge. Later on we will also derive expressions which permit the investigation of the local charge fluctuations.

We have already pointed out in the preceding discussion, the fact that interaction effects are very important for the discussion of charge and potential fluctuations. For a moment, however, we now consider non-interacting carriers and only later on we introduce screening. First we want to derive an operator for the total charge on a mesoscopic conductor. To this end we use the continuity equation and integrate it over the total volume of the conductor. This gives a relation between the charge in this volume and the particle currents entering this volume,

$$\sum_{\alpha} \hat{I}_{\alpha}(\omega) = i\omega e\hat{\mathcal{N}}(\omega), \tag{14.30}$$

where $\hat{I}_{\alpha}(\omega)$ is the current operator in lead α (see Appendix A) and $e\hat{\mathcal{N}}$ is the operator of the charge in the mesoscopic conductor. From the current operator (see Appendix A) we obtain [3] for the density operator

$$\hat{\mathcal{N}}(\omega) = \hbar \sum_{\alpha\beta\gamma} \sum_{mn} \int dE \hat{a}^{\dagger}_{\beta m}(E) \mathcal{N}_{\beta\gamma mn}(E, E+\hbar\omega) \hat{a}_{\gamma n}(E+\hbar\omega), \tag{14.31}$$

whrere $\hat{a}^{\dagger}_{\beta m}(E)$ (and $\hat{a}_{\beta m}(E)$) creates (annihilates) an incoming particle with energy E in lead β and channel m and where the diagonal and non-diagonal density of states elements $\mathcal{N}_{\beta\gamma mn}$ are

$$\mathcal{N}_{\beta\gamma mn}(E, E') = \frac{-i/2\pi}{(E-E')} \sum_{\alpha} \left[\delta_{mn}\delta_{\alpha\beta}\delta_{\alpha\gamma} - \sum_{k} s^{\dagger}_{\alpha\beta mk}(E) s_{\alpha\gamma kn}(E') \right]. \tag{14.32}$$

In particular, in the zero-frequency limit, we find [3] (in matrix notation),

$$\mathcal{N}_{\beta\gamma}(E) = \frac{1}{2\pi i} \sum_{\alpha} \mathbf{s}^{\dagger}_{\alpha\beta}(E) \frac{d\mathbf{s}_{\alpha\gamma}(E)}{dE}, \tag{14.33}$$

Eq. (14.33) is nothing but the multi-lead generalization of Eq. (14.3). Note that here we have implicitly assumed that the integration volume also enters the scattering matrices used: The integration volume intersects the leads of the conductor and we define the scattering matrix in such a way that it relates incoming and outgoing waves at these intersections. Proceeding as for the case of current fluctuations [6, 7, 10] we find for the fluctuation spectrum of the total charge

$$S_{NN}(\omega)=h\sum_{\gamma\delta}\int dE \mathrm{Tr}\left[\mathcal{N}^{\dagger}_{\gamma\delta}(E,E+\hbar\omega)\mathcal{N}_{\delta\gamma}(E+\hbar\omega,E)\right]F_{\gamma\delta}(E,\omega) \quad (14.34)$$

where the trace is over quantum channels and

$$F_{\gamma\delta}(E,\omega)=\{f_{\gamma}(E)\,[1-f_{\delta}(E+\hbar\omega)]+[1-f_{\gamma}(E)]\,f_{\delta}(E+\hbar\omega)\}\,, \quad (14.35)$$

is a combination of frequency-dependent Fermi functions. Eq. (14.34) is, in the absence of interactions, the general fluctuation spectrum of the charge on a mesoscopic conductor. The true charge on a mesoscopic conductor cannot be found from an analysis of free-particle fluctuations. As we have pointed out the charge operator and the fluctuation spectra depend on the integration volume. Screening is necessary, if only to eliminate this dependence. In some particular cases, the volume of integration can be physically motivated, for instance in the case of a cavity connected to contacts, the main effect we are interested in comes from the charging of the cavity itself and thus we limit the integration volume to that of the cavity.

Our next goal is to find the true charge, *i.e.* the charge that can build up on a mesoscopic conductor in the presence of interaction. Let us consider the low-frequency limit. Then the charge and the potential for the conductor shown in Fig. 14.2 are, on the one hand, related by $Q = CU$. Now we write this equation in operator form with a charge operator $\hat{Q}$ and a potential operator $\hat{U}$. We thus have $\hat{Q} = C\hat{U}$. On the other hand, the charge on the mesoscopic conductor can also be expressed in terms of the charge $e\hat{\mathcal{N}}$ injected from the reservoir (in response of an increase of the the reservoir voltage or due to a fluctuation), and a screened charge $e^2N\hat{U}$ which is the charge due to the response of the electrostatic potential to the injected charge. Here N is the average density of states. Thus in terms of the operators, we have

$$\hat{Q} = C\hat{U} = e\hat{\mathcal{N}} - e^2N\hat{U}. \quad (14.36)$$

This equation now permits us to express $\hat{U}$ in terms of the capacitance and the operator for the bare injected charges, $\hat{U} = e(C+e^2N)^{-1}\hat{\mathcal{N}}$ and gives

$$\hat{Q} = \frac{Ce\hat{\mathcal{N}}}{(C+e^2N)} = e\frac{C_{\mu}}{e^2}\frac{\hat{\mathcal{N}}}{N} \quad (14.37)$$

for the operator of the true charge. Here C_{μ} is the electrochemical capacitance, Eq. (14.5). Eq. (14.37) states that a charge injected into the conductor due to the population of incident states in an energy interval dE is not the bare charge $eNdE$ but only $eNdE/N(e^2/C_\mu) = C_\mu dE/e$.

Next we can use these expression to evaluate the (zero-frequency) fluctuation spectrum of the charge. We find

$$S_{QQ}(0) = \frac{C_\mu^2}{e^2}\frac{S_{NN}(0)}{N^2}. \tag{14.38}$$

Comparison with Eq. (14.11) gives us an expression for the charge relaxation resistance in terms of the fluctuation spectrum of the bare charges,

$$R_q = \frac{1}{2kTe^2}\frac{S_{NN}(0)}{N^2}. \tag{14.39}$$

At equilibrium the particle fluctuation spectrum is

$$S_{NN}(0) = 2kTh\mathrm{Tr}(\hat{\mathcal{N}}^\dagger\hat{\mathcal{N}})$$

and we recover for R_q the expression, Eq. (14.6).

As an example let us again consider the QPC as specified by the scattering matrix, Eq. (14.21). For the elements of the density of states matrix, Eq. (14.33), a little algebra leads to

$$\mathcal{N}_{11} = \mathcal{N}_{22} = \frac{1}{2\pi}\frac{d\phi_n}{dE}, \tag{14.40}$$

$$\mathcal{N}_{12} = \mathcal{N}_{21} = \frac{1}{4\pi}\frac{1}{\sqrt{R_n T_n}}\frac{dT_n}{dE}. \tag{14.41}$$

With these density of states elements, we can determine the particle fluctuation spectrum, Eq. (14.34) in the white-noise limit, and R_q with the help of Eq. (14.39). That leads again to R_q as given by Eq. (14.23) and Eq. (14.27).

7. THE MESOSCOPIC CAPACITOR: MESOSCOPIC GATE

Consider next the case of a system in which all components are mesoscopic. An example is the conductor shown in Fig. 14.1. A mesoscopic cavity is connected (via a single lead) to a reservoir and in proximity to a QPC. These two conductors can again be viewed as the two plates of a small capacitor. We assume that each electric field line emanating from the cavity ends up either again on the cavity or on the QPC. Now the fluctuations of the charge Q_1 on the cavity and the charge Q_2 on the QPC are related to the electrostatic potentials U_1 and U_2 on these two conductors via $Q_1 = C(U_1 - U_2)$ and $Q_2 = C(U_2 - U_1)$. Note, overall there is no charge accumulation, $Q_1 + Q_1 = 0$. We now present a discussion of the fluctuations of the charge fluctuations of this conductor, generalizing the approach outlined above.

The charge on conductor i can also be written in terms of the bare fluctuating charges $e\hat{\mathcal{N}}_i$, counteracted by a screening charge $eN_i ed\hat{U}_i$,

$$\hat{Q}_1 = C(\hat{U}_1 - \hat{U}_2) = e\hat{\mathcal{N}}_1 - e^2 N_1 \hat{U}_1, \qquad (14.42)$$

$$\hat{Q}_2 = C(\hat{U}_2 - \hat{U}_1) = e\hat{\mathcal{N}}_2 - e^2 N_2 \hat{U}_2. \qquad (14.43)$$

Note that the charge conservation immediately leads to $\hat{Q}_2 = -\hat{Q}_1$. We have a dipole and $\hat{Q}$ is the charge operator of the dipole. We solve these equations for $\hat{U}_1$ and $\hat{U}_2$. Using $D_i \equiv e^2 N_i$ for the density of states we find that the effective interaction G_{ij} between the two systems is

$$\mathbf{G} = \frac{C_\mu}{D_1 D_2 C} \begin{pmatrix} C + D_2 & C \\ C & C + D_1 \end{pmatrix}. \qquad (14.44)$$

The electrochemical capacitance is the series capacitance of the geometrical contribution and the quantum contribution of the two conductors [1],

$$C_\mu^{-1} = C^{-1} + (e^2 N_1)^{-1} + (e^2 N_2)^{-1}. \qquad (14.45)$$

With Eq. (14.44) we find for the potential operators

$$\hat{U}_i = e \sum_j G_{ij} \hat{\mathcal{N}}_j. \qquad (14.46)$$

In the low frequency limit of interest here the elements of the density of states matrix $\mathcal{N}^{(i)}_{\gamma\delta}$ are specified by Eq. (14.33). Using Eqs. (14.46) and (14.34) we find that at equilibrium the low frequency fluctuations of the potential in conductor 1 are given by

$$S_{U_1 U_1}(0) = 2kT \left(\frac{C_\mu}{C}\right)^2 \left[\left(\frac{C + D_2}{D_2}\right)^2 R_q^{(1)} + \left(\frac{C}{D_1}\right)^2 R_q^{(2)}\right] \qquad (14.47)$$

with $R_q^{(i)}$ determined by Eq. (14.6) using the scattering matrix of conductor i. Similarly,

$$S_{U_2 U_2}(0) = 2kT \left(\frac{C_\mu}{C}\right)^2 \left[\left(\frac{C}{D_2}\right)^2 R_q^{(1)} + \left(\frac{C + D_1}{D_1}\right)^2 R_q^{(2)}\right]. \qquad (14.48)$$

The voltage fluctuations are correlated,

$$S_{U_1 U_2}(0) = 2kT \left(\frac{C_\mu}{C}\right)^2 \left[\frac{(C + D_2)C}{D_2^2} R_q^{(1)} + \frac{(C + D_1)C}{D_1^2} R_q^{(2)}\right]. \qquad (14.49)$$

With the help of these spectra we also obtain the spectrum $S_{UU}(0)$ of the potential difference $U = U_1 - U_2$, which is simply given by

$$S_{UU}(0) = 2kT \left(\frac{C_\mu}{C}\right)^2 \left[R_q^{(1)} + R_q^{(2)}\right]. \tag{14.50}$$

The charge relaxation resistances of the individual systems add [1]. In the non-interacting limit (infinite capacitance C) the spectra, Eqs. (14.47-14.49), all tend to the same limit,

$$S_{U_iU_j}(0) = 2kT \frac{1}{(D_1 + D_2)^2} \left[D_1^2 R_q^{(1)} + D_2^2 R_q^{(2)}\right], \tag{14.51}$$

whereas the spectrum of the voltage difference S_{UU} vanishes. In the physically more important, charge neutral limit, we have for the auto-correlation spectra

$$S_{U_iU_i}(0) = 2kTR_q^{(i)} \tag{14.52}$$

and the correlation spectrum Eq. (14.49) vanishes. The fluctuation spectrum of the voltage difference is given by Eq. (14.50) with $C_\mu/C = 1$, *i.e.* it is determined just by the sum of the two charge relaxation resistances. With the help of these spectra, we can now find the dephasing rates in the two conductors. A carrier in conductor 1 is subject to the fluctuating potential U_1 and a carrier in conductor 2 sees the fluctuating potential U_2. Thus the dephasing rates [21] are $\Gamma_\phi^{(i)} = (e^2/\hbar^2 t)\langle \int dt' \int dt'' U_i(t') U_i(t'')\rangle$. The fluctuation spectra $S_{U_1U_1}$ and $S_{U_2U_2}$ each contain two contributions which are additive. For instance, the first term in $S_{U_1U_1}$ represents a contribution to the fluctuation spectrum which can be viewed as being predominantly due to fluctuations in the conductor 1 itself, whereas the second term is due to the presence of the second conductor. We can separate the dephasing rate of each conductor into two contributions, $\Gamma_\phi^i = \Gamma_\phi^{i1} + \Gamma_\phi^{i2}$ where the first (upper) index i indicates the conductor and the second upper index indicates whether this rate is due to the conductor itself (index ii) or due to the presence of the other conductor (index ij with $i \neq j$). We thus obtain the following dephasing rates

$$\Gamma_\phi^{11} = \frac{e^2}{\hbar^2} kT \left(\frac{C_\mu}{C}\right)^2 \left(\frac{C + D_2}{D_2}\right)^2 R_q^{(1)}, \tag{14.53}$$

$$\Gamma_\phi^{12} = \frac{e^2}{\hbar^2} kT \left(\frac{C_\mu}{D_1}\right)^2 R_q^{(2)}, \tag{14.54}$$

$$\Gamma_\phi^{21} = \frac{e^2}{\hbar^2} kT \left(\frac{C_\mu}{D_2}\right)^2 R_q^{(1)}, \tag{14.55}$$

$$\Gamma_\phi^{22} = \frac{e^2}{\hbar^2} kT \left(\frac{C_\mu}{C}\right)^2 \left(\frac{C + D_1}{D_1}\right)^2 R_q^{(2)}. \tag{14.56}$$

In the large capacitance limit, this gives for the dephasing rates,

$$\Gamma_{\phi}^{11} = \frac{e^2}{\hbar^2} kT \left(\frac{D_1}{D_1 + D_2} \right)^2 R_q^{(1)}, \tag{14.57}$$

$$\Gamma_{\phi}^{12} = \frac{e^2}{\hbar^2} kT \left(\frac{D_2}{D_1 + D_2} \right)^2 R_q^{(2)}. \tag{14.58}$$

In this limit the rates depend on the ratio of the density of states of the two conductors. In contrast, in the zero-capacitance limit, Eq. (14.54) gives for the dephasing rate,

$$\Gamma_{\phi}^{11} = \frac{e^2}{\hbar^2} kT R_q^{(1)} \tag{14.59}$$

whereas Γ_{ϕ}^{12} becomes proportional C^2 and thus vanishes in the charge-neutral limit. In the important case that C is small compared to the quantum capacitance (the Coulomb energy is large compared to the level spacing) we find

$$\Gamma_{\phi}^{12} = \frac{e^2}{\hbar^2} kT \left(\frac{C}{D_1} \right)^2 R_q^{(2)}. \tag{14.60}$$

We have seen that for a single mesoscopic conductor, coupled to a back gate, the treatment of the single lead case, can immediately be generalized to the case of a many lead conductor. Similarly, the case of two coupled mesoscopic conductors, each of them connected via several leads to different reservoirs, is in fact contained in the expressions given above. We only need to use the full scattering matrix of each conductor to evaluate the density of states and the charge relaxation resistance. This is true only for the equilibrium fluctuations. The non-equilibrium situation requires special thought.

8. NONEQUILIBRIUM CHARGE FLUCTUATIONS AND DEPHASING

Let us consider two mesoscopic conductors as above but let us consider the case that one of these conductors, say conductor 2, is connected to two reservoirs. An example of such a conductor is shown in Fig. 14.5. One of the conductors, a quantum dot or chaotic cavity, is located either in position A (to be discussed below) or in position B (to be discussed later on). A second conductor consists of a wire patterned into a two dimensional electron gas. It contains a quantum point contact (QPC). The sample is subject to a high magnetic field which generates edge states (indicated by lines with arrows). In situation A it is the charge on the quantum dot and the charge in the center of the QPC which

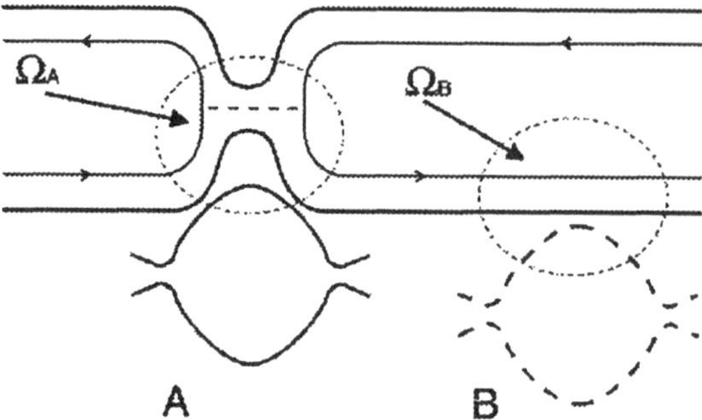

Figure 14.5 Quantum point contact coupled to a quantum dot either in position A or B. After Ref. [21].

are Coulomb coupled. In situation B the charge on the quantum dot is Coulomb coupled with the charge on the edge state far away from the QPC.

Let us now consider the case, where conductor 2 (the wire with the QPC) is brought into a transport state. For simplicity, we consider a voltage $eV \gg kT$ so large that the thermal noise in this conductor can be neglected. To investigate the effect which the presence of conductor 2 has on conductor 1, we can thus take conductor 2 to be in the zero-temperature limit. Even at zero temperature, there is noise generated in conductor 2 which is shot noise due to the granularity of the charge [10]. We are interested in the charge fluctuations of this conductor and reconsider Eq. (14.34). In the zero-temperature limit, Eq. (14.34) is now evaluated with a Fermi function for contact 1 of this conductor at an electrochemical potential μ_1 and a Fermi function for contact 2 at an electrochemical potential μ_2. To be definite, we take $\mu_1 > \mu_2$ such that a voltage $eV = \mu_1 - \mu_2$ is established. The resulting bare particle number fluctuation spectrum, Eq. (14.34), is now determined only by a subset of the density of states matrix elements and is proportional to the applied voltage,

$$S_{NN}^{(2)}(0) = 2h\,\mathrm{Tr}\left[(\mathcal{N}_{21}^{(2)})^{\dagger}\mathcal{N}_{21}^{(2)}\right] e|V|. \tag{14.61}$$

The upper index (2) is to indicate that this spectrum and the matrix elements are evaluated for conductor 2. Eq. (14.61) is valid to linear order in the applied voltage only. To linear order in the applied voltage, we can also in the presence of transport use Eqs. (14.42) and (14.43), since screening is determined by the equilibrium density of states N evaluated at the Fermi energy. The nonequilibrium particle density fluctuation spectrum leads to a novel resistance

$$R_v^{(2)} = \frac{h}{e^2} \frac{\mathrm{Tr}\left[(\mathcal{N}_{21}^{(2)})^\dagger \mathcal{N}_{21}^{(2)}\right]}{N^2}. \tag{14.62}$$

Formally we can define this novel resistance via $R_v = e^2 S_{NN}(0)/2eV$. With this resistance we now find for the voltage fluctuation spectra

$$S_{U_1U_1}(0) = 2\left(\frac{C_\mu}{C}\right)^2 \left[\left(\frac{C+D_2}{D_2}\right)^2 R_q^{(1)} kT + \left(\frac{C}{D_1}\right)^2 R_v^{(2)} e|V|\right] \tag{14.63}$$

with $R_q^{(i)}$ determined by Eq. (14.6). For the voltage fluctuation spectrum in conductor 2 we find,

$$S_{U_2U_2}(0) = 2\left(\frac{C_\mu}{C}\right)^2 \left[\left(\frac{C}{D_2}\right)^2 R_q^{(1)} kT + \left(\frac{C+D_1}{D_1}\right)^2 R_v^{(2)} e|V|\right]. \tag{14.64}$$

The voltage fluctuations are correlated,

$$S_{U_1U_2}(0) = 2\left(\frac{C_\mu}{C}\right)^2 \left[\frac{(C+D_2)C}{D_2^2} R_q^{(1)} kT + \frac{(C+D_1)}{D_1^2} R_v^{(2)} e|V|\right]. \tag{14.65}$$

With the help of these spectra we also obtain the spectrum $S_{UU}(0)$ of the potential difference $U = U_1 - U_2$, which is simply given by

$$S_{UU}(0) = 2\left(\frac{C_\mu}{C}\right)^2 \left[R_q^{(1)} kT + R_v^{(2)} e|V|\right]. \tag{14.66}$$

Whereas the dephasing rate Γ_ϕ^{11} remains unchanged and is given by Eq. (14.54) the dephasing rate Γ_ϕ^{12} is now determined by $R_v eV$ and is given by

$$\Gamma_\phi^{12} = \frac{e^2}{\hbar^2}\left(\frac{C_\mu}{D_1}\right)^2 R_v^{(2)} e|V| \tag{14.67}$$

In the large capacitance limit $(C_\mu/D_1)^2$ approaches $[D_2/(D_1+D_2)]^2$ whereas in the small capacitance limit $(C_\mu/D_1)^2$ tends to $(C/D_1)^2$.

9. THE RESISTANCE R_V OF A QUANTUM POINT CONTACT

We now discuss R_v for the specific example of a QPC. The calculation applies to the geometry of Fig. 14.1 or to the geometry of Fig. 14.5 with the quantum dot in position A. Using the density matrix elements for a symmetric QPC given by Eqs. (14.41), we find [3]

$$R_v = \frac{h}{e^2} \frac{\sum_n \frac{1}{4R_nT_n}\left(\frac{dT_n}{dE}\right)^2}{\left[\sum_n (d\phi_n/dE)\right]^2} = \frac{h}{e^2}\left(\frac{\pi}{\hbar\omega_x}\right)^2 \frac{\sum_n T_nR_n}{\left[\sum_n (d\phi_n/dE)\right]^2}, \quad (14.68)$$

where we have made use of the fact that the transmission probabilities have the form of Fermi functions, Eq. (14.26). Note that the resistance R_v is proportional to the shot noise power $S(0) = 2(e^2/h)\sum_n T_nR_n eV$.

This statement is in contrast to the experimental papers [16, 17] and to theoretical works [19] where arguments are presented which lead to a dephasing rate proportional to $(\Delta T)^2/4TR$. The arguments which are advanced take the factor TR as being due to the shot noise and ΔT as the change in the transmission probability due to the variation of the charge on the QPC. According to these arguments, the dephasing rate is inversely proportional to the shot noise power S. Of course we are not forbidden to relate TR to the zero temperature current noise spectrum. But this identification breaks down, if we are not strictly at zero temperature, if the channels of the QPC do not open in a well separated way (and several channels contribute) *etc.* More importantly, as the quantum dot is moved from position A in Fig. 14.5 to position B the resulting dephasing rate is, as we will see, unambiguously proportional to the shot noise. Clearly arguments which hold that the dephasing rate is inversely proportional to the shot noise in configuration A but proportional to the shot noise in configuration B lead to a paradox.

Also Eq. (14.68) is valid for a saddle point constriction, the adiabatic model of Glazman *et al.* [41] leads to the same conclusion. Thus it stands to reason that Eq. (14.26) is in fact much more general than the simple model used indicates.

To evaluate Eq. (14.68) further, Ref. [3] calculates the phase derivatives $d\phi_n/dE$ semi-classically in WKB approximation. The resulting resistance R_v is shown in Fig. 14.6. Application of a magnetic field changes R_v only qualitatively (see Ref. [21]).

Without screening R_v would exhibit a bell shaped behavior as a function of energy, *i.e.* it would be proportional to $T_n(1 - T_n)$ in the energy range in which the n-th transmission channel is partially open. Screening, which in R_v is inversely proportional to the density of states squared, generates the dip at the threshold of the new quantum channel at the energy which corresponds to $T_n = 1/2$. In the semi-classical evaluation used here the density of states is singular at this point. Quantum corrections will reduce the density of states and will tend to diminish the dip shown in Fig. 14.6. It is interesting to note that the experiment [16] does indeed show a double hump behavior of the dephasing rate.

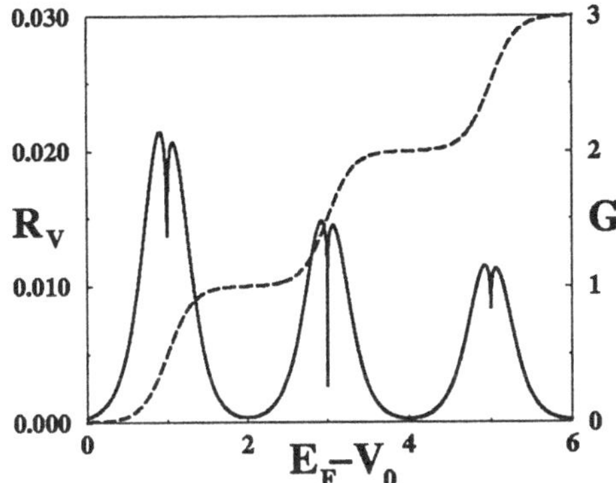

Figure 14.6 R_v (solid line) for a saddle QPC in units of h/e^2 and G (dashed line) in units of e^2/h as a function of $E_F - V_0$ in units of $\hbar\omega_x$ with $\omega_y/\omega_x = 2$ and a screening length $m\omega_x\lambda^2/\hbar = 25$. R_v and G are for spinless electrons. After Ref. [42].

10. LOCAL CHARGE FLUCTUATIONS

Thus far we discussed situations in which the total charge fluctuation determine the potential fluctuations and the dephasing rates. Many additional geometries, however, require a discussion not only of the total charge fluctuations, but a consideration of the local charge.

To describe the charge distribution due to carriers in an energy interval dE in our conductor, we consider the Fermi-field [7]

$$\hat{\Psi}(\mathbf{r},t) = \sum_{\alpha m} \int dE \left[\frac{1}{hv_{\alpha m}(E)}\right]^{1/2} \psi_{\alpha m}(\mathbf{r},E)\hat{a}_{\alpha m}(E)e^{-iEt/\hbar}, \quad (14.69)$$

which annihilates an electron at point $\mathbf{r}$ and time t. The Fermi field Eq. (14.69) is built up from all scattering states $\psi_{\alpha m}(\mathbf{r},E)$ which have unit incident amplitude in contact α in channel m. The operator $\hat{a}_{\alpha m}(E)$ annihilates an incident carrier in reservoir α in channel m. $v_{\alpha m}$ is the velocity in the incident channel m in reservoir α. The local carrier density at point $\mathbf{r}$ and time t is determined by $\hat{n}(\mathbf{r},t) = \hat{\Psi}^{\dagger}(\mathbf{r},t)\hat{\Psi}(\mathbf{r},t)$. We will investigate the density operator in the frequency domain, $\hat{n}(\mathbf{r},\omega)$. Using the Fermi-field we find,

$$\begin{aligned}\hat{n}(\mathbf{r},\omega) \;=\; & \sum_{\alpha m\beta n} \int dE\, [1/hv_{\alpha m}(E)]^{1/2}\, [1/hv_{\beta n}(E+\hbar\omega)]^{1/2} \\ & \psi^{*}_{\alpha m}(\mathbf{r},E)\psi_{\beta n}(\mathbf{r},E+\hbar\omega)\hat{a}^{\dagger}_{\alpha m}(E)\hat{a}_{\beta n}(E+\hbar\omega). \quad (14.70)\end{aligned}$$

This equation defines a density matrix operator with elements

$$n_{\gamma\delta mn}(\mathbf{r},E,E+\hbar\omega) = \frac{\psi^{*}_{\gamma m}(\mathbf{r},E)\psi_{\delta n}(\mathbf{r},E+\hbar\omega)}{h[v_{\gamma m}(E)v_{\delta n}(E+\hbar\omega)]^{1/2}}. \quad (14.71)$$

It is now very convenient and instructive to consider an expression for the density operator not in terms of wave functions but more directly in terms of the scattering matrix. In the zero-frequency limit, in matrix form, we can connect the scattering states and the scattering matrix with the density of states matrix [25],

$$\mathbf{n}_{\beta\gamma}(\alpha, \mathbf{r}) = -(1/4\pi i)[\mathbf{s}^{\dagger}_{\alpha\beta}(\delta \mathbf{s}_{\alpha\gamma}/\delta eU(\mathbf{r})) - (\delta \mathbf{s}^{\dagger}_{\alpha\beta}/\delta eU(\mathbf{r}))\mathbf{s}_{\alpha\gamma}]. \quad (14.72)$$

All quantities in this expression are evaluated at the energy E. The matrix elements of Eq. (14.71) are connected to the matrices $\mathbf{n}_{\beta\gamma}(\alpha, \mathbf{r})$ via

$$\mathbf{n}_{\gamma\delta}(\mathbf{r}) = \sum_{\alpha} \mathbf{n}_{\beta\gamma}(\alpha, \mathbf{r}) \quad (14.73)$$

The price we have to pay to gain local information is to use functional derivatives of the scattering matrix with respect to the local potential.

Eq. (14.73) was given in Ref. [25] without proof. Appendix B outlines how this result is obtained. We are not interested in the microscopic local charge fluctuations but we will consider the fluctuating charge in a certain volume Ω. To make the problem treatable, we will make the assumption that in the volume of interest the potential can be described by a single variable. Thus the quantities of interest are obtained by integrating the density operator Eq. (14.73) over a volume Ω,

$$\mathcal{N}_{\beta\gamma} = \int_{\Omega} d^3\mathbf{r}\, \mathbf{n}_{\beta\gamma}(\mathbf{r}). \quad (14.74)$$

With the help of the charge density matrix the low frequency limit of the bare charge fluctuations can be obtained [1, 3, 21]. It is again given by Eq. (14.34) but now with the local density of states matrix: the elements of $\mathcal{N}_{\gamma\delta}$ are in the zero-frequency limit of interest here given by Eq. (14.74). Thus we are now in a position to find the local charge fluctuations in any volume element of interest.

11. CHARGE FLUCTUATIONS OF AN EDGE STATE

Consider the conductor shown in Fig. 14.5 but with conductor 1 located at B. Now the charge fluctuations on the quantum dot couple to the charge fluctuations on the edge state. To simplify the problem, we assume that the relevant charge fluctuations in conductor 2 are of importance only near the quantum dot (in the region Ω_B of Fig. 14.5). The rest of the conductor is treated as charge neutral.

The scattering matrix of the QPC alone can be described by $r \equiv s_{11} = s_{22} = -i\mathcal{R}^{1/2}$ and $t \equiv s_{21} = s_{12} = \mathcal{T}^{1/2}$ where $\mathcal{T} = 1 - \mathcal{R}$ is the transmission probability through the QPC. Here the indices 1 and 2 label the reservoirs (see Fig. 14.5). We can entirely neglect the phase accumulated by carriers traversing the QPC. However, a carrier traversing the region underneath the gate acquires a phase $\phi(U)$ which depends on the electrostatic potential U in this region. Since we consider only the charge pile up in this region all additional phases in the scattering problem are here without importance. The total scattering matrix of the QPC and the traversal of the region Ω is then simply

$$\mathbf{s} = \begin{pmatrix} r & t \\ te^{i\phi} & re^{i\phi} \end{pmatrix}. \tag{14.75}$$

If the polarity of the magnetic field is reversed the scattering matrix is given by $s_{\alpha\beta}(B) = s_{\beta\alpha}(-B)$, *i.e.* in the reversed magnetic field it is only the second column of the scattering matrix which contains the phase $\phi(U)$. In what follows, the dependence of the scattering matrix on the phase ϕ is crucial. We emphasize that the approach presented here can be generalized by considering all the phases of the problem and by considering these phases and the amplitudes to depend on the entire electrostatic potential landscape [25]. In the evaluation of the density of states matrix elements , we make use of the fact that $d\phi/edU = -d\phi/dE$ in the WKB-limit. However, $d\phi/dE = 2\pi N$ where N is just the density of states of the edge state underneath the gate (in region Ω_B of Fig. 14.5). Simple algebra gives [21, 42] $\mathcal{N}_{11} = \mathcal{T}N$,

$$\mathcal{N}_{21} = \mathcal{N}_{12}^{*} = r^{*}tN, \tag{14.76}$$

and $\mathcal{N}_{22} = \mathcal{R}N_2$. Using only the zero-frequency limit of the elements of the charge operator determined above gives,

$$S_{NN}(\omega) = hN^2 \Big[\mathcal{T}^2 \textstyle\int dE\ F_{11}(E,\omega) + \mathcal{T}\mathcal{R} \int dE\ F_{12}(E,\omega) \\ + \mathcal{T}\mathcal{R} \int dE\ F_{21}(E,\omega) + \mathcal{R}^2 \int dE\ F_{22}(E,\omega)\Big], \tag{14.77}$$

which in the zero-frequency limit is,

$$S_{NN}(\omega) = 2hN^2 \Big[\mathcal{T} \textstyle\int dE\ f_1(1-f_1) + \mathcal{T}\mathcal{R} \int dE\ (f_1 - f_2)^2 \\ + \int dE\ \mathcal{R} f_2(1-f_2)\Big]. \tag{14.78}$$

The first term represents the equilibrium noise which is transmitted from contact 1 through the QPC into the edge channel adjacent to the quantum dot. The last term is the equilibrium noise of reservoir 2 which is fed into this edge state through reflection at the QPC. The middle term is the non-equilibrium, (two particle) shot noise contribution. In the zero temperature limit, the equilibrium noise terms do not contribute, and the shot noise term is proportional to the applied voltage and given by

$$S_{NN}(0) = 2hN^2 \mathcal{R}\mathcal{T} e|V|. \tag{14.79}$$

To proceed, we can again consider Eqs. (14.42) and (14.43). The charge operator $d\hat{Q}_1$ describes as before the charge fluctuations on the quantum dot, $d\hat{Q}_2$ the charge fluctuations on the edge state in proximity to the quantum dot. Now $\hat{\mathcal{N}}_2$ is the bare particle fluctuation operator on the edge state, and N_2 and $\hat{\mathcal{U}}_2$ are the density of states of the edge state in Ω_B and the potential operator in the region Ω_B. The potential fluctuations in conductor 1 are again given by Eq. (14.63) but with a resistance $R_v^{(2)}$ which is [21]

$$R_v^{(2)} = (h/e^2)\mathcal{R}\mathcal{T}. \tag{14.80}$$

For a single edge state the resistance R_v is independent of the density of states. The resistance R_v is again proportional to the shot noise generated by the QPC. It is maximal for a semi-transparent QPC, $\mathcal{T} = \mathcal{R} = 1/2$. The resulting dephasing rate is [21]

$$\Gamma_\phi^{12} = \frac{e^2}{\hbar^2}\left(\frac{C_\mu}{D_1}\right)^2 R_v^{(2)} eV = 4\pi^2 \left(\frac{C_\mu}{D_1}\right)^2 \mathcal{R}\mathcal{T}e|V|. \tag{14.81}$$

Suppose now that the quantum dot is at resonance. Then its density of states is given by $D_1 = 2e^2/(\pi\Gamma)$ where Γ denotes the width of the quantum level due to decay of carriers into the leads. We then obtain

$$\Gamma_\phi^{12} = \pi^4\Gamma^2 \left(\frac{C_\mu}{e^2}\right)^2 \mathcal{R}\mathcal{T}e|V|. \tag{14.82}$$

In the limit $e^2/C \gg \pi\Gamma/2$ and $e^2/C \gg N_2^{-1}$ the capacitance C_μ can be replaced by the geometrical capacitance C. In the opposite limit, if $e^2/C \ll \pi\Gamma/2$ and $e^2/C \ll N_2^{-1}$ we find a dephasing rate

$$\Gamma_\phi^{12} = \pi^4(N_2\Gamma)^2\mathcal{R}\mathcal{T}e|V|. \tag{14.83}$$

The limit described by Eq. (14.82) is realistic for mesoscopic samples.

The dephasing rate given by Eq. (14.82) is inversely proportional to the square of the coupling constant e^2/C. It vanishes as this coupling constant tends to infinity, since charge fluctuations become energetically prohibitive. In contrast, a perturbation treatment which considers only the bare charge fluctuations would lead to a result that is proportional to the square of the coupling constant. Second, the rate given by Eq. (14.82) is proportional to the square of the width of the resonance in the quantum dot.

Ref. [21] presents two additional results. First if there are more than two edge states which contribute to transport the charge resistance R_v is modified. As an example, consider the case, where one of the edge states is transmitted perfectly through the conductor. Transport in this edge state (in the zero-temperature limit) generates no current noise. But it can contribute to screening if the two edge states are not to far apart near the gate. If both edge states see the same potential U_2, the resistance R_v becomes,

$$R_v = \frac{h}{e^2}\left(\frac{N_2}{N_1+N_2}\right)^2 \mathcal{T}\mathcal{R}e|V|, \tag{14.84}$$

where N_1 is the density of states of the outer edge state in Ω_B and N_2 is the density of states of the the inner only partially transmitted edge state in region Ω_B. Eq. (14.84) is valid if there is no population equilibration (due to elastic or inelastic scattering) among the two edge channels between the QPC and the dot. Thus in the presence of additional edge states, the resistance R_v is reduced below that of a single edge state.

Another result given in Ref. [21] is of experimental interest [17]. If a voltage probe is inserted between the QPC and the quantum dot in position B, and if the voltage probe is such that every carrier enters it, it acts as a complete phase randomizer [43]. Ref. [21] shows that for a single edge state R_v and thus the dephasing rate, remains unchanged by the presence of the voltage probe, if the voltage probe is ideal. An ideal voltmeter has infinite impedance. To derive this result, it is necessary to know the ac-conductance matrix of the conductor, since the voltage at the probe is now a time-dependent fluctuating quantity. Its Fourier amplitude acts thus like an ac-voltage applied at this contact. Ref. [21] only stated the result, for a more detailed discussion we must refer the reader to Ref. [23].

12. DISCUSSION

In this work we have discussed the relationship between the dephasing rate and quantities which determine the RC-time in Coulomb coupled structures. The RC-time reflects a collective behavior of the electrons in these two conductors. In fact, as we have pointed out, as far as the total charge is concerned, the Coulomb interaction, imposes, via the requirement of charge neutrality over large distances, a complete correlation between the charge fluctuations on the two conductors. The RC-time is determined by an electrochemical capacitance and at equilibrium by a charge relaxation resistance R_q and in the presence of transport by a resistance R_v.

We have emphasized the simple case, where each conductor is only described by one potential. There is then only one dipole whose fluctuations govern the charge dynamics of the coupled conductors. The theory is not limited to such a simplified discussion. A formal expression for the electrochemical capacitance [44] and for the equilibrium charge relaxation resistance [25] using the microscopic potential landscape have already been derived. However, for the geometries which do not allow for any geometrical symmetries, such general expressions are difficult to evaluate. More fruitful is an approach, which proceeds like the discussion of ac-conductance, by dividing the sample into a limited series of volumes whose potentials and charges are related by a geometrical capacitance matrix. In such a discrete potential model it is then possible with some finite algebraic effort, to present a theory which contains not only the dynamics of one fluctuating dipole, but that of a number of dipoles.

We conclude by mentioning that recently progress has also been made to extend the theory presented here to normal-superconducting, hybrid structures [45, 46]. This is interesting since charge and particle fluctuations due to the presence of electron and hole quasi-particles are not simply proportional to one another as in a normal conductor.

Appendix A: The current operator

In this appendix we recall some results form the scattering theory of electrical conduction which are needed in the main part of this work. We consider a conductor and assume that its internal electrostatic potential is fixed at its static equilibrium value. The reservoirs are represented as perfect conductors which permit a sequence of transverse states with threshold energy below the Fermi energy. These states form the quantum channels which permit the definition of in-going and outgoing particle fluxes. The current operator in such a reservoir is,

$$\hat{I}_{\alpha}(\omega) = e \int dE [\hat{a}^{\dagger}_{\alpha}(E)\hat{a}_{\alpha}(E+\hbar\omega) - \hat{b}^{\dagger}_{\alpha}(E)\hat{b}_{\alpha}(E+\hbar\omega)] \qquad (14.85)$$

where $\hat{a}^{\dagger}_{\alpha}(E)$ creates an incoming particle flux in reservoir α. $\hat{a}_{\alpha}(E)$ is an M_{α} component vector: one component for each transverse channel with threshold below the Fermi energy. $\hat{b}^{\dagger}_{\alpha}(E)$ creates out-going fluxes in lead α. Eq. (14.85) is just the Fourier transform in frequency-space of the time-dependent occupation number of the incoming currents minus the occupation number of the outgoing currents. The operators $\hat{a}_{\alpha}(E)$ and $\hat{b}_{\alpha}(E)$ are not independent but related by a unitary transformation [7]:

$$\hat{b}_{\alpha} = \sum_{\beta} \mathbf{s}_{\alpha\beta} \hat{a}_{\beta}. \qquad (14.86)$$

Here $\mathbf{s}_{\alpha\beta}$ is the scattering matrix with dimensions $M_{\alpha} \times M_{\beta}$ which relates the incoming (current) amplitudes to the outgoing current amplitudes.

Using Eq. (14.86) to eliminate the occupation numbers of the outgoing channels in terms of the ingoing channels yields a current operator [7]

$$\hat{I}_\alpha(\omega) = e \int dE \sum_{\beta\gamma} \hat{a}^\dagger_\beta(E) \mathbf{A}_{\beta\gamma}(\alpha, E, E+\hbar\omega) \hat{a}_\gamma(E+\hbar\omega). \qquad (14.87)$$

with a *current matrix*

$$\mathbf{A}_{\delta\gamma}(\alpha, E, E') = \delta_{\alpha\delta}\delta_{\alpha\gamma}\mathbf{1}_\alpha - \mathbf{s}^\dagger_{\alpha\delta}(E)\mathbf{s}_{\alpha\gamma}(E'). \qquad (14.88)$$

Here $\mathbf{1}_\alpha$ is the unit matrix with dimensions $M_\alpha \times M_\alpha$. The properties of the current matrix are discussed in detail in Ref. [7].

Appendix B: Scattering matrix expression for local densities

To derive Eq. (14.73), we start from the Schrödinger equation for carriers in an electrostatic potential $U(\mathbf{r})$. The scattering state $\psi_\delta(\mathbf{r}, E)$ is an exact solution of this equation. Now we add a small complex valued potential $-i\Gamma(\mathbf{r})$ to the real electrostatic potential $U(\mathbf{r})$. The complex valued potential is non-vanishing only in a small region around $\mathbf{r}$. The complex valued potential changes the scattering state $\psi_\delta(\mathbf{r}, E)$ due to absorption at $\mathbf{r}$ into a state $\Psi_\delta(\mathbf{r}, E, \Gamma(\mathbf{r}))$. Now we proceed by multiplying the Schrödinger equation for $\Psi_\delta(\mathbf{r}, E, \Gamma(\mathbf{r}))$ by $\Psi^*_\gamma(\mathbf{r}, E, \Gamma(\mathbf{r}))$. Next we consider the Schrödinger equation for the state $\Psi^*_\gamma(\mathbf{r}, E, \Gamma(\mathbf{r}))$. The potential has a small positive imaginary part at $\mathbf{r}$, and thus a total potential $eU(\mathbf{r}) + i\Gamma(\mathbf{r})$. We multiply this equation with $\Psi_\delta(\mathbf{r}, E, \Gamma(\mathbf{r}))$. Subtraction of the two equations gained in this manner from one another gives

$$-(\hbar^2/2m)[\Psi^*_\gamma \Delta^2 \Psi_\delta - \Psi_\delta \Delta^2 \Psi^*_\gamma] = -2i\Gamma(\mathbf{r})\Psi^*_\gamma \Psi_\delta. \qquad (14.89)$$

The left hand side of this equation is, apart from a factor $i\hbar$ just the divergence of the local current density $\tilde{j}_{\gamma\delta}(\mathbf{r})$. Thus we can also write

$$\hbar \nabla \tilde{j}_{\gamma\delta}(\mathbf{r}) = -2\Gamma(\mathbf{r})\Psi^*_\gamma \Psi_\delta. \qquad (14.90)$$

The current-matrix elements which are related to the scattering matrix correspond to a carrier flux not at some definite energy but to the current in some small energy interval,

$$j_{\gamma\delta}(E', E, \mathbf{r}) = [1/hv_\gamma(E')]^{1/2}[1/hv_\delta(E)]^{1/2}\tilde{j}_{\gamma\delta}(E', E, \mathbf{r})dE'dE.$$

Using this (for $E = E'$) and integrating the resulting equation over the volume of the conductor, gives

$$\sum_\alpha d\mathbf{I}_{\gamma\delta}(\alpha) = 2\int d^3r (1/hv_\gamma)^{1/2}(1/hv_\delta)^{1/2}\Gamma(\mathbf{r})\Psi^*_\gamma \Psi_\delta. \qquad (14.91)$$

where we have taken currents which enter the volume to be positive. Here we have used the freedom in the integration volume to make it large enough such that the surface of the volume intersects the reservoirs. If these intersections coincide with the intersections which are used to define the scattering matrices, we can use the expressions which relate the currents to the scattering matrix elements

$$\sum_{\alpha} \mathbf{A}_{\gamma\delta}(\alpha, \Gamma) = 2 \int d^3r (1/hv_\gamma)^{1/2} (1/hv_\delta)^{1/2} \Gamma(\mathbf{r}) \Psi^*_\gamma \Psi_\delta. \quad (14.92)$$

Now we take on both sides the functional derivative with respect to Γ. At $\Gamma = 0$, on the right hand side, this derivative gives us back the original scattering states which exist in the absence of absorption,

$$(1/h) \sum_{\alpha} \delta \mathbf{A}_{\gamma\delta}(\alpha)/\delta\Gamma(\mathbf{r})|_{\Gamma=0} = (4\pi/h^2)(v_\gamma v_\delta)^{-1/2} \psi^*_\gamma \psi_\delta. \quad (14.93)$$

It remains to re-write the functional derivative $\delta\mathbf{A}_{\gamma\delta}(\alpha)/\delta\Gamma(\mathbf{r})$ in terms of functional derivatives with respect to the potential. To do this we note that in the presence of absorption the scattering matrix $\mathbf{s}$ is a functional of $eU(\mathbf{r}) - i\Gamma(\mathbf{r})$ and the scattering matrices $\mathbf{s}^\dagger$ are a functional of $eU(\mathbf{r}) + i\Gamma(\mathbf{r})$. It follows that

$$\delta \mathbf{A}_{\gamma\delta}(\alpha)/\delta\Gamma(\mathbf{r}) = i(\delta \mathbf{s}^*_{\alpha\gamma}/e\delta U(\mathbf{r}))\mathbf{s}_{\alpha\delta} - \mathbf{s}^*_{\alpha\gamma}(\delta \mathbf{s}_{\alpha\delta}/\delta U(\mathbf{r})) \quad (14.94)$$

and hence

$$\sum_{\alpha} n_{\gamma\delta}(\alpha, \mathbf{r}) = (1/h)(v_\gamma v_\delta)^{-1/2} \psi^*_\gamma \psi_\delta. \quad (14.95)$$

Acknowledgements

I have benefited from discussions with B. L. Altshuler, Y. M. Blanter, C. W. J. Beenakker, A. M. Martin and C. Texier. This work is supported by the Swiss National Foundation and the European Network on Dynamics of Hybrid Nanostructures.

References

[1] M. Büttiker, H. Thomas, and A. Prêtre, Phys. Lett. A **180**, 364 (1993).
[2] Ya. M. Blanter and M. Büttiker, Phys. Rev. B **59**, 10217 (1999).
[3] M. H. Pedersen, S. A. van Langen and M. Büttiker, Phys. Rev. B **57**, 1838 (1998).
[4] I. O. Kulik and A. N. Omel'yanchuk, Sov. J. Low Temp. Phys. **10**, 158 (1984).

[5] V. A. Khlus, Sov. Phys. JETP **66**, 1243 (1987).
[6] G. B. Lesovik, JETP Lett. **49**, 592 (1989).
[7] M. Büttiker, Phys. Rev. Lett. **65**, 2901 (1990); Phys. Rev. B**46**, 12485 (1992).
[8] M. Henny, S. Oberholzer, C. Strunk, T. Heinzel, K. Ensslin, M. Holland, and C. Schönenberger, Science **284**, 296 (1999).
[9] W. D. Oliver, J. Kim, R. C. Liu, and Y. Yamamoto, Science **284**, 299 (1999).
[10] Ya. M. Blanter and M. Büttiker, Physics Reports, (unpublished). cond-mat/9910158
[11] B. J. van Wees, H. van Houten, C. W. J. Beenakker, J. G. Williamson, L. P. Kouwenhoven, D. van der Marel and C. T. Foxon, Phys. Rev. Lett. **60**, 848 (1988).
[12] D. A. Wharam, T. J. Thornton, R. Newbury, M. Pepper, H. Ahmed, J. E. F. Frost, D. G. Hasko, D. C. Peacock, D. A. Ritchie and G. A. C. Jones, J. Phys. C **21**, L209 (1988).
[13] B. L. Altshuler, A. G. Aronov and D. Khmel'nitskii, J. Phys. C**15**, 7367 (1982).
[14] U. Sivan, Y. Imry and A. G. Aronov, Europhys. Lett. **28**, 115 (1994).
[15] Y. M. Blanter, Phys. Rev. B **54**, 12807 (1996).
[16] E. Buks, R. Schuster, M. Heiblum, D. Mahalu and V. Umansky, Nature **391**, 871 (1998).
[17] D. Sprinzak, E. Buks, M. Heiblum and H. Shtrikman, (unpublished). cond-mat/9907162
[18] S. A. Gurvitz, Phys. Rev. B **56**, 15215 (1997).
[19] I. L. Aleiner, N. S. Wingreen, and Y. Meir, Phys. Rev. Lett. **79**, 3740 (1997).
[20] Y. B. Levinson, Europhys. Lett. **39**, 299 (1997).
[21] M. Büttiker and A. M. Martin, Phys. Rev. B **61**, 2737 (2000).
[22] M. Büttiker, in *Quantum Physics at mesoscopic scale*, edited by D.C. Glattli, M. Sanquer and J. Tran Thanh Van (Editions Frontieres, 1999). cond-mat/9906386
[23] M. Büttiker, in *Statistical and Dynamical Aspects of Mesoscopic Systems*, XVI Sitges Conference, (Lecture Notes in Physics, Springer Verlag). cond-mat/9908116
[24] Y. B. Levinson, Phys. Rev. B **61**, 4748 (2000).
[25] M. Büttiker, J. Math. Phys., **37**, 4793 (1996).
[26] F. T. Smith, Phys. Rev. **118** 349 (1960).
[27] R. Dashen, S. -k Ma and H. J. Bernstein, Phys. Rev. **187**, 345 (1969).

[28] As emphasized below, it is potential derivatives, not energy derivatives, which are fundamental. This distinction is also important in the discussion of tunneling times: M. Büttiker, *Electronic Properties of Multilayers and low Dimensional Semiconductors*, edited by J. M. Chamberlain, L. Eaves, and J. C. Portal, (Plenum, New York, 1990). p. 297-315.
[29] V. A. Gopar, P. A. Mello, and M. Büttiker, Phys. Rev. Lett. **77**, 3005 (1996).
[30] P. W. Brouwer and M. Büttiker, Europhys. Lett. **37**, 441-446 (1997).
[31] P. W. Brouwer, S. A. van Langen, K. M. Frahm, M. Büttiker, and C. W. J. Beenakker, Phys. Rev. Lett. **79**, 914 (1997).
[32] A. G. Huibers, M. Switkes, C. M. Marcus, K. Campman and A. C. Gossard Phys. Rev. Lett. **81**, 200 (1998).
[33] M. Büttiker, Physica Scripta, **T54**, 104 - 110, (1994); M. Büttiker and C. A. Stafford, Phys. Rev. Lett. **76**, 495 (1996).
[34] R. Deblock, Y. Noat, H. Bouchiat, B. Reulet, D. Mailly, (unpublished). cond-mat/9910035
[35] M. Büttiker, J. Korean Phys. Soc. **34**, 121-130 (1999).
[36] C. W. J. Beenakker, private communication, (1999); C. Texier and M. Büttiker, (unpublished).
[37] Ya. M. Blanter, F.W.J. Hekking, and M. Büttiker, Phys. Rev. Lett. **81**, 1925 (1998).
[38] W. Chen, T. P. Smith, M. Büttiker, and M. Shayegan, Phys. Rev. Lett. **73**, 146 (1994).
[39] M. Büttiker and T. Christen, in *High Magnetic Fields in the Physics of Semiconductors*, edited by G. Landwehr and W. Ossau, (World Scientific, Singapur, 1997). p. 193 - 202. cond-mat/9607051
[40] M. Büttiker, Phys. Rev. B **41**, 7906 (1990).
[41] L. I. Glazman, G. B. Lesovik, D. E. Khmel'nitskii, and R. I. Shekhter, JETP Lett. **48**, 238 (1988).
[42] A. M. Martin and M. Büttiker, (unpublished).
[43] M. Büttiker, Phys. Rev. B **32**, 1846 (1985).
[44] M. Büttiker, J. Phys.: Condens. Matter **5**, 9361 (1993).
[45] T. Gramespacher and M. Büttiker, Phys. Rev. B **61**, 8125 (2000).
[46] A. M. Martin, T. Gramespacher and M. Büttiker, Phys. Rev. B **60**, R12581 (1999).

Chapter 15

TRANSPORT AND NOISE IN MULTI-TERMINAL DIFFUSIVE CONDUCTORS

E. V. Sukhorukov and D. Loss
Department of Physics and Astronomy, University of Basel, Klingelbergstrasse 82, CH-4056 Basel, Switzerland

1. INTRODUCTION

Shot-noise induced by current or voltage fluctuations in electron transport is a striking manifestation of charge quantization (for a review, see [1] and [2]). It serves as a sensitive tool to study correlations in conductors: while shot-noise assumes a maximum value (with Poissonian distribution) in the absence of correlations, it becomes suppressed when correlations set in as e.g. imposed by the Pauli principle [3]-[6].

In diffusive mesoscopic two-terminal conductors where the inelastic scattering lengths exceed the system size the shot-noise suppression factor was predicted [7]-[11] to be 1/3. The 1/3-suppression of shot-noise in diffusive conductors is now experimentally confirmed [12]. While some derivations [8]-[11] are based on a scattering matrix approach and thus a priori include quantum phase coherence, no such effects are included in the semiclassical Boltzmann-Langevin equation approach, which nevertheless leads to the same result [7]. However, while in the quantum approach for a two-terminal conductor the factor 1/3 was even shown to be universal [10], the semiclassical derivations given so far [7, 13] are restricted to quasi-one-dimensional conductors.

We present here the theory of transport and noise in multiterminal diffusive conductors [14, 15]. This problem has been recently addressed by Blanter and Büttiker in Ref. [11], where they use the scattering matrix formulation followed by an impurity averaging procedure. Having the advantage of including quantum phase coherence, this approach is somewhat cumbersome to generalize to an arbitrary geometry and arbitrary disorder. In contrast to this, our approach is based on a semiclassical

I. O. Kulik and R. Ellialtioğlu (eds.),
Quantum Mesoscopic Phenomena and Mesoscopic Devices in Microelectronics, 243–250.

Boltzmann-Langevin equation, which greatly simplifies the calculations. In Section 2. we formulate the problem and give a formal solution of it, while in Section 3. we present some applications of our theory.

2. FORMALISM

2.1 Boltzmann-Langevin method

We consider a multiterminal mesoscopic diffusive conductor of arbitrary 2D or 3D shape which is connected to N perfect metallic contacts of area L_n, $1 \leq n \leq N$, where the voltages V_n or outgoing currents I_n are measured. Our goal is to calculate the multiterminal spectral density S_{nm} of current fluctuations $\delta I_n(t)$ at zero frequency $\omega = 0$, defined by

$$S_{nm} = \int dt \langle \delta I_n(t) \delta I_m(0) \rangle. \tag{15.1}$$

We start from the (semiclassical) Boltzmann-Langevin equation for the distribution function [16]. Using standard diffusion approximations and neglecting the accumulation of charge (we are interested in $\omega = 0$ limit) we get the following equations for the symmetric $f_0(\varepsilon, \mathbf{r})$ and antisymmetric $\mathbf{f}_1(\varepsilon, \mathbf{r})$ parts of the average distribution function $f(\varepsilon, \mathbf{r}) = f_0(\varepsilon, \mathbf{r}) + \mathbf{n} \cdot \mathbf{f}_1(\varepsilon, \mathbf{r})$,

$$\nabla[\sigma(\mathbf{r}) \nabla f_0(\varepsilon, \mathbf{r})], \quad \mathbf{f}_1(\varepsilon, \mathbf{r}) = -v_F \tau_{im}(\mathbf{r}) \nabla f_0(\varepsilon, \mathbf{r}), \tag{15.2}$$

and for the fluctuations $\delta \mathbf{j}(\mathbf{r}, t)$ and $\delta V(\mathbf{r}, t)$ of the current density and of the potential, respectively,

$$\delta \mathbf{j}(\mathbf{r}) + \sigma(\mathbf{r}) \nabla \delta V(\mathbf{r}) = \delta \mathbf{j}^s(\mathbf{r}), \quad \nabla \cdot \delta \mathbf{j}(\mathbf{r}) = 0, \tag{15.3}$$

and where the correlator of the Langevin fluctuation sources is given by[1]

$$\langle \delta j^s_\alpha(\mathbf{r}, t) \delta j^s_\beta(\mathbf{r}', t') \rangle = \delta_{\alpha\beta} \delta(t - t') \delta(\mathbf{r} - \mathbf{r}') \sigma(\mathbf{r}) \Pi(\mathbf{r}),$$
$$\Pi(\mathbf{r}) = 2 \int d\varepsilon f_0(\varepsilon, \mathbf{r})[1 - f_0(\varepsilon, \mathbf{r})]. \tag{15.4}$$

Here we introduced the density of states ν_F, and the local conductivity $\sigma(\mathbf{r}) = \gamma e^2 \nu_F v_F^2 \tau_{im}(\mathbf{r})$, $(\gamma_{3D,2D} = 1/3, 1/2)$.

Now we specify the boundary conditions to be imposed on Eqs. (15.2) and (15.3). First, we assume that for a given energy there is no current

[1] We note that the function Π describes the local broadening of the distribution function and can be thought of as an effective (noise) temperature. Then we see that the correlator (15.4) takes an equilibrium-like form of the fluctuation-dissipation theorem. This is a direct consequence of our diffusion approximation.

through the surface S. Second, since the contacts with area L_n are perfect conductors the average potential V is fixed and independent of the position $\mathbf{r}$ on L_n. Third, the contacts are assumed to be in thermal equilibrium with outside reservoirs. The boundary conditions for (15.2) and (15.3), respectively, then read explicitly

$$f_0(\varepsilon, \mathbf{r})\,|_{L_n} = f_T(\varepsilon - eV_n), \quad d\mathbf{s}\cdot\nabla f_0(\varepsilon, \mathbf{r})\,|_S = 0, \tag{15.5}$$

$$d\mathbf{s}\cdot\delta\mathbf{j}(\mathbf{r}, t)\,|_S = 0, \quad \delta V(\mathbf{r}, t)\,|_{L_n} = 0, \tag{15.6}$$

where $f_T(\varepsilon)$ is the equilibrium distribution function at temperature T, and ds is a vector area element perpendicular to the surface.

2.2 Multiterminal spectral density of noise

We derive now the exact solution of Eqs. (15.2) and (15.3) with boundary conditions (15.5, 15.6) and use it to evaluate S_{nm} for a multiterminal conductor of arbitrary shape and distribution of impurities. For this it is convenient to follow Ref. [17] and to introduce N characteristic potentials $\phi_n(\mathbf{r})$ which satisfy the diffusion equation with associated boundary conditions:

$$\nabla[\sigma(\mathbf{r})\nabla\phi_n(\mathbf{r})] = 0, \quad d\mathbf{s}\cdot\nabla\phi_n(\mathbf{r})\,|_S = 0, \quad \phi_n(\mathbf{r})\,|_{L_m} = \delta_{nm}, \tag{15.7}$$

with $\phi_n(\mathbf{r}) \geq 0$ and the sum rule $\sum_{n=1}^{N}\phi_n(\mathbf{r}) = 1$. The potential V can then be expressed in terms of ϕ_n, $V(\mathbf{r}) = \sum_n \phi_n(\mathbf{r})V_n$, and the conductance then follows as

$$G_{nm} = -\int_{L_n} d\mathbf{s}\cdot\sigma(\mathbf{r})\nabla\phi_m(\mathbf{r}) = -\int d\mathbf{r}\,\sigma(\mathbf{r})[\nabla\phi_n\cdot\nabla\phi_m]. \tag{15.8}$$

Next we multiply the first part of (15.3) by $\nabla\phi_n$ and integrate it over the volume. Subsequent partial integration of the *lhs* gives the solution of (15.3) in terms of ϕ_n: $\delta I_n - \sum_m G_{nm}\delta V_m = \delta I_n^s = \int d\mathbf{r}\nabla\phi_n\cdot\delta\mathbf{j}^s$. Using the correlator (15.4) and the boundary condition (15.6) we obtain a general result for the multiterminal spectral density

$$S_{nm} = \int d\mathbf{r}\,\sigma(\mathbf{r})\Pi(\mathbf{r})[\nabla\phi_n(\mathbf{r})\cdot\nabla\phi_m(\mathbf{r})], \tag{15.9}$$

with the properties: $S_{nm} = S_{mn}$ and $\sum_n S_{nm} = 0$.

3. APPLICATIONS

3.1 Universality of noise

To prove the universality we calculate the effective temperature Π given by (15.4). For simplicity we work now in the $T = 0$ limit (the generalization to $T > 0$ is straightforward [15]). First we note that f_0 satisfies the diffusion equation (15.2) with boundary conditions (15.5). The solution in terms of ϕ_n then is, $f_0(\varepsilon, \mathbf{r}) = \sum_n \phi_n(\mathbf{r})\theta(\varepsilon - eV_n)$. Substituting this f_0 into Π in (15.4) we obtain

$$\Pi(\mathbf{r}) = e \sum_{k,l} \phi_k(\mathbf{r})\phi_l(\mathbf{r})|V_k - V_l|, \tag{15.10}$$

which in combination with Eq. (15.9) gives the general expression for S_{nm} in the case of elastic scattering.

Now we are in the position to prove the universality of the 1/3-suppression of shot-noise. To be specific we choose $V_n = \delta_{n1}V_1$, **i.e.** only contact one has a non-vanishing voltage. Then, using $\sum_n \phi_n = 1$, we get $\Pi = 2e|V|\phi_1[1-\phi_1]$. Substituting this into (15.9) and integrating by parts we obtain, $S_{1n} = 2e|V| \oint d\mathbf{s}\cdot\nabla\phi_n\sigma[\phi_1^2/2 - \phi_1^3/3] = -\frac{1}{3}e|I_n|$. For $n = 1$ we have $S_{11} = \frac{1}{3}e|I|$, where $I \equiv -I_1 > 0$ is the incoming current. Thus, the factor $\frac{1}{3}$ is indeed universal in the sense that it does not depend on the shape of the conductor nor on its impurity distribution. This generalizes the known universality of a two–terminal conductor [10] to a multiterminal geometry.

3.2 Exchange effects

Next we discuss exchange effects [18] in a four terminal conductor. They can be probed by measuring S_{13} in three ways: $V_n = V_0\delta_{n2}$ (A), $V_n = V_0\delta_{n4}$ (B), and $V_n = V_0\delta_{n2} + V_0\delta_{n4}$ (C). Then we take $\Delta S_{13} = S_{13}^C - S_{13}^A - S_{13}^B$ as a measure of the exchange effect. It measures the interference of electrons coming from mutually incoherent sources, which is caused by the indistinguishability of the electrons. Naively, one might expect that this interference effect averages to zero in diffusive conductors. However, it comes now as some surprise that in our semi-classical Boltzmann approach ΔS_{13} turns out to be non-zero in general and can even be of the order of the shot noise itself.

Indeed, for an arbitrary four–terminal geometry of the conductor the exchange effect can be expressed in terms of characteristic potentials:

$$\Delta S_{nm} = -4eV_0 \int d\mathbf{r} \nabla\phi_n \cdot \hat{\sigma} \nabla\phi_m \phi_k \phi_l, \tag{15.11}$$

where all indices are different. From this general formula it follows that $\Delta S_{nm} = \Delta S_{kl}$, and $\Delta S_{nm} + \Delta S_{nl} + \Delta S_{nk} = 0$. The last equation means that the exchange effect can change sign, *i.e.* cross correlations can be either suppressed or enhanced. On the other hand, the set-up can be slightly modified: instead of cross correlations, the noise density in one of the contacts of a multiterminal ($N > 2$) diffusive conductor is measured, say S_{11}, in all three experiments. Then, it follows from (15.11) that

$$\Delta S_{11} = -4eV_0 \int d\mathbf{r} \nabla\phi_1 \cdot \hat{\sigma} \nabla\phi_1 \phi_2 \phi_3 < 0, \tag{15.12}$$

i.e. the correlations are always suppressed, which is a direct manifestation of the Pauli principle. This experiment, which is analogous to the experiment of Hanbury Brown and Twiss in optics [19], was recently realized with a ballistic electron beam splitter [20, 21]. We have shown here that this effect is also observable in mesoscopic diffusive conductors.

3.3 Noise induced by thermal transport

Here we address a new phenomenon, namely the current noise in multiterminal diffusive conductors in the presence of thermal transport. We assume no energy relaxation in the conductor due to phonons, and no voltage is applied to the contacts, which are kept in equilibrium at different temperatures T_n. In this regime only possible transport is the thermal transport caused by diffusion of hot electrons from one lead to another [15]. We turn now to the calculation of the spectral density of noise.

To calculate Π we need to know the distribution function f_0. It obeys the Eq. (15.2) with the boundary conditions (15.5) in the contacts. The solution then reads explicitly, $f_0 = \sum_n \phi_n f_{T_n}$, and with the help of the sum rule for ϕ_n we get,

$$\Pi = \sum_{kl} \phi_k \phi_l Z_{kl}, \quad Z_{kl} = T_k T_l \int_{-\infty}^{\infty} ds \left[1 - \tanh(T_k s) \tanh(T_l s)\right]. \tag{15.13}$$

This together with Eq. (15.9) gives the spectral density of noise S_{nm}.

In equilibrium $T_n = T$, $Z_{kl} = 2T$ and (15.13) and we find for the equilibrium noise, $S_{nm} = -2G_{nm}T$. On the other hand, if for example $T_k \gg T_l$, then $Z_{kl} = (2\ln 2)T_k$. We consider then two situations, where e.g. either $T_1 = T$ and $T_{n\neq 1} = 0$, or $T_1 = 0$ and $T_{n\neq 1} = T$. In other words, only one contact is either heated up to high enough temperature T or cooled down to zero temperature and the other contacts are kept at the same temperature. Using the sum rule for ϕ_n and carrying out the integration in (15.9) we obtain for both cases,

$$S_{1n} = -\frac{2}{3}(1 + \ln 2) G_{1n} T. \tag{15.14}$$

Using the Wiedemann-Franz law in the form $Q_n = (\pi^2/6e^2) G_{1n} T^2$ [15] we can express the noise power (15.14) in terms of thermal currents Q_n,

$S_{1n} = \pm 4(1+\ln 2)(e/\pi)^2 T^{-1} Q_n$, with the sign depending on the sign of Q_n. This expressions reflects the universality of the noise in the presence of thermal transport.

3.4 Noise induced by spin transport

Next we discuss the current noise induced by transport of spin in a diffusive conductor attached to Fermi leads with spin-dependent chemical potentials $\mu_{n\sigma} = eV_{n\sigma}$. Assuming no spin relaxation and interaction processes (for short enough conductor) we can consider the two spin subsystems of the electron gas in the conductor as being independent. Then, we can write for the charge current $I_n^c = I_{n\uparrow} + I_{n\downarrow} = \frac{1}{2}\sum_m G_{nm}(V_{n\uparrow}+V_{n\downarrow})$ and for the spin current $I_n^s = I_{n\uparrow} - I_{n\downarrow} = \frac{1}{2}\sum_m G_{nm}(V_{n\uparrow} - V_{n\downarrow})$. Because of no correlations, $\langle \delta I_{n\uparrow}(t)\delta I_{m\downarrow}(0)\rangle = 0$, the total charge current noise is given by the sum of the contributions from two spin subsystems, $S_{nm} = S_{nm}(\uparrow) + S_{nm}(\downarrow)$. Now we can choose $V_{n\uparrow} = -V_{n\downarrow}$, so that the charge current through the conductor vanishes, $I_n^c = 0$, while the spin current I_n^s is non-zero in general. In particular, if the bias is applied to only one contact, say $V_{1\uparrow} = -V_{1\downarrow} \neq 0$, while $V_{n\sigma} = 0$ for $n \neq 1$, then we have

$$S_{1n} = -\frac{1}{3}e(|I_{n\uparrow}| + |I_{n\downarrow}|) = -\frac{1}{3}e|I_n^s|, \quad n \neq 1, \quad S_{11} = \frac{1}{3}e|I_1^s|. \tag{15.15}$$

Thus, one can observe a non-vanishing noise in the charge current induced by spin transport only (*i.e.* in the absence of charge transport).

3.5 Noninvasive voltage probe

We propose here an experiment [15] in which one can measure the effective noise temperature Π at some point $\mathbf{r} = \mathbf{r}_0$ on the surface of the conductor by measuring the voltage fluctuations in a noninvasive voltage probe. This is an open contact with a small area on the surface of the conductor around the point $\mathbf{r} = \mathbf{r}_0$. The contact is not attached to the reservoir so that it does not cause an equilibration of the electron gas, and as a result we have $\Pi = const$ around the point $\mathbf{r}_0$. Then, (15.9) can be rewritten as follows: $S = \int d\mathbf{r} \nabla\phi \cdot \hat{\sigma} \nabla\phi \Pi = \Pi(\mathbf{r}_0) \int d\mathbf{r} \nabla\phi \cdot \hat{\sigma} \nabla\phi$, where ϕ is the characteristic potential corresponding to the noninvasive probe. Using Eqs. (15.8) we get $S = R^{-1}\Pi(\mathbf{r}_0)$, where R is the resistance of the contact which comes from the volume around $\mathbf{r}_0$. Finally, taking into account $\delta I_n - \sum_m G_{nm}\delta V_m = \delta I_n^s$ and the fact that there is no current through the voltage probe, $\delta I = 0$, we obtain

$$S^v \equiv \int dt \langle \delta V(t) \delta V(0) \rangle = R\Pi(\mathbf{r}_0). \tag{15.16}$$

Thus, Π can be directly measured which gives important information about nonequilibrium processes in the conductor. Although our consideration is restricted to the diffusive regime, one can show [22] that the formula (15.16) can also be applied to the case where the probe is tunnel coupled to the conductor, with R being the tunneling resistance. An experiment to measure shot noise at local tunneling contacts is discussed in Ref. [23].

References

[1] M. J. M. de Jong, and C. W. J. Beenakker, Shot noise in mesoscopic systems, in L. P. Kouwenhoven, G. Schön, and L. L. Sohn (eds.), *Mesoscopic electron transport*, NATO ASI, Series E: Applied Sciences, Vol. 345, Kluwer, Dordrecht, 225–258 (1997).

[2] Ya. M. Blanter and M. Büttiker, Shot Noise in Mesoscopic Conductors, cond-mat/9910158, (1999).

[3] V. A. Khlus, Current and voltage fluctuations in microjunctions between normal metals and superconductors, *Sov. Phys. JETP* **66**, 1243–1249 (1987).

[4] G. B. Lesovik, Excess quantum noise in 2D ballistic point contact, *JETP Lett.***49**, 592–594 (1989).

[5] B. Yurke and G.P. Kochanski, Momentum noise in vacuum tunneling transducers *Phys. Rev. B***41**, 8184–8194 (1990).

[6] M. Büttiker, Scattering theory of thermal and excess noise in open conductors, *Phys. Rev. Lett.***65**, 2901–2904 (1990).

[7] K. E. Nagaev, On the shot noise in dirty metal contacts, *Phys. Lett. A* **169**, 103–107 (1992).

[8] C. W. J. Beenakker and M. Büttiker, Suppression of shot noise in metallic diffusive conductors, *Phys. Rev.* B**46**, 1889–1892 (1992).

[9] M. J. M. de Jong and C. W. J. Beenakker, Mesoscopic fluctuations in the shot-noise power of metals, *Phys. Rev.* B**46**, 13400–13406 (1992).

[10] Yu. V. Nazarov, Limits of universality in disordered conductors, *Phys. Rev. Lett.***73**, 134–137 (1994).

[11] Ya. M. Blanter and M. Büttiker, Shot-noise current-current correlations in multiterminal diffusive conductors, *Phys. Rev.* B**56**, 2127–2136 (1997).

[12] M. Henny, S. Oberholzer, C. Strunk and C. Schönenberger, The 1/3-shot noise suppression in diffusive nanowires, *Phys. Rev.* B**59**, 2871-2880 (1999).

[13] M. J. M. de Jong and C. W. J. Beenakker, Semiclassical theory of shot noise in mesoscopic conductors, *Physica* A**230**, 219–248 (1996).

[14] E. V. Sukhorukov and D. Loss, Universality of shot noise in multiterminal diffusive conductors, *Phys. Rev. Lett.* **80**, 4959–4962 (1998).

[15] E.V. Sukhorukov and D. Loss, Noise in multiterminal diffusive conductors: Universality, nonlocality, and exchange effects, *Phys. Rev.* B**59**, 13054–13066 (1999).

[16] Sh. M. Kogan and A. Ya. Shul'man, Theory of fluctuations in a nonequilibrium electron gas, *Sov. Phys. JETP* **29**, 467–474 (1969).

[17] M. Büttiker, Capacitance, Admittance and Rectification of Small Conductors, J. Phys.: Condens. Matter **5**, 9361–9378 (1993).

[18] M. Büttiker, Scattering theory of current and intensity noise correlations in conductors and wave guides, *Phys. Rev.* B**46**, 12485–12507 (1992).

[19] R. Hanbury Brown and R. Q. Twiss, Correlation between photons in two coherent beams of light, *Nature (London)* **177**, 27–29 (1956).

[20] R. C. Liu, B. Odom, Y. Yamamoto and S. Tarucha, Quantum interference in electron collision, *Nature* **391**, 263–265 (1998).

[21] M. Henny, S. Oberholzer, C. Strunk, T. Heinzel, K. Ensslin, M. Holland and C. Schönenberger, The fermionic Hanbury Brown and Twiss experiment, *Science* **284**, 296–298 (1999).

[22] E. V. Sukhorukov and D. Loss, Measuring effective noise temperature in mesoscopic conductor using noninvasive voltage probe, (1998) unpublished.

[23] Th. Gramespacher and M. Büttiker, Quantum shot noise at local tunneling contacts on mesoscopic multiprobe conductors, *Phys. Rev. Lett.* **81**, 2763–2766 (1998).

Chapter 16

MEMORY EFFECTS IN STOCHASTIC RATCHETS

B. Tanatar, E. Kececioglu and M. C. Yalabik
Department of Physics, Bilkent University
06533 Bilkent, Ankara, Turkey

1. INTRODUCTION

The motion of a particle under the influence of random forces and an asymmetric periodic potential has been attracting a lot of interest in recent years. [1] The so-called ratchet models exhibit a variety of interesting phenomena stemming from nonequilibrium fluctuations such as the occurrence of a net macroscopic current [2, 3] which depends on the spatial asymmetry of the external potential as well as the statistical properties of the fluctuating forces. The stochastic ratchets are important in understanding biological systems such as molecular motors and they also find application for electrons in superlattices, quantum dots, Josephson junctions, and atomic systems. Recent experiments [4] making use of nanolithography techniques provide a wealth of possibilities in the study of quantum properties.

We consider the effects of memory friction for a Brownian particle subject to spatial and temporal forces. When the dissipative effects of the fluctuating (random) force depend on the system's past behavior, the memory function replaces the constant friction term and a generalized Langevin equation results. Previous studies of thermal ratchets making use of the memory function or correlated noise relied on the simulation of Langevin equation to obtain transport properties. [5] A corresponding Fokker-Planck equation[6] (FPE) describing the dynamics of the system is difficult to construct owing to the non-Markovian nature of the underlying process. Introducing some simplifying assumptions we form an integro-differential equation satisfied by the probability density which may be construed as a generalized FPE. The non-Markovian nature of

I. O. Kulik and R. Ellialtioğlu (eds.),
Quantum Mesoscopic Phenomena and Mesoscopic Devices in Microelectronics, 251–256.

the problem is preserved in the final FPE we obtain. It has been known that for a non-Markovian system, the is no generic way of obtaining a FPE if the system does not exhibit steady state solutions. In our case, we know from the outset that steady state solutions exist through Langevin simulations, even in the non-Markovian limit. We demonstrate by numerical calculations that the same qualitative and quantitative results can be obtained as those from the generalized Langevin equation. Our application of the generalized FPE to the ratchet problem shows interesting properties resulting from the memory effects. In particular, we find a regime in the current characteristics exhibiting instabilities.

2. THEORETICAL BACKGROUND

In order to construct a FPE for stochastic ratchets including the memory effects, we begin with the basic equation of motion governing the dynamics of a Brownian particle, *viz.* the generalized Langevin equation

$$\frac{dv}{dt} = -\int_{-\infty}^{t} \mu(t-t')\, v(t')\, dt' + f(t) - \frac{dV(x)}{dx} + \xi(t)\,, \tag{16.1}$$

where $x(t)$ and $v(t) = dx/dt$, respectively, are the position and velocity of the particle, $V(x)$ is the asymmetric periodic potential, $f(t)$ is a time-dependent external driving force, and $\xi(t)$ is the stochastic force with zero mean. Our aim is to obtain an equation for the probability density function $P(v, x, t)$. Since we can express v' and x' at time $t + dt$ when we know v and x at time t, it should also be possible to find a solution for $P(v', x', t + dt)$ when we know $P(v, x, t)$. In other words, by using the fact that the total probability is conserved, we aim at finding how $P(v, x, t)$ transforms to $P(v', x', t + dt)$. To this end, we first discretize the integral in Eq. (16.1), so that

$$-\int_{-\infty}^{t} \mu(t-t')\; v(t')\; dt' = -\sum_{n=0}^{\infty} \mu(n\Delta t')\, v(t - n\Delta t')\Delta t'\,, \tag{16.2}$$

with $\mu(0)\Delta t'$ approaching γ as $\Delta t' \to 0$. Here γ is the friction constant in the absence of memory correlations in the dissipation term. The discretized Langevin equation is cast in a form which includes the probability density P_n for the random variable v_n. The product over $P_n(v_n)$ indicates an assumption of statistical independence of the velocities v_n at different times, although such velocities are clearly expected to be correlated. Our approximation is expected to be valid for velocity distributions near steady state, and enables us to proceed with the algebra. Performing the integrations over v and x, keeping only terms of order

Δt, and considering the limits $\Delta t' \to 0$ and $\Delta t \to 0$, we find

$$\frac{\partial P}{\partial t} = \frac{\partial P}{\partial v}\left[\int_{-\infty}^{t} \mu(t-t')\bar{v}(t')\,dt' + \frac{dV(x)}{dx} - f(t) + \gamma(v-\bar{v})\right] \\ -\frac{\partial P}{\partial x}v + D\,\frac{\partial^2 P}{\partial v^2} + \gamma P(v,x,t)\,. \tag{16.3}$$

Here $\bar{v}(t)$ is the average velocity at a given time, γ is equal to $\mu(0)\Delta t'$ as $\Delta t' \to 0$ as defined earlier, and D is the diffusion constant. Our main result, Eq. (16.3), describes the dynamics of stochastic ratchets when the memory effects are included. We recover the conventional FPE if the memory function is a delta function. Incorporation of memory effects in the FPE has been investigated by Fonseca *et al.* [7] in the context of Kramers model for reaction rates.

3. RESULTS AND DISCUSSIONS

We have numerically solved the Fokker-Planck and Langevin equations to test how well our proposed FPE describes the features of thermal ratchets. We have found that the FPE reproduces all the established results such as current reversal and noise rectification. In the results below, we have specifically used $f(t) = A\sin(\omega t)$ with $A = 1.0$ and $\omega = 0.2$ for the driving force. The ratchet potential is taken to be $V(x) = b_0 \sin(2\pi x/L) + b_1 \sin(4\pi x/L)$, with b_0, b_1, and L are constants. [5] We model the memory function as $\mu(t) = (\gamma/\tau)\exp(-|t|/\tau)$, where τ is the correlation time. Finally, $\xi(t)$ is a Gaussian random variable with zero average and time correlation is given by the fluctuation-dissipation theorem $\langle\xi(t)\xi(t')\rangle = 2D\mu(t-t')$, for $t > t'$, and approaches $\langle\xi(t)\xi(t')\rangle = 2\gamma D\delta(t-t')$ as τ approaches zero.

We first show that our proposed FPE and Langevin equation yield the same current value for the ratchet problem. A simple finite element algorithm is used to solve the equations. In Fig. 16.1 we illustrate our results for the time dependence of the velocity. The average velocity of the particle corresponding to long time behavior is shown in Fig. 16.1*b*. A simple running average algorithm is used to find the average velocities. In fact, when properly averaged over the periods of oscillations, the long-time behaviors are just straight lines. Nevertheless, it is seen that the average velocity, or equivalently the current which is the physical observable, is the same for both methods. It is evident in that the FPE results approach the Langevin equation values with increasing time. We now turn to the effects of memory correlation in the ratchet problem. We have analyzed the ratchet problem for various values of the correlation time τ and the strength of the correlations γ. We find that as τ is

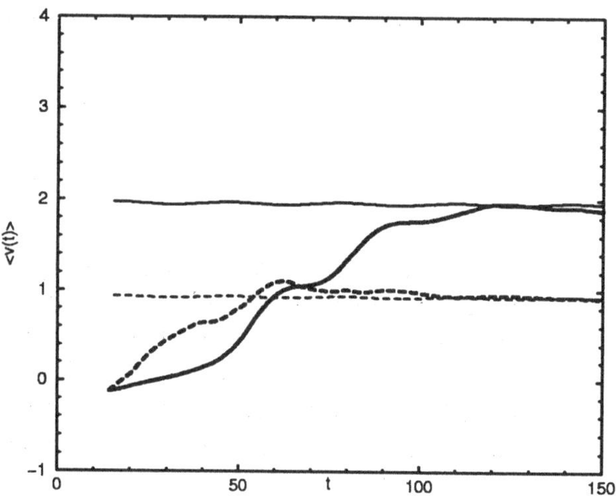

Figure 16.1 The average velocity obtained from the Langevin simulation (thin lines) and from the solution of generalized FPE (thick lines). The solid and dashed lines indicate the amplitude $A = 1$ and $A = 0.5$, respectively, of the driving force $f(t)$.

increased, because of negative feedback of the correlation to the system, the current decreases (current is proportional to the slope of the curves). When we explore the large τ regime such that it becomes comparable to the period of oscillations in $x(t)$, we find a different behavior. In Fig. 16.2 we again show $x(t)$ for different values of τ. The topmost curve ($\tau = 0$) is enlarged many times in order to show the differences between the case when $\tau = 0$ and when τ is large. It is observed that as τ becomes comparable to the period of oscillations of the $\tau = 0$ case, there appears some instabilities in the system stemming from the positive feedback of the memory. The memory term in the definition of the friction now behaves as if it is a driving term in resonance with the actual frequency of the oscillations. Stochastic resonances in ratchet problems or in a class of FPE have been reported. [8] When $f(t)$ is switched off no instabilities are observed, consistent with the known results. We have checked that similar instabilities are also obtained by direct Langevin simulations. We have also observed that as the magnitude of γ increases, the period of oscillations decreases under the resonant conditions.

In short, we have developed a FPE corresponding to a generalized Langevin equation to describe the dynamics of a Brownian particle under the influence of external potentials and memory effects. The proposed FPE reproduces the known results and reduces to the white-noise limit when the memory effect is absent. The correlated fluctuating force in combination with the memory effects give rise to stochastic resonances.

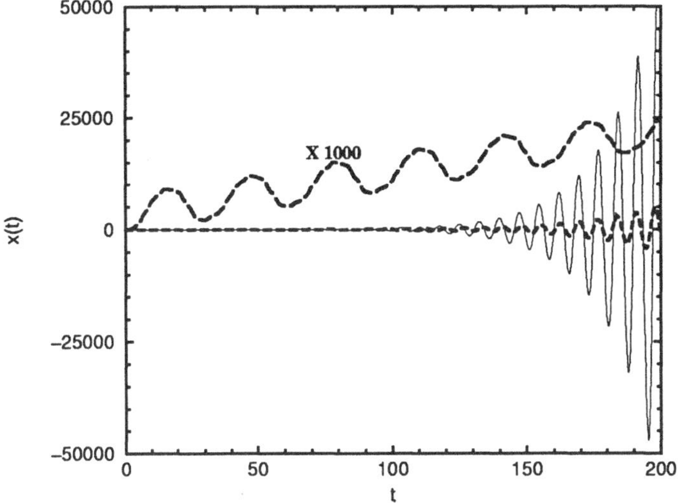

Figure 16.2 The position variable $x(t)$ for various values of the correlation time τ at $\gamma = 1$. The dotted and solid lines indicate $\tau = 18$ and $\tau = 30$, respectively. $\tau = 0$ (dashed line) case is also shown for reference.

Acknowledgement

This work was partially supported by the Scientific and Technical Research Council of Turkey (TÜBİTAK) under Grant No. TBAG-1662.

References

[1] P. Hänggi and R. Bartussek, Brownian rectifiers: How to convert Brownian motion into directed transport, in *Nonlinear Physics of Complex Systems-Current Status and Future Trends*, J. Parisi, S. C. Müller, and W. Zimmermann, eds., (Springer, Berlin, 1996) pp. 294-308.

[2] M. O. Magnasco, Forces thermal ratchets *Phys. Rev. Lett.* **71**, 1477-1481 (1993); R. D. Astumian and M. Bier, Fluctuation driven ratchets: molecular motors *Phys. Rev. Lett.* **72**, 1766-1769 (1994); C. R. Doering, W. Horsthemke and J. Riordan, Nonequilibrium fluctuation-induced transport *Phys. Rev. Lett.* **72**, 2984-2987 (1994).

[3] R. Bartusek, P. Hänggi and J. G. Kissner, Periodically rocked thermal ratchets *Europhys. Lett.* **28**, 459-464 (1994); B. Lindner, L. Schimansky-Geier, P. Reimann, P. Hänggi and M. Nagaoka, Inertia ratchets: a numerical study versus theory *Phys. Rev. E* **59**, 1417-1424 (1999).

[4] H. Linke, W. Sheng, A. Löfgren, H. Xu, P. Omling and P. E. Lindelof, A quantum dot ratchet: experiment and theory *Europhys. Lett.* **44**, 343-349 (1998); A. Lorke, S. Wimmer, B. Jager, J. P. Kotthaus, W. Wegscheider and M. Bichler, Far infrared and transport properties of antidot arrays with broken symmetry *Physica B* **249-251**, 312-316 (1998).

[5] L. Ibarra-Bracamontes and V. Romero-Rochin, Stochastic ratchets with colored thermal noise *Phys. Rev. E* **56**, 4048-4051 (1997); P. Jung, J. G. Kissner and P. Hänggi, Regular and chaotic transport in asymmetric periodic potentials: inertia ratchets *Phys. Rev. Lett.* **76**, 3436-3439 (1996); C. M. Arizmendi and F. Family, Memory correlation effect on thermal ratchets *Physica A* **251**, 368-381 (1998).

[6] H. Risken, *The Fokker-Planck Equation*, Springer Series in Synergetics Vol. 18 Springer, Berlin, 1984.

[7] T. Fonseca, J. A. N. F. Gomez, P. Grigolini and F. Marchesoni, The theory of chemical reaction rates *Adv. Chem. Phys.* **62**, 389 (1985).

[8] A. Neiman and S. Wokyung, Memory effects on stochastic resonance *Phys. Lett. A* **223**, 341-347 (1996).

IV

AHARONOV-BOHM EFFECT AND VORTICES

Chapter 17

NON-DECAYING CURRENTS IN NORMAL METALS

I. O. Kulik
Department of Physics, Bilkent University
06533 Bilkent, Ankara, Turkey

Abstract In a present paper, we review cases when stable, time-independent currents may flow in normal metals, both single-connected or multiple-connected, in presence of magnetic field or the field of vector potential. These include the spatially oscillating currents in narrow metallic stripes originating due to Landau diamagnetism (the Landau persistent currents), the currents in hollow normal-metal or semiconducting cylinders and rings induced by the Aharonov-Bohm flux threading the conducting loop (the Aharonov-Bohm persistent currents), and currents in hollow cylinders subject to radial flux and existing even in the absence of the longitudinal, or Aharonov-Bohm flux (the "transverse" persistent currents), as well as the non-decaying currents in axial magnetic field, modulated via the Berry phase effect by the transverse or azimuthal components of the field and oscillating as a function of the latter.

1. INTRODUCTION

It is commonly believed that current in a normal (nonsuperconducting) metal can only flow if the voltage is applied to the sample, and that current transport is necessarily related to the Joule heat dissipation inside the sample. This is however a prejudice. To start with, the diamagnetic properties of metals (the Landau diamagnetism) can be explained by currents flowing around the sample near its surface, as was first demonstrated by Teller [1] in his interpretation of the Landau theory [2]. The other case is the Aharonov-Bohm effect [3] in double connected metallic samples which produces currents created by the magnetic flux inside the sample orifice. It was long debated, especially by Byers and Yang [4] and by Bloch [5] whether stable currents may exist in the noninteract-

I. O. Kulik and R. Ellialtioğlu (eds.),
Quantum Mesoscopic Phenomena and Mesoscopic Devices in Microelectronics, 259–282.

ing electron gas without the off-diagonal long range order specific to superconductivity. Bloch, in particular, concluded that

"Except for abnormally small radius and irrespective of the specific properties of the system one is thus led to the exclusion of stable flux trapping in a one dimensional ring".

The "small radius" corresponds to what we call now mesoscopic sample. In the latter, stable current and flux trapping is possible. It was first clearly stated by the author [6] that the non-decaying current DOES exist in a small metallic loop provided that scattering of electrons (both elastic and inelastic) is not too strong. In the paper [6] it was concluded that

"The current state corresponds in this case to a minimum of the free energy, so that allowance for dissipation does not lead to its decay".

Later, Buttiker, Imry and Landauer [7] reached the same conclusion by considering Aharonov-Bohm current in a dirty ring, showing that the magnetic flux Φ in a ring serves as a quasi-momentum if the ring is unwrapped into an effective periodic structure with a period equal to the ring circumference L. Therefore, the energy becomes a periodic function of Φ with a period of a normal-metal flux quantum $\Phi_0 = hc/e$, and the current originates as a nonzero derivative $J = -c\,\partial E/\partial\Phi$.

The purpose of the present paper is to review and extend the situations in which non-decaying ("persistent") current may exist in non-superconducting metals. The cases considered include the spatially oscillating currents in narrow metallic stripes in a magnetic field responsible for the Landau diamagnetism (the Landau persistent currents), the currents driven by the Aharonov-Bohm flux in hollow metallic cylinders and rings (the Aharonov-Bohm persistent currents), and the non-decaying charge flow in hollow disordered cylinders subject to radial magnetic flux (the "transverse" persistent currents), as well as the non-decaying currents originating due to Berry's phase [8] effects for electron spins interacting with the radial and azimuthal magnetic fields in the ring. Part of the material presented can also be found in the review papers related to the Aharonov-Bohm effect in solids [9]-[15], *etc.*

2. PERSISTENT CURRENT IN A LONG METALLIC STRIPE

Consider a situation when magnetic field **B** is applied perpendicular to the surface of two-dimensional metallic stripe with electron density n and width d (Fig. 17.1). The sample will show the Landau effect; *i.e.*, the dependence of its energy E on B and thus attaining the magnetic moment $M_z = -\partial E/\partial B$. We will assume that the width of the stripe

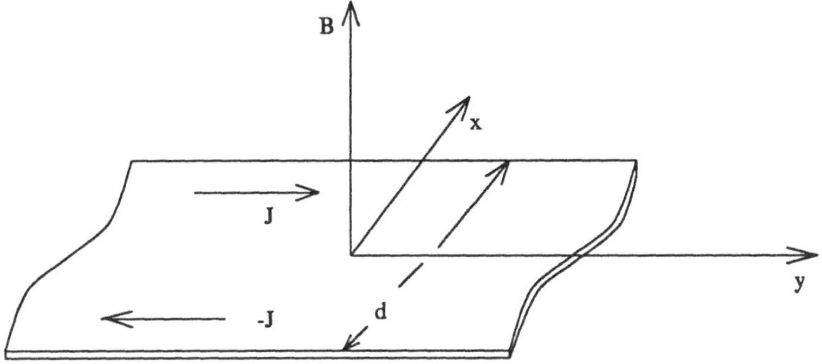

Figure 17.1 Sketch of a two-dimensional metallic stripe of width d in a perpendicular magnetic field $\mathbf{B}$. Arrows show the average current direction in the stripe.

d is much smaller than the cyclotron radius $r_H = mv_F c/eB$ which will allow to include the effect of magnetic field as a perturbation, and to find the current density from the Schrödinger equation expanded in powers of B. Neglecting the spin, Schrödinger equation reads

$$-\frac{\hbar^2}{2m}\frac{\partial^2\psi}{\partial x^2} - \frac{\hbar^2}{2m}\left(\frac{\partial}{\partial y} - \frac{ieBx}{\hbar c}\right)^2 \psi = E\psi. \tag{17.1}$$

In zero magnetic field, the eigenstates of Eq. 17.1 are

$$\psi^0_{nk} = \frac{1}{\sqrt{L}} e^{iky} \left(\frac{2}{d}\right)^{1/2} \sin\frac{\pi n}{d}(x + d/2) \tag{17.2}$$

corresponding to energies

$$\varepsilon^0_n = \frac{\hbar^2}{2m}\left(k^2 + \pi^2 n^2/d^2\right).$$

The perturbation Hamiltonian is

$$H_1 = \frac{\hbar^2}{2m}\left(2kk_0\frac{x}{d} + k_0^2\frac{x^2}{d^2}\right) \tag{17.3}$$

where $k_0 = eBd/\hbar c$. To first order in B, it has matrix elements

$$(H_1)_{mn} = \frac{2kk_0\hbar^2}{m}\int_0^1 \sin\pi n\left(\xi - \frac{1}{2}\right)\sin\pi m\left(\xi - \frac{1}{2}\right)\xi d\xi =$$

$$= \begin{cases} -\frac{8\hbar^2 kk_0}{\pi^2 m}\frac{nm}{(n^2-m^2)^2} & \text{if } n-m \text{ is odd} \\ 0 & \text{otherwise} \end{cases} \tag{17.4}$$

It is then a straightforward to solve for the correction to the wave function and to the current. The current density is

$$j(\xi) = -j_0 \sum_{n=1}^{\infty} \int_{-\infty}^{\infty} \frac{d\kappa}{e^{(\kappa^2+n^2-n_F^2)/2n_F\tau}+1}$$

$$\times \left[\left(\xi - \frac{1}{2}\right) \sin^2 \pi n\xi - \frac{32\kappa^2}{\pi^2} \sum_{m=1}^{\infty} \frac{nm\delta_{n-m,2k+1}}{(n^2-m^2)^3} \sin \pi n\xi \sin \pi m\xi \right] \quad (17.5)$$

where $\xi = x/d$, $j_0 = 2e\hbar k_0/md^2 = 2e^2B/mcd$, and $\tau = T/\Delta E$ where $\Delta E = hp_F/md$ is the distance between discrete energy levels in a ring at the Fermi energy. The dependence $j(\xi)$ is calculated numerically and shown in Fig. 17.2 at various temperatures T.

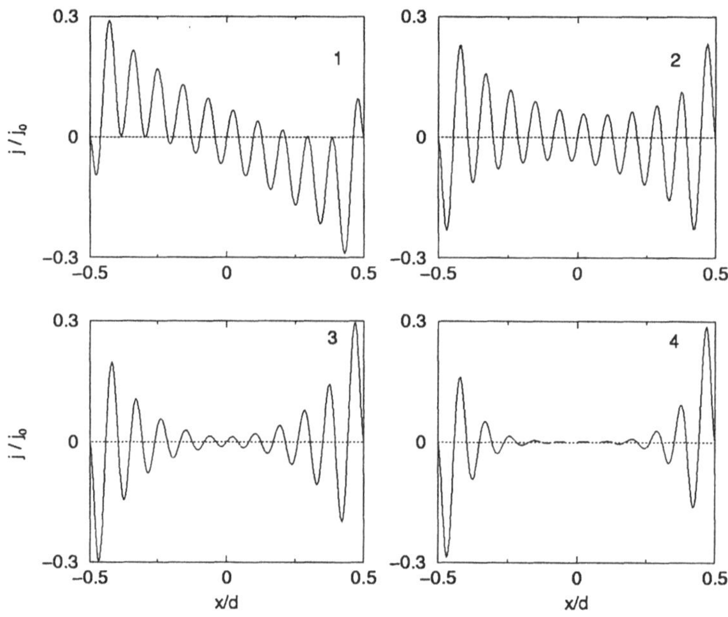

Figure 17.2 Persistent current in a stripe at various temperatures. Panel 1: $\tau = 0$; panel 2: $\tau = 0.15$; panel 3: $\tau = 0.45$; panel 4: $\tau = 0.9$ where τ is the temperature in units of the level spacing ΔE at the Fermi energy ε_F. The value of the electron concentration corresponds to the "Fermi number" n_F defined according to $\varepsilon_F = h^2n_F^2/2md^2$ and equal to $n_F = 10.9$.

First of all, our calculation shows that the current does exist. Since it corresponds to minimum of free energy (at given magnetic field) it does not decay in time, even if the scattering is included to the calculation. Integration of the current $j(\xi)$ with a factor $(\xi - 1/2)$ gives magnetic moment M of the sample which in our case (in thin sample) proves to be temperature dependent, unlike in the standard Landau diamagnetism in bulk samples.

The current will not vanish (but may decrease in amplitude) if scattering is included, in the same way as the Landau diamagnetism persists in dirty metals. This argument does not show need for any further justification of the persistent current. (We can find proof even on the experimental basis, referring to the innumerable papers studying the manifestations of the Landau diamagnetism, the De Haas - Van Alphen effect [16].)

The current in a stripe shows periodic oscillation at low temperature with a spatial period

$$\lambda = \frac{2\pi}{Q_0}, \quad Q_0 = 2k_F \tag{17.6}$$

where k_F is the Fermi momentum. These oscillations is nothing else than the known Friedel oscillation in a degenerate Fermi liquid [17]. At elevated temperature, the amplitude of oscillation decreases in the middle of the sample, and current is pushed to the specimen edges. Mention that the amplitude of oscillation is much greater than the average current. At higher temperature, amplitude of oscillation follows the exponential law with a characteristic decay length

$$\xi_T = \frac{\hbar v_F}{2\pi T} \tag{17.7}$$

which reveals also in such phenomena as the Andreev reflection [18] from the normal metal backed to superconductor [19].

3. PERSISTENT CURRENTS IN METALLIC RINGS AND CYLINDERS

3.1 Clean Rings

In a one-dimensional metallic ring pierced by solenoid creating Aharonov-Bohm flux $\Phi = A_\varphi L$ where A_φ is the vector potential and L is the ring circumference (Fig. 17.3), the Schrödinger equation

$$\frac{-\hbar^2}{2mR^2}\left(\frac{\partial}{\partial\varphi} - \frac{ieAL}{hc}\right)^2 \psi = E\psi \tag{17.8}$$

has a solution

$$\psi_n = \frac{1}{\sqrt{L}} e^{in\varphi} \tag{17.9}$$

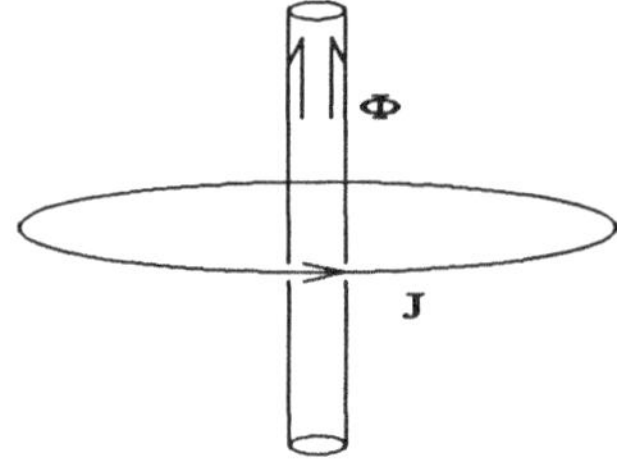

Figure 17.3 One-dimensional metallic ring in the field of vector potential created by the solenoid piercing the ring.

corresponding to electron energy

$$\varepsilon_n = \frac{\hbar^2}{2mR^2}\left(n - \frac{\Phi}{\Phi_0}\right)^2 \tag{17.10}$$

where $\Phi_0 = hc/e$ is the flux quantum.

The thermal averaged current in a ring is calculated as

$$J = \frac{eh}{mL^2}\sum_{n=-\infty}^{\infty}\frac{n - \Phi/\Phi_0}{e^{[(n-\Phi/\Phi_0)^2 - n_F^2]/2n_F\tau} + 1} \tag{17.11}$$

where we introduced τ as a relative temperature $\tau = T/\Delta E$ where ΔE is the distance between two successive levels of the discrete spectrum (Eq. 17.10) at the energy equal to Fermi energy ($\varepsilon_F = h^2 n_F^2/2mL^2$)

$$\Delta E = h^2 n_F/mL^2 = hv_F/L. \tag{17.12}$$

Straightforward calculation using the Poisson formula

$$\sum_{n=-\infty}^{\infty} f(n) = \sum_{-\infty}^{\infty}\int_{-\infty}^{\infty} f(n)e^{2\pi i n s}dn$$

gives the current as an oscillating function of flux with the period Φ_0

$$J = \frac{ev_F}{\pi L}\sum_{s=1}^{\infty}\frac{2\pi^2 s\tau}{\sinh(2\pi^2 s\tau)}\cos(k_F L s)\sin\left(2\pi s\frac{\Phi}{\Phi_0}\right). \tag{17.13}$$

At low temperature, amplitude of oscillation is of the order

$$J_{\max} \sim \frac{ev_F}{\pi L} \text{ at } T \to 0. \tag{17.14}$$

At higher temperature, only the lowest harmonic ($s = 1$) survives with an amplitude

$$J_{\max}^{1d} \simeq \frac{ev_F}{L}\frac{4\pi T}{\Delta E}e^{-2\pi^2 T/\Delta E},\ T \geq \Delta E. \tag{17.15}$$

In a two-dimensional sample (thin hollow cylinder of radius R), persistent current is proportional to the number of transverse conducting channels [6]

$$J_{\max}^{2d} \simeq \frac{4\sqrt{2\pi}}{\sqrt{k_F L}} \frac{ev_F}{L} N_\perp \frac{4\pi T}{\Delta E} e^{-2\pi^2 T/\Delta E} \tag{17.16}$$

where $N_\perp = L_z k_F/2\pi$, whereas in the three-dimensional ring of cross section $S = d_1 d_2$ it increases with the number of perpendicular channels $N_\perp = k_F^2 S/4\pi$ as

$$J_{\max}^{3d} \sim \frac{ev_F}{L} N_\perp^{1/2}. \tag{17.17}$$

3.2 Ring Interrupted By A Barrier

Assume that one-dimensional ring is interrupted by a barrier $V\delta(x)$ where x is a coordinate along the ring, $x = R\varphi$ such that $0 \le x \le L = 2\pi R$. The solution to the Schrödinger equation is of the form

$$\psi = C_1 e^{ik_1 x} + C_2 e^{-ik_2 x} \tag{17.18}$$

where $k_{1,2} = k \pm eA/\hbar c$ and $k = \sqrt{2m\varepsilon}/\hbar$, with C_1, C_2 found from the boundary condition

$$\psi(0) = \psi(L) \text{ and } -\frac{\hbar^2}{2m}[\psi'(0) - \psi'(L)] + V\psi(0) = 0. \tag{17.19}$$

The allowed energy values are found from the equation equivalent to the $1d$ Kronig-Penney model

$$\cos kL + \frac{mV}{\hbar^2 k} \sin kL = \cos 2\pi \frac{\Phi}{\Phi_0}. \tag{17.20}$$

The wave function (Eq. 17.18), with the constriction implied by this equation, can be rewritten in the form

$$\psi = Ce^{i\alpha\frac{x}{L}} \left(e^{-i\alpha/2} \sin \kappa \frac{x}{L} + e^{i\alpha/2} \sin \kappa \frac{L-x}{L} \right) \tag{17.21}$$

where

$$C = \frac{1}{\sqrt{L}} \left[1 + \frac{\sin\kappa}{\kappa}(\cos\kappa - \cos\alpha) - \cos\kappa\cos\alpha \right]^{-\frac{1}{2}}. \tag{17.22}$$

By introducing the dimensionless momentum $\kappa = |k|L$, Eq. (17.20) is rewritten as

$$\cos\kappa + g\frac{\sin\kappa}{\kappa} = \cos\alpha \tag{17.23}$$

where $g = mVL/\hbar^2$ and $\alpha = 2\pi\Phi/\Phi_0$. Persistent current in the ring is found as an average of the current operator $\hat{j} = \frac{e\hbar}{m}(\frac{1}{i}\frac{\partial}{\partial x} - \frac{eA}{\hbar c})$ and equals

$$j = -\frac{e\hbar}{mL}\sum_n \frac{\kappa_n \sin\kappa_n \sin\alpha}{1 + \frac{\sin\kappa_n}{\kappa_n}(\cos_n - \cos\alpha) - \cos\kappa_n \cos\alpha}. \tag{17.24}$$

In case of no barrier ($V = 0$), the solutions of Eq. (17.23) are $\kappa_n = 2\pi n + \alpha$, $n = 0, \pm 1, \pm 2, ...$ which of course bring us back to the previous formula (Eq. 17.10). In the opposite limit of a strong barrier, $g_o \gg 1$ (where $g_o = g/k_F L = mV/\hbar^2 k_F \sim Vk_F/\varepsilon_F$; Vk_F is an effective barrier height of a potential with an effective width Δx of the order of the Fermi wavelength $2\pi/k_F$) the solution to Eq. (17.23) is

$$\kappa_n \simeq \pi n + \frac{1}{g_o}\frac{(-1)^n \cos\alpha - 1}{k_F L}. \tag{17.25}$$

We receive in this limit the formula for the persistent current

$$J = J_{\max} \sin 2\pi \frac{\Phi}{\Phi_0} \tag{17.26}$$

with the maximal amplitude

$$J_{\max} \simeq \frac{ev_F}{L}\left(\frac{l}{L}\right)^{1/2} \sum_{n=1}^{\infty} \frac{(-1)^n n^2}{e^{(n^2 - n_F^2)/2n_F T} + 1} \tag{17.27}$$

In the latter expression, we introduced an effective mean free path of electron in the ring

$$l \simeq L\frac{D}{1-D} \tag{17.28}$$

where D is the barrier transmissivity $D \sim 1/g_o^2$. It then follows from Eq. (17.27) that persistent current decreases with the transmissivity of the barrier slower than the normal-state conductance of the ring $G \sim (ne^2 S/p_F)(l/L)$ does.

3.3 Dirty Rings

Elastic scattering of electrons decreases the amplitude of persistent current oscillation. The impurities can be viewed as barriers which increase the effective length needed for electron to make a full round up of the ring and to interfere with the initial state. The interference between the initial and the final states showing the effect of the Aharonov-Bohm flux persists until the length becomes of the order of the localization length $\zeta \sim lN_\perp$ in a bulk sample [20] where $N_\perp = k_F^2 S_\perp / 4\pi$ is the number of transverse channels in a ring of cross section $S_\perp$. In the diffusive regime,

$$l \ll L \ll \zeta, \tag{17.29}$$

conductance of the ring can be calculated as a scattering problem, by using the Landauer formula [21] (see also [10]) relating ring conductance G to the transmission amplitudes $t_{\alpha\beta}$ between the incoming and outgoing channels α and β

$$G = \frac{2e^2}{h} \sum_{\alpha,\beta} |t_{\alpha\beta}|^2. \tag{17.30}$$

We therefore estimate the ring conductance as

$$G \sim N_\perp \langle |t|^2 \rangle G_0 \tag{17.31}$$

where $G_0 = 2e^2/h = 1/12.9\mathrm{k}\Omega$ is the quantum of conductance, *i.e.*, G is proportional to square of transmission amplitude. The Aharonov-Bohm effect, on the contrary, is expected to relate amplitude of the persistent current *linearly* to $|t_{\alpha\beta}|$, as was shown above in case of model problem with a δ-functional barrier. Translated from the subsection 2, the value of persistent current is expected to be, by the order of magnitude

$$J_{\max} \sim \frac{ev_F}{L} \left(\frac{G}{G_0} \right)^{1/2}. \tag{17.32}$$

We may expect that similar dependences may hold also for a diffusive ring. And indeed, the numeric simulation of the Aharonov-Bohm current in disordered ring seems to support this hypothesis.

In Fig. 17.4, we show the mean free path dependence of the current amplitude at zero temperature, which was received by calculating the energy E of N electrons in a $3d$ ring of volume $d_1 d_2 L$ with varying defect concentration, as function of vector potential A, and then calculating the current as a derivative dE/dA. Within the tight-binding approximation, the Hamiltonian of the model is

$$H = -t \sum_{\mathbf{n}} \sum_{\mathbf{m}} a_{\mathbf{n}+\mathbf{m}}^+ a_{\mathbf{n}} e^{i\alpha m_x} + \sum_{\mathbf{n}} V_{\mathbf{n}} a_{\mathbf{n}}^+ a_{\mathbf{n}} \tag{17.33}$$

where $\alpha = 2\pi\Phi/N\Phi_0$ is the phase difference between the near lattice sites along the ring circumference, $a_{\mathbf{n}}^+$ ($a_{\mathbf{n}}$) the creation (annihilation) operators at site $\mathbf{n}$, and $\mathbf{m}$ is the vector pointing from $\mathbf{n}$ to the nearest site. Potential at the site $V_{\mathbf{n}} = V\xi_{\mathbf{n}}$ is a random quantity depending on the value of $\xi_{\mathbf{n}} = 0, 1$ where 1 is occurring with a probability c. This probability determines the impurity concentration and the electron

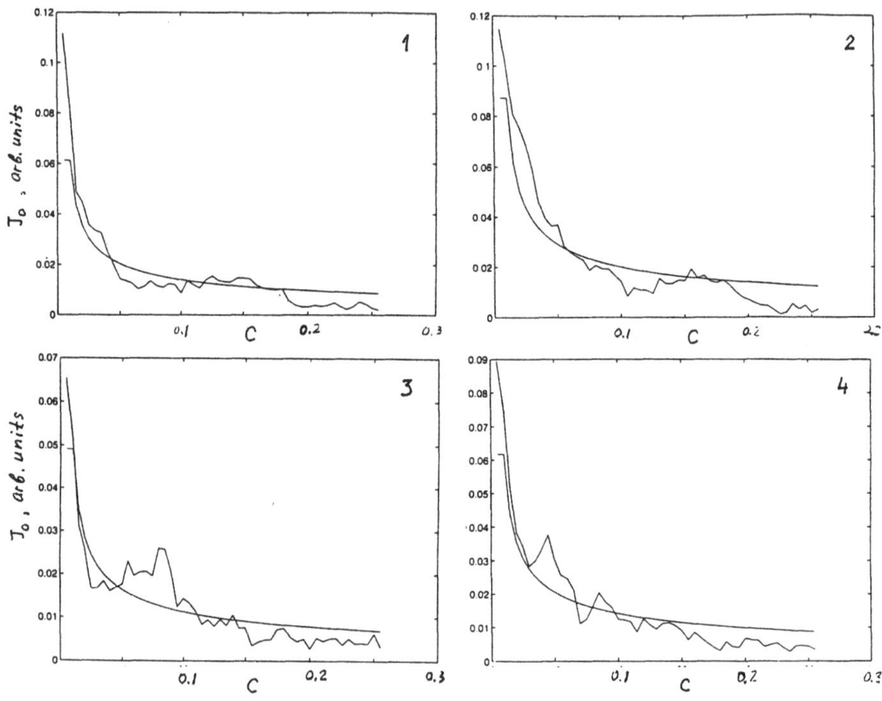

Figure 17.4 Persistent current amplitude J versus defect concentration c in a disordered ring (zigzag line) compared to the empirical $J(c)$ dependence according to Eq. (17.31) (continuous line). The dependences have been received in the various runs of numeric experiment with different number of electrons N_e and gradually increasing concentration c. Panel 1: $N_e = 25$; panel 2: $N_e = 35$; panel 3: $N_e = 45$; panel 4: $N_e = 50$. The sample has a cross section of 4×4 sites and a perimeter length of 30 sites. The "theoretical" curves have been received by estimating electron mean free path from the estimate $l \sim a/c$ where a is intersite spacing) and fitting the magnitude of the current at smaller concentration used ($c = 0.05$) to the "experimental" value at that concentration.

mean free path $l \sim a/c$ (a is the intersite distance taken as unity in our units). The limit $V \to \infty$ is equivalent to breaking of all connections of a site having $\xi_{\mathbf{n}} = 0$ to its neighbors.

The irregular curves in Fig. 17.4 represent the change in the magnitude of persistent current at addition of new defect sites (thus increasing c and decreasing l) whereas the smooth curves show the empirical law (17.32) with G estimated from Eq. (17.31) and corrected to the value of conductance at $c = 0$. The above calculations are not fully convincing since the expected localization limit is almost equal to the ring length, in our relatively small rings, and show disagreement with other theories [22] predicting smaller persistent currents. Mention however that some experimental data [23] in which persistent current, if have been properly estimated, was found to be *larger* than the value calculated on basis of Ref. [22]. Other experimental works [24, 25] show smaller oscillation amplitude.

3.4 Weak Localization Effects

Disordered metals show the effect of quantum interference in the Aharonov-Bohm field in an another way, as was first pointed out by Aronov, Altshuler and Spivak [26], and observed in an experiment of Sharvin and Sharvin [27]. The mesoscopic rings display oscillatory dependence of their kinetic rather than thermodynamic properties on the flux piercing the ring, with the half of the Aharonov-Bohm period, $hc/2e$. The origin of these oscillations is in that the time-reversed paths of electrons traversing the ring circumference in the clockwise and in the counterclockwise directions interfere with one another giving rise to conductance oscillations with the period equal to $\Phi_0/2$

$$\sigma \sim \sin\left(\frac{2e}{\hbar c}\oint \mathbf{A}dl\right) \sim \sin\left(4\pi\frac{\Phi}{\Phi_0}\right). \qquad (17.34)$$

Extensive reviews on that subject are presented in Refs. [9, 11, 28].

4. FLUCTUATIONS OF PERSISTENT CURRENT

Persistent current is not a macroscopic phenomenon like, *e.g.*, supercurrent in a ring. The amplitude of persistent current in a $1d$ ring is of the order of the current produced by a *single* electron orbiting the ring with a *high* velocity, that of the order of the Fermi velocity $v_F \sim 10^8$ cm/s. This makes a current to be of a sizeable amplitude

$$J \sim 1\ \mu\text{A}$$

in a ring of radius $R \sim 10^{-4}$ cm. More than that, unlike the superconducting currents, persistent current in mesoscopic loop is subject to quantum and thermal fluctuations. The average current equals to

$$\langle \hat{J} \rangle = \sum_n \langle \hat{J} \rangle_n N_n \qquad (17.35)$$

where $\hat{J}$ is the current operator and N_n is the thermal average of the occupation probability of a quantum state n. The root mean square

(RMS) fluctuation of the current, $\langle\delta\hat{J}^2\rangle^{1/2}$, is found from the identity

$$\delta\bar{J}^2 = \langle(\hat{J} - \langle\hat{J}\rangle)^2\rangle = \sum_n \langle\hat{J}^2\rangle_n N_n - \langle\hat{J}\rangle_n^2 N_n^2 \tag{17.36}$$

Consider a ring with a δ-barrier from Section 3.2. The average value of the current density j is given by Eq. (17.24), whereas the average value of j^2 is

$$\langle\hat{j}^2\rangle_n = \langle\hat{j}\rangle_n^2 = \left(\frac{e\hbar}{mL^2}\right)\kappa_n^2. \tag{17.37}$$

In a ballistic ring, we receive

$$\delta\hat{J}^2 = \left(\frac{ev_F}{L}\right)^2 \frac{1}{(4\pi n_F)^2} \sum_{n=-\infty}^{\infty} \frac{(n-f)^2}{\cosh^2 \frac{(n-f)^2 - n_F^2}{4\pi n_F}} \tag{17.38}$$

which gives

$$\langle\delta\hat{J}^2\rangle \simeq \left(\frac{ev_F}{L}\right)^2 \frac{T/\Delta E}{2\pi k_F L}. \tag{17.39}$$

In a poorly connected ring ($V_{max} \gg \varepsilon_F$ where $V_{max} = Vk_F$ is an effective height of the barrier) the average value of the persistent current becomes smaller than the RMS fluctuations of the current. The motion of electrons is then more like the "persistent drift" rather than the regular flow of charge similar to that in a superconducting metal.

5. TRANSVERSE PERSISTENT CURRENT

Persistent current in ring appears due to violation of the time-reversal symmetry created by the "longitudinal" flux directed along the ring symmetry axis. The nonzero φ-component of vector potential, A_φ, determines the asymmetry between the clockwise and counterclockwise directions, and allows for the nonzero φ-component of the current. Assume however that in a double ring (Fig. 17.5a) the total longitudinal flux is zero but the radial flux (the one due to B_r component of the field) is not. If the double-ring is asymmetric, the phase gradients in two rings will not be equal to one another which is equivalent to the appearance of an effective phase gradient in the φ-direction. The current in a system (termed in [29] as the "transverse" persistent current), may then appear even if the longitudinal flux is zero. The experimental observation of persistent currents in strong transverse magnetic fields was also reported in Ref. [30].

Assume that hopping amplitudes t_1, t_2 in a double ring are not equal, and that the hopping amplitude t_{12} couples the rings. Such model, in a

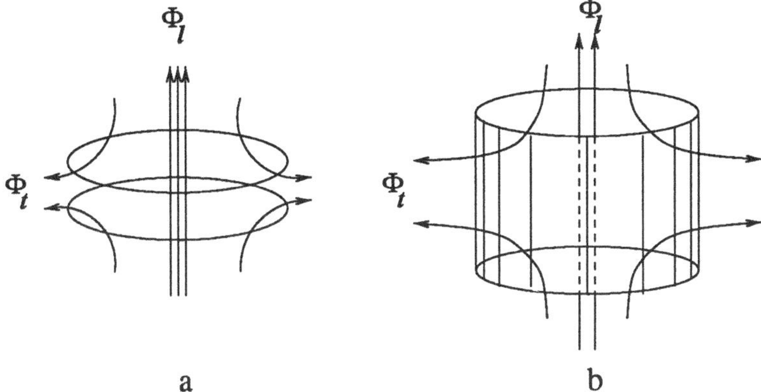

Figure 17.5 Sketch of a setup for observation of the Aharonov-Bohm effect in a strong perpendicular field. (a) Two coupled rings; (b) Hollow cylinder with walls traversed by the transverse flux Φ_t created by magnetic field in the radial direction. Shown are lines of force of magnetic field around the cylinder.

tight-binding approximation, is described by the Hamiltonian

$$H = -\sum_{n=1}^{N}\left(t_1 a_n^+ a_{n+1} e^{i(\alpha+\beta)} + t_2 b^+ +_n b_{n+1} e^{i(\alpha-\beta)} + t_{12} a_n^+ b_n\right) + \text{H.c.} \tag{17.40}$$

in which α stands for the longitudinal and β for the transverse flux

$$\alpha = 2\pi\Phi/N\Phi_0,\ \beta = \pi\Phi_r/N\Phi_0. \tag{17.41}$$

Solving for the plane wave state

$$\psi = \sum_{n=1}^{N} e^{ikn}(Aa_n^+ + Bb_n^+)|0\rangle$$

we receive the energy

$$\varepsilon_{k\sigma} = -t_1\cos(k+\alpha+\beta) - t_2\cos(k+\alpha-\beta)$$

$$+\sigma\sqrt{(t_1\cos(k+\alpha+\beta) - t_2\cos(k+\alpha-\beta)^2 + t_{12}^2} \tag{17.42}$$

and then calculate the current

$$J = -\frac{e}{N\hbar}\sum_k\sum_{\sigma=-1}^{1}\left[e^{(\varepsilon_{k\sigma}-\zeta)/T}+1\right]^{-1}\frac{\partial}{\partial\alpha}\varepsilon_{k\sigma}(\alpha) \tag{17.43}$$

which may be nonzero even if the longitudinal flux $\Phi = 0$. Such "transverse" currents appear if system is not central symmetric, *e.g.*, when

$t_1 \neq t_2$. In a hollow dirty cylinder (Fig. 17.5*b*), asymmetry is caused by random distribution of impurities with the potential $V_{nm} = V\xi_{nm}$, $0 < \xi_{nm} < 1$

$$H = -t \sum_{n=1}^{N} \sum_{m=1}^{M} \left(a_{nm}^{+} a_{n+1,m} e^{i[\alpha+\beta(m-M/2)]} + a_{nm}^{+} a_{n,m+1} \right) + \text{H.c.}$$

$$+ \sum_{nm} V_{nm} a_{nm}^{+} a_{nm} \qquad (17.44)$$

where α is given by first Eq. (17.41) and $\beta = 2\pi\Phi_r/NM$.

Fig. 17.6 shows an example of persistent current versus transverse flux dependence in a cylindrical shell with 10×10 sites and varying amplitude V of the impurity potential. The chaotic quasi-oscillatory behavior is hard to interpret quantitatively, it corresponds most probably to the flux quantization in the local "loops" formed by the impurity islands. The dependence, if properly inverted, may serve as an information on the inhomogeneous state in mesoscopic structure.

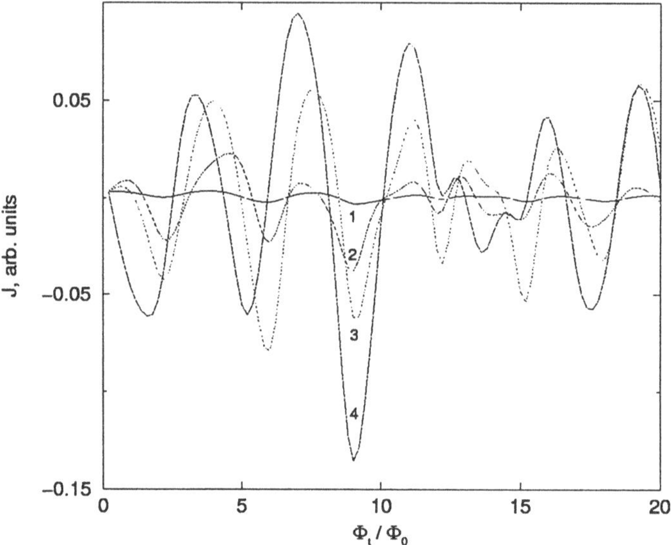

Figure 17.6 Azimuthal current in a disordered cylindrical shell with 10×10 sites versus transverse flux through wall Φ_t at the zero Aharonov-Bohm flux Φ_l at various amplitudes of the disorder potential V. Line 1: $V/t = 0$; line 2: $V/t = 0.05$; line 3: $V/t = 0.1$; line 4: $V/t = 0.15$. The small but nonzero value of current at $V = 0$ is related to numeric procedure of calculating current as the derivative of energy with respect to flux and then putting flux $\Phi_l \to 0$.

6. BERRY'S PHASE AND OSCILLATORY SPIN DYNAMICS IN MESOSCOPIC RINGS

Aharonov-Bohm effect represents the first but not the last example of more general concept of quantal phase accumulated by electron in its motion in a slowly varying field which was considered by Berry [8] (see also [31]) in the context of molecular dynamics. When electron moves adiabatically (slower than the field changes) in the field of the vector potential, its wave function accumulates a phase

$$\Delta\varphi = \frac{e}{\hbar c}\int_{\mathbf{r}_1}^{\mathbf{r}_2} \mathbf{A}dl = \frac{e}{\hbar c}A_\varphi(\theta_2 - \theta_1) \tag{17.45}$$

where θ_1 and θ_2 are azimuthal angles of the initial and final locations $\mathbf{r}_1$ and $\mathbf{r}_2$ on a ring. This phase difference is the Aharonov-Bohm phase discussed in previous sections. Consider now the effect of magnetic field on the electron spin. Assume that magnetic field at any point of the ring makes fixed orientation with the local tangential vector on the contour (Fig. 17.7). The tangential component of the field, B_φ, can in principle be created by a current-carrying wire inserted to the ring. The radial component, B_r, formally corresponds to the field produced by the line of magnetic monopoles inserted into the ring. In reality, such field can be created by a proper combination of solenoids around the ring, as was explained in Section 5 (see Fig. 17.6).

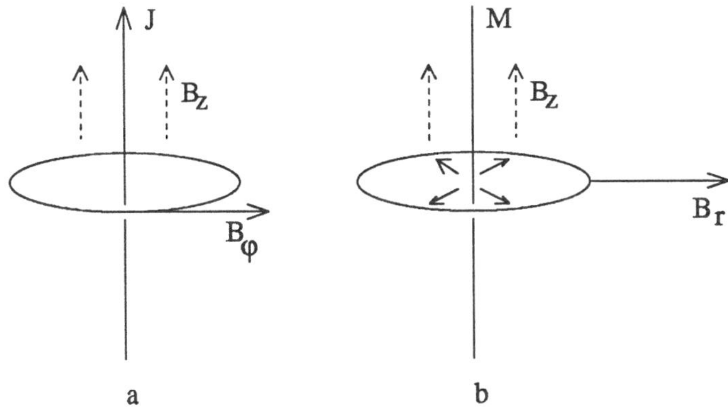

Figure 17.7 Sketch of the thought Berry-phase experiments with an azimuthal field created by a current-carrying wire piercing the ring (a), and radial field generated by the line of magnetic monopoles inside the ring (b).

When electron is slowly rotating along the ring, its spin function will accumulate phase due to spinor transformation [32]

$$\exp\left[\frac{1}{2}i\mathbf{n}\cdot\vec{\sigma}(\theta_2 - \theta_1)\right]$$

where $\vec{\sigma}$ is the Pauli matrix vector and $\mathbf{n}$ is the unit vector along the rotation axis. With $\mathbf{n}$ being in the direction of the vector $\mathbf{B}$, this produces a phase difference between the points θ_1, θ_2 on the ring

$$\Delta\varphi = \frac{1}{2}(\theta_2 - \theta_1)\sin\alpha \tag{17.46}$$

where α is an angle between $\mathbf{B}$ and its projection onto the plane of the ring. Such effects which are additive to the topological effects of the Aharonov and Bohm, have been considered by Stern [33] and by Loss, Goldbart and Balatsky [34]. Actually, the condition for adiabaticity explains the origin of the effect rather is strictly required. The theory of the Berry's phase effects for electron spin, as well as the Aharonov-Bohm effects due to the orbital motion of electrons in a ring, will be presented below in a form which covers both the adiabatical ($\omega_H T_0 \gg 1$) as well as nonadiabatical ($\omega_H T_0 \leq 1$) regimes where ω_H is the cyclotron frequency eB/mc and T_0^{-1} the frequency of electron rotation v_F/L.

Hamiltonian of particle in a ring including the Zeeman energy in the nonrelativistic approximation (the Pauli Hamiltonian),

$$H = \frac{1}{2m}\left(\hat{\mathbf{p}} - \frac{e}{c}\mathbf{A}\right)^2 - \mu\mathbf{B}\cdot\vec{\sigma}, \tag{17.47}$$

is presented in a matrix form

$$H = -\varepsilon_0\left(\frac{\partial}{\partial\varphi} - i\Phi/\Phi_0\right)^2 + \varepsilon_\perp\begin{pmatrix} 0 & e^{-i(\varphi+\gamma)} \\ e^{i(\varphi+\gamma)} & 0 \end{pmatrix} + \varepsilon_{||}\begin{pmatrix} 1 & 0 \\ 0 & -1 \end{pmatrix} \tag{17.48}$$

where ε_0 is the ring quantization energy, and $\varepsilon_\perp$ and $\varepsilon_{||}$ are the components of the Zeeman energy $\varepsilon_Z = g\mu_B B$

$$\varepsilon_0 = \hbar^2/2mR^2, \quad \varepsilon_\perp = \varepsilon_Z\cos\alpha, \quad \varepsilon_{||} = \varepsilon_Z\sin\alpha. \tag{17.49}$$

γ is the angle between the radial and azimuthal components of magnetic field $\gamma = \arctan(B_\varphi/B_r)$. Φ is the total flux comprising that from the z-component of magnetic field and from the solenoid which can in principle be inserted into the ring. Eigenstates of Eq. (17.48) can easily be found if we present the wave function in the form

$$\Psi = \sum_{n=-\infty}^{\infty}\begin{pmatrix} u_n \\ v_n \end{pmatrix} e^{in\varphi} \tag{17.50}$$

and write down the equations for u_n, v_n

$$\begin{aligned} \varepsilon_0(n - \Phi/\Phi_0)^2 u_n + \varepsilon_\perp e^{-i\gamma} v_{n+1} + \varepsilon_{||} u_n &= \varepsilon u_n, \\ \varepsilon_0(n - \Phi/\Phi_0)^2 v_n + \varepsilon_\perp e^{-i\gamma} u_{n-1} - \varepsilon_{||} v_n &= \varepsilon v_n. \end{aligned} \tag{17.51}$$

Then we receive an equation

$$\left[\varepsilon_0(n-f)^2+\varepsilon_{\|}-\varepsilon+\frac{\varepsilon_{\perp}^2}{\varepsilon-\varepsilon_0(n+1-f)^2+\varepsilon_{\|}}\right]u_n=0 \qquad (17.52)$$

where $f=\Phi/\Phi_0$, and the relation between u_n and v_{n+1}

$$v_{n+1}=\frac{\varepsilon_{\perp}e^{i\gamma}}{\varepsilon-\varepsilon_0(n-f)^2+\varepsilon_{\|}}u_n. \qquad (17.53)$$

The energy eigenvalues are

$$\varepsilon_n^{\pm}=\frac{1}{2}(\varepsilon_1+\varepsilon_2)\pm\frac{1}{2}\sqrt{(\varepsilon_1-\varepsilon_2)^2+4\varepsilon_{\perp}^2} \qquad (17.54)$$

where

$$\varepsilon_1=\varepsilon_0(n-f)^2+\varepsilon_{\|},\quad \varepsilon_2=\varepsilon_0(n+1-f)^2-\varepsilon_{\|}$$

The current in the ring is found by differentiating the energy with respect to the Aharonov-Bohm flux Φ (and putting the latter to zero if the *total* flux vanishes, due to possible compensation between the solenoid flux and the flux created by the Zeeman field, B_z)

$$J=-\frac{e}{h}\sum_{n=-\infty}^{\infty}\sum_{\sigma=-1}^{1}\left\{\exp\left[\frac{\varepsilon_{n\sigma}(f)-\zeta}{T}\right]+1\right\}^{-1}\frac{\partial}{\partial f}\varepsilon_{n\sigma}(f)\,. \qquad (17.55)$$

Expression (17.55) can be easily evaluated. Skipping the corresponding lengthy formula, we show below the representative dependences of the persistent currents on the longitudinal flux Φ and on the tilt angle α between the perpendicular field $B_{\perp}=\sqrt{b_r^2+B_{\varphi}^2}$ and the Zeeman field $B_{\|}$.

Fig. 17.8 shows the current as a function of flux in the ring Φ in case when the ratio of the Zeeman energy ε_Z to the distance between the discrete energy levels at the Fermi energy $\Delta E=2n_F\varepsilon_0$ equals to 5.

ΔE is the representative energy scale for the Aharonov-Bohm effect whereas the projections of ε_Z to the ring plane and on the symmetry axis determine the spin-related Berry-phase energy scales $\varepsilon_{\perp}$, $\varepsilon_{\|}$. In case when $\alpha=0$ or $\alpha=\pi/2$, the current vanishes at zero Aharonov-Bohm flux, as was mentioned in [13]. Persistent current is an oscillating function of the Aharonov-Bohm flux at any value of α, it does not depend on the angle γ between the azimuthal and radial field components. The shift of position of the minima in $J(f)$ dependences is in accord with the expected Berry phase shift from Eq. (17.46)

$$\Delta f=\frac{1}{2}\sin\alpha. \qquad (17.56)$$

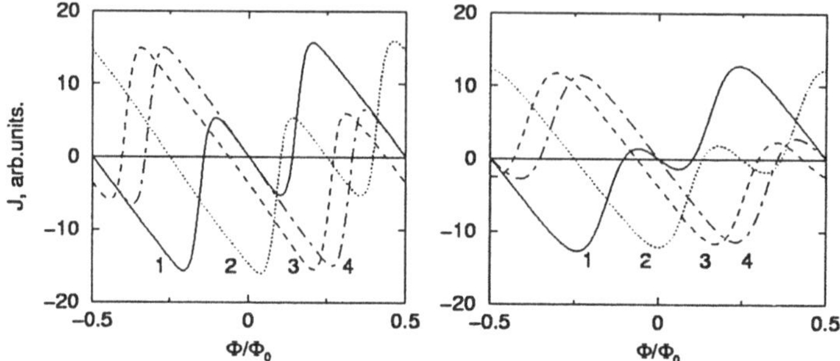

Figure 17.8 Aharonov-Bohm current versus magnetic flux at various values of the Berry angle α and the temperature τ. Left panel: $\tau = 0.02$, right panel: $\tau = 0.05$. Line 1 in both panels corresponds to $\alpha = 0$, line 2 to $\alpha = \pi/6$, line 3 to $\alpha = \pi/3$ and line 4 to $\alpha = \pi/2$. Electron concentration is chosen according to the value of parameter $n_F = 10.25$.

There remains a mystery on the origin of persistent current at $f = 0$, in particular this current shows the non-monotonic dependence on temperature with a maximum at low temperature.

Fig. 17.9 shows the dependence of the persistent current on $\varepsilon_\perp$ at fixed longitudinal flux. The oscillation displayed are the manifestation of an another effect, similar to the De Haas-Van Alphen effects in metals, and can be accounted for by the passage by the Zeeman-split Fermi energy through the set of quantized energy states in a ring in the vicinity of the Fermi energy. Since the energy of the states depends on Φ, persistent current is also an oscillating function of Φ with a period Φ_0.

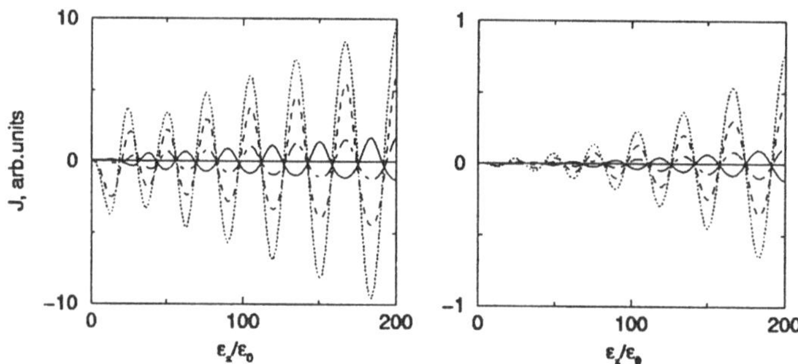

Figure 17.9 Aharonov-Bohm current versus the Zeeman splitting ε_Z at temperature $\tau = 0.2$ (left panel) and $\tau = 0.5$ (right panel). In both panels, solid line corresponds to $\alpha = 0$, dotted line to $\alpha = \pi/6$, dashed line to $\alpha = \pi/3$ and dot-dashed line to $\alpha = \pi/2$. Parameter $n_F = 10.25$.

7. EXOTIC AHARONOV-BOHM AND BERRY-PHASE EFFECTS

The understanding of phase coherence in mesoscopic specimens stimulated search for various generalizations of the generic situation depicted with the Aharonov-Bohm phase. We mention some of those within the limits of our competence.

The phase shift of the electron wave function can be produced by spin-orbit interaction [35]

$$H_{so} = -\frac{\hbar}{4m^2c^2}\vec{\sigma}\cdot(\mathbf{p}\times\nabla)V(\mathbf{r}) = \frac{\hbar^2}{4m^2c^2}\left(\frac{1}{r}\frac{dV}{dr}\right)\mathrm{l}\cdot\vec{\sigma} \qquad (17.57)$$

where $V(\mathbf{r})$ is an electrostatic potential. In case when V is created by a charged line with a linear charge density τ inserted into the ring (Fig. 17.10a), the phase shift (known as the Aharonov-Casher effect [36]) becomes

$$\delta = \frac{e\tau}{mc^2}(n_\uparrow - n_\downarrow). \qquad (17.58)$$

The interaction Hamiltonian describing this effect, to be added to the Hamiltonian in Eq. (17.48), is

$$H_{int} = -\frac{ie\hbar^2\tau}{2m^2c^2R^2}\sigma_z\frac{\partial}{\partial\varphi}. \qquad (17.59)$$

The phase shift is very small a compared to the Aharonov-Bohm shift $(\Delta\varphi)_{AB}$,

$$\delta/(\Delta\varphi)_{AB} \sim \frac{V}{mc^2}\times\frac{g\mu_B}{\varepsilon_F}. \qquad (17.60)$$

Even for the semiconducting crystals with a large g-factors [37], δ remains much smaller than $(\Delta\varphi)_{AB}$.

For the atomic potentials of randomly distributed spins, according to [38], Aharonov-Bohm persistent current shifts in phase as

$$J = \frac{1}{2}[J_{AB}(2\pi\Phi/\Phi_0 + \delta) + J_{AB}(2\pi\Phi/\Phi_0 - \delta)]. \qquad (17.61)$$

Similar spin-orbit effects have been considered by a number of authors [39], *etc.* The Aharonov-Casher effect has been suggested for neutral particles such as atoms in liquid helium [40].

It was proposed [41, 42] and possibly, found in an experiment [43], that persistent current may exist in the nonmetallic materials such as Peierls insulators.

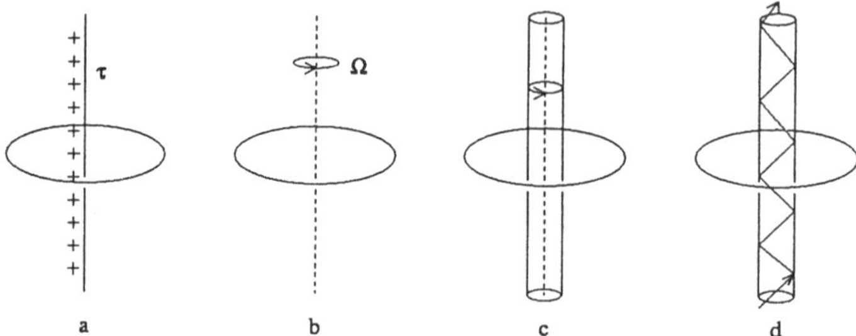

Figure 17.10 Possible geometrical and topological Aharonov-Bohm configurations: (a) Ring pierced by a charged dielectric line (Aharonov-Casher geometry); (b) Rotating ring; (c) Ring in the Lense-Thirring field of a rotating massive body; (d) Ring enclosing an optical fiber transmitting the electromagnetic radiation through its orifice.

Geometrical phase may appear in a rotating metal (Fig. 17.10*b*), and even cause "persistent rotation" of the latter in the Aharonov-Bohm field [44]. This is extremely weak effect requiring, for its observation, temperatures in the range of μK even in nanoscopic samples ($R \leq 10^{-6}$ cm). Even smaller are the effects of gravitational interaction with the rotating massive bodies (Fig. 17.10*c*) producing the Lense-Thirring field [45] which results in the phase shift

$$\alpha \sim \frac{GmMR^2}{\hbar c^2}\Omega$$

where G is gravitational constant, R is radius and M the mass (per unit length) of a cylinder rotating with an angular velocity Ω. The latter effect is certainly out of reach to any terrestrial experiment.

The effects of time-dependent fluxes (Fig. 17.10*d*) have been considered by Aronov *et al.* [46] and by the author and Shumovsky [47]. The specific case of the so-called 2nd Aharonov-Bohm effect (the phase shift due to scalar potential) have been addressed in [13] and nominated as the "persistent charge", *i.e.*, periodic in the VT charge accumulation on plates of a mesoscopic capacitor subject to voltage pulses of amplitude V and duration T. A specific effect of inelastic backscattering in a "clean" ring is shown to restore the Aharonov-Bohm oscillation otherwise suppressed by a time varying magnetic flux [48, 49].

Strong coupling effects (Wigner crystallization [50]), Luttinger liquid [51]) make change to the amplitude of persistent current and its temperature dependence, in comparison to noninteracting electrons.

We mention also earlier works on flux quantization in bulk mesoscopic cylinders [52, 53, 54]. Due to surface electron states accumulating the Aharonov-Bohm phase in the same way as the rotating electrons in a hollow cylinder do, magnetic moment of cylinder have been shown to oscillate in function of flux [52]. The oscillations are also seen in the longitudinal conductivity of Bi cylinders [53, 54]. These experiments have been the first demonstration of the quantum oscillations in mesoscopic specimens with the period of a single flux quantum hc/e.

References

[1] E. Teller, Der diamagnetismus von freien electronen, Zs. Physik **67**, 311 (1931).

[2] L. D. Landau, Zs. Physik **64**, 629 (1930).

[3] Y. Aharonov and D. Bohm, Phys. Rev. **115**, 485 (1959).

[4] N. Byers and C. N. Yang, Theoretical considerations concerning quantized magnetic flux in superconducting cylinders, Phys. Rev. Lett. **7**, 46 (1961).

[5] F. Bloch, Flux quantization and dimensionality, Phys. Rev. **166**, 415 (1968);
Off-diagonal long-range order and persistent currents in a hollow cylinder, Phys. Rev. **137**, A787 (1965).

[6] I. O. Kulik, Flux quantization in a normal metal, JETP Lett. **11**, 275 (1970).

[7] M. Buttiker, Y. Imry and R. Landauer, Josephson behavior in small normal one-dimensional ring, Phys. Lett. **A96**, 365 (1983).

[8] M. V. Berry, Quantal phase factors accompanying adiabatic changes, Proc. Roy. Soc. (London) **A392**, 45 (1984).

[9] S. Washburn, Aharonov-Bohm effects in loops of gold, in: Mesoscopic Phenomena in Solids, *eds.* B. L. Altshuler, P. A. Lee and R. A. Webb, (Elsevier, 1991) p.1.

[10] Y. Imry, Physics of mesoscopic systems, in: Directions in Condensed Matter Physics, *eds.* G. Grinstein and G. Mazenko, (World Scientific, Singapore, 1986) p.101.

[11] S. Washburn and R. A. Webb, Aharonov-Bohm effect in normal metal. Quantum coherence and transport. Adv. Phys. **35**, 375 (1986).

[12] Y. Imry, Introduction to Mesoscopic Physics, (Oxford Univ. Press, Oxford, 1997).

[13] I. O. Kulik, Magnetic and electric Aharonov-Bohm effects in nanostructures, Physica B, **218**, 252 (1996).

[14] I. O. Kulik, Persistent current and persistent charge in nanostructures, in: Quantum Optics and Spectroscopy of Solids, *eds.* T.

Hakioglu and A. S. Shumovsky, (Kluwer Acad. Publ., Dordrecht, 1997) p.45.

[15] I. V. Krive and A. S. Rozhavsky, Non-traditional Aharonov-Bohm effects in condensed matter, Int. J. Mod. Phys. **B6**, 1255 (1992).

[16] D. Shönberg, Magnetic Oscillations in Metals, (Cambridge Univ. Press, Cambridge, 1984).

[17] J. M. Ziman, Principles of the Theory of Solids, (Cambridge Univ. Press, Cambridge, 1972).

[18] A. F. Andreev, Sov. Phys. JETP **19**, 1228 (1964); **22**, 455 (1966).

[19] I. O. Kulik, Macroscopic quantization and the proximity effect in SNS junctions, Zh. Eksp. Teor. Fiz. **57**, 1745 (1969) [Sov. Phys. JETP **30**, 944 (1969)].

[20] A. A. Abrikosov, Fundamentals of the Theory of Metals, (North-Holland, Amsterdam, 1988).

[21] R. Landauer, Electrical resistance of disordered one-dimensional lattices, Phil Mag. **21**, 863 (1970).

[22] H. F. Cheung, E. K. Riedel and Y. Gefen, Persistent current in mesoscopic rings and cylinders, Phys. Rev. Lett. **62**, 587 (1989).

[23] V. Chandrasekhar, R. A. Webb, M. J. Brady, M. B. Ketchen, W. J. Gallagher and A. Kleinsasser, Magnetic response of a single, isolated gold loop, Phys. Rev. Lett. **67**, 3578 (1991).

[24] D. Mally, C. Chapelier and A. Benoit, Experimental observation of persistent current in a GaAs-AlGaAs single loop, Phys. Rev. Lett. **70**, 2020 (1993).

[25] L. P. Levy, G. Dolan, J. Dunsmuir and H. Bouchiat, Magnetization of mesoscopic copper rings: Evidence for persistent currents, Phys. Rev. Lett. **64**, 2074 (1990).

[26] B. L. Altshuler, A. G. Aronov and B. Z. Spivak, JETP Lett. **33**, 94 (1981).

[27] D. Yu. Sharvin and Yu. V. Sharvin, JETP Lett. **34**, 272 (1982).

[28] B. L. Altshuler and A. G. Aronov, in: Electron-Electron Interactions in Disordered Systems, *eds.* A. L. Efros and M. Pollak, (Elsevier, Amsterdam, 1985) p.1.

[29] I. O. Kulik, "Transverse" persistent currents in mesoscopic cylinders and rings, To appear in Physica B, 2000.

[30] D. Eliahu, R. Berkovits, M.Abraham and Y. Avishai, Mesoscopic persistent-current correlations in the presence of strong magnetic fields, Phys. Rev. **B49**, 14448 (1994).

[31] C. A. Mead, The geometric phases in molecular systems, Rev. Mod. Phys. **64**, 51 (1992).

[32] L. D. Landau and E. M. Lifshitz, Quantum Mechanics, (Pergamon, New York, 1965).

[33] A. Stern, Berry's phase, motive forces, and mesoscopic conductivity, Phys. Rev. Lett. **68**, 1022 (1992).

[34] D. Loss, P. Goldbart and A. V. Balatsky, Berry's phase and persistent charge and spin currents in textured mesoscopic rings, Phys. Rev. Lett. **65**, 1655 (1990).

[35] R. Shankar, Principles of Quantum Mechanics, Plenum, New York, 1987.

[36] Y. Aharonov and A. Casher, Topological quantum effects for neutral particles, Phys. Rev. Lett. **53**, 319 (1984).

[37] E. N. Bogachek and U. Landman, Aharonov-Bohm and Aharonov-Casher tunneling effects and edge states in double-barrier structures, Phys. Rev. **B50**, 2678 (1994).

[38] Y. Meir, Y. Gefen and O. Entin-Wohlman, Universal effects of spin-orbit scattering in mesoscopic systems, Phys. Rev. Lett. **63**, 798 (1989).

[39] Y. Lyanda-Geller, Quantum interference and electron-electron interactions at strong spin-orbit coupling in disordered systems, Phys. Rev. Lett. **80**, 4273 (1998).

[40] A. V. Balatsky and B. L. Altshuler, Persistent spin and mass currents and Aharonov-Casher effect, Phys. Rev. Lett. **70**, 1678 (1993).

[41] I. O. Kulik, A. S. Rozhavsky and E. N. Bogachek, Magnetic flux quantization in dielectrics, Zh. Eksp. Teor. Fiz. Pis'ma **47**, 251 (1988) [JETP Lett. **47**, 304 (1988)].

[42] E. N. Bogachek, I. V. Krive, I. O. Kulik and A. S. Rozhavsky, Instanton Aharonov-Bohm effect and macroscopic quantum coherence in charge-density-wave systems, Phys. Rev. B42, 7614 (1990).

[43] Yu. I. Latyshev. O. Laborde, P. Monceau and S. Klaumunzer, Aharonov-Bohm effect on charge density wave (CDW) moving through columnar defects in $NbSe_3$, Phys. Rev. Lett. **78**, 919 (1997).

[44] I. O. Kulik, Small rigid rotators as qubits, in: Quantum Mesoscopic Phenomena and Mesoscopic Devices in Microelectronics, Program and Abstracts, p.108, *eds.* I. O. Kulik, R. Ellialtıoğlu, B. Tanatar and C. Yalabık., Bilkent Univ., Ankara, 1999.

[45] L. D. Landau and E. M. Lifshitz, Theory of Fields, (North-Holland, Amsterdam, 1988).

[46] I. E. Aronov, A. Grincwajg, M. Jonson, R. I. Shekhter and E. N. Bogachek, Sol. St. Communs. **91**, 75 (1994).

[47] I. O. Kulik and A. S. Shumovsky, Aharonov-Bohm effect induced by light in a fiber, Appl. Phys. Lett. **69**, 2779 (1996).

[48] T. Swahn, E. N. Bogachek, Yu. M. Galperin, M. Jonson and R. I. Shekhter, Phys. Rev. Lett. **73** 162 (1994).

[49] E. N. Bogachek, Yu. M. Galperin, M. Jonson, R. I. Shekhter and T. Swahn, J. Phys.: Cond. Matter **8**, 2603 (1996).

[50] I. V. Krive, P. Sandström, R. I. Shekhter, S. M. Girvin and M. Jonson, Phys. Rev. B **52**, 16451 (1995).

[51] M. V. Moskalets, Physica E **5**, 124 (1999).

[52] E. N. Bogachek and G. A. Gogadze, Sov. Phys. JETP **36**, 973 (1973).

[53] N. B. Brandt, D. V. Gitsu, A. A. Nikolaeva and Ya. G. Ponomarev, JETP Lett. **24**, 272 (1976).

[54] N. B. Brandt, E. N. Bogachek, D. V. Gitsu, G. A. Gogadze, I. O. Kulik, A. A. Nikolaeva and Ya. G. Ponomarev, Flux quantization effects in metal microcylinders in a tilted magnetic field, Fiz. Nizk. Temp. **8**, 718 (1982) [Sov. J. Low Temp. Phys. **8**, 358 (1982)].

Chapter 18

PERSISTENT CURRENT IN A MESOSCOPIC RING WITH STRONGLY COUPLED POLARONS

M. Bayindir and I. O. Kulik
Department of Physics, Bilkent University
06533 Bilkent, Ankara, Turkey

Abstract We investigate influence of the electron-phonon interaction on persistent current in a mesoscopic ring threaded by a Aharonov-Bohm flux Φ. We find that all thermodynamic quantities are still periodic in Φ with a period $\Phi_0 = hc/e$. In both cases, weak- and strong-coupling, effective mass of the electrons is increased and amplitude of persistent current is suppressed. In the latter case the amplitude decreases exponentially as a function of the electron-phonon coupling constant.

1. INTRODUCTION

In classical electrodynamics, motion of a charged particle is completely determined by local electric field $\mathbf{E}$ and magnetic field $\mathbf{B}$ that act upon it. The scalar potential ϕ and vector potential $\mathbf{A}$ were first introduced as a mathematical tool for calculation concerning electromagnetic fields. However, in quantum theory these potentials appear in the Schrödinger equation explicitly and therefore they affect all physical quantities directly. Aharonov and Bohm [1] have shown that contrary to the conclusion of classical electrodynamics, there exist effects of the potentials on the charged particles, even in the region where all fields vanish (hence, there is no force acting on the particles). The experiment of Chambers [2] proved the existence of Aharonov-Bohm (AB) effect and stimulated experimental and theoretical studies in this field [3, 4, 5]. This effect has purely quantum mechanical origin because it comes from the interference phenomenon. The most well-known example of the AB effect is

I. O. Kulik and R. Ellialtioğlu (eds.),
Quantum Mesoscopic Phenomena and Mesoscopic Devices in Microelectronics, 283–292.

the persistent current in the normal-metal rings threaded by a magnetic flux [6, 7, 8].

Sensitivity of the equilibrium properties, such as average energy or magnetization of a small free electron systems to a magnetic field was noticed rather early [9, 10, 11]. Existence of non-decaying (persistent) current in normal-metal rings (where mean free path L_ϕ exceeds the circumference of the ring $L = 2\pi R$) enclosing a magnetic flux was predicted by Kulik [6] in 1970. Later Büttiker, Imry, and Landauer [8] proposed the persistent current in the normal one-dimensional disordered ring. In 1990's, experimental works [12, 13] confirmed the existence of persistent current in mesoscopic rings. This current arises due to the boundary conditions imposed on the wave function by the doubly connected nature of the loop. As a consequences of the boundary conditions, all physical properties of the ring are periodic in magnetic flux Φ with a period $\Phi_0 = hc/e$ [14, 15].

Scattering mechanisms result in decreasing, but not diminishing, of the persistent current. It is believed that inelastic, *i.e.* electron-phonon interaction, should be small to make the electronic states in the ring long-lived (phase conserving). In the literature, the effects of the electron-phonon interaction were studied in the mesoscopic AB rings in metallic [16, 17] and semiconducting cases [18].

In this paper we investigate the electron-phonon interaction within two limiting cases (weak- and strong-coupling) in a metallic AB ring and show that the polaronic effect doesn't exclude the existence of flux quantization and persistent current in the ring.

2. PERSISTENT CURRENTS IN MESOSCOPIC RINGS

In this section we give a brief summary of the persistent current in mesoscopic AB rings in the absence of electron-phonon interaction (See Refs. [19, 20, 21]). Let us consider a one-dimensional ring with the circumference $L = Na$, where N is the number of lattice sites and a is the lattice constant ($a = 1$ is taken throughout this paper), threaded by a magnetic flux Φ as illustrated in Fig. 18.1.

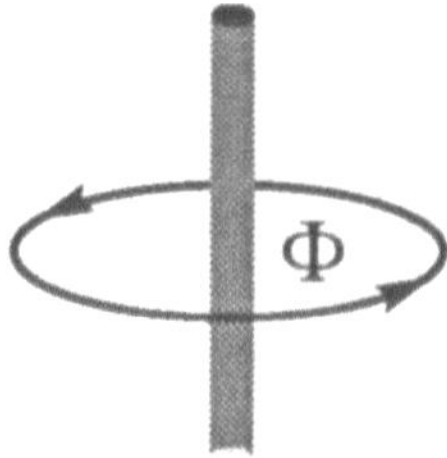

Figure 18.1 One-dimensional ring of circumference L threaded by an AB flux Φ.

The existence of persistent current in the ring requires that there should not be strong localization, so that overlapping of the atomic orbitals is much important rather themselves. In the tight binding approximation, we can write the Hamiltonian as

$$H_e = -t_0 \sum_{n=1}^{N} (a_n^+ a_{n+1} e^{i\alpha} + a_{n+1}^+ a_n e^{-i\alpha}) \,, \tag{18.1}$$

where t_0 is hopping amplitude between the nearest-neighbors for an undistorted lattice, the operators a_n (a_n^+) annihilates (creates) an electron at site n and α is the corresponding phase change which can be expressed in terms of AB flux Φ

$$\alpha = \frac{e}{\hbar c} \int_n^{n+1} \mathbf{A} \cdot dl = 2\pi \frac{\Phi}{N\Phi_0} \,, \tag{18.2}$$

where $\Phi_0 = hc/e \simeq 4.1 \times 10^{-7}$ G cm^2 is the flux quantum. We can express the wave function in terms of the atomic orbitals:

$$\Psi_k = \frac{1}{\sqrt{N}} \sum_{n=1}^{N} e^{ikn} a_n^+ |0\rangle \,. \tag{18.3}$$

where $|0\rangle$ is the vacuum state. From property of Fermi operators, $[a_m^+, a_n]_+ = \delta_{mn}$, and the periodic boundary condition, the Schrödinger equation $H\Psi_k = \epsilon_k \Psi_k$ gives us

$$\epsilon_k = -2t_0 \cos(k + \alpha), \qquad k = \frac{2\pi}{N} n, \qquad n = 0, 1, 2 \ldots N - 1 \,. \tag{18.4}$$

The corresponding eigenvalue spectrum and the persistent current are given by

$$\epsilon_n(\Phi) = -2t_0 \cos\left[\frac{2\pi}{N}(n + \frac{\Phi}{\Phi_0})\right] \,, \tag{18.5}$$

$$I_n(\Phi) = -c \frac{d\epsilon_n}{d\Phi} = -I_0 \sin\left[\frac{2\pi}{N}(n + \frac{\Phi}{\Phi_0})\right] \,, \tag{18.6}$$

where $I_0 = 4c\pi t_0 / N\Phi_0$ is the current amplitude. The energy is periodic in Φ with a period Φ_0, so are all thermodynamic properties. The number of electrons, $\mathcal{N}$, is an obvious restriction for total energy which is the sum of all eigenvalues corresponding to an occupied state. Ground state energy (we ignore spin of the electron) is given by

$$E(\Phi) = \sum_n \epsilon_n(\Phi) \,, \tag{18.7}$$

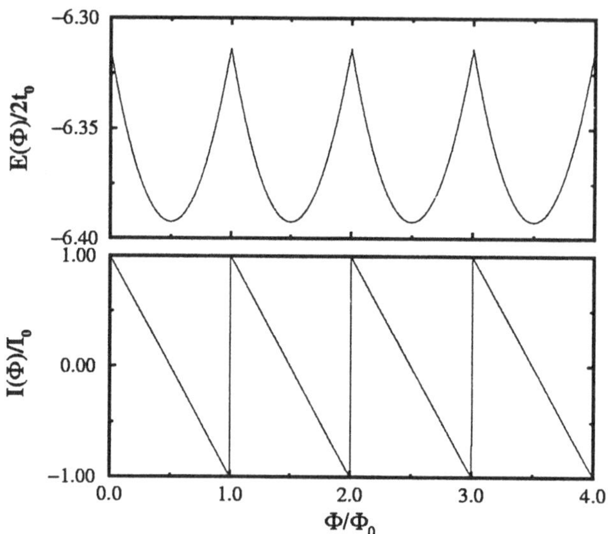

Figure 18.2 The normalized total energy $E(\Phi)/2t_0$ and the current $I(\Phi)/I_0$ as a function of flux Φ/Φ_0 for $N = 20$ and $\mathcal{N} = 10$ in a one-dimensional ring.

where the summation is carried out over $\mathcal{N}$ lowest energies for each value of flux Φ. The total current flowing along the ring can be written as

$$I(\Phi) = \sum_n I_n = -I_0 \sum_n \sin\left[\frac{2\pi}{N}(n + \frac{\Phi}{\Phi_0})\right] . \tag{18.8}$$

Figure 2 shows the periodic nature of the total energy and current as a function of flux Φ/Φ_0 for $N = 20$ and $\mathcal{N} = 10$ in the ring.

3. ELECTRON-PHONON INTERACTION IN THE RING

3.1 Weak-Coupling Polarons

The hopping amplitudes $t_{n,n+1}$ are changed when the atoms are displaced by u_n from their positions in the undistorted lattice, and which can be expanded to the first order about the equilibrium position [22]:

$$t_{n,n+1} = t_0 + g(u_n - u_{n+1}) , \tag{18.9}$$

where g is the electron-lattice displacement (electron-phonon) coupling constant. In present case, the Hamiltonian is modified as

$$H = -\sum_{n=1}^{N} t_{n,n+1}(a_n^+ a_{n+1} e^{i\alpha} + a_{n+1}^+ a_n e^{-i\alpha}) . \tag{18.10}$$

If we substitute the Eq. (18.9) into (18.10), the second term can be treated as the electron-phonon interaction Hamiltonian which is given by

$$H_{e-ph} = g \sum_{n=1}^{N} (u_n - u_{n+1})(a_n^+ a_{n+1} e^{i\alpha} + a_{n+1}^+ a_n e^{-i\alpha}) , \qquad (18.11)$$

where u_n can be written in terms of the phonon operators b and $b^\dagger$

$$u_n = \sum_q M_q (b_q + b_{-q}^+) e^{iqn} , \qquad (18.12)$$

where $M_q = (\hbar/2MNw_q)^{1/2}$, $\omega_q = 2\omega_0 |\sin(q/2)|$ is the frequency of phonon with momentum q, and b_q (b_q^+) annihilates (creates) a phonon with momentum q. The operator a_n can be written in the momentum representation as

$$a_n = \frac{1}{\sqrt{N}} \sum_k a_k e^{ikn} . \qquad (18.13)$$

By using Eqs. (18.12) and (18.13), H_{e-ph} becomes

$$\begin{aligned} H_{e-ph} &= -g \sum_{k,q} M_q (1 - e^{iq})(b_q + b_{-q}^+) \\ &\times \left[e^{i(k+\alpha-q)} a_k^+ a_{k-q} + e^{-i(k+\alpha+q)} a_{k+q}^+ a_k \right] . \end{aligned} \qquad (18.14)$$

For small values of coupling constant g, we can use the second order perturbation theory to determine contribution to the energy due to electron-phonon interaction:

$$\Delta E_k^{(2)} = \sum_q \frac{|\langle \phi_0 | H_{e-ph} | \phi_q \rangle|^2}{E_k^{(0)} - E_{k+q}^{(0)} - \hbar\omega_q} , \qquad (18.15)$$

where $\phi_0 = a_k^+ |0\rangle$ is the ground state and $\phi_q = a_{k+q}^+ b_{-q}^+ |0\rangle$ is the excited state, and $E_k^{(0)} = -2t_0 \cos(k + \alpha)$ are the energies of electron in the absence of electron-phonon interaction. Converting the summation in the Eq. (18.15) into integration, we get

$$\Delta E_k^{(2)} = -\gamma \int_{-\pi/2}^{\pi/2} dq \frac{\cos^2(p+q)}{\sin(p+q)\sin q/|\sin q| + \lambda} , \qquad (18.16)$$

where $\gamma = \hbar g^2/\pi M t_0 \omega_0$, and $\lambda = \hbar\omega_0/2t_0$ and $p = k + \alpha$. For small values of p, we can expand the integrand around $p = 0$ by using the Taylor expansion formula

$$\Delta E_k^{(2)} = -\gamma(J_0 + J_1 p + J_2 p^2/2 + \cdots) , \qquad (18.17)$$

where

$$J_0 = \int_0^{\pi/2} dx \frac{2\cos^2 x}{\sin x + \lambda}, \qquad J_1 = 0, \qquad J_2 = \frac{1}{\lambda^2}. \tag{18.18}$$

Hence, the second order correction to the energy due to electron-phonon coupling is given by

$$\Delta E_k^{(2)} = -\gamma J_0 - \frac{\gamma}{2\lambda^2}(k+\alpha)^2 . \tag{18.19}$$

The total energy becomes

$$\begin{aligned} E_k &= E_k^0 + \Delta E_k^{(2)} \\ &= E_0 - 2t_0 \cos(k+\alpha) - \frac{\gamma}{2\lambda^2}(k+\alpha)^2 , \end{aligned} \tag{18.20}$$

where $E_0 = -\gamma J_0$. For small values of $k+\alpha$, we can use the expansion $\cos(k+\alpha) \simeq 1 - (k+\alpha)^2/2$, then

$$E_k = \tilde{E}_0 + \frac{t^\star}{2}(k+\alpha)^2 , \tag{18.21}$$

where $\tilde{E}_0 = E_0 - 2t_0$ and $t^\star = t_0(1 - \gamma/2\lambda^2 t_0)$. Since the effective mass of the electron is related to the hopping amplitude via $m^\star \propto 1/t^\star$, we conclude that the effective mass of the electron is increased due to electron-phonon interaction:

$$m^\star = m\left(1 + \frac{\gamma}{2\lambda^2 t_0}\right) . \tag{18.22}$$

In addition to the increasing of effective mass since $I_0 \propto t^\star$, the amplitude of persistent current is suppressed due to electron-phonon coupling.

3.2 Strong-Coupling Polarons

We have studied the weak-coupling polarons in the previous section. In this section we will consider the strong-coupling case. The strong coupling polaron is the self-trapped state of the electron. If we denote by l the characteristic size of the region which traps the electron (in reality, this region may move over the crystal as a whole), all of our calculations so far have related to case of large-radius polarons, $l \gg a$. We now consider small-radius polarons where $l \ll a$. The Hamiltonian of the system can be written as

$$H = H_e + H_{ph} + H_{e-ph} , \tag{18.23}$$

where H_e is the electronic part of the Hamiltonian which is given by the Eq. (18.1), H_{ph} represents the phononic part which is given by

$$H_{ph} = \sum_q \hbar\omega_q b_q^+ b_q \,, \tag{18.24}$$

and H_{e-ph} denotes the electron-phonon interaction which can be written as

$$H_{e-ph} = \sum_{k,q} V_q a_{k+q}^+ a_k (b_q + b_{-q}^+) \,. \tag{18.25}$$

Using the Eq. (18.13), H_{e-ph} can be written in the site representation as

$$H_{e-ph} = \sum_{n,q} V_q a_n^+ a_n (b_q + b_{-q}^+) e^{iqn} \,. \tag{18.26}$$

where V_q is the electron-phonon coupling constant which depends only on magnitude of q. We can take the wave function as

$$\Psi = \sum_m C_m a_m^+ \left|0\right\rangle \,, \tag{18.27}$$

where C_m is a functions of the phonon operators. In the extreme strong-coupling limit one can neglect the hopping term ($t_0 = 0$) in the Eq. (18.23). This approximation leads to a simplified equation

$$\sum_q \left[\hbar\omega_q b_q^+ b_q + V_q(b_q + b_{-q}^+) e^{iqm}\right] C_m = E C_m \,. \tag{18.28}$$

Introducing a transformation, $b_q = B_q + \Lambda_q$, in such a way that the electron-phonon interaction term in the Eq. (18.28) can be eliminated, yields

$$\sum_q \left[\hbar\omega_q B_q^+ B_q - \frac{V_q^2}{\hbar\omega_q}\right] C_m = E C_m \,. \tag{18.29}$$

Notice that the operators b_q and B_q have the same commutation relations, $[b_q, b_{q'}^+] = \delta_{q,q'} \Longrightarrow [B_q, B_{q'}^+] = \delta_{q,q'}$, and $\Lambda_q = -(V_q/\omega_q)\exp(-iqm)$. The eigenvalues of the Eq. (18.29) are given by

$$E^{(0)} = \sum_q (\hbar\omega_q - \frac{V_q^2}{\hbar\omega_q}) \,. \tag{18.30}$$

Thus, the transformation leads to a decrease in phonon energy at each site by an amount $\Delta = V_q^2/\hbar\omega_q$, which is known as the polaron shift. For each values of q, the ground state wave function must satisfy the following relation:

$$b_q \psi_q^{(0)} = \Lambda_q \psi_q^{(0)} \,. \tag{18.31}$$

The solution of the Eq. (18.31) is given by

$$\psi_q^{(0)} = \exp(-\frac{1}{2}|\Lambda_q|^2) \sum_{n=0}^{\infty} \frac{(\Lambda_q)^n}{n!} (b_q^+)^n \; |0\rangle \; . \qquad (18.32)$$

The same result can be received with a canonical transformation of Lang and Firsov [23]. For all allowed values of q, the ground state wave function becomes

$$\Psi_m^{(0)} = \prod_q \psi_q^{(0)} \; . \qquad (18.33)$$

To understand the consequences of the strong electron-phonon coupling, we calculate matrix elements of H_e

$$\langle \Psi_n^{(0)} | H_e | \Psi_m^{(0)} \rangle = -t^\star e^{i\alpha} \delta_{n,m+1} - t^\star e^{-i\alpha} \delta_{n,m-1} \; . \qquad (18.34)$$

Here, the hopping amplitude is modified as

$$t^\star = t_0 e^{-\Gamma} \; , \qquad (18.35)$$

where

$$\Gamma = 2 \int_0^\pi \frac{dq}{2\pi} \frac{V_q^2}{\hbar^2 \omega_q^2} (1 - \cos q) \; . \qquad (18.36)$$

Thus, in the strong-coupling case ($\Gamma \gg 1$ and $l \ll a$), the hopping amplitude decreases exponentially as a function of electron-phonon coupling constant V_q. The effective mass increases exponentially because of $m^\star \propto 1/t^\star$. More important result is that, in the present case, amplitude of the persistent current is suppressed exponentially since $I_0 \propto t^\star = t_0 e^{-\Gamma}$. Similar result was also obtained in Ref. [17] within a variational approach.

4. CONCLUSIONS

We investigated effects of electron-phonon interaction in the mesoscopic metal rings threaded by a magnetic flux. In both cases, weak- or strong-coupling, in spite of the fact that magnitude of the persistent current is suppressed, the current is still periodic in flux Φ with a period Φ_0. In the strong-coupling case, the magnitude is decreased exponentially as a function of the electron-phonon coupling strength. Finally, it is worth noting that the experiments were performed with a finite width rings [12]. Another effect of the electron-phonon interaction can be in the fluctuation pairing. This may increase I_0 since it was found quite early [24] that superconductivity restores due to Aharonov-Bohm effect in a small ring. The polaronic effect in coupled mesoscopic rings, as well as superconducting pairing and finite temperature effects will be subject of future investigation.

References

[1] Y. Aharonov and D. Bohm, Phys. Rev. **115**, 485 (1959).

[2] R. G. Chambers, Phys. Rev. Lett. **5**, 3 (1960).

[3] A. Tonomura, Rev. Mod. Phys. **59**, 639 (1987); Adv. Phys. **41**, 59 (1992).

[4] M. Peshkin and A. Tonomura, *The Aharonov-Bohm Effect* (Springer, Berlin Heidelberg, 1989), and references therein.

[5] J. Hamilton, *Aharonov-Bohm and Other Cyclic Phenomena* (Springer, Berlin Heidelberg, 1997), and references therein.

[6] I. O. Kulik, Pis'ma Zh. Eksp. Teor. Fiz. **11**, 407 (1970) [JETP Lett. **11**, 275 (1970)].

[7] I. O. Kulik, in *Quantum Optics and the Spectroscopy of Solids*, edited by T. Hakioğlu and A. S. Shumovsky (Kluwer, Netherlands,1997), and references therein.

[8] M. Büttiker, Y. Imry, and R. Landauer, Phys. Lett. **96A**, 365 (1983).

[9] L. Pauling, J. Chem. Phys. **4**, 673 (1936).

[10] F. London, J. Phys. Rad. **8**, 397 (1937).

[11] R. B. Dingle, Proc. Phys. Soc. A **212**, 47 (1952).

[12] L. P. Lévy, G. Dolan, J. Dunsmuir, and H. Bouchiat, Phys. Rev. Lett. **64**, 2074 (1990).

[13] V. Chandrasekhar, R. A. Webb, M. J. Brady, M. B. Ketchen, W. J. Gallagher, and A. Kleinsasser, Phys. Rev. Lett. **67**, 3578 (1991).

[14] N. Byers and C. N. Yang, Phys. Rev. Lett. **7**, 46 (1961).

[15] F. Bloch, Phys. Rev. B **2**, 109 (1970).

[16] A. Ferretti, I. O. Kulik, and A. Lami, Phys. Rev. B **45**, 5486 (1992).

[17] Yi-Chang Zhou, Xin-E Yang, and Huo-Zhong Li, Phys. Lett. **190A**, 123 (1994).

[18] Tong-zhong Li, Ke-lin Wang, Gan Qin, and Jin-long Yang, J. Phys.: Condens. Matter **10**, 319 (1998).

[19] S. Washburn and R. A. Webb, Adv. Phys. **35**, 375 (1986).

[20] Y. Imry, *Introduction to Mesoscopic Physics* (Oxford University Press, New York-Oxford, 1997).

[21] *Mesoscopic Physics and Electronics*, edited by T. Ando *et al.* (Springer, Berlin Heidelberg, 1998).

[22] L. N. Bulaevskii, Ups. Fiz. Nauk. **115**, 263 (1975) [Sov. Phys. Ups. **18**, 131 (1975)].

[23] I. G. Lang and Yu. A. Firsov, Zh. Eksp. Teor. Fiz. **43**, 1843 (1962) [Sov. Phys. JETP **16**, 1301 (1963)].

[24] L. Gunther and Y. Imry, Solid State Commun. **7**, 1391 (1969).

Chapter 19

SUPERFLUIDITY AND PLANAR VORTICES IN SYSTEMS WITH PAIRING OF SPATIALLY SEPARATED ELECTRONS AND HOLES

S. I. Shevchenko

B.I. Verkin Inst. for Low Temp. Phys. and Engineering, Natl. Acad. Sci. of Ukraine
Lenin ave. 47, 310164 Kharkov, Ukraine

1. INTRODUCTION

A possibility of pairing of spatially separated electrons and holes (PSSEH), and transition of the formed pairs into a rather unusual superconducting state, in which electron supercurrent is accompanied by hole supercurrent that is equal to the former one in value and opposite in direction, was predicted by Lozovik and Yudson [1], and independently by Shevchenko [2]. An opportunity of Bose-condensation of electron-hole pairs in the systems without spatially separated electrons and holes and various manifestation of such condensation, in particular, a possibility of superfluidity of pairs was discussed long before the appearance of the works by Lozovik, Yudson and Shevchenko. However, in such systems the interband transitions which are always exist in real physical systems lead to fixation of phase of an order parameter due to electron-hole pairing. This induces the appearance of a gap in an energy spectrum and, as a result, the transition takes place into a dielectric state and not into superfluid one.

Two fundamental moments differ the system with PSSEH from those in which unseparated electrons and holes are paired. In system with PSSEH interband transitions coincide with interlayer ones and the probability of interlayer transitions can be easily changed varying a dielectric layer thickness d which separates layers with electron and hole conductivity. Since the probability of interlayer transitions decreases exponentially

I. O. Kulik and R. Ellialtioğlu (eds.),
Quantum Mesoscopic Phenomena and Mesoscopic Devices in Microelectronics, 293–298.

with the growth of d, and the binding energy of electrons with holes falls off algebraically with d then at $d \approx 10^{-6}$ cm the binding energy can reach 10^2 K, but interband transitions can be already completely neglected. As a result in systems with PSSEH electron-hole pairs can transit into a truly superfluid state. Moreover, in systems with PSSEH no local compensation of electron current by hole one takes place that makes it possible to observe equal and opposite in direction supercurrents in layers with electron and hole conductivity. This means that in systems with PSSEH a rather unusual superconductivity mechanism can be realized, and in this connection I offered to call such superconductivity "the capacitor superconductivity". For shortness I'll use sometimes this term.

It is worth noting that the capacitor superconductivity was recently rediscovered in a number of theoretical studies [3, 4, 5], in which the behavior of double quantum wells in a strong magnetic field normal to the structure plane was studied. For such two-layered systems with half-occupied lowest Landau level of each layer the idea of spontaneous interlayer coherence was introduced in [4, 5]. This is equivalent to the idea of superconductivity in systems with PSSEH which was discussed in [1, 2]. In these systems the unoccupied states of the Landau level are considered as holes. However, the problem on the nature of a ground state of the two-layer system with the half-filled Landau level in each of layers seems not finally solved. More clear from the physical point of view is the case when filling factors of Landau level $\nu_i = 2\pi n_i l_H^2$ (where n_i — the electron density in the layer i, $l_H = (c\hbar/eH_z)^{1/2}$) are connected by the following relationship: $\nu_1 + \nu_2 = 1$ with $\nu_1 \ll 1$. In this case a distance between pairs is much larger than a pair dimension and the problem of pairing is reduced to two-particle one.

There are a few experimental works [6, 7, 8] where the authors believe that they observe not only PSSEH but also the superfluidity of the electron-hole pairs. And though such an interpretation is not unique these experiments stimulate further theoretical research concerning the systems with PSSEH.

2. PLANAR VORTICES IN SYSTEMS WITH SPATIALLY PAIRED ELECTRONS

The features of the systems with PSSEH below the temperature of superconducting transition are investigated in this report. A new phenomenon in these systems is predicted. It is shown that magnetic field with two-dimensional divergence not equal to zero can create the macroscopic number of planar vortices having the same circulation. This essentially differs the studied structures from usual superconductors in which the vortices can be created by the homogeneous magnetic field. It will be considered the low density limit when the size of an electron-hole pair is less than the distance between them. At the temperature less than the binding energy of the pairs all electrons will be paired with holes, and we go over from a system of two-dimensional fermions to a system of two-dimensional bosons. Since the pairs are bosons they can pass to a superfluid state on lowering the temperature. Below the temperature of superfluid transition one can describe the behavior of a superfluid component with an order parameter. The part of the energy depending on the phase of the order parameter φ has a form [9]:

$$E = \int \left[\frac{n_s}{2M} \left(\hbar \nabla \varphi\right)^2 + n_s \frac{\hbar e d}{Mc} \left(\hat{z} \times \vec{H} \right) \cdot \nabla \varphi \right] d^2\rho. \tag{19.1}$$

The second term in (19.1) may cause a rise of a macroscopic amount of vortices with identical circulation. Indeed, assuming that the field $\vec{H}$, besides H_z, has a plane component which can be imagined as a two-dimensional "hedgehog" (*i.e.*, the planar component of magnetic field diverges radially from a certain center), we find that the second term makes rising the circular currents to be advantageous from the energy point of view. However, since the phase φ can acquire only a quantized increment $2\pi n$ (where n is an integer), the mechanism of realization of these circular currents consists in the rising the quantized planar vortices with identical circulation.

To find the energy of the system of vortices having identical circulation it is necessary to substitute a phase φ into Eq. (19.1) in the form

$$\varphi = \sum_i \arctan \frac{y - y_i}{x - x_i}, \tag{19.2}$$

where x_i and y_i are the coordinates of the i-th vortex. It is necessary, however, to keep in mind that the velocity related to a vortex falls off slowly. This makes us take into account finite size of a sample. Below I shell consider only the case when an electron-hole system is a disk of radius R and magnetic field excited by the two-dimensional azimuth current circulating in a plane parallel to the disk plane. In this case the energy of vortices is equal to:

$$E = \sum_{i=1}^{N} E_v(\rho_i) + \frac{1}{2} \sum_{i \neq j} U(\vec{\rho_i}, \vec{\rho_j}), \tag{19.3}$$

where E_v is the energy of one vortex and U is the interaction energy of vortices with each other:

$$E_v(\rho) = \frac{\pi\hbar^2 n_s}{M}\left(\ln\frac{R^2-\rho^2}{R\xi} - \frac{R-\rho}{\lambda}\right), \tag{19.4}$$

$$U(\vec{\rho_i},\vec{\rho_j}) = \frac{\pi\hbar^2 n_s}{M}\ln\frac{R^2 - 2\rho_i\rho_j\cos(\theta_i-\theta_j) + \rho_i^2\rho_j^2/R^2}{\rho_i^2 - 2\rho_i\rho_j\cos(\theta_i-\theta_j) + \rho_j^2} \tag{19.5}$$

The first term in Eq. (19.4) is the inherent energy of a vortex in the absence of an external magnetic field. The second term is the energy of interaction of a vortex with the magnetic field which enters into Eq. (19.4) through a multiplier

$$\lambda^{-1} = \frac{4\pi Ied}{\hbar c^2} \equiv \frac{2H_\rho ed}{\hbar c} \tag{19.6}$$

To find the critical current I_{c1} at which an appearance of vortices reduces the energy of the system, we should put the number of vortices $N = 1$ in Eq. (19.3), and the vortex coordinate $\rho_1 = 0$. Setting $E_v = 0$ one can obtain

$$\lambda_{c1}^{-1} \equiv \frac{4\pi I_{c1} ed}{\hbar c^2} = \frac{1}{R}\ln\frac{R}{\xi} \tag{19.7}$$

Substituting $d = 10^{-6}$ cm into Eq. (19.7) and taking into account that $H_\rho = 2\pi I/c$ we find $H_\rho^{c1} \approx 1$ G for $R = 1$ cm.

At $I > I_{c1}$ the system has a finite number of vortices which can be easily found for currents $I \gg I_{c1}$. In this case, vortices may be considered continuously distributed. Then, by introducing the vortex density $n_{v(\vec{\rho})}$ we may rewrite expression (19.3) in the form:

$$E = \int E_v(\vec{\rho})n_v(\vec{\rho})\, d^2\rho + \frac{1}{2}\int n_v(\vec{\rho_1})U(\vec{\rho_1},\vec{\rho_2})n_v(\vec{\rho_2})\, d^2\rho_1\, d^2\rho_2 \tag{19.8}$$

Varying Eq. (19.8) in $n_{v(\vec{\rho})}$ and setting the result equal to zero, we obtain the equation:

$$\int U(\vec{\rho},\vec{\rho_1})n_v(\vec{\rho_1})\, d^2\rho_1 = -E_v(\rho) \tag{19.9}$$

The solution of this equation has the form:

$$n_v(\rho) = \frac{1}{4\pi\lambda\rho} - \frac{1}{\pi}\frac{R^2}{(R^2-\rho^2)^2} \tag{19.10}$$

This expression for $n_v(\rho)$ differs drastically from analogous result for neutral superfluid systems and for ordinary superconductors. In these cases the equilibrium concentration of vortices is spatially uniform. In the case considered $n_v(\rho)$ decreases with increasing ρ in spite of the fact that the external H_ρ does not depend on ρ. Although the planar vortices resemble Onsager–Feynman and Abrikosov vortices they form a structure fundamentally different from the one the above vortices form. This structure is ordered but non-invariant under translation.

We can show [9] that the cores of vortices must be situated on the circumferences whose centers coincide with the disk center. If denote the radii of these circumferences as ρ_i and a number of vortices on the circumference i as N_i, then:

$$N_i = \frac{\rho_{i+1} - \rho_i}{2\lambda}, \tag{19.11}$$

and

$$(\rho_{i+1} - \rho_i)^2 = 5.55\lambda\rho_i \tag{19.12}$$

In this case the vortices occupy not the whole disk but only a part of it near the center, and the radius of the maximum circumference occupied by vortices is equal to [10]

$$\rho_m = \frac{\lambda_{c1} - \lambda}{\lambda_{c1}} \frac{R}{1 + \ln(R/\rho_m)} \tag{19.13}$$

Expressions (19.11, 19.12) are true for vortices created in the disk by magnetic field of the azimuthal current. However, one can show that in the arbitrary magnetic field which two-dimensional divergence differs from zero in the limit when vortices can be considered continuously distributed, their density is equal to

$$n_v(\vec{\rho}) = \frac{1}{2\pi} \frac{ed}{\hbar c} \frac{\partial H_z}{\partial z} \tag{19.14}$$

It follows from this expression that planar vortices in superfluid PSSEH systems are excited by a non-uniform magnetic field of rather general form. This means that no special experimental conditions will be required for planar vortices to appear in a sample. Rather, a problem will consist of removing vortices.

References

[1] Yu. E. Lozovik and V. I. Yudson, A novel mechanism of superconductivity: pairing of spatially separated electrons and holes, *Zh. Exp. Teor. Fiz.* **71**, 389–399 (1976).

[2] S. I. Shevchenko, Theory of superconductivity of systems with pairing of spatially separated electrons and holes, *Low Temp. Phys.***2**, 251–260 (1976).

[3] X. G. Wen and A. Zee, Neutral superfluid modes and "magnetic" monopoles in multilayer quantum Hall systems, *Phys. Rev. Lett.* **69**, 1811–1814 (1992).

[4] K. Moon, H. Mori, Kung Yang, S. M. Girvin, A. H. MacDonald, L. Zheng, D. Yoshioka and Shon-Cheng Zhang, Spontaneous interlayer coherence in double-layer quantum Hall systems: charged vortices and Kosterlitz-Thouless phase transitions, *Phys. Rev.* **B51**, 5138–5170 (1995).

[5] Z. F. Ezawa, Quantum coherence and Skyrmions in a bilayer quantum Hall system, *Phys. Rev.* **B55**, 7771–7790 (1997).

[6] S. Q. Murphy, J. P. Eisenstein, G. S. Boebinger, L. N. Pfeiffer and K.W. West, Many-body integer quantum Hall effect: evidence for new phase transitions, *Phys. Rev. Lett.* **72**, 728–731 (1994).

[7] L. V. Butov, A. Z. Zrener, G. Abstrein, G. Böhm and G. Weinmann, Condensation of indirect excitons in coupled AlAs/GaAs quantum wells, *Phys. Rev. Lett.* **73**, 304–307 (1994).

[8] L. J. Cooper, N. K. Patel, V. Drouot, E. H. Linfield, D. A. Ritchie and M. Pepper, Resistance resonance induced by electron-hole hybridization in a strongly coupled InAs/GaSb/AlSb heterostructure, *Phys. Rev.* **B57**, 11915–11918 (1998).

[9] S. I. Shevchenko, Vortices and vortex structures in systems with pairing of spatially separated electrons and holes, *Phys. Rev.* **B56**, 10355–10368 (1997).

[10] S. I. Shevchenko and V. A. Bezugly, On vortex phase of systems with pairing of spatially separated electrons and holes, *Low. Temp. Phys* **25**, 366–375 (1999).

V

JOSEPHSON EFFECT

Chapter 20

WEAKLY COUPLED MACROSCOPIC QUANTUM SYSTEMS: LIKENESS WITH DIFFERENCE

"Analogy denotes a resemblance not between things but between the relations of things"

—W.S. Javons, The principles of Science

A. Barone
INFM - Dipartimento di Scienze Fisiche, Università di Napoli "Federico II"
Piazzale Tecchio 80 - 80125 Napoli, Italy

1. INTRODUCTION

The underlying physics of phenomena such as superconductivity, superfluidity and Bose Einstein Condensation (BEC) is particularly fascinating also in connection with the fundamental issue of the Josephson Effect [1]-[4]. Indeed, the great complexity of weakly coupled macroscopic quantum systems deserves great attention also for the deep diagnostic power these structures offer as a probe of intriguing aspects of the states of the matter. The objective of the paper is a qualitative discussion of evidences, clues and ambiguities in the analogies of Josephson phenomena in such different systems.

The Josephson effect should not be seen just as a sort of "spin off" of the superconducting state or as a tool for a variety of stimulating applications, indeed it represents a unique powerful probe of fundamental aspects of the superconducting state and of the underlying physics of phenomena which go far beyond the field of superconductivity. The investigations, of both theoretical and experimental nature, on new states of the matter, on Macroscopic Quantum Tunneling (MQT) and Coherence (MQC) are concrete examples of paramount importance [5].

In Section 2 some basic aspects of superconductivity in metals, superfluidity in liquid helium and Bose Einstein Condensation (BEC) in

I. O. Kulik and R. Ellialtioğlu (eds.),
Quantum Mesoscopic Phenomena and Mesoscopic Devices in Microelectronics, 301–320.

alkali gases, are briefly considered stressing remarkable similarities and differences from both theoretical and experimental point of view. The Josephson effect occurring in systems of superfluid phases of the two helium isotopes, ^{4}He and ^{3}He and in Bose condensed alkali gases will be discussed in Section 3 and compared with the "archetype" superconductive Josephson coupling. Recent experiments of Josephson effect in coupled atomic Bose Einstein condensates will be briefly reviewed. Additional considerations on the analogies and differences between superconducting and BEC weak link systems are discussed in Section 4.

I hope that a qualitative approach, though sacrifying some rigor, can favor a more general picture which is suitable for a short overview of the matter and for a ground of possible new speculations. Obviously a demand of greater profundity or unavoidable useful scepticism can be gratified by the listed bibliography.

2. QUANTUM LIQUIDS

In this section we shall briefly recall aspects of macroscopic quantum systems such as superconductivity, superfluidity and Bose Einstein Condensation (BEC). Rather than following an historical track of the observed phenomena we shall start from the BEC and than proceed with a possible comparison with the other systems.

The achievement of Bose Einstein condensation in the spin polarized alkali gases [6] can be considered as a corner stone in the fundamentals of quantum mechanics [7].[1]

It is characterized by the onset of a "new" state in which, under thermodynamic equilibrium (or close to it), a macroscopic number of particles occupy a single one-particle state. Such an "unicum" we know to be characterized, from the quantum statistics point of view, by the indistinguishability of the particles.

Loosely speaking we can include in the frame of such systems the class of the so-called "quantum liquids" such as the spin polarized atoms stabilized at $T \cong 0.1$ μK and sufficiently high densities (say up to 10^{15} cm^{-3}), the liquid helium (both ^{4}He and ^{3}He) [8], the electrons in metals, ("s-d" excitons in the cuprates) [9].

In ^{4}He and spin polarized alkali gases the bosons are "preformed" at temperatures T_B much larger than that of condensation T_c.

Indeed, in considering superconductivity and superfluidity of ^{3}He, the main difference from the "archetype" BEC lies in the circumstance that, in these Fermionic systems, condensation occurs simultaneously to (it is

[1] Actually more recently BEC has been achieved also in atomic hydrogen (see below).

allowed by) Cooper pairing. According to Leggett, we can label these processes as "pseudo Bose condensation". It can be included within this definition also the circumstance that, in these systems, condensation takes place in the momentum space while BEC in the spin polarized alkali gases occurs both in the momentum and in the real space or, more correctly, it is evident also in the real space (see below).

In weakly interacting BEC systems almost 100% of the atoms are condensed at T=0 (actually in Bose condensed alkali atoms it is possible to have a fraction down to ~5% of atoms not in the ground state) while for fermions systems only a small fraction of electrons (fermions) are involved in Cooper pairs. Despite this difference, even in the latter case we can treat the whole system as behaving in a superfluid way for most purposes.

An important aspect emerging from the comparison between the ^{4}He and alkali spin polarized atoms systems (for both is $T_c \ll T_B$) lies in the dramatic difference of their atomic densities. In the former case we are dealing with $\sim 10^{23}$ cm^{-3}, in the latter with $\sim 10^{15}$ cm^{-3}. This circumstance leads to a drastic difference in the effect of intrinsic interactions which is of great importance in characterizing the fundamental state. Namely, for a weak interaction the many body ground state is given by the product $\prod_{i=1}^{N} \varphi_0(x_i)$ of the particle wave functions over all particles in the condensate. In the ^{4}He the ground state is far more complicated but, still we have a finite fraction of atoms occupying a single one particle state. Of course, weak interactions allow also an easier theoretical handling of the problems.

The weakness of the interactions inside these BEC gases renders these much more susceptible to "deformations" by external actions than the quite "rigid" He system. This peculiarity, together with the possibility of keeping a BEC gas in a "virtual" container, that is, confined by magnetic and or laser trapping, allows a great variety of possible manipulations including shaping, timing, creation of potential wells, atoms displacement and, maybe, more.

As stated by Leggett [10], a further important issue which is a consequence of these considerations, is the presence of different kinetic effects in the two systems and the peculiar aspects of energy and momentum conservation. In this context we can indeed include the consequences of the interaction between the He and the rigid walls of the physical box necessary to contain the system.

Let us now observe that, in BEC experiments since smaller volume produces larger zero point energy (E $\sim V^{-2/3}$), it is possible, just acting on the confinement of the BEC system (for a given number of atoms), to balance the attractive interaction creating thereby the metastable condensate [11]. This is believed to occur in ^{7}Li which has a negative scattering length (attractive interaction). Experiments carried out by Bradley *et al.* [12] seem to confirm such a picture.

Near the critical point, the condensate of bosons with attractive interaction is "kept" in its metastable state by a quite low energy barrier. This circumstance allows the possibility for such a state to decay via Macroscopic Quantum Tunneling (MQT) to a state with infinitely low energy.

Expressions of the decay rate via such a process have been obtained by Kagan *et al.* [13] and by Shuryak [14] through the calculation of the overlap integral between the metastable and the collapsed state, while Stoof [15] used the WKB method and, subsequently, Ueda and Leggett [16] applied a variational method and the instanton technique. The results are reported in Table. 20.1 N_o is the number of condensate bosons, N_c the critical number of condensate bosons (beyond N_c the system is absolutely unstable); l_o is the amplitude of zero-point oscillations of the trap and $L^* = 3|a|N_o$. a is the s-wave scattering length.

Table 20.1 MQT decay rate expressions

Kagan *et al.* [13]	$\tau^{-1} \sim \exp[-\frac{3}{2}N_0 \log(l_0/L^*)]$
Shuryak [14]	$\tau^{-1} \sim \exp[-0.57(N_c - N_o)]$
Stoof [15]	$\tau^{-1} \sim \exp(N_o)$
Ueda and Leggett [16]	$\tau^{-1} \sim \exp(N_c - N_o)^{5/4}$

While all expressions are quantitatively different from each other, the result obtained by Ueda and Leggett leads to a quite different qualitative conclusion from the others. Namely, the MQT process results in the dominant decay mechanism of the condensate system near the critical point of collapse at T=0. As stated by the authors, in principle, effects of higher order interactions, not taken into account in their analysis, could prevent the collapse of the system via MQT.

Before concluding this section, let us observe that, as far as a comparison between traditional superfluids and BEC systems is concerned, a dramatic difference lies in the intrinsic time scale significantly longer for the latter, combined with the above discussed possibilities for BEC alkali gases of optical manipulation of significant physical parameters on a very short time scale. We shall come back to this point in the next section.

3. JOSEPHSON EFFECT IN DIFFERENT MACROSCOPIC QUANTUM SYSTEMS

In this section we shall confine our attention to structures of weakly coupled macroscopic quantum systems. The physics governing such structures, namely the Josephson effect, [3] allows a great insight in the underlying "nature" of the different macroscopic systems.

3.1 Weakly Coupled Superconductors

Let us briefly recall the essence of the Josephson effect resorting to rather simple models (for detailed and extensive analysis the reader is addressed to specialized bibliography (*e.g.* [3]). We can refer to a structure of two superconducting films separated by a very thin (~10 Å) dielectric barrier. Such a sandwich type junction is just one of the possible configuration of coupled superconductors structure exhibiting the Josephson effect. In Fig. 20.1 the schematic of a Josephson junction is reported together with a typical dc current-voltage characteristics.

A very simple derivation of the Josephson effect due to Feynmann [17, 3] is based on the "two state system" in which the two weakly coupled superconductors can be described in terms of the amplitudes Ψ_L and Ψ_R of the left ($|L\rangle$) and right ($|R\rangle$) states as

$$\begin{aligned} i\hbar\frac{\partial \Psi_R}{\partial t} &= -eV\Psi_R + K\Psi_L \\ i\hbar\frac{\partial \Psi_L}{\partial t} &= -eV\Psi_L + K\Psi_R \end{aligned} \tag{20.1}$$

where V is a dc potential difference across the junction corresponding to an energy shift $E_L - E_R = 2eV$ (zero energy is chosen halfway between the energies on the two sides).

By substituting Ψ_L and Ψ_R with their expressions $\Psi_L = \sqrt{\rho_L}e^{i\varphi_L}$ and $\Psi_R = \sqrt{\rho_R}e^{i\varphi_R}$ in Eq. 20.1 and separating real and imaginary terms in each equation, the Josephson relations follow :

$$\begin{aligned} J &= J_c \sin\varphi && (a) \\ \frac{\partial \varphi}{\partial t} &= \frac{2e}{\hbar}V && (b) \end{aligned} \tag{20.2}$$

where φ is the relative phase $\varphi = \varphi_L - \varphi_R$ and $J_c = 2K/\hbar\rho$ $(\rho = \rho_L = \rho_R)$. We see that to $V = 0$ corresponds a constant value of the phase difference φ leading thereby to the possibility of finite (non

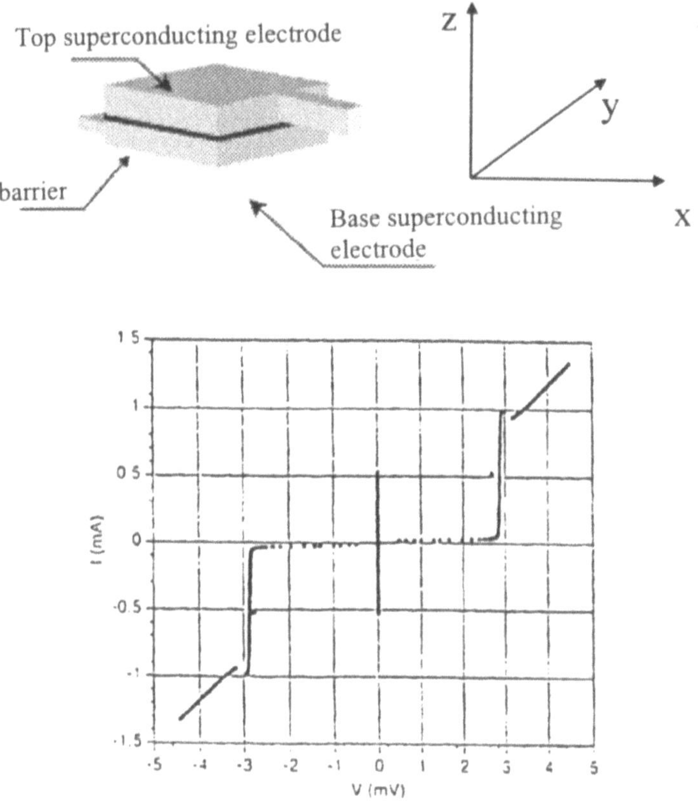

Figure 20.1 Sketch of the junction structure.

zero) supercurrent density J (dc Josephson effect). Moreover, applying a voltage $V \neq 0$ across the junction, by integration of Eq. 20.2*b*, it follows $\varphi = 2eVt/\hbar + \varphi_0$ and therefore an ac current occurs given by

$$J = J_c sin\left(\varphi_0 + \frac{2e}{\hbar}Vt\right) \tag{20.3}$$

characterized by a frequency $\omega = 2\pi\nu = 2eV/\hbar$ (ac Josephson effect).

The Josephson relations can be also obtained [18] observing that for an isolated superconductor the number of pairs, N, and the phase of the macroscopic superconducting wave function (say α) are conjugate variables and therefore an uncertainty principle $\Delta\alpha\Delta N \geq \pi$ holds which corresponds to the circumstance of ΔN=0 (isolated superconductor) and $\Delta\varphi = \infty$.

If, instead, we consider two weakly linked superconductors (or two parts of the same superconductor) S_L and S_R, each characterized by a number of pairs N_L (N_R) and a phase α_L (α_R), than the relative variables $n = N_L - N_R$ and $\varphi = \alpha_L - \alpha_R$ are conjugate variables. n can fluctuate and, accordingly, φ becomes a meaningful parameter.

The total Hamiltonian of the system of two coupled superconductors (*i.e.* the junction) can be written [18]

$$H = E_1(1 - \cos\varphi) + \frac{(2en)^2}{2C} \tag{20.4}$$

where the first term represents the whole junction coupling energy and the second term accounts for the electrostatic energy due to the difference in charge between the two sides of the junction (C being the junction capacitance). The Hamilton equations for the two variables n and φ are given by

$$\begin{aligned} \frac{dn}{dt} &= -\frac{i}{\hbar}[H, n] = -\frac{1}{\hbar}\frac{\partial H}{\partial \varphi} \\ \frac{d\varphi}{dt} &= -\frac{i}{\hbar}[H, \varphi] = \frac{1}{\hbar}\frac{\partial H}{\partial n} \end{aligned} \tag{20.5}$$

from which the constitutive Josephson relations (20.2) follow.

3.2 Weak Coupling Of Liquid He Systems

Although quite different in nature, I would consider Josephson effect in superfluid ^{3}He coupled reservoirs as a link between the phenomenology of superconductive Cooper pairs tunneling and that of overlapping Bose-Einstein condensates.

Assuming such a "strategy of comparison", let us begin to look at similarities and differences between superconducting and ^{3}He systems.

It is well known that in the BCS theory the superconducting transition is treated as the instability of the Fermi liquid with respect to Cooper pairing and the simultaneous pair condensation at $T = T_c$. From this point of view ^{3}He behaves analogously. The main difference lies in the Josephson effect where, instead of Cooper pairs tunneling through the barrier separating two superconductors, a "massive" atomic tunneling occurs through weakly linked ^{3}He reservoirs.

Evidence of weakly coupled phase-coherent systems in the case of superfluid Helium dates back years ago (*e.g.* [19]). The more recent works we refer to are those carried out in the last years by the Group of the University of California at Berkeley [12, 13]. In these experiments the system consists of two weakly coupled reservoirs of superfluid ^{3}He B phase of density ρ. The link is provided by a chip with an array of over 4200 holes of small diameter (comparable to the superfluid healing length ξ_{hl} at the operation temperature) separating the two reservoirs. Josephson type oscillations of frequency f were produced by applying a pressure difference ΔP created by a soft elastic membrane (Fig. 20.2).

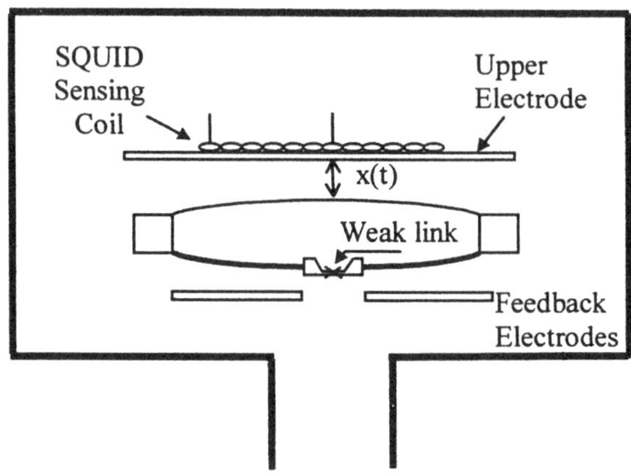

Figure 20.2 Weakly coupled superfluid ^{3}He reservoirs.

The analogy between the Superconducting Josephson Junction (SJJ) and the ^{3}He Josephson Junction (^{3}He JJ) systems can be cast in a simple form as summarized in Table. 20.2.

Table 20.2 Analogy Superconductor–^{3}He B

SJJ	^{3}He JJ
$\xi \simeq 10 \div 10^3$ Å	$\xi_{hl} \simeq 500 \div 1000$ Å
$I = I_c \sin\varphi$	$I^{(m)} = I_c^{(m)} \sin\varphi$
$\frac{\partial\varphi}{\partial t} = \frac{2e}{\hbar}V$	$\frac{\partial\varphi}{\partial t} = \frac{m\Delta p}{\rho h}$
$\nu = 4.82 \times 10^{14}$ Hz/V	$f = 183.7 \times 10^3$ Hz/Pa
t $\approx$10 Å	Ø $\leq$1000 Å
$W, L \leq \xi$	

t (barrier thickness) $\Rightarrow$ SiN chip with 4000 holes of diameter Ø $\cong 10^3$ Å
$\Delta p \Rightarrow$ realized by elastic membrane.

This scheme can be kept also for a comparison with the ^{4}He isotope, just by identifying m with the mass of the ^{4}He atom, rather than twice the atomic mass of ^{3}He and using f=69 kHz/Pa. In the case of ^{4}He, however, the experiment would be much more difficult due to the small critical healing length (more than 2 orders of magnitude smaller than that of ^{3}He).

In the experiment on ^{3}He B reported in [20] the array of apertures correctly behaves as a single aperture, the holes acting coherently as a single quantum weak link. The same group in a subsequent paper [21] has also shown the occurrence of a metastable superfluid state in the ^{3}He weak link corresponding to a quantum phase difference of π. Pushing further the analogy, we can say that this reproduces the superconducting "π-junction" observable in superconductors in connection with extrinsic paramagnetic impurities [22] or, intrinsically, related to the unconventional symmetry of the superconducting order parameter, as first proposed [23] in connection with half flux quantization in heavy Fermion systems[2] and widely investigated in connection with the d-wave symmetry in high T_c superconductors [25]. Similar π-oscillations related to geometrical factors were discussed in the context of Bose Josephson Junction (BJJ) by Raghavan *et al.* [26]. In that work such a phase mode was ascribed to the momentum dependent length (non rigidity) of the pendulum analog. Indeed such a pendulum can support an inverted orientation oscillating also about $\varphi = \pi$.

3.3 Weakly Coupled Bose-Einstein Condensates

In the context of BEC, the Josephson effect can be investigated in double well tunneling (or other coupling mechanism) that is, realizing coupling between initially separated condensates or producing a "barrier" in a simple condensate creating there by the two weakly coupled condensates systems. An important work toward such an objective has been performed at MIT [27]. In these experiments, the two condensates prepared in the double-well potential (realized by focusing an off-resonant laser light in the middle of the magnetic trap) were allowed, by switching off the confining potential, to freely expand for 40 ms and subsequently overlap.

Well defined atomic interference pattern resulting from the overlapping region were detected as a clear consequence of the macroscopic quantum nature and coherence of the BEC system. The double condensate was observed by non destructive phase-contrast imaging; the "coupling", namely the distance between the two condensates was of about 40 μm. Such a distance was modified by playing on the power of

[2] Half quantum vortices in the superfluid ^{3}He A-phase were discussed by Volovik and Salomaa [24]

the argon ion laser sheet. A thin barrier down to 1 μm was stated by the authors to be necessary for an actual Josephson coupling.

Josephson coupling can be envisaged also between two-component condensates, namely two internal states of the atom. This is indeed the case of two condensates in a single trap but in different hyperfine states (the so called "internal Josephson effect") [28]. In the brilliant experiments carried out at JILA a condensate of ^{87}Rb was created in the $|F = 1, m_f = -1\rangle$ state and, applying a short ($\sim$400 μs) two-photon pulse, a transfer of 50% of the atoms from the original to the $|F = 2, m_f = -1\rangle$ state was realized. The double condensate system prepared by such $\pi/2$ pulse was studied in its time evolution. After released from the trap, the condensates expand preserving the initial phase difference. Actually, stable interference patterns demonstrate that each realization of the experiment reproduces the same relative phase evolution keeping memory of the original one. It is interesting to observe that such a phenomenon ("internal Josephson effect") has a quite close analogy with the longitudinal magnetic resonance in ^{3}He A [29].

As for the "external" Josephson effect we can refer to a system consisting of a double-well trap (see Fig. 20.3) with a suitable barrier separating the two condensates, where, as usual, $\mu_L - \mu_R$ is the difference between the chemical potential in the left and right trap.

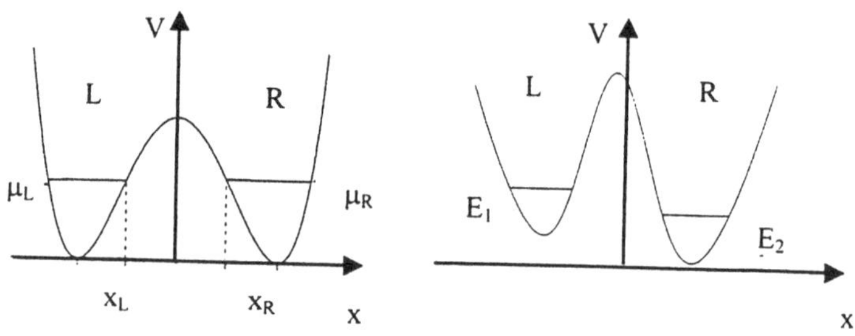

Figure 20.3 Double well trap for two Bose-Einstein condensates.

Let us observe that such a difference can be realized by filling the two traps with a different number of atoms, by asymmetric positioning of the barrier, *etc.* At zero temperature, the BEC system can be described by the theory of weakly interacting Bose gas due to Gross and Pitaevskii [30]. The Gross–Pitaevskii equation (GPE) for the order parameter can be written as:

$$i\hbar\frac{d\Psi(r,t)}{dt} = \frac{\hbar^2}{2m}\nabla^2\Psi(r,t) + \left[V(r) + g|\Psi(r,t)|^2\right]\Psi(r,t) \qquad (20.6)$$

where $V(r)$ is the confining potential, $g = 4\pi\hbar^2 a/m$, m the atomic mass and a the s-wave scattering length. $a > 0$ ($a < 0$) corresponds to an effective repulsion (attraction) between the atoms. This is a nonlinear Schrödinger equation. The nonlinearity stems from the local mean-field interaction energy proportional to the particle density and therefore to $|\Psi(r,t)|^2$. In the absence of interactions between atoms (*i.e.*, g=0) the usual Schrödinger equation is essentially recovered.

For a high barrier between the two wells the solutions in the two traps have the form

$$\Psi_{L,R}(r,t) = \Phi_{L,R}(r) \exp\left(-i\frac{\mu_{L,R}}{\hbar}t\right).$$

The "weak coupling" between the two traps, realized by the overlap of the two condensates, can be well described (indeed, in that region, nonlinearities can be neglected) by the linear combination

$$\Psi(r,t) = \Phi_1(r) \exp\left(-i\frac{\mu_L}{\hbar}t\right) + \Phi_2(r) \exp\left(-i\frac{\mu_R}{\hbar}t\right) \tag{20.7}$$

which provides a time dependent solution of the GPE. Dalfovo, Pitaevskii and Stringari [31] have discussed the problem in the one-dimensional limit reaching the typical Josephson relation

$$I = I_c \sin\left(\frac{\mu_L - \mu_R}{\hbar}t\right). \tag{20.8}$$

The extension to the 3D model of the Josephson type configuration has been given by Zapata, Sols and Leggett [32]. This analysis is of great interest also because it is the first, to my knowledge, which takes into account the "normal tunneling" namely the incoherent exchange of thermally exited atoms. Starting from Eq. 20.7 the corresponding two state dynamical equations are [33]:

$$\begin{aligned} i\hbar\frac{\partial\Psi_L}{\partial t} &= (E_L + U_L N_L)\Psi_L - K\Psi_R \\ i\hbar\frac{\partial\Psi_R}{\partial t} &= (E_R + U_R N_R)\Psi_R - K\Psi_L \end{aligned} \tag{20.9}$$

where $\Psi_{L,R}(t) = \sqrt{N_{L,R}}\, e^{i\varphi_{L,R}(t)}$ and $\mathrm{N}_L+\mathrm{N}_R = |\Psi_L|^2 + |\Psi_R|^2 = N_T$ is the total number (constant) of atoms. E_L (E_R) is the zero-point energy in the well L (R). K represents the "weak coupling" amplitude between the two condensates. Equations (20.9) are formally similar to (20.1) of Section 3.1, however, in this case, they are nonlinear due to the atomic self interaction energies. Using the same notations as in [33] we can write:

$$n(t) = \frac{N_L(t) - N_R(t)}{N_T} \quad \text{and} \quad \varphi = \varphi_L - \varphi_R. \tag{20.10}$$

The Hamiltonian can be written

$$H = \frac{\Lambda}{2}n^2 + \Delta E n - \sqrt{1-n^2}\cos\varphi \tag{20.11}$$

With $\Lambda = N_T(U_L + U_R)/4K$ and $\Delta E = [(E_L - E_R) + (U_L - U_R)]/2K$. Since n and φ are conjugate variables we can write the Hamiltonian equations:

$$\begin{aligned} \frac{dn}{dt} &= -\frac{\partial H}{\partial \varphi} \\ \frac{d\varphi}{dt} &= \frac{\partial H}{\partial n} \end{aligned} \tag{20.12}$$

and, therefore:

$$\begin{aligned} \frac{dn}{dt} &= -\sqrt{1-n^2}\sin\varphi(t) \\ \frac{d\varphi}{dt} &= \Delta E + \Lambda n(t) + \frac{n(t)}{\sqrt{1-n^2(t)}}\cos\varphi(t). \end{aligned} \tag{20.13}$$

In the above quoted paper Smerzi *et al.* [33], using such two-mode nonlinear GPE, gave a description of the time oscillation of the population imbalance of the two condensates in terms of elliptic functions discussing different aspects of the dynamical regime. As for the dramatic differences in the time scales involved in BJJ systems compared with SJJ structures let us observe that the frequency of small amplitude oscillations of the pendulum analog of the former, ω_L is of the order of kHz while the corresponding plasma frequency of the latter is of the order of GHz.

It is interesting to recall that in the original Feynmann two level system description the Josephson effect in the coupled superconductors the feeding current (the source) was not included in the equations [3, 34]. In the case of weakly coupled BEC systems the "physiological" absence of feeding current (at least in today's experiments) is, in my opinion, an intriguing issue. This point has been considered in [35] where a time dependent barrier moving at a constant velocity inside the trap has been proposed to "mimic" the current source of an SSJ structure.

A more subtle subject of analogy between SJJ and BJJ structures has been recently proposed by the very interesting experiment carried out by Anderson and Kasevich [35]. In this case standing waves are realized by a retro-reflected laser beam (wavelength λ) producing a periodicity of $\lambda/2$. The traps are created at the anti-nodes of the optical standing wave oriented along z and loaded by Bose condensed atoms of mass m. The difference in the chemical potential, $\mu_{i+1} - \mu_i$, is produced by the gravitational potential $U = mgz$ (see Fig. 20.4). $E_R = \hbar^2 k^2/2m$ is the photon recoil energy ($k = 2\pi/\lambda$). The depth of the wells is such

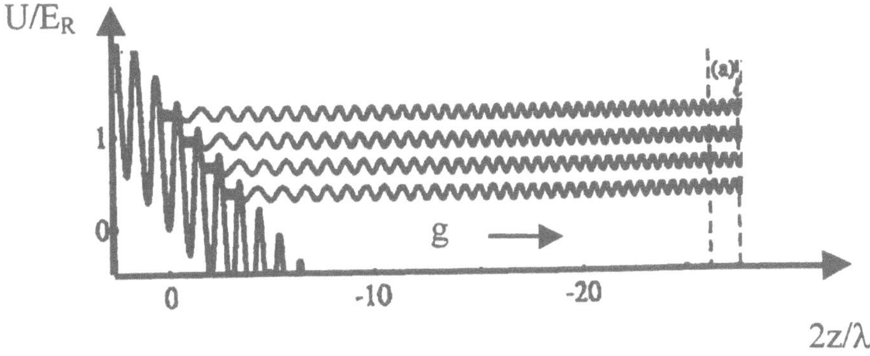

Figure 20.4 Standing waves realized by retro-reflected laser beam.

that each contains one bound state only. The coherent tunneling of atoms from the array of traps leads to an atomic current modulated at frequency $mg\lambda/2\hbar$ in analogy with the Josephson effect in that, for a SJJ structure is

$$\Delta\mu = 2eV, \quad \frac{\partial\varphi}{\partial t} = \frac{2eV}{\hbar}, \tag{20.14}$$

while in this BEC system is

$$\Delta\mu = mgz, \quad \frac{\partial\varphi}{\partial t} = \frac{mg\lambda}{2\hbar}. \tag{20.15}$$

The sinusoidal dependence on φ of the atomic supercurrent, $\mathrm{I}^{(m)}(\varphi)$, is however not demonstrated by the experiment.

4. MORE ON SUPERCONDUCTING *VS* BEC JOSEPHSON WEAK LINKS

As for the Josephson coupling, we would expect that the number of atoms in the two trapped BEC cannot be too (or arbitrarily) small. In other words, the mean number of atoms in the two noninteracting trapped BEC (bulk region) has to be quite large with respect to the number of atoms involved in the interaction (weak link region). Indeed, from this point of view, the best configuration for a superconducting weak link bridge is the VTB (Variable Thickness Bridge) type in which the banks are, largely, thicker than the link (see Chapter 7 of Ref. [3]).

More formally, in the tunneling frame, it implies a non trivial choice of a set of single electron wave functions for the left and right superconductor.

In the case of superconducting tunneling, the possibility of writing the total Hamiltonian by adding to the Hamiltonians of the two noninteracting metals, H_L and H_R, the interaction term H_T implies the existence of a set of single electron wave functions ϕ_k and χ_q for the left and right metal respectively. Such functions should have the following properties: as a first ϕ_k and χ_q should form together a complete orthonormal set and, on the other hand, the wave function for an electron in the left (right) metal should be expressed only in terms of the φ_k's (χ_q's). Since these two requirements cannot be simultaneously satisfied it is possible to proceed by considering the states $\varphi_k(\chi_q)$ defined by assuming that the barrier extends to $+\infty$ $(-\infty)$ (following Bardeen). With this assumption the ϕ_k's and χ_q's possess an exponential tail in the barrier region and are not orthogonal. Therefore the c_k and d_q operators do not commute in a rigorous sense. As discussed by Prange (see Chapter 2 of Ref. [3] and references reported therein), under the assumption of specular tunneling between states of equal energy, the anticommutation relations $\{c_k^+, d_q\} = \{c_k, d_q\} = 0$ may be assumed to hold to the lowest order in H_T. We expect that "analog" considerations could be proper for a "coupled BEC's" system. A different approach is followed in Ref. [33] where the left and right wave functions are considered as sum and difference of exact eigenstates of the Gross-Pitaevskii equation.

Moreover, in the case of superconductors, using the full Hamiltonian formalism for the derivation of the Josephson constitutive relations, it is possible to derive the complete expression of the current which includes three terms [3]:

$$I(t, V_o, T) = I_{qp}(V_o, T) + I_{j1}(V_0, T)\sin\varphi(t) + I_{j2}(V_o, T)\cos\varphi(t). \tag{20.16}$$

At V_o=0 the only non vanishing contribution to the total current is given by $I_{j1}(V_o, T)\sin\varphi(t)$ which is the dc Josephson current, while for $V \neq 0$ the phase dependent terms give the ac Josephson current of frequency $\omega_j = 2eV/\hbar$.

The analog of the quasiparticle tunneling current I_{qp} in the case of BEC represents an intriguing point. Indeed, the equivalent "normal component" also in BEC systems could be of importance though the obtained condensates in current experiments show a negligible small fraction of "normal" particles. On the other hand, the Gross-Pitaevski equation, in the usual formulation, does ignore the effect of excited quasiparticles. It is possible to handle the mean field effects by adding an extra term into the effective potential, but a more serious issue relates to the effect of quasiparticles creation by the time-dependent field of the condensate, which is outside the objective of the Gross-Pitaevski equation. This remains to date an open problem.

As already mentioned in Section 3.2, Zapata, Sols and Leggett [32] have discussed the effect of dissipation related to the incoherent exchange of normal atoms. The analysis was accomplished in the limits of high and low potential barriers. The result is that the "today" observability of coherence between BEC systems requires low barrier and a further temperature reduction combined with higher number of particles in the condensate.

Even more subtle, but maybe of potential interest, could be to investigate on the analog of the interference (quasi particle-pairs) $\cos\varphi$ term of equation (20.16) assuming that such a term is not peculiar of Cooper electrons paired systems, or more generally, of pseudo-BEC. As previously discussed, in dealing with weakly coupled BEC systems, the way of realizing the link is that of "cutting" a single trapped BEC into two condensates by a far off-resonant laser sheet, providing, thereby, the "tunnel barrier". In the literature the phenomena of weakly coupled BEC systems are usually discussed in terms of tunneling, in analogy with superconducting Josephson tunnel junctions. Actually, as it is well known, the tunneling is a sufficient, not a necessary condition, for the occurrence of the Josephson effect. Smooth "weak links" (without tunneling barrier) of dimensions smaller than the coherence length (at the operation temperature) provide overlapping of the macroscopic wave functions describing as well the two weakly coupled systems. The phenomenological Ginzburg-Landau theory can be therefore applied to describe the whole structure.

As we have seen straightforward oversimplified theoretical frame to obtain the Josephson equations is that of the "two state systems" proposed by Feynmann [17]. Let us follow here a different approach in which we do not refer to the two coupled equations describing the junction system but to a set of equations each describing the system with 1, 2,..., n pairs transferred. The ground state of two isolated superconductors, S_1 and S_2, with N_1 and N_2 electrons respectively is

$$\Psi_0 = \Psi^{(1)}(N_1)\Psi^{(2)}(N_2) \tag{20.17}$$

with energy

$$E_o = E^{(1)}(N_1) + E^{(2)}(N_2). \tag{20.18}$$

Let us consider the generic state $|n\rangle$ corresponding to the transfer of n pairs from superconductor 1 to superconductor 2. We can generate an infinite set of base states

$$\begin{aligned} |n-1\rangle\Psi_{n-1} &= \Psi^{(1)}[N_1 - 2(n-1)]\Psi^{(2)}[N_2 + 2(n-1)] \\ |n\rangle\Psi_n &= \Psi^{(1)}(N_1 - 2n)\Psi^{(2)}(N_2 + 2n) \\ |n+1\rangle\Psi_{n+1} &= \Psi^{(1)}[N_1 - 2(n+1)]\Psi^{(2)}[N_2 + 2(n+1)]. \end{aligned} \tag{20.19}$$

Antisymmetrization operator is omitted. Let us consider the case of equal chemical potentials (no energy is necessary for the transfer of pairs) in the two superconductors so that all states Ψ_n will be degenerate in energy with Ψ_o. The transfer of Cooper pairs can occur via tunneling effect when the two superconductors are no longer isolated but separated by a microscopic distance (say 10 Å). Indeed the weak coupling, as we know, of the two superconductors can be realized in different ways. The coupling of S_1 and S_2 is assumed as a perturbation which transfers one pair at a time. We assume that the time evolution of the system can be described by an infinite number of equations given, for each value of n, by

$$i\hbar\frac{\partial\Psi_n}{\partial t} = E_0\Psi_n - M\Psi_{n+1} - M\Psi_{n-1} \tag{20.20}$$

where M is the matrix element connecting the coupled states.

Let us look for solutions such as $\Psi_n = a(n)e^{-iEt/\hbar}$, $a(n)$ being the time independent amplitude of having the state with n transferred pairs. Thus

$$Ea(n) = E_0 a(n) - Ma(n+1) - Ma(n-1) \tag{20.21}$$

These are linear differential equations with constant coefficients, so we can consider solutions of the form $a(n) = e^{in\varphi}$. Then we get $E_0 e^{in\varphi} - Me^{i(n+1)\varphi} - Me^{i(n-1)\varphi}$. That is

$$E_\varphi = E_0 - 2M\cos\varphi. \tag{20.22}$$

In the scheme of a one-dimensional lattice of atoms this gives the eigenvalues in the tight binding approximation with $\hbar\varphi$ (crystal momentum) and n being conjugate variables. In this analogy the phase change φ from one site to the next in the atomic chain corresponds to the relative phase between the two superconductors. Thus Hamiltonian equation apply giving the Josephson relations:

$$I = I_o \sin\varphi$$

$$\frac{\partial\varphi}{\partial t} = \frac{2e}{\hbar}V$$

The basic equations considered connect states $n \to n+m$ with $m = \pm 1$. For $m = 2, 3, \ldots$, we should obtain higher order harmonics, that is

$$E_\varphi = E_0 - 2\sum_{j=1}^{l} M_j \cos j\varphi. \tag{20.23}$$

And thus

$$I = I_0 \sin\varphi + I_0' \sin 2\varphi + \ldots \tag{20.24}$$

Let us observe that in (20.22) it appears the matrix element M linearly (rather than quadratically) due to the coherence involved in the problem. Furthermore, it could be of interest to push further the analogy and relate expression (20.24) with the more general current-phase relation observed in different kinds of weakly-coupled superconductors systems.

It could be maybe of interest to use such an approach for a BJJ structure, referring to atoms rather than Cooper pairs and including in the equations the nonlinear term due to the atomic self-interaction energies, namely, resorting once again to the Gross-Pitaevskii equation.

The set of equations (20.20) could also describe the dynamics of the atomic traps array in the experiments by Anderson and Kasevich [36].

Acknowledgements

I am indebted with Tony Leggett for reading the manuscript and for the stimulating discussions and precious suggestions. I wish also to thank Augusto Smerzi, Guglielmo Tino and Andrej Varlamov for useful comments.

References

[1] B. D. Josephson, Possible new effects in superconductive tunneling, Phys. Lett. **1**, 251–253 (1962).

[2] I. O. Kulik and I. K. Yanson, The Josephson effect in superconductive tunneling structures, Izdatel'stuo Nauka, Moscow, 1970. Israel Program for Scientific Translations, Jerusalem, 1972.

[3] A. Barone and G. Paternò, Physics and Applications of the Josephson Effect, John Wiley, New York, 1982.

[4] K. K. Likharev, Dynamics of Josephson Junctions and Circuits, Gordon and Breach Science Publisher, New York, 1984.

[5] For a recent account see P. Silvestrini, B. Ruggiero, F. Petraccione and A. Barone, eds. Macroscopic Quantum Tunneling and Coherence, Journal of Superconductivity **12**, n.6 (1999).

[6] M. H. Anderson, J. R. Ensher, M. R. Matthews, C. E. Wieman and E. A. Cornell, Observation of Bose-Einstein Condensation in

a dilute atomic vapor, Science **269**, 198–201 (1995); K. B. Davis, M. O. Mewes, M. R. Andrews, N. J. van Druten, D. S. Durfee, D. M. Kurn and W. Ketterle, Bose-Einstein Condensation in a gas of sodium atoms, Phys. Rev. Lett. **75**, 3969–3972 (1995).

[7] For comprehensive reviews on BEC see: A. Griffin, D. W. Snoke and S. Stringari, eds., Bose Einstein Condensation, Cambridge University Press, Cambridge, 1995; F. Dalfovo, S. Giorgini, L. P. Pitaevskii and S. Stringari, Theory of Bose-Einstein Condensation in trapped gases, Rev. Mod. Phys. **71**, 463–513 (1999); G. M. Tino and M. Inguscio, Experiments in Bose-Einstein Condensation, Rivista del Nuovo Cimento **22**, 1–43 (1999).

[8] A. J. Leggett, A theory of the new phases of liquid ^{3}He, Rev. Mod. Phys. **47**, 331–414 (1975).

[9] P. V. Shevchenko and O. P. Sushkov, Phase oscillations between two superconducting condensates in cuprate superconductors, Phys. Lett. A **236**, 137–142 (1997); see also the early paper by A. J. Leggett, Number-phase fluctuations in two-band superconductors, Prog. Theor. Phys. **36**, 901–930 (1966).

[10] A. J. Leggett, BEC: The alkali gases from the perspective of research on liquid helium, Proc. 16th, International Conference on Atomic Physics (W. Bayliss, ed.) Windsor, Ontario, 1998.

[11] P. A. Ruprecht, M. J. Holland, K. Burnett and M. Edward, Time-dependent solution of the nonlinear Schrödinger equation for Bose-Condensed trapped neutral atoms, Phys. Rev. A **51**, 4704 (1995).

[12] C. C. Bradley, C. A. Sackett, J. J. Tollett and R. G. Hulet, Evidence of Bose-Einstein Condensation in an atomic gas with attractive interactions, Phys. Rev. Lett. **75**, 1687–1690 (1997); Bradley *et al.*, Bose-Einstein Condensation of lithium: observation of limited condensate number, Phys. Rev. Lett. **78**, 985-989 (1997); see also A. Parola, L. Salasnich and L. Reatto, structure and stability of bosonic clouds: Alkali-metal atoms with negative scattering length, Phys. Rev. A **57**, R3180-R3183 (1998).

[13] Yu. Kagan, G. Shlyapnikov and J. Walraven, Bose-Einstein Condensation in trapped atomic gases Phys. Rev. Lett. **76**, 2670-2673, (1996).

[14] E. V. Shuryak, Metastable Bose condensate made of atoms with attractive interaction, Phys. Rev. A **54**, 3151–3154 (1996).

[15] H. T. C. Stoof, cond-mat/9601150.

[16] Ueda and A. J. Leggett, Macroscopic Quantum Tunneling of a Bose Condensate with attractive interaction, Phys. Rev. Lett. **80**, 1756–1759 (1988).

[17] R. P. Feynmann, R. B. Leighton and M. Sands, The Schrödinger equation in a classical context: A seminar on superconductivity in the Feynmann Lectures on Physics, Vol. III Addison-Wesley, 1965, Chap. 21.

[18] P. W. Anderson, Special effects in superconductivity in Lectures on the Manybody Problem Ravello (E. R. Caianiello ed., Vol. II, Academic Press, 1963), pp.113–135.

[19] O. Avenel and E. Varoquaux, Josephson effect and quantum phase slippage in superfluids, Phys. Rev. Lett. **60**, 416-419 (1988); Varoquaux *et al.*, Phase slippage in superfluid ^{3}He-B, Physica B **178**, 309–316 (1990).

[20] S. V. Pereverzev, A. Loshak, S. Backaus, J.C. Davis and R. E. Packard, Quantum oscillations between two weakly coupled reservoirs of superfluid ^{3}He, Nature **388**, 449–451 (1997).

[21] S. Backaus, S. Pereverzev, R. W. Simmonds, A. Loshak, J. C. Davis and R. E. Packard, Discovery of metastable π-state in a superfluid ^{3}He weak link, Nature **392**, 687–690 (1998).

[22] L. N. Bulaevskii, V. V. Kuzii and A. A. Sobyanin, Superconducting system with weak coupling to the current in the ground state, JETP Lett. **25**, 290–293 (1977).

[23] V. B. Geshkenbein, A. I. Larkin and A. Barone, Vortices with half magnetic flux quanta in Heavy Fermion superconductors, Phys. Rev. B **36**, 235–238 (1987).

[24] M. M. Salomaa and G. E. Volovik, "half quantum vortices in superfluid ^{3}He A", Phys. Rev. Lett. **55**, 1184–1187 (1985).

[25] D. A. Wollman, D. J. Van Harlingen, W. C. Lee, D. M. Ginsberg and A. J. Leggett, Experimental determination of the superconducting pairing state in YBCO-Pb dc SQUIDS, Phys. Rev. Lett. **74**, 797–801 (1993). For an extensive review see: D. J. Van Harlingen, Phase-sensitive tests of the symmetry of the pairing states in high-temperature superconductors: evidence for $d_{x^2-y^2}$ symmetry, Rev. Mod. Phys. **47**, 515–535 (1997).

[26] S. Raghavan, A. Smerzi, S. Fantoni and S. R. Shenoy, Coherent oscillations between two weakly coupled Bose-Einstein condensates: Josephson effects, π oscillations and macroscopic quantum self trapping, Phys. Rev. A **59**, 620–633 (1999).

[27] M. R. Andrews, C. G. Towsend, H. J. Miesner, D. S. Durfee, D. M. Kurn and W. Ketterle, Observation of interference between two Bose Condensates, Science **275**, 637–641 (1997).

[28] D. S. Hall M. R., Matthews, C. E. Wieman and E. A. Cornell, Measurements of relative phase in two component Bose Condensates; Phys. Rev. Lett. **81**, 1543–1546 (1998); J. Williams, R. Walser,

J. Cooper, E. Cornell and M. Holland, Nonlinear Josephson-type oscillations of a driven, two component Bose-Einstein Condensate, cond-mat/9806337, 27 June 1998; D. S. Hall, M. R. Matthews, J. R. Ensher, C. E. Wieman and E. A. Cornell, The dynamics of component separation in binary mixture of Bose-Einstein Condensates, cond-mat/9804138, 14 April 1988; J. William, R. Walser, J. Cooper, E. A. Cornell and M. Holland, Excitation of an antisymmetric collective mode in a strongly coupled two-component Bose-Einstein Condensate.

[29] *e.g.* R. A. Webb, R. L. Kleinberg and J. C. Wheatly, Phys. Rev. Lett. **33** 145–148 (1974); see also A. J. Leggett in Refs. [8] and [10].

[30] L. P. Pitaevskii, Sov. Phys. JETP **13**, 451 (1961); E. P. Gross, Nuovo Cimento **20**, 454 (1961).

[31] F. Dalfovo, L. Pitaevskii and S. Stringari, Order parameter at the boundary of a trapped Bose gas, Phys. Rev. A **54**, 4213–4217; see also early work by Javanainen J. (1986), Oscillatory exchange of atoms between traps containing Bose condensates, Phys. Rev. Lett. **57**, 3164–3166 (1996).

[32] I. Zapata and A. J. Leggett, Josephson effect between trapped Bose-Einstein condensates, Phys. Rev. A **57**, R28–R31 (1998).

[33] A. Smerzi, S. Fantoni, S. Giovanazzi and S. R. Shenoy, Quantum coherent atomic tunneling between two trapped Bose-Einstein condensates, Phys. Rev. Lett. **79**, 4950–4953 (1997).

[34] H. Ohta, A self-consistent model of the Josephson junction IC-SQUID 76 (H. D. Hahlbohm and H. Lubbig, eds.) W. De Grujter, Berlin, 1976.

[35] S. Giovannazzi, PhD Thesis, Macroscopic Quantum, Coherence Phenomena in Bose Einstein Condensates, ISAS-International School for Advanced Studies, 1998.

[36] B. P. Anderson and M. A. Kasevich, Macroscopic Quantum Interference from Atomic Tunnel Arrays Science **282**, 1686–1689 (1998).

Chapter 21

MACROSCOPIC QUANTUM PHENOMENA IN JOSEPHSON SYSTEMS

P. Silvestrini
Istituto di Cibernetica del CNR
I-80072 Arco Felice, Italy
and
Macroscopic Quantum Coherence group, INFN
I-80126 Napoli, Italy

1. INTRODUCTION

How does quantum mechanics survive in macroscopic systems in thermal equilibrium with the external environment [1, 2, 3]? Previous experiments show quantum effects at temperatures below a certain crossover temperature [4]-[7], in order to guarantee the condition that the thermal energy be very low with respect to the energy level spacing [3]. We present evidence of macroscopic quantum effects in Josephson junction at temperature well above the crossover temperature. The clear observation of quantum effects up to high temperatures has been obtained by measuring the tunneling rate out of a metastable state while the external energy was changing with time. Thermal mixing of eigenstates does not destroy quantum effects if the rate of change of the external energy is fast with respect to the rate of thermally induced energy transitions between levels. In our experiments this condition has been fulfilled up to temperatures much higher than the crossover temperature by a suitable choice of sweeping frequency of the external bias, demonstrating that quantum mechanics does not need to be supplemented by new principles at high temperatures.

2. OUTLINE OF THE THEORY

A Josephson tunnel junctions consists of two superconductors separated by a thin insulating barrier (10-20 Å). A macroscopic quantum

I. O. Kulik and R. Ellialtioğlu (eds.),
Quantum Mesoscopic Phenomena and Mesoscopic Devices in Microelectronics, 321–328.

variable ϕ describes the collective behaviour of the junction [8]. A static supercurrent can pass through the junction with no voltage drop up to a maximum value I_c, known as the critical current. The supercurrent state is metastable and its lifetime is a function of the external energy given to the system, modulated by the bias current I. When the external current is increased from zero a transition from supercurrent state to the non-zero voltage state (V=2-3 mV) is observed for a current value smaller than I_c. This switching current value is a random variable whose probability distribution $P(I)$ is measured by repeating the observation many times. From the measured histogram (10^5 events recorded) it is straightforward to derive the escape rate out of the supercurrent state, in the current range for which transitions occur. The switching dynamics of a Josephson junction has a correspondence with the motion of a Brownian particle in a washboard potential $U(\phi) = -U_o(\alpha\phi + cos\phi)$. $U_o = \hbar I_c/2e$ is the Josephson coupling energy, and α is the bias current I normalized to the critical one, $\alpha = I/I_c$. The supercurrent state is separated from the voltage state by the energy barrier E_o which is decreasing with increasing the bias current: $E_o = U_o\{-\pi\alpha + 2[\alpha \arcsin\alpha + (1-\alpha^2)^{1/2}]\}$. The voltage switching is related to the escape from the metastable state via either tunneling through the potential barrier or thermal activation over the potential barrier. Within this mechanical analogue, the friction coefficient is modeled by an effective junction resistance $\eta = \hbar^2/Re^2$ while the distributed capacitance C of the structure forming the junction has a correspondence with the particle mass $M = \hbar^2 C/e^2$. An action of the system inside the well is defined as a function of the energy [9]:

$$S(E) = \int_{R(E)} d\phi \left(2M\sqrt{U(\phi) - E}\right) \tag{21.1}$$

where $R(E)$ is the classically accessible region defined as $U(\phi) \leq E$. The energy levels are determined from the condition $S(E_j) = \hbar\pi(j + 1/2)$, $(j = 0, 1, \ldots, N)$, so the "total" action $\tilde{S}(I) = S(E_o)$ determines the number N of energy levels inside the well and is modulated by the external current. Incoherent transitions from the ith into the jth level will occur at a rate w_{ij}, due to the interaction with the thermal bath [9].

The escape rate Γ is the sum of the contributions of tunneling from the various levels, $\Gamma = \sum_j \gamma_j \rho_j$, where ρ_j is the probability of finding the system in j^{th} energy state, plus the thermal hopping. The tunneling rate γ_j from single levels increases exponentially with the energy E_j [9], while ρ_j decreases exponentially with the ratio E_j/k_BT. Therefore there will be a crossover temperature T_c dividing the temperature range in two regions, depending on which exponent dominates. It turns out that T_c

is related to the oscillation frequency at the bottom of the potential well [9], known as plasma frequency $\omega_j = (2eI_c/\hbar C)^{1/2}(1-\alpha^2)^{1/4}$, as $k_B T_c = h\omega_j/2\pi$. It is easy to figure out that the main contribution to the tunneling will occur from the ground state at temperatures below the crossover temperature, or from the upper level for $T > T_c$. For $T < T_c$, in thermal equilibrium, the system is frozen in the ground state and the escape rate is just due to the tunneling. For $T > T_c$, the thermal diffusion process from the bottom levels towards the top of the barrier fills the upper level and the escape can occur. No quantum effect can be observed as long as the characteristic time of the escape process will be dominated by the thermal diffusion to reach the top of the barrier. This is the picture in quasi-stationary conditions, which are realized for a steady current or for sweeping frequency very low with respect to the characteristic time of the thermal diffusion. The presence of discrete energy levels can be revealed by a fast sweep of the external current. In this case the process is no longer stationary. At the initial time ($I = 0$) the system is in the supercurrent state in one of the energy levels, in a deep potential well. While the current is increased with time at the rate dI/dt the number of levels, and the total action $\tilde{S}$ as well, decreases. If the rate of change of $\tilde{S}$ satisfies the condition $d\tilde{S}/dt > \hbar \mathrm{w}_{ij}$, the thermal transitions are negligible during the time t for the action to change of $\hbar\pi$, namely for the upper level to reach the barrier top. Therefore the escape out of the metastable state will occur in the current range for which the barrier has been reduced enough for the system to stay in the upper level. It is straightforward to figure that this leads to an oscillatory behaviour of the escape rate as a function of I. The process is well described by the kinetic equation for the probability ρ_j of finding the system in the jth energy level [9]:

$$d\rho_j/dt = \Sigma_i(\mathrm{w}_{ij}\rho_i - \mathrm{w}_{ji}\rho_j) - \gamma_j\rho_j, \qquad i,j = 0,...,N \tag{21.2}$$

In statistical terms, the escape dynamics is summarized as follows: during the escape process the population of the upper levels undergoes to a fast tunneling rate, which is high for levels close to the barrier top. If the sweep rate is so high that the refilling from the lowest levels by thermal diffusion has not enough time to take place, the upper levels will be depopulated and, once empty, can no longer contribute to the escape. So that until the next level approaches energy values close to the top, with high tunneling rate, there is a rapid decrease of the decay rate. This occurs periodically as a level is emptied and leads to a distribution modulation related to the presence of quantized energy levels. Many oscillations are observed for $T \gg T_c$, since the initial Boltzmann population is dealt out over many levels. Following this idea we have planned the experiment.

3. EXPERIMENTS

A typical experimental configuration has been used to take the experimental histograms (10^5 switching events recorded) equivalent to the switching current distributions $P(I)$ [10]. Our electronics must be fast enough to allow high sweeping frequencies of the external bias (resulting in dI/dt up to 100 A/s), in order to induce non-stationary conditions in junctions. The time resolution of the experiment τ_r, namely the time uncertainty in recording the switching current, was the ultimate limit determining the maximum significant sweeping frequency. In fact Γ oscillates as a function of I, and therefore the time t elapsed between two peaks depends on dI/dt. We obviously must satisfy the condition $\tau_r < t$, in order to resolve the peaks. Indeed a critical experimental task is to achieve a time resolution τ_r=100-200 ps. This allows us to increase the sweeping frequency up to have t=1-2 ns. It is worth noting that from the point of view of quantum mechanics the potential energy describing the junction is changing adiabatically, since the highest sweeping frequency ω was several order of magnitude smaller than the plasma frequency, which determines the level spacing ($\omega/\omega_j = 10^{-6}$). The other important point is that the system dissipation must be very low, since the w_{ij} transition rates are proportional to the junction conductance [9]. In building our experimental setup, great care has been devoted to try to have the system dissipation dominated by the intrinsic mechanism, namely, the quasiparticle tunneling. The quasiparticle resistance R_{qp} depends exponentially on the temperature [11]. This leads to a very low intrinsic damping level below T=2 K ($R_{qp} \cong 100$ kΩ at $T \leq 2$ K; $R_{qp} \cong 100$ Ω at T=4.2 K). It is worth to stress that the effective resistance in this kind of experiment may be limited by any external shunting impedance [5]. In our case we had a 87.3 kΩ resistor located close to the junction, while a great care has been devoted to reduce any stray capacitance, which may determine the real part of the complex impedance at frequencies of the order of the levels spacing. The experiment used Nb-AlO$_x$-Nb Josephson junctions with high quality factor V_m >80 mV [8]. In order to have different energy level spacing, junctions with different critical current densities J_c are chosen. Here we present data on a sample with $J_c \cong 400$ A/cm^2. The junction parameters independently measured at 1.2 K are: I_c=80±1 μA, and C=1.2±0.2 pF. The system dissipation is determined from the fitting of data in the pure thermal limit [11], namely from the low frequency measurements, and results in an effective resistance of R=10±5 kΩ. This way of measuring the average dissipation of the whole system includes the contribution from the shunting (frequency dependent) impedance of the load line, as well as the intrinsic one (quasiparticle resistance) [11]. These independent measurements allows us to determine all the relevant quantum parameters [9], namely energy levels, tunneling rates, thermal rates, the crossover temperature and to compare data by theory. The crossover temperature is related to the oscillation frequency at the bottom of the well and therefore to the level spacing. This very important parameters to fit data by theory has been also independently determined by measuring at low temperature (down to 40 mK) the well known transition of the stationary distributions in the quantum regime [5]. All the measurements lead to the same result for the relevant junction parameters, confirming the correctness of our quantum picture of the junction.

In Fig. 21.1 we report both the experimental histograms $P(I)$ and escape rate as a function of the external current biasing the junction, Γ vs I, at T=1.2 K. The sweeping frequency is high enough to induce the observation of quantum effects, $dI/dt > 20$ A/s. The row experimental data already give evidence of energy quantization. Two different sweeping frequencies are plotted to stress that energy change is in fact the cause of the observed oscillation: time period of oscillations is quite different in measurements at different sweeping frequencies, but energy spacing is definitely the same. It is worth noting that different junctions present different energy spacing, which correctly shifts the expected and measured energy level quantization for the different junction parameters. As expected, the tunneling rates do not depend on T, provided that the dissipation is small enough. Furthermore we have seen that quantum effects cannot be revealed by our apparatus at T=4.2 K, since the number of thermally activated quasiparticles and the correlated dissipation increase exponentially with the temperature (R=100 Ω at T=4.2 K). The theoretical curves are obtained by solving Eq.21.2 [12]. The agreement between data and theory is quite convincing, taking into account the complexity of the system we are modeling and the absence of free fitting parameters.

4. CONCLUSIONS

In conclusion, our experiments give a direct evidence that quantum effects are observable in macroscopic systems at temperature higher than the crossover one. In our system the incoherent mixing of eigenstates is related to the diffusion process whose characteristic time is of the order of $1/RC$. By a fast sweeping of the external bias we have been

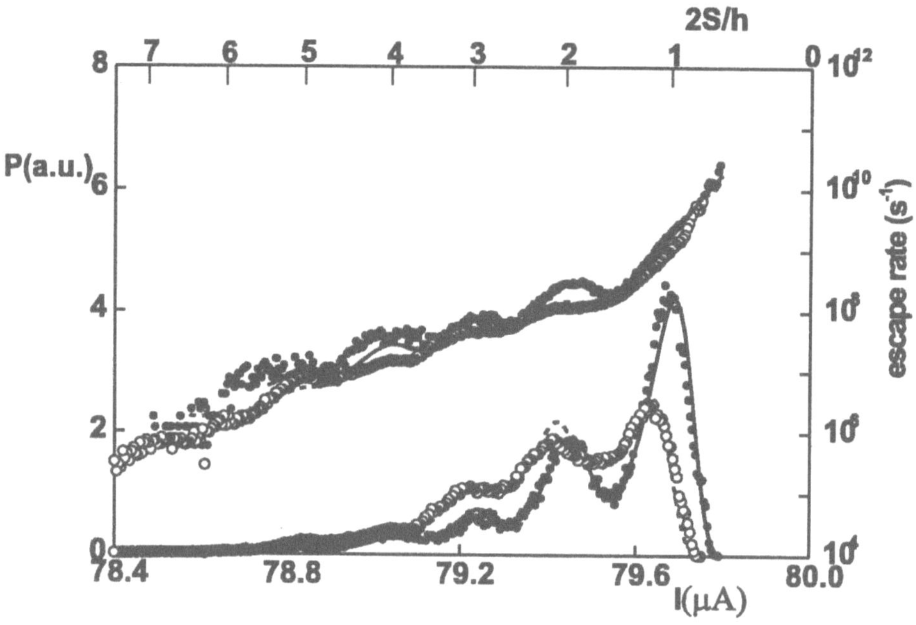

Figure 21.1 Switching current distribution P (left axis, lower curves) and the decay rate of the supercurrent state Γ, as functions of the current I (right axis, upper curves). Dots refer to data taken at T=1.2 K and dI/dt=102 A/s; open circles refer to data taken at T=1.2 K and dI/dt=33 A/s. The theoretical predictions from Eq. (21.2) refer to I_c=80 μA, C=1.2 pF, R=10 kΩ, and T=1.2 K with two different sweeping frequencies: dI/dt=102 A/s (solid line) and dI/dt=33 A/s (dashed line). The upper horizontal axis shows the conversion of the bias current to the total action S of the system. In the inset the experimental current spacing $\Delta I_{exp} = I_{min}^{N} - I_{min}^{N-1}$ normalized to the expected one [9], ΔI_{the}, for different sweeping rates, is shown. The error bars include the uncertainty on the time resolution and the statistical one as well.

able to obtain that the time t for the action to change of $\hbar\pi$ is of the same order than $1/RC$. Therefore the incoherent mixing of eigenstates is negligible while we are measuring the escape rate from the upper level. This condition can be in principle realized at any temperature, provided that the macroscopic degree of freedom is sufficiently decoupled from the environment (small dissipation), so that the lifetime of the quantum states is much longer than the characteristic time of the system (level width much smaller than the level separation). The presented effects could be relevant in view of experiments on macroscopic quantum coherence [13], and quantum computing [14].

Acknowledgements

I am grateful to Anthony Leggett, Antonio Barone, and Berardo Ruggiero for useful discussions and hints. Thanks are due to Michel Devoret, Dan Flees, Joanathan Friedman, James Lukens, and John Martinis for useful discussions on our experimental set-up at CNR-IC laboratories (Arco Felice, Italy).

References

[1] J. A. Wheeler and W. H. Zurek, *Quantum Theory of Measurements*, (Princeton Univ.Press, Princeton, 1983).

[2] E. Schrödinger, Naturwissenschaften **23**, 807, 823, 844 (1935); reprinted in English in (1).

[3] A. D. Caldeira and A. J. Leggett, "Influence of dissipation on Quantum Tunneling in Macroscopic Systems", Phys. Rev. Lett. **46**, 211 (1981).

[4] S. Washburn, R. A. Webb, R. F. Voss and S. Faris, Phys. Rev. Lett. **54**, 2712-2715 (1985).

[5] J. Clarke, A. N. Cleland, M. H. Devoret, D. Esteve and J. M. Martinis, Quantum Mechanics of a Macroscopic Variable: the Phase Difference of a Josephson Junction, Science **239**, 992-997 (1988).

[6] D. B. Schwartz, B. Sen, C. N. Archie and J. Lukens, Quantitative Study of the effect of the Environment on the Macroscopic Quantum Tunneling, Phys. Rev. Lett. **57**, 1547-1550 (1985).

[7] R. Rouse, S. Han and J. E. Lukens, Observation of resonant Tunneling between Macroscopically distinct quantum levels, Phys. Rev. Lett. **75**, 1614-1617 (1995).

[8] A. Barone and G. Paternò, *Physics and Applications of the Josephson Effect*, (Wiley, New York, 1982).

[9] A. I. Larkin and Yu. N. Ovchinnikov, Effects of level quantization on the lifetime of metastable states, Sov. Phys. JETP **64**, 185-189 (1986) [Zh. Eksp. Teor. Fiz. **91**, 318-325 (1986)].

[10] P. Silvestrini, V. G. Palmieri, B. Ruggiero and M. Russo, Observation of energy levels quantization in underdamped Josephson junction above the classical-quantum regime crossover temperature, Phys. Rev. Lett. **79**, 3046-3049 (1997); P. Silvestrini, B. Ruggiero, C. Granata and E. Esposito, Supercurrent decay of Josephson junctions in non-stationary conditions: experimental evidence of macroscopic quantum effects, Phys. Lett. (1999), submitted.

[11] P. Silvestrini, R. Cristiano, S. Pagano, O. Liengme and K. E. Gray, Effect of Dissipation on Thermal Activation in an Underdamped Josephson Junction: First Evidence of a Transition between Different Damping Regimes, Phys. Rev. Lett. **60**, 844-847 (1988); B.

Ruggiero, C. Granata, V. G. Palmieri, A. Esposito, M. Russo and P. Silvestrini, Supercurrent decay in extremely underdamped junctions, Phys. Rev. B **57**, 134-137 (1998); B. Ruggiero, C. Granata, E. Esposito, M. Russo and P. Silvestrini, Extremely underdamped Josephson junctions for low noise applications, Appl. Phys. Lett. **75**, 121-123 (1999).

[12] P. Silvestrini, Yu. N. Ovchinnikov and R. Cristiano, Effects of level quantization on the supercurrent decay in Josephson junctions: the non stationary case, Phys. Rev. B **41**, 7341-7344 (1990); P. Silvestrini, Temperature dependence of macroscopic quantum effects in non-stationary conditions, Phys. Lett. A **152,** 306-310 (1991); P. Silvestrini, B. Ruggiero and A. Esposito, The role of quantized energy levels in the macroscopic quantum behavior of Josephson junctions, Low Temp Phys. **22**, 195-207 (1996) [Fiz. Nik. Temp. **22**, 252-266 (1996)]

[13] D. V. Averin, Solid State qubits under control, Nature **398**, 748-749 (1999); Y. Nakamura, C. D. Chen and J. S. Tsai, Coherent control of macroscopic quantum states in a single-Cooper-pair box, Nature **398**, 786-788 (1999); see also *Proc. of International Workshop on "Macroscopic Quantum Tunneling and Coherence"*, Special Issue of J. of Supercond. Vol.**12.**, No.6 (1999)

[14] D. P. DiVincenzo, Quantum computation, Science **270**, 255-261 (1995); Y. Makhlin, G. Schön and A. Shnirman, Josephson-junction qubits with controlled couplings, Nature **398**, 305-309 (1999).

Chapter 22

VORTEX CONFINEMENT PHENOMENA IN MESOSCOPIC SUPERCONDUCTORS

V. V. Moshchalkov, V. Bruyndoncx, L. Van Look,
J. Bekaert, M. J. Van Bael and Y. Bruynseraede
Laboratorium voor Vaste-Stoffysica en Magnetisme, K. U. Leuven
Celestijnenlaan 200 D, B-3001 Leuven, Belgium

S. J. Bending
Department of Physics, University of Bath
Claverton Down, Bath BA2 7AY, United Kingdom

Abstract We report on vortex confinement phenomena in mesoscopic superconducting samples of different connectivity: singly connected dot, dot with two microholes ("antidots") and dot with four antidots. For these structures, we study the crossover from the "network" regime in low fields to the giant vortex state in high fields. Flux dynamics and pinning are investigated in superconductors with huge regular arrays of artificial pinning centers (antidots or magnetic dots). In films with an antidot lattice, Shapiro steps, induced by rf irradiation, have been observed. In films with a lattice of magnetic dots, stable vortex configurations have been identified using scanning Hall probe microscopy.

1. INTRODUCTION

Superconductivity is a remarkable example of a macroscopic quantum phenomenon. The absence of any resistance to the flow of a dc current in a superconductor is typical for a quantum mechanical stationary non-dissipative state, like Bohr orbits in a hydrogen atom. The fact that quantum mechanics dominates the behavior of electrons in atoms is considered as normal, since the relevant length scale of the electron orbits lies in the angström range. Much less expected is a quantum behavior of a whole macroscopic superconducting sample. Interestingly, the form

I. O. Kulik and R. Ellialtioğlu (eds.),
Quantum Mesoscopic Phenomena and Mesoscopic Devices in Microelectronics, 329–345.

of the sample itself defines the confinement geometry for the superconducting condensate. The superconducting order parameter Ψ (analogue of the wave function in the Schrödinger equation), obeys the coupled Ginzburg-Landau (GL) equations which play the same fundamental role for superconductivity as the Schrödinger equation for electrons in quantum mechanical systems. Moreover, the linearized first GL equation is formally identical to the Schrödinger equation. Therefore, the shift of the transition temperature with magnetic field, $T_c(H=0)-T_c(H)$, corresponds to the lowest Landau level $E_{LLL}(H)$, found as a solution of the Schrödinger equation with the proper "superconducting" boundary conditions [1]. Using nanostructuring, the confinement geometry for superconducting samples can be tuned, thus leading to a substantial modification of $E_{LLL}(H)$ and to an enhancement of $T_c(H)$ (for a review, see Ref. [2]). Moreover, nanoengineered regular pinning arrays also provide the conditions necessary for a dramatic increase of the critical current j_c up to its theoretical limit, the depairing current [2].

In the present work, we focus on several characteristic samples, clearly demonstrating how nanostructuring is affecting the two important parameters, critical current and critical field. We begin with superconducting mesoscopic dots of different connectivity: singly connected dots and dots with two or four openings. The role of connectivity on the normal/superconducting phase boundary $T_c(H)$ is investigated. Further on, we describe the dynamic properties of the driven interstitial vortex lattice in a film with an antidot lattice in the presence of rf irradiation. Finally, we present results on the critical current of a superconducting film covering a regular array of magnetic dots, acting as efficient pinning centers. Stable vortex configurations have been identified in these films by studying the field distribution with high resolution scanning Hall probe microscopy.

2. MESOSCOPIC SUPERCONDUCTING DOTS OF DIFFERENT CONNECTIVITY

We present here the measured phase boundaries $T_c(\Phi)$ of superconducting samples with three different topologies, which are shown in Fig. 22.1. The flux is defined as $\Phi = \mu_0 HS$, with S the outer sample area. The three structures studied are a filled microsquare, and two squares with 4 and 2 openings (antidots) respectively.

The main idea is to study the influence of connectivity, varied by introducing antidots inside a microsquare, on the crossover from the "network" behavior at low fields to a "giant vortex state" [3, 4] at high fields, and whether eventually the two configurations (vortices pinned

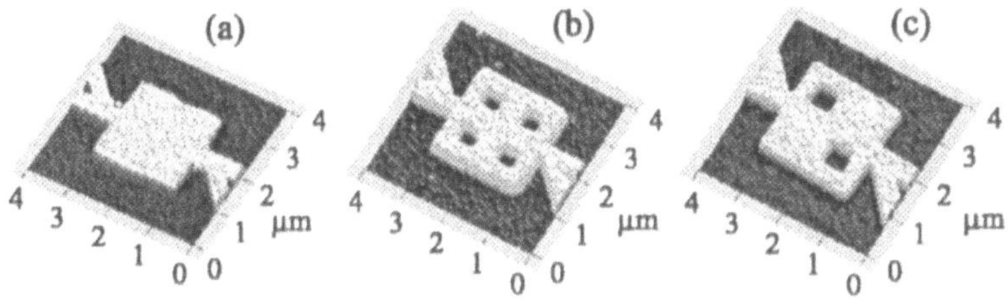

Figure 22.1 AFM images of the three structures: (*a*) full, (*b*) 4-antidot, and (*c*) 2-antidot microsquares.

by the antidots and the giant vortex state) can coexist. We will mainly focus on the high magnetic field regime.

Three different microstructures, shown in Fig. 22.1, have been studied. A square dot, with side $a = 2.04$ μm is taken as a reference sample (a); a square of side $a = 2.04$ μm, with four 0.46×0.46 μm^2 square antidots (b); and a square, side $a = 2.14$ μm, with two 0.52×0.52 μm^2 antidots, placed along a diagonal (c). The structures were characterized by X-ray, SEM and AFM (Fig. 22.1). The $T_c(\Phi)$ measurements are done in a continuous run, keeping the sample resistance at 50% of the normal state value and sweeping the magnetic field slowly while recording the temperature. The magnetic field was applied perpendicular to the structures. More experimental details can be found in Ref. [5].

In Fig. 22.2 we present the experimental phase boundary $T_c(\Phi)$ of the three structures. For the reference full square, we observe pseudoperiodic oscillations in $T_c(\Phi)$ superimposed with an almost linear background, where the period of the oscillations slightly decreases with increasing field, in agreement with previous studies [6]-[9]. *These observations are characteristic for the presence of the giant vortex state.* For the

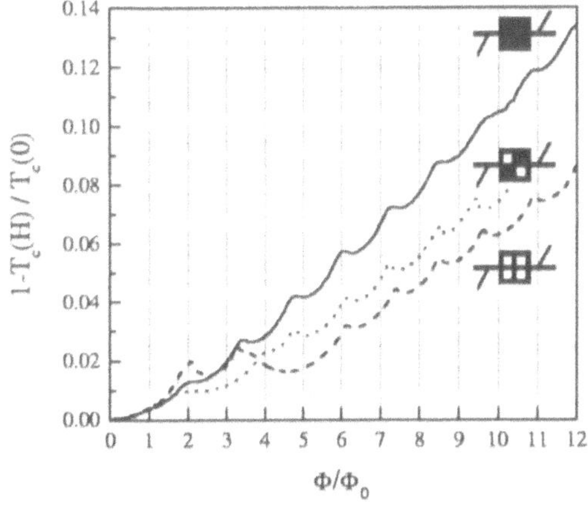

Figure 22.2 $T_c(\Phi)$ phase boundaries in reduced units of critical temperature and flux. For $\Phi/\Phi_0 > 5$ the peaks in $T_c(\Phi)$ appear at the same flux Φ/Φ_0 in all the structures.

perforated microstructures, *two different magnetic field regimes* can be distinguished. At *high magnetic flux*, the oscillations in $T_c(\Phi)$ are pseudoperiodic as well. This is similar to the 'single object' regime, as it was found by Bezryadin and Pannetier [10], but here, at $T_c(\Phi)$, a surface superconducting sheath develops near the sample's outer boundary only. The comparison of the $T_c(\Phi)$ data obtained on the perforated Al microstructures with that of the reference microsquare without antidots confirms the presence of a giant vortex state in the three structures in the high magnetic flux regime. For the *low flux* part of the phase diagram, distinct features appear (*i.e.*, below $\sim 5\Phi_0$): for the 2-antidot sample we observe the same number of peaks compared to the full square, but with a considerable shift of the positions of the first peaks. For the 4-antidot structure extra peaks can be clearly seen below $\sim 5\Phi_0$.

The $T_c(\Phi)$ curve measured for the full square is quite similar to the result obtained from a calculation [4, 9] for a mesoscopic disk in the presence of a magnetic field. In that model the linearized first GL theory equation is solved with the boundary condition for an ideal superconductor/insulator interface:

$$(-i\hbar\nabla - 2e\mathbf{A})\Psi\Big|_{\perp,b} = 0. \tag{22.1}$$

The vector potential $\mathbf{A}$ is related to the applied magnetic field $\mathbf{H}$ through μ_0 $\mathbf{H}$=rot $\mathbf{A}$. The series of peaks in the $T_c(\Phi)$ curve correspond to transitions between states with different angular momenta $L \to L+1$ of Ψ as successive flux quanta, $\Phi = L\Phi_0$, enter the superconductor. A comparison with the experimental result for a square structure was made in Ref. [6]. In a recent paper by Jadallah *et al.* [11] the $T_c(\Phi)$ phase boundary is studied theoretically and is compared to the experimental $T_c(\Phi)$ curve for the full square, described in the present paper.

It is important to note that here we defined the flux as $\Phi = \mu_0 H S_{\text{eff}}$, with S_{eff} the effective area of the whole microsquare. It is close to the exact outer sample area S, and was introduced in order to fit the peak positions to the calculated $T_c(\Phi)$ for a circular dot. Doing so, we obtain an effective area of 3.9 μm^2, close to the actual size of the structure, 4.2 μm^2. The introduction of this 'effective area' is obviously not needed if the $T_c(\Phi)$ is compared with a calculation performed for a square [11]. From the parabolic shape of the L=0 state, we find the coherence length, $\xi(0) = 92$ nm.

In contrast to the experimental result presented in Ref. [7], which was obtained for a substantially larger, but circular dot, the field period for the full square can be matched to the theoretical predictions in the whole field interval. When a sufficiently high magnetic field is applied to the sample, a superconducting edge state is formed, where superconductivity only nucleates within a surface layer of thickness w_H. The remaining area acts like a normal core of radius $R_{\text{eff}} \approx R - w_H$, and carries L flux quanta in its interior. Due to the expanding normal core, the sample can then be seen topologically as a loop of variable radius. For this reason, the $T_c(\Phi)$ of the dot shows *nonperiodic* Little-Parks-like oscillations. In comparison to the loop, which has a parabolic background on $T_c(\Phi)$, the background for the dot is quasi-linear, because of the additional energy cost (*i.e.* extra reduction of T_c) for suppressing superconductivity in the sample interior.

All the structures, in the high flux regime, have peaks in $T_c(\Phi)$ at the same Φ values. How can we understand this striking coincidence of the peak positions at high fields? For this, we have to look how the order parameter Ψ nucleates along a curved superconductor/insulator boundary. Fig. 22.3 shows the calculated $T_c(\Phi)$ curves for a single circular dot and for an antidot (see also Ref. [10]) in an infinite film, both of radius R. For comparison, we plotted H_{c3}= 1.695 H_{c2}, for a plain normal/insulator interface as a dashed line [12].

Since the dot has a larger $H_{c3}(T)$ than the antidot, Ψ is expected to grow initially at the outer sample boundary, as the temperature drops below T_c. At slightly lower temperatures, surface superconductivity should as well nucleate around the antidots. In the mean time, however, Ψ has reached already a finite value over the whole width of the strips. The resistively measured $T_c(\Phi)$ curves, probably because of this substantially different H_{c3} for a dot and an antidot, only show peaks related to the switching of the angular momentum L, associated with a closed contour along the outer sample boundary. At the $T_c(\Phi)$ boundary, in the high magnetic field regime, there is no such closed superconducting path around each single antidot, and therefore the fluxoid

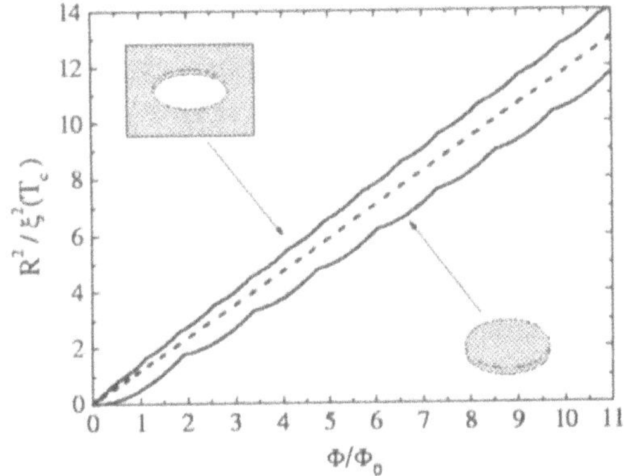

Figure 22.3 Calculated phase boundaries (*i.e.*, the $H_{c3}(T)$ curves) for a circular antidot and a dot in normalized units of temperature and magnetic flux. Superconductivity always nucleates initially near the dot/insulator boundary (the dot has the highest H_{c3}). The dashed line gives the $H_{c3}(T)$ = 1.695 $H_{c2}(T)$ curve for a semi-infinite plane [12].

quantization condition does not need to be fulfilled for a closed contour encircling each single antidot.

The background depression of T_c is different for the three structures studied (Fig. 22.2). The larger the perforated area (in other words the smaller the area exposed to the perpendicular magnetic field), the less $T_c(\Phi)$ is pushed to lower temperatures. Another clear example of a similar behavior is given in Ref. [6], where the $T_c(\Phi)$ of the (square) dot is shown to be lower than the $T_c(\Phi)$ of the loop, when exposed to a perpendicular magnetic field. This general rule applies, for instance, also to simple strips for which $T_c(\Phi)$ is suppressed more when the width w increases.

Summarizing this section, we have presented the experimental superconducting/normal phase boundaries $T_c(\Phi)$ of a mesoscopic full square and two perforated mesoscopic aluminum squares. The flux interval was divided in two regimes by comparing the results with the behavior of a full square microstructure: for *low magnetic flux* the 4-antidot structure behaves like a network consisting of quasi-one-dimensional strips, giving rise to extra peaks in $T_c(\Phi)$ in comparison to the full square. In the 2-antidot structure the peak positions are only shifted compared to the full square. As soon as each antidot confines one flux quantum, the giant vortex develops, resulting in pseudoperiodic oscillations in the $T_c(\Phi)$ and a quasi-linear background on $T_c(\Phi)$ at high magnetic fields. In this regime, the peak positions coincide for all three structures studied when the phase boundaries are plotted in flux quanta units (where flux is referred to the total sample area). For *high magnetic flux*, the presence of the antidots apparently does not change the phase winding number L for a closed contour around the outer perimeter of the whole square.

3. SHAPIRO STEPS IN A SUPERCONDUCTING FILM WITH AN ANTIDOT LATTICE

The number of introduced antidots can be dramatically increased (up to tens of millions) in macroscopically large superconducting films with an antidot lattice. These films with a regular array of microholes are convenient models to study the effects of irradiation with a radio-frequency signal on vortex dynamics. In this case, as in Josephson junctions, Shapiro steps [13] appear in the voltage-current $V(I)$ characteristics. Moreover, the height of the steps is proportional to the number of moving vortices. The appearance of Shapiro steps is not surprising, since the damped pendulum equation [14], describing the evolution in time of $\Delta\varphi$, the phase difference between the two electrodes of a Josephson junction in the framework of the Resistively Shunted Junction (RSJ) model, is formally identical to the equation of motion of a driven particle in a tilted washboard potential. Shapiro steps are therefore expected in every system where an object, moving in a periodic potential, is driven by superposed dc and ac forces. Two interesting examples of such systems are the dc flux transformer [15, 16] and a superconducting film with a laterally modulated thickness, studied by Martinoli *et al.* [17].

We will focus on the observation of Shapiro steps in a 50 nm thick superconducting Pb film with a periodic potential created by a square lattice of antidots (size 0.4×0.4 μm^2 and period 2 μm) (see Fig. 22.4*a*). The preparation of these samples is discussed elsewhere [1]. The films have a strip geometry of 0.3×3 mm^2 with two current and voltage contacts. The distance between the voltage contacts is 2 mm. From the $T_c(H)$ phase boundary of a coevaporated reference film, the superconducting coherence length was determined to be $\xi(0) = 38$ nm [18].

Choosing the appropriate temperature and magnetic field, this system can be tuned into the "matching" configuration in which every antidot is occupied by a single vortex and an interstitial vortex lattice is formed at the centers of the cells, caged by the surrounding occupied antidots (see Fig. 22.4*b*). This configuration can be realized at the second matching field [19]. The weakly pinned interstitial sublattice is easily moved through the potential, created by the strongly pinned vortices at the antidots. Due to the applied rf current and the associated Lorentz force, the pinning potential produced by the singly occupied antidots is tilted periodically, and when the resulting flow of the vortex lattice is in phase with the rf modulation of the pinning potential, steps occur in the $V(I)$ characteristics at well defined voltages V_n. For a square lattice of moving vortices with k moving vortices per unit cell of the antidot array, these

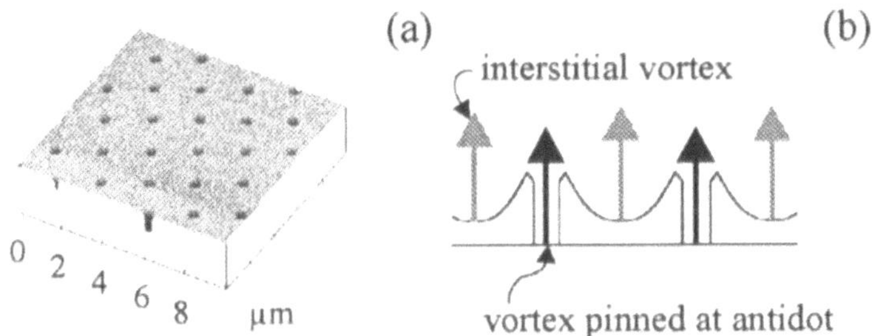

Figure 22.4 (*a*) Atomic force micrograph of the Pb (50 nm) film with a square antidot array. The antidot size is 0.4 μm and the lattice period is 2 μm. (*b*) Schematic representation of the free energy along a diagonal cut of the square antidot lattice for the second matching field. The black and gray arrows represent the vortices strongly pinned at the antidots and the weakly pinned interstitial vortices, respectively.

voltages are given by [17] :

$$V_n = Nk\frac{h}{2e}\frac{\langle v\rangle}{d} = n\left(Nk\frac{h}{2e}\nu\right) = n\left(Nk\Phi_o\nu\right) \equiv nV_o \qquad (22.2)$$

where n is an integer, N is the number of antidot rows between the voltage contacts (N= 1000 for this sample), $\langle v\rangle$ is the average velocity of the coherent motion of the interstitial vortex lattice, ν is the frequency of the applied rf signal and d is the period of the periodic potential created by the singly occupied antidots. By comparing the observed step height V_0 with the one expected from Eq. (22.2), the number k of moving vortices per unit cell can be determined. This technique can therefore be used to detect the presence of interstitial vortices in superconducting films with an artificial periodic pinning array.

In Fig. 22.5 we show the $V(I)$ curve at T = 7.151 K= 0.995 T_c obtained for a sinusoidally modulated current: $I = I_{dc} + I_{rf}\sin(2\pi\nu t)$, where $I_{rf} \approx I_c$, the critical current of the sample, and ν= 40 MHz. The field was fixed at the second matching field $\mu_0 H_2$= 1.03 mT. The $V(I)$ curve is nearly a straight line through the origin, with smooth periodic steps superimposed on it. In the derivative $\delta I/\delta V$ versus V, the voltage separation $V_o = N\nu h/2e = 81.3\ \mu$V can be determined, which is within 2% of the value 82.8 μV found from Eq. (22.2) using k= 1. Here we have assumed that the antidots trap only single-quantum vortices.

Indeed, the calculated saturation number n_s [20] indicates that, at the temperature used in our experiment, only one vortex can be pinned per antidot, thus forcing the excess vortices to occupy the much weaker pinning sites at interstices (see Fig. 22.4*b*) [21]-[23]. Recent numerical simulations [24] have shown that for a range of driving forces, the

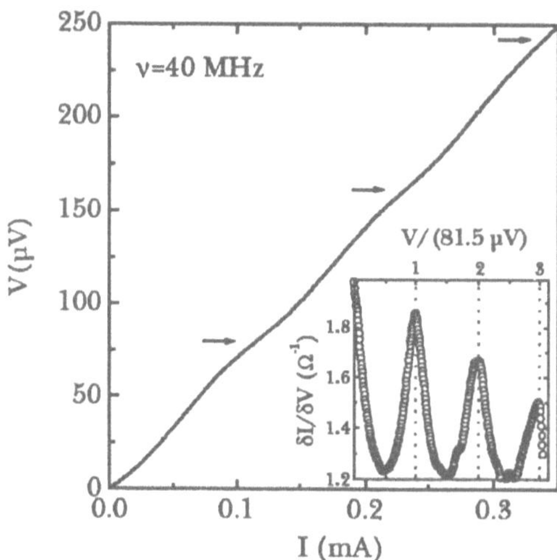

Figure 22.5 $V(I)$-curve of a Pb (50 nm) film with a square antidot lattice at T= 7.151 K and $\mu_0 H = \mu_0 H_2$= 1.03 mT. The inset shown in the right corner shows the derivative $\delta I/\delta V$ *vs.* V. The horizontal arrows mark the voltage steps expected at multiples of $V_o = 81.3\ \mu$V (Eq. 22.2).

motion of interstitial vortices is confined to 1D channels between the antidot rows. With all the interstitial positions filled, one can expect that the repulsive interaction within the 1D channel and the shear between neighboring rows will lead to a coherent motion of all interstitial vortices, as we observe in our experiments.

Concluding this section, we have shown that Shapiro steps are present in the $V(I)$ curves of a superconducting film with a square antidot lattice when a rf current is superposed on the dc transport current. The height of these voltage steps depends on the number of moving vortices per antidot lattice unit cell. These observations open new possibilities to control the Josephson effect in superconducting films through nanostructuring.

4. PINNING PHENOMENA IN PB FILMS WITH A REGULAR LATTICE OF MAGNETIC AU/CO/AU DOTS

Regular pinning arrays can also be created by a periodic repetition of a cell containing a magnetic dot instead of an antidot. Square arrays of magnetic Au/Co/Au dots are prepared on SiO_2 substrates by means of electron beam lithography and lift-off techniques. The deposition of the Au/Co/Au trilayer was done at room temperature by molecular beam evaporation in an MBE apparatus. The resulting polycrystalline structure of both materials, Au and Co, is confirmed by X-ray diffraction measurements. On top of the dot lattices, a 500 Å thick Pb layer is deposited by electron beam evaporation at 77 K. The sample is finally covered with a 200 Å protecting Ge layer.

The topography and magnetic domain state of the dot lattices are characterized by means of atomic force microscopy (AFM) and magnetic force microscopy (MFM), respectively, before deposition of the superconducting layer. These measurements are performed on a Nanoscope III system from Digital Instruments using tapping mode.

The topography and the magnetic signal are separated by a two-step lift mode procedure, in which the phase detection technique is used for the magnetic signal. All images are taken at room temperature and in zero magnetic field. The superconducting flux pinning properties of the hybrid systems (Pb film on top of a lattice of dots) are studied by performing magnetization measurements in a Quantum Design SQUID magnetometer. The magnetic field in the $M(H)$ measurements is always oriented perpendicular to the substrate plane.

The AFM experiments indicate a uniform shape, size and period of the dots over the entire sample. An AFM topograph, showing a well-defined

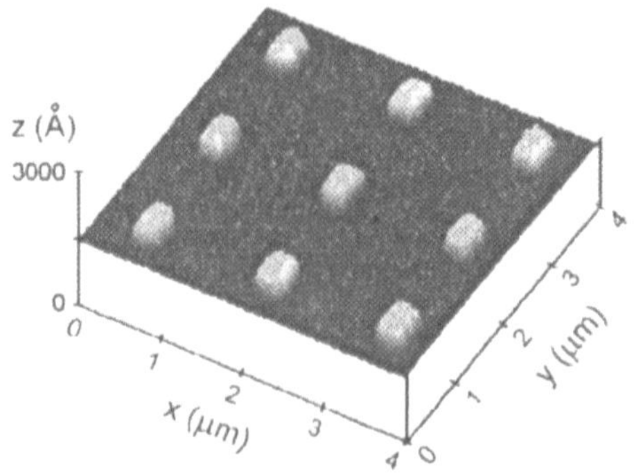

Figure 22.6 AFM topograph of a square lattice of rectangular Au/Co/Au dots, showing well-defined and equally shaped dots with lattice period $d = 1.5$ μm.

square lattice of rectangular Au/Co/Au dots is presented in Fig. 22.6. The lattice period equals $d = 1.5$ μm while the lateral dimensions of the rectangular dots are 0.36 and 0.54 μm for their short (L_s) and long (L_l) axis, respectively. The total thickness of the Au/Co/Au trilayer is 380 Å.

The magnetic domain structure of the dots was investigated by MFM at room temperature and in zero field. Due to the polycrystalline structure, the magnetocrystalline anisotropy is averaged out and the magnetic properties of the rectangular dots are dominated by the shape anisotropy, resulting in a preferential magnetization direction parallel to L_l. In the

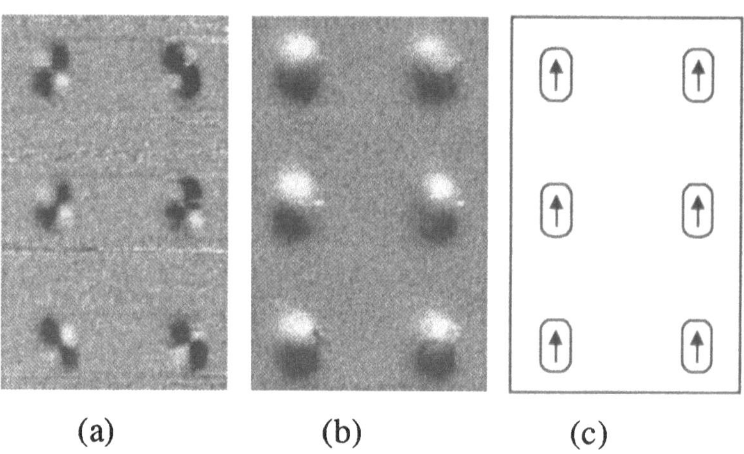

Figure 22.7 (*a*), (*b*) MFM images recorded at room temperature in zero field of a 2.7×4.0 $\mu\mathrm{m}^2$ area of the dot lattice, containing 6 magnetic dots. (*a*) shows the magnetic response of the dots before applying a magnetic field. The magnetic image of a similar area of the sample in the remanent state, after magnetic saturation in a magnetic field of 1 T pointing upwards, is given in (*b*). The suggested remanent magnetic domain configuration of the dots after magnetization is schematically presented in (*c*).

as-grown state, the magnetic contrast of the MFM images resembles a small 2×2 checkerboard pattern, as is shown in Fig. 22.7*a*.

In contrast with the *two-domain* as-grown state, the dots reveal a single domain structure in the remanent state ($H = 0$) after magnetic saturation along their easy axis of magnetization L_l in a field of 1 T, well above the saturation field. This is shown in the MFM image of Fig. 22.7*b*, taken after magnetizing the dots. The magnetic contrast of each dot shows a dark and a bright spot on the opposite sides of the dot, which is the typical contrast for a *single domain* object with magnetic moment in the substrate plane [25, 26]. The corresponding domain structure is displayed in Fig. 22.7*c*.

It has been shown recently that a lattice of sub-micron ferromagnetic dots acts as a strong periodic pinning potential for the flux lines in a superconducting Nb film that is grown on top of it [27]. In the following experiment, we will study to what extent a periodic lattice of magnetic Au/Co/Au dots can inhibit the vortex motion and increase j_c in a continuous superconducting Pb film that is deposited on top of the lattice. The possibility to change the magnetic domain structure and hence the stray field of these Au/Co/Au dots enables us to study *the influence of the magnetic stray field* on the pinning effects. Magnetization measurements are performed on a Pb film on a square lattice of Au/Co/Au dots, before and after magnetizing the dots.

Fig. 22.8 shows the upper branch of the magnetization loops measured at T/T_c= 0.97 for a Pb (500 Å) film on top of a square lattice of Au/Co/Au dots as function of H/H_1. $\mu_0 H_1 = \Phi_0/(1.5\ \mu\text{m})^2 = 0.92$ mT represents the first matching field for which the density of flux lines equals the density of dots. The presence of the lattice of Au/Co/Au dots results in a strong enhancement of the width of the magnetization loop $M \sim j_c$ compared to the reference Pb (500 Å) film without dots (line).

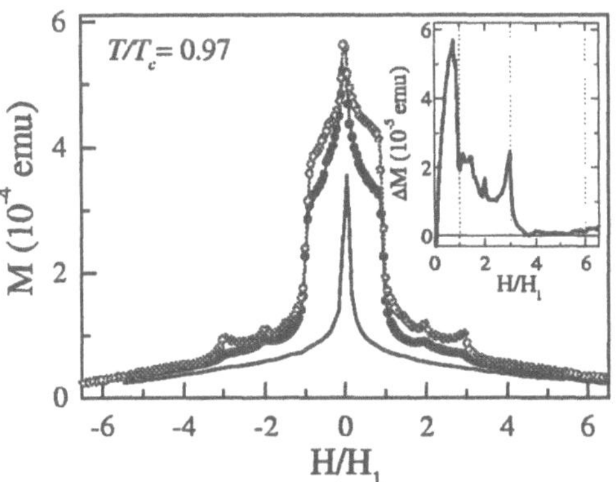

Figure 22.8 Upper half of the $M(H/H_1)$ hysteresis curves at $T/T_c = 0.97$, for a 500 Å Pb film on a square lattice of Au/Co/Au dots ($d = 1.5\ \mu$m) before (filled symbols) and after (open symbols) magnetizing the dots, and for a reference Pb (500 Å) film (line). The inset shows the difference in magnetization for a superconducting film with a single domain (after magnetization) and two-domain (before magnetization) magnetic state of the dots forming a regular pinning array.

Moreover, pronounced anomalies are observed for T close to T_c at certain integer and rational multiples of H_1, *e.g.* a sudden drop of $M(H)$ at $H/H_1 = 1$ and a maximum at $H/H_1 = 2$ and 3. These observations indicate that the lattice of Au/Co/Au dots creates a strong periodic pinning potential for the flux lines in the Pb film that is deposited on top of it, similar to the thin films or multilayers with an antidot lattice [21, 22, 23]. A clear influence of the stray field of the dots on the pinning properties can be noticed when comparing the magnetization curves for the magnetized (open symbols) and the non-magnetized dots (filled symbols). It can be seen that after magnetizing the dots (i) $M(\sim j_c)$ is further increased, (ii) the matching anomalies are much more pronounced, and (iii) additional matching anomalies are observed (see *e.g.* Fig. 22.8 at $H/H_1 = 3/2$).

When the temperature is decreased, the matching anomalies are less pronounced and they disappear at $T/T_c < 0.85$. On the other hand, very close to T_c, a significant number of *fractional* matching anomalies can be observed at certain rational multiples of H_1.

Crucial for the observation of matching effects is the long-range repulsive interaction of the vortices, and the presence of a uniform flux density in large parts of the sample. This leads to the requirements that (i) the penetration depth $\lambda(T)$ should be large compared to the period of the vortex- and pinning lattice and (ii) no strong spatial flux density gradient may exist in the sample. Both these conditions are only fulfilled at sufficiently high temperatures.

All observed matching anomalies can be associated with the formation of specific stable vortex configurations in the imposed square pinning potential. These configurations can be identified using local scanning probe techniques such as Scanning Hall Probe Microscopy (SHPM).

Moreover, $M(H)$ is further enhanced and the matching effects are more pronounced after magnetization of the dots. Our experiments hence indicate that *the dots with the largest stray field provide the most effective pinning in the broadest field range.*

A possible explanation for these experimental results is closely related to the stray field of the magnetic dots, which locally suppresses or even destroys superconductivity around the magnetic dots. These regions with a suppressed Ψ around the magnetic dots provide an additional contribution to the pinning. Since the stray field for the single domain dots after magnetization is stronger and more extended than for the two-domain dots, a more effective local destruction of Ψ creates a more effective pinning center for the flux lines. On the other hand, after magnetizing the dots, all dots exhibit exactly the same stray field pattern, which enhances the order of the pinning potential when compared to the situation before magnetizing the dots. This improved order of the pinning array provides an additional explanation for the enhanced matching effects after the dots are magnetized [28].

In the regular arrays of *ferromagnetic* dots studied here, additional contributions besides the core pinning arise due to the interaction of the vortices with the local magnetic fields generated by the dots [27, 29, 30]. Using SHPM, we directly visualized the commensurate vortex configurations simultaneously with the local stray fields of the magnetic Au/Co/Au dots. We estimate that the active Hall sensor was about 200-300 nm above the sample during the scans shown here. A more detailed description of the SHPM can be found elsewhere [31, 32].

At a temperature of 77 K, SHPM reveals the dipole stray fields characteristic for an ordered array of single domain particles (*cfr.* MFM image in Fig. 22.7*b*). In what follows, we present measurements at a temperature of 6.5 K, well below the superconducting critical temperature of the Pb film (T_c = 7.16 K) with an applied field normal to the sample plane.

SHPM results at H/H_1= 1/2, 1, 3/2 and 2 after field cooling are shown in Fig. 22.9*a*, *b*, *c* and *d*, respectively. In order to identify the vortex locations, we subtracted the measured dipole contribution at $T > T_c$ from the obtained image at $T < T_c$, which gives Fig. 22.9, containing

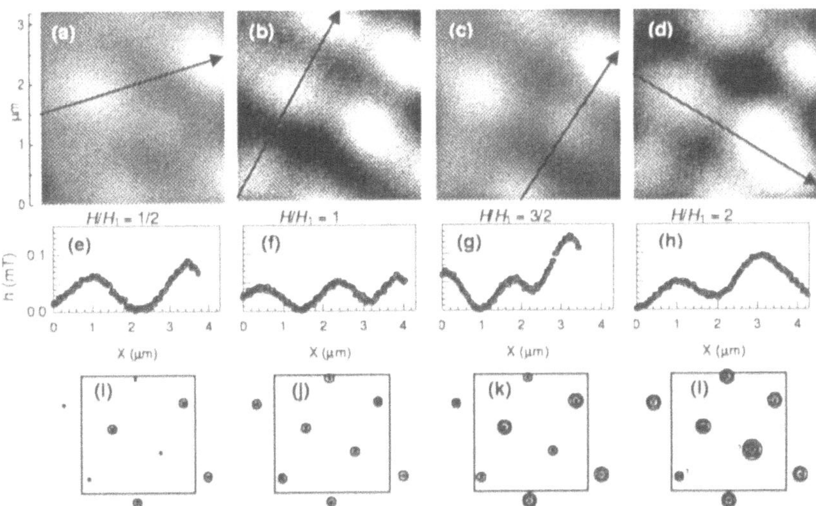

Figure 22.9 (*a*)-(*d*) SHPM images at T= 6.5 K after subtraction of the dipole contribution at different values of H/H_1 as indicated. (*e*)-(*h*) the local field distribution along the arrow directions in (*a*), (*b*), (*c*), and (*d*) respectively. (*i*)-(*l*) schematic representation of the dot array (points) and configuration of pinned vortices (circles) for the different H/H_1 values.

information about the flux distribution only. The white spots are to be associated with the pinned positive flux lines, and appear in an ordered vortex configuration commensurate with the underlying square pinning array. At $H/H_1 = 1/2$, we observe the 'checkerboard' structure where every second dot is occupied by a flux line, forming a square vortex lattice rotated by 45° with respect to the pinning array. At H_1, every dot traps a single flux line and the vortex structure mirrors that of the dot array. For $H/H_1 = 3/2$, the structure is a direct superposition of that for $H/H_1 = 1/2$ and it $H/H_1 = 1$, where half of the sites pin two flux quanta (these are the brighter spots with a higher local field value) while the rest trap just one. Finally for $H/H_1 = 2$, every dot has trapped two flux lines, except for one dot in the bottom right corner which has three flux quanta associated with it, possibly due to a small sample inhomogeneity or because the applied field slightly exceeds $2H_1$.

Up to $2H_1$ we see no evidence for vortices occupying interstitial positions at these low temperatures ($T \leq 6.5$ K). We note that, due to the divergence of the penetration depth $\lambda(T)$ at T_c, a very different behavior can indeed be expected at temperatures closer to T_c. The SHPM results are only valid in that low temperature regime, where $\lambda(T)$ is smaller than the size of a Au/Co/Au dot.

Acknowledgements

The authors are thankful to the Fund for Scientific Research Flanders (FWO), the Flemish Concerted Action (GOA), the Belgian Inter-University Attraction Poles (IUAP), and the ESF program VORTEX for the financial support. MJVB is Research Fellow of the FWO. SB acknowledges the British EPSRC and MOD Grant No. GR/J03077, and the University of Bath Initiative Fund. Discussions with K. Temst, J. G. Rodrigo, T. Puig, V. Fomin, J. Devreese, and C. Strunk are gratefully acknowledged.

References

[1] V. V. Moshchalkov *et al.*, Quantum interference and confinement phenomena in mesoscopic superconducting systems, Physica Scripta **T55**, 168-176 (1994).

[2] V. V. Moshchalkov, V. Bruyndoncx, L. Van Look, M. J. Van Bael and Y. Bruynseraede, in *Handbook of Nanostructured Materials and Nanotechnology*, H. S. Nalwa, ed., Volume 3, Chapter 9, (Academic Press, San Diego, 1999).

[3] H. J. Fink and A. G. Presson, Magnetic irreversible solution of the Ginzburg-Landau equations, Phys. Rev. **151**, 219-228 (1966).

[4] V. V. Moshchalkov, X. G. Qiu and V. Bruyndoncx, Paramagnetic Meissner effect from the self-consistent solution of the Ginzburg-

Landau equations, Phys. Rev. B **55**, 11793-11801 (1997).

[5] V. Bruyndoncx, J. G. Rodrigo, T. Puig, L. Van Look, V. V. Moshchalkov and R. Jonckheere, Giant vortex state in perforated aluminum microsquares, Phys. Rev. B **60**, (1999).

[6] V. V. Moshchalkov, L. Gielen, C. Strunk, R. Jonckheere, X. Qiu, C. Van Haesendonck and Y. Bruynseraede, Effect of sample topology on the critical fields of mesoscopic superconductors, Nature (London) **373**, 319-321 (1995).

[7] O. Buisson, P. Gandit, R. Rammal, Y. Y. Wang and B. Pannetier, Magnetic oscillations of a superconducting disk, Phys. Lett. A **150**, 36-42 (1990).

[8] A. K. Geim, I. V. Grigorieva, S. V. Dubonos, J. G. S. Lok, J. C. Maan, A. E. Filippov and F. M. Peeters, Phase transitions in individual sub-micrometer superconductors, Nature (London) **390**, 259-261 (1997).

[9] D. Saint-James, Etude du champ critique H_{c3} dans une géometrie cylindrique, Phys. Lett. **15**, 13-15 (1965).

[10] A. Bezryadin and B. Pannetier, Nucleation of superconductivity in a thin film with a lattice of circular holes, J. Low Temp. Phys. **98**, 251-268 (1995).

[11] H. T. Jadallah, J. Rubinstein and P. Sternberg, Phase transition curves for mesoscopic superconducting samples, Phys. Rev. Lett. **82**, 2935-2938 (1999).

[12] D. Saint-James and P.-G. de Gennes, Onset of superconductivity in decreasing fields, Phys. Lett. **7**, 306 (1963).

[13] S. Shapiro, Josephson currents in superconducting tunneling; the effect of microwaves and other observations, Phys. Rev. Lett. **11**, 80-82 (1963).

[14] J. R. Waldram, A. B. Pippard and J. Clarke, Theory of the current-voltage characteristics of SNS junctions and other superconducting weak links, Phil. Trans. Roy. Soc. Lond. A. **268**, 265-287 (1970).

[15] M. D. Sherril and W. A. Lindstrom, Superconducting DC transformer coupling, Phys. Rev. B **11**, 1125-1130 (1975).

[16] A. Gilabert, I. K. Schuller, V. V. Moshchalkov and Y. Bruynseraede, New Josephson-like effect in a superconducting transformer, Appl. Phys. Lett. **64**, 2885-2887 (1994).

[17] P. Martinoli, O. Daldini, C. Leemann and E. Stocker, A.C. Quantum interference in superconducting films with periodically modulated thickness, Solid State Commun. **17**, 205-209 (1975).

[18] E. Rosseel, *Critical Parameters of superconductors with an antidot lattice*, Ph.D. Thesis, Katholieke Universiteit Leuven, Leuven, 1998.

[19] E. Rosseel, M. J. Van Bael, M. Baert, R. Jonckheere, V. V. Moshchalkov and Y. Bruynseraede, Depinning of caged interstitial vortices in superconductors with periodic pinning, Phys. Rev. B **53**, R2983-R2986 (1996).

[20] A. Buzdin and D. Feinberg, Electromagnetic pinning of vortices by non-superconducting defects and their influence on screening, Physica C **256**, 303-311 (1996).

[21] M. Baert, V. V. Metlushko, R. Jonckheere, V. V. Moshchalkov and Y. Bruynseraede, Composite flux-line lattices stabilized in superconducting films by a regular array of artificial defects, Phys. Rev. Lett. **74**, 3269-3272 (1995).

[22] V. V. Moshchalkov, M. Baert, V. V. Metlushko, E. Rosseel, M. J. Van Bael, K. Temst, R. Jonckheere and Y. Bruynseraede, Magnetization of multiple-quanta vortex lattices, Phys. Rev. B **54**, 7385-7393 (1996).

[23] V. V. Moshchalkov, M. Baert, V. V. Metlushko, E. Rosseel, M. J. Van Bael, K. Temst, Y. Bruynseraede and R. Jonckheere, Pinning by an antidot lattice: The problem of the optimum antidot size, Phys. Rev. B **57**, 3615-3622(1998).

[24] C. Reichhardt, C. J. Olson and F. Nori, Dynamic phases of vortices in superconductors with periodic pinning, Phys. Rev. Lett. **78**, 2648-2651 (1997).

[25] G. A. Gibson and S. Schultz, Magnetic force microscopy study of the micromagnetics of submicrometer magnetic particles, J. Appl. Phys. **73**, 4516-4521 (1993).

[26] R. M. H., New, R. F. W. Pease and R. L. White, Physical and magnetic properties of submicron lithographically patterned magnetic islands, J. Vac. Sci. Technol. B **13**, 1089-1094 (1995).

[27] J. I. Martín, M. Vélez, J. Nogués and I. K. Schuller, Flux Pinning in a Superconductor by an Array of Submicrometer Magnetic Dots, Phys. Rev. Lett. **79**, 1929-1932 (1997).

[28] C. Reichhardt, J. Groth, C. J. Olson, S. B. Field and F. Nori, Spatiotemporal dynamics and plastic flow of vortices in superconductors with periodic arrays of pinning sites, Phys. Rev. B **54**, 16108-16115 (1996).

[29] D. J. Morgan and J. B. Ketterson, Asymmetric Flux Pinning in a Regular Array of Magnetic Dipoles, Phys. Rev. Lett. **80**, 3614-3617 (1998).

[30] M. J. Van Bael, K. Temst, V. V. Moshchalkov and Y. Bruynseraede, Magnetic properties of submicron Co islands and their use as artificial pinning centers, Phys. Rev. B **59**, 14674-14679 (1999).

[31] A. Oral, S. J. Bending and M. Henini, Real-time scanning Hall probe microscopy, Appl. Phys. Lett. **69**, 1324-1326 (1996).

[32] A. Oral, J. C. Barnard, S. J. Bending, I. I. Kaya, S. Ooi, T. Tamegai and M. Henini, Direct Observation of Melting of the Vortex Solid in $Bi_2Sr_2CaCu_2O_{8+\delta}$ Single Crystals, Phys. Rev. Lett. **80**, 3610-3613 (1998).

VI
MESOSCOPIC SUPERCONDUCTIVITY

Chapter 23

SUPERCONDUCTING NANOPARTICLES AND NANOWIRES

M. Tinkham
Harvard University
Cambridge, MA 02138, USA

1. INTRODUCTION

The properties of macroscopic samples of the classic, phonon-mediated superconductors are now rather well understood (although the situation is much less clear with the high-temperature superconductors). In this chapter, we address the question of how superconductivity is affected if the sample size is reduced to the nanometer scale. We distinguish two cases: nanoparticles and nanowires, in which 3 and 2 dimensions of the sample, respectively, are at the nanoscale, so that the system can be approximated as being governed by the physics of zero or one dimension, respectively.

First, we shall discuss the case of nanograins, where the small volume of the sample implies a finite spacing δ between the quantized states of the conduction electron gas. As pointed out by Anderson [1] in 1959, when the spacing δ becomes larger than the superconducting energy gap Δ, the BCS theory implies that superconductivity should be excluded. When experiments of Black *et al.* [2] finally allowed the superconducting energy gap to be measured in single nanoparticles of Al, a number of theoretical groups were stimulated to reexamine this issue. We shall first briefly survey some of the more accessible theoretical analyses and then describe the experimental results. For a detailed overview of the current theoretical situation, see the next Chapter by von Delft.

Secondly, we shall review the experimental and theoretical situation concerning superconducting nanowires. Here again there is a long-standing theorem, which states that long-range order, and hence superconductivity, is not possible in a strictly one-dimensional system. But since even

I. O. Kulik and R. Ellialtioğlu (eds.),
Quantum Mesoscopic Phenomena and Mesoscopic Devices in Microelectronics, 349–360.

10 nm diameter wires contain $\sim 10^4$ channels, the operational question is: under what conditions is superconductivity effectively destroyed in real nanowire samples. In this case, we shall start with a brief review of the well-understood thermally activated resistance in superconducting filaments, and then discuss the low temperature quantum regime in which neither theory nor experiment were on firm ground at the time of this ASI. In the interest of bringing this volume up to date, we will also describe some new experimental results that dramatically illuminate the situation.

2. SUPERCONDUCTIVITY IN NANOPARTICLES: THE SUPERCONDUCTING SIZE EFFECT

In the early days of the BCS theory of superconductivity, P. W. Anderson [1] pointed out that grains of an ordinarily superconducting metal should *not* become superconducting if the grain were small enough that the level spacing δ exceeded the width of the (low-temperature) superconducting gap Δ of a bulk sample. With the availability of tunneling measurements on single nanoparticles, it became possible to test this qualitative prediction and other more quantitative recent theories.

For a first orientation, we examine the implications of standard BCS results, when they are applied to small samples. The energy of the superconducting state at $T = 0$ relative to the normal state is usually written as $-N(0)\Delta^2/2$, where $N(0)$ is the normal density of states in the sample volume. Rewriting $N(0)$ as $1/\delta$, we obtain $-\Delta^2/2\delta$. This "condensation energy" will be washed out by fluctuations unless it exceeds k_BT. Thus, we expect the superconducting transition temperature to be strongly suppressed when a sample is small enough that $\delta > \Delta$. Another suggestive result is obtained by comparing the BCS results for the cases of odd *vs.* even numbers of electrons. If the number is odd, there must be an unpaired quasiparticle excitation above the energy gap, which raises the system energy by at least Δ. Thus the energy lowering becomes only $\Delta - \Delta^2/2\delta$, which is even more easily overcome by thermal fluctuations. In fact, this formula predicts that there is no energy gain at all in the superconducting state if $\delta > \Delta/2$. In other words, for a certain range of particle size (and hence δ) we expect that the grain would be superconducting with an even number of electrons, but not for an odd number!

Although these conclusions are qualitatively correct, they obviously lack rigor. A straightforward improvement is simply to solve the BCS gap equation, retaining a discrete density of states rather than making the continuum approximation. The problem is easily solved if one assumes that the states have a uniform spacing δ. This was done in an elementary way by Black [3] and more carefully by von Delft and collaborators [5] with similar results. They found that superconductivity is excluded, even at T=0, if δ/Δ exceeds 3.56 or 0.89 for even or odd numbers of electrons, respectively. The problem was taken a step further by Smith and Ambegaokar [6] who replaced the assumption of a uniform level density by the sort of spectrum predicted by random matrix theory. In this case, there is only a probabilistic relationship between the existence of a superconducting state and the *mean* level spacing relative to the value of Δ in bulk samples. Physically, this result reflects the fact that the spacing between the few levels nearest the Fermi energy dominates in BCS, and that these *particular* spacings may be quite different from the *mean* spacing.

It is difficult to test how quantitatively these predictions are borne out. Single superconducting grains this small have a diamagnetic moment less than a Bohr magneton, which is too small to be detected even by SQUID magnetometry. Thus, we must rely on tunnel spectroscopy to probe for the existence of a superconducting energy gap. This technique works well [2] for grains large enough that the level separation is much less than Δ, because then there is an obvious gap of width Δ (which can be destroyed by applying a strong magnetic field) in the closely spaced sequence of discrete levels. This is illustrated by data of Black *et al.* shown in Fig. 23.1. But for the most interesting particle size range, where $\delta \sim \Delta$, all levels are sufficiently widely spaced that a superconducting gap can not be distinguished by inspection from a simple interval between discrete levels, and all the levels shift with a magnetic field. The best one can do is plot the observed superconducting gap for those grains in which a gap can be identified *vs.* estimated grain size. When this is done as in Fig. 23.2, one finds that if a superconducting gap can be identified, its width is close to that in the bulk superconductor, but a superconducting gap can no longer be identified when the grain is small enough (diameter < 10 nm) that $\delta > \Delta$. This is consistent with the theoretical prediction, but hardly a quantitative check of the theory.

3. SUPERCONDUCTING NANOWIRES: A DISSIPATIVE PHASE TRANSITION

Resistive voltages appear between two contacts (1 and 2) to a superconductor when there is a time variation of the phase difference φ_{12} of the Ginzburg-Landau wave function between those points. Specifically, $2eV_{12} = \hbar\partial\varphi_{12}/\partial t$. Such a changing phase difference arises in a bulk

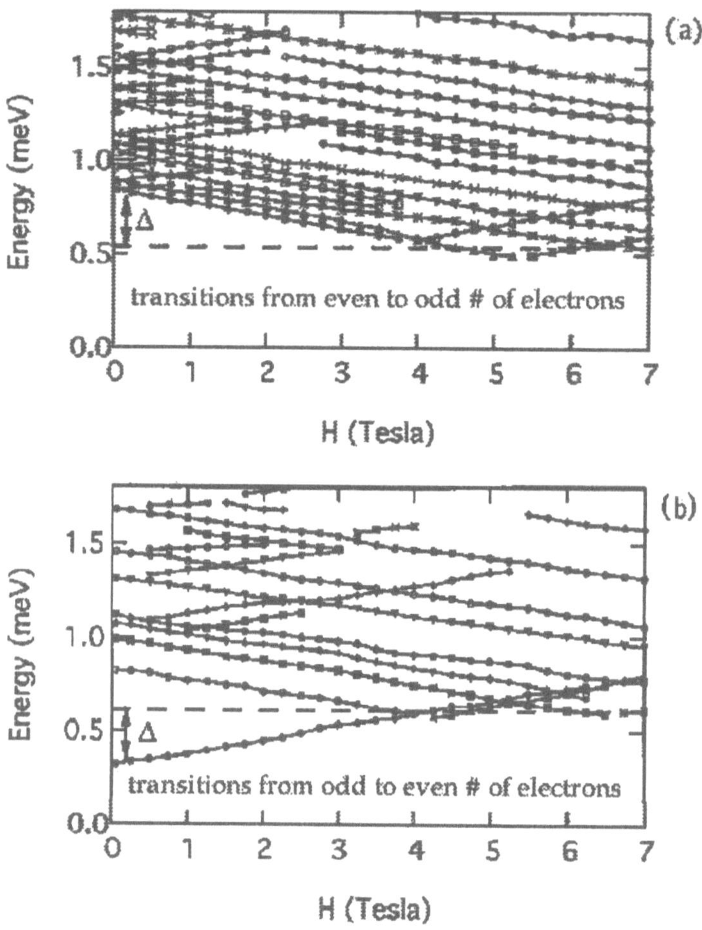

Figure 23.1 Magnetic field dependence of tunneling resonance energies in an aluminum grain large enough to have a mean level spacing δ which is small compared to the superconducting energy gap Δ. The absence of tunneling at lower energies reflects the Coulomb blockade. The upper panel shows the data for a grain with an even number of electrons going to an odd number, which requires an extra energy Δ. The lower panel shows the odd to even case, where the added electron pairs with an existing quasiparticle, releasing a pairing energy Δ per electron, which makes the tunneling possible at a lower bias voltage. (After Ref. [2].)

superconductor from the motion of 3D vortex lines across a line connecting the points 1 and 2, and in thin superconducting films from the similar motion of 2D vortex spots. In both cases, the motion of a single quantum vortex carrying flux $\Phi_o = h/2e$ causes a phase slip by 2π, and

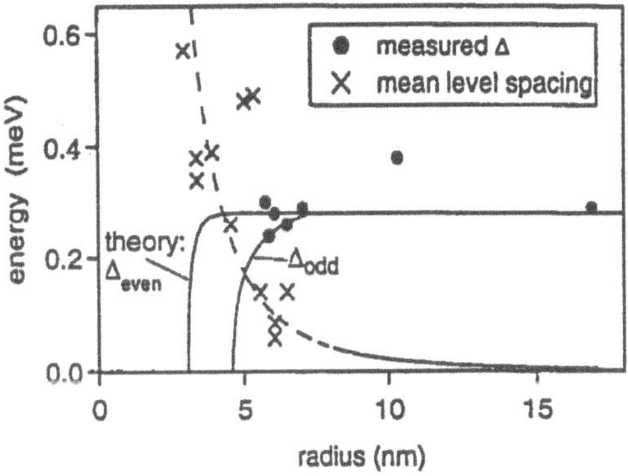

Figure 23.2 Dependence of the measured superconducting tunneling gap as a function of particle size, compared to the measured mean level spacing and the predicted dependence of the parity-dependent order parameter. (After Ref. [4].)

hence a voltage pulse satisfying

$$\int V_{12} dt = \Phi_0 \tag{23.1}$$

The superposition of these pulses from many vortex lines gives rise to a dc voltage. On the other hand, in a quasi-one-dimensional superconducting wire [*i.e.*, one with diameter $< \xi(T)$], the phase slip by 2π can only occur by having the amplitude of the order parameter ψ at some point x drop to zero, so that the phase is undefined, and then having $|\psi(x)|$ build back up to its usual value, but with a phase differing by 2π from the value before the phase-slip event. (Only phase slips by integer multiples of 2π are allowed, since only those leave the phase unchanged modulo 2π.) If one writes V_{12} as $\int_1^2 E dx$, Eq. (23.1) becomes $\int\int E dx dt = \Phi_0$. In other words, the electric field associated with a phase slip obeys a quantum constraint and is confined in a localized area in the x-t plane, analogous to a localized vortex containing a quantum of magnetic flux in the x-y plane in a 2D superconducting film.

The minimum energy cost to depress $|\psi|$ to zero at some point along the filamentary conductor to allow a phase slip can readily be estimated as the loss of superconducting condensation energy $H_c^2/8\pi$ per unit volume over a volume given by the product of the cross-sectional area of the wire A and the coherence length $\xi(T)$. A precise evaluation of this free energy cost by Langer and Ambegaokar [7] showed that it is

$$\Delta F = \frac{8\sqrt{2}}{3}\frac{H_c^2}{8\pi} A\xi \; . \tag{23.2}$$

Inserting the known temperature dependences near T_c of the superconducting parameters, $H_c \sim (1-t)$ and $\xi \sim (1-t)^{-1/2}$, we see that $\Delta F \sim (1-t)^{3/2}$, where $t = T/T_c$. Since thermally activated phase slips, causing a resistive voltage, will occur at a rate dominated by the factor

$$\Omega \exp(-\Delta F/k_B T), \tag{23.3}$$

where Ω is an attempt frequency, this ΔF must not exceed $k_B T$ by too large a factor if thermally-activated phase slips are to occur sufficiently frequently to give rise to a measurable dc average voltage.

The theoretical prediction for the resistive transition was confirmed experimentally by Newbower *et al.* [8], for example. He measured the resistive transition in tin "whiskers" $\sim 0.5\ \mu$m in diameter, and found that the drop in resistance below T_c followed the theoretically predicted exponential form, and became immeasurably small as soon as T was about 1 mK below T_c. (For this detailed quantitative comparison, it is necessary to use a corrected value of the pre-exponential factor Ω, found by McCumber and Halperin [9].) Thus, for all practical purposes, such micron-diameter superconducting filaments show a sharp transition from the normal state to a state in which resistive, thermally-activated phase-slip events occur only on a time scale like the age of the universe!

How does this situation change when one considers filaments with diameters on the nanometer scale instead of the micrometer scale? We can make an estimate by scaling from the data for the tin whiskers. Focusing on the exponential dependence (23.3) of resistance on ΔF, and the dependence (23.2) of ΔF on parameters, we see that, for a given material, the product $A(1-t)^{3/2}/t$ controls the amount of resistance. Obviously, the thermally-activated resistance always goes to zero in the unattainable limit T=0. A more practical question is: how thin a filament would be required to broaden the thermally-activated resistive transition from 1 mK below T_c, as in tin whiskers, down to $T \ll T_c$, say $T_c/3$. That is, if the $(1-t)^{3/2}/t$ factor is increased from 4.4×10^{-6} to 1.6, how small a cross-sectional area would be needed to give a resistance comparable to the lowest measurable value seen in the tin whiskers. This leads to an estimate of $A \sim 1$ nm^2, which is smaller than any samples studied to date. From this simple estimate, it is clear that thermally-activated processes will be quickly frozen out as one goes significantly below T_c, even in filaments as narrow as a few nm in diameter. As a result, we infer that the resistance of available superconducting nanowires at low temperatures will be dominated by *quantum* fluctuations, as opposed to thermal fluctuations, in which case resistance will persist even at T=0. Before discussing the developing theory of quantum phase slips (QPS) in more detail, let us first examine the available experimental data for guidance.

At the time of this ASI, the principal experiments in the literature that were interpreted as giving evidence for quantum phase slips were those of Giordano, carried out some years earlier [10]. He studied wires of several different superconductors, 50-150 μm long with diameters in the range 16-100 nm, produced by step-edge lithographic techniques. Near T_c, these samples showed a sharp fall in resistance similar to that found earlier in tin whiskers, but with a width of ~0.1 K instead of ~0.001 K. This increase is transition width for a narrower wire is generally consistent with the argument in the previous paragraph. But on cooling further, the decrease in resistance was found to change to a slower exponential drop which Giordano found could be fitted to a phenomenological model based on quantum tunneling. Specifically, he argued that in an overdamped system the probability of MQT or QPS could be estimated by replacing the k_BT in the denominator of the exponent of the thermally activated model (23.3) by the energy $\hbar/\tau_{GL} = 8k_BT_c(1-t)/\pi$. Unlike k_BT, this energy does not go to zero at T=0, but actually has its maximum value there. Physically, this energy can be thought of as the energy uncertainty introduced by the rapid quantum fluctuations on the time scale τ_{GL}. Note that at T=0, this energy uncertainty exceeds the energy gap 1.76 k_BT_c so that the gap is largely ineffective, and the conductivity is essentially the same as in the normal state. The factor of $(1-t)$ partially cancels the $(1-t)^{3/2}$ factor in ΔF, leaving a much slower temperature dependence than for the thermally-activated process. For typical parameter values, the dominant resistive process below T/T_c ~0.9 was inferred to be the quantum tunneling process, in which the temperature dependence arises only from the temperature dependence of superconducting parameters, not from the k_BT in the exponential thermal activation expression. Despite this slower drop in resistance, in all samples but one the resistance dropped below detectability at his lowest temperature. The exception was the very narrowest sample, with a nominal diameter of 16 nm, in which the resistance appeared to bottom out at a finite value $\sim 0.2\ R_n$.

Shortly after this ASI, new data were obtained by Bezryadin *et al.* [11], which markedly illuminate the experimental situation. They produced their nanowires by sputter depositing a MoGe amorphous alloy on free-standing carbon nanotubes. In this way they were able to make rather uniform superconducting nanowires of length ~150 nm, thickness ~5 nm, and widths ~5-15 nm. The normal state resistances ranged from 4.5 to 23 kΩ. Films of this alloy had T_c ~5.5 K, comparable with the results of a careful study of this system by Graybeal [12]. The key data from this new work, obtained up to the time of this writing, are the plots of measured (linear) resistance of the wire *vs.* temperature shown in Fig. 23.3. A clear dichotomy is evident. The samples i1, i2, and i3, with normal resistances of 23, 15, and 10 kΩ remain resistive at low temperatures,

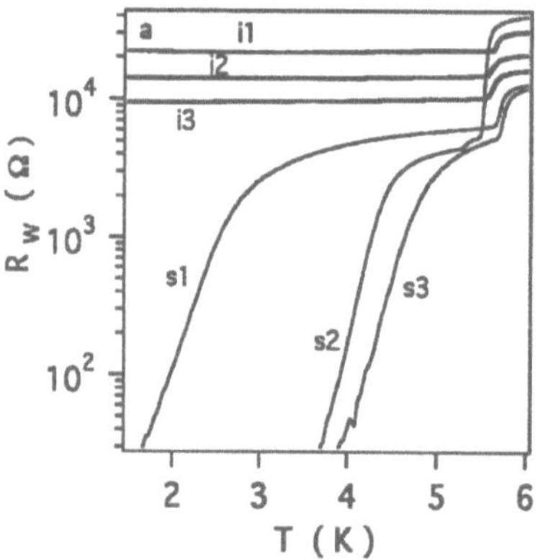

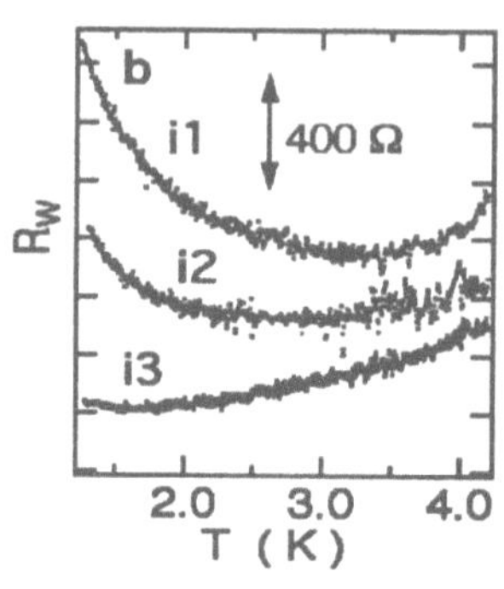

Figure 23.3 (*a*) Resistance *vs.* temperature curves for six different samples, measured in series with broad thin film leads. The resistance drop at ~5.5 K corresponds to the temperature at which these thin film leads of the same material and thickness become superconducting. All wires are 5 nm thick but their widths and hence resistances are different as described in the text. Curves i1, i2, and i3 show insulating behavior as $T \to 0$, while curves s1, s2, and s3 show superconducting behavior. The behavior observed appears to depend on whether the normal resistance at T_c is above or below the quantum resistance ~6.5 kΩ. (*b*) Expanded version of the R vs T curves of the insulating samples, showing the upturn in resistance at lower temperatures. The curves have been displaced vertically for clarity. (After Ref. [11].)

even showing a negative temperature dependence (see Fig. 23.3*b*) suggesting insulating behavior at low temperatures. On the other hand, samples s1, s2, and s3, with normal resistances of 6.4, 4.5, and 5.7 kΩ, all show resistances which drop decisively upon cooling, presumably approaching zero at T=0. Although at present no samples with resistances in the range 6.5-10 kΩ have been measured, these results strongly suggest that (within experimental error) wires with normal resistance above the superconducting quantum of resistance $R_Q = h/4e^2 \approx$6.5 kΩ remain resistive, tending toward insulating at low temperatures, while wires with $R < R_Q$ tend toward zero resistance at T=0. This behavior is referred to as a *dissipative phase transition* at T=0, because the controlling parameter in the phase transition is not the temperature but the level of dissipation set by the wire conductance.

With this clear experimental guidance, we now review the theoretical considerations thought to underlie the superconductor-insulator dissipative quantum phase transition in the T=0 limit. A pioneer paper in treating the effect of dissipation on macroscopic quantum tunneling is that of Caldeira and Leggett [13]. They showed that linear friction suppresses quantum tunneling by an exponential factor $\exp[-A\eta(\Delta q)^2/\hbar]$

where A is of order unity, η is the classical linear friction coefficient, and Δq is the distance tunneled. More specifically relevant to the problem at hand, Schmid [14] analyzed the properties of an ideal Josephson junction, shunted by a damping resistance R. The static energy of a Josephson junction is given by $-E_J \cos \varphi$, where φ is the phase difference across the device and $E_J = \hbar I_c/2e$. This energy is periodic with period 2π in the phase variable. If there is no damping, the problem can be solved in analogy with the periodic electron potential in a crystal, with the familiar energy band solutions. In these energy band solutions, the particle is not localized in any potential minimum, but occupies Bloch-wave states, which extend over the crystal. These solutions respond to any electric field, or, in the absence of a field, they describe the diffusion of the electron through the crystal. Returning to the Josephson junction case, the analog of the Bloch waves describe a situation in which the phase variable is delocalized, and responds freely to even slight driving forces. (Currents tilt the "washboard" just as a dc E-field tilts the periodic potential in a crystal.) In other words, even a slight tilt in the potential induces phase evolution, which implies a voltage. That is, the Josephson junction does not act like a superconductor. What Schmid's work showed was that the damping produced by a conductance $1/R_Q$ was the minimum necessary to localize the phase, and hence to turn the device into a superconductor. This result was recently confirmed experimentally on a single classic Josephson tunnel junction by Penttila *et al.* [15]. (Their work also showed that, in order to fit the experimental phase diagram in detail, the simple theoretical predictions need to be adjusted to take account of the limited sensitivity of electrical measurements.) The data of Bezryadin *et al.*, suggest that the same criterion also describes the behavior of nanowires, even when they are as much as 20 coherence lengths long. The open question is: why, and under what conditions, is this true? Work by Zaikin *et al.* [16] and by Demler [17] indicates that the *diameter* of the wire relative to the London penetration depth (~15-35 nm) is an important additional parameter. Physically speaking, for thicker wires the damping for high-frequency currents associated with phase slips will be governed by a skin layer of the wire, while the normal resistance at T_c is governed by the whole cross-sectional area. Thus there would no longer be a simple equality between these two resistances, and hence the condition for the phase transition might be less simple. Another important significance of this diameter is that for smaller diameters the Mooij-Schön [18] plasma wave velocity c_{MS} is reduced well below c by the increased kinetic inductance. It is less clear at what *length* scale considerations other than the total resistance of a narrow wire will become important.

An important aspect of the result of Schmid (as refined by Bulgadaev [19]) is that the criterion for localization of the phase by damping is *universal*, independent of the magnitudes of the reactive terms (Josephson energy E_J and charging energy E_c) which determine the barrier height through which the phase variable must tunnel. In other words, while these energies may determine the *temperature* scale at which the $T=0$ results are approached, and the sensitivity needed to detect them, they should not affect the ideal results governed by dissipation at $T=0$. If one can generalize this result to the present case of the nanowire, this would imply that the critical condition there should *not* depend on the critical current of the wire, as determined by its diameter and the superconducting parameters H_c, *etc.* This point underlines a clear distinction between a dissipative phase transition and one determined by considerations of free energy.

In the absence of a definitive theoretical consensus, we conclude with a few qualitative remarks that may be helpful for understanding why the *normal* resistance of the *entire* wire should be the crucial parameter, even when the wire is many coherence lengths long. The central distinction is that *dissipated* energy dominates for a quantum tunneling process, while an *activation* energy dominates a thermally-activated process. The minimum energy ΔF allowing a phase slip is given by (23.2), which implies a process localized within a coherence length. On the other hand, at $T=0$ there is no thermal source of energy to compare with ΔF, and instead, one needs to minimize the energy *dissipated.* This dissipated energy for a phase slip will be $\int IV dt$. Given the quantum constraint (23.1), we see that the energy dissipated scales with I, or with $1/R$ for Ohmic damping. Thus, for minimum damping, and hence the highest probability of a quantum phase slip, the system will take advantage of the maximum available resistance, namely that of the *entire* wire. Any process involving a shorter length (such as the coherence length) would lead to more current and more dissipation, and hence be less likely to occur. Thus, the maximum resistance, *i.e.*, that of the entire wire, can be expected to control the process, as is observed.

Why does the *normal* resistance of the wire play a dominant role? As remarked earlier, insofar as the Ginzburg-Landau relaxation time τ_{GL} governs the time scale of these quantum fluctuations, the characteristic frequencies of the voltage pulse associated with a phase slip are comparable with or exceed the energy gap. At such high frequencies, the electrodynamic properties of the superconducting state are essentially the same as in the normal state [20]. Another perspective is that the effective T_c of these nanowires is depressed, as suggested by the $R(T)$ curves in Fig. 23.3*a*, so that the superconducting state is "weakened" and the normal resistance is a reasonably self-consistent approximation in the temperature range not too far below the bulk T_c.

Taking this idea further, if the voltage pulse described by (23.1) occurs in a nanowire of normal resistance R, the instantaneous current is $\sim V(t)/R$, and (23.1) implies that the integrated charge flow associated with a single phase slip is $\delta Q = \Phi_o/R = h/2eR = 2eR_Q/R$, where $R_Q = h/4e^2$ is the quantum of resistance for Cooper pairs. Thus, if $R > R_Q$, less than two electrons can be transferred by a single phase-slip process, precluding the transfer of a Cooper pair, and normal electrons will dominate, and vice versa. This simple heuristic argument suggests that $R = R_Q$ might represent an important threshold value, as in fact turns out to be the case for quantum phase slips. Clearly a much more rigorous analysis will be required to completely illuminate the behavior of this novel and interesting system.

Acknowledgements

The author is happy to acknowledge the crucial contributions of his coworkers D. C. Ralph, C. T. Black, A. Bezryadin and C. N. Lau to the work reported here, as well as valuable discussions with A. D. Zaikin, E. Demler and Y. Oreg. The financial support of this research by NSF grants DMR-97-01487, DMR 98-09363, and PHY98-71810, and by ONR grant N00014-96-0108 is gratefully acknowledged.

References

[1] P. W. Anderson, Theory of dirty superconductors, J. Phys. Chem. Solids **11**, 26-30 (1959).

[2] C. T. Black, D. C. Ralph and M. Tinkham, Spectroscopy of the superconducting gap in individual nanometer-scale aluminum particles, Phys. Rev. Lett. **76**, 688-691 (1996); D. C. Ralph, C. T. Black and M. Tinkham, Gate-voltage studies of discrete electronic states in Al nanoparticles, Phys. Rev. Lett. **78**, 4087-4090 (1997).

[3] C. T. Black, Tunneling spectroscopy of nanometer-scale metal particles, Ph.D. Dissertation, (Physics Dept., Harvard University, 1996).

[4] D. C. Ralph, C. T. Black, J. M. Hergenrother, J. G. Lu and M. Tinkham, Ultrasmall superconductors, in L. L. Sohn, L. P. Kouwenhoven, and G. Schön, (eds.), Mesoscopic Electron Transport, (Kluwer, Dordrecht, 1997) pp. 447-468.

[5] J. Von Delft, A. D. Zaikin, D. S. Golubev and W. Tichy, Parity-affected superconductivity in ultrasmall metallic grains, Phys. Rev. Lett. **77**, 3189-3192 (1996).

[6] R. A. Smith and V. Ambegaokar, Effects of level statistics on superconductivity in ultrasmall metallic grains, Phys. Rev. Lett. **77**, 4962-4965 (1996).

[7] J. S. Langer and V. Ambegaokar, Intrinsic resisitive transition in narrow superconducting channels, Phys. Rev. **164**, 498-510 (1967).
[8] R. S. Newbower, M. R. Beasley and M. Tinkham, Fluctuation effects on the superconducting transition of tin whisker crystals, Phys. Rev. B**5**, 864-868 (1972); also M. Tinkham *Introduction to Superconductivity*, 2nd ed., (McGraw-Hill, New York, 1996) p. 288.
[9] D. E. McCumber and B. I. Halperin, Time scale of intrinsic resistive fluctuations in thin superconducting wires, Phys. Rev. B **1**, 1054-1070 (1970).
[10] N. Giordano, Evidence for macroscopic quantum tunneling in one-dimensional superconductors, Phys. Rev. Lett. **61**, 2137-2140 (1988); also Superconducting fluctuations in one dimension, Physica **B203**, 460-466 (1994).
[11] A. Bezryadin, C. N. Lau and M. Tinkham, Nature **404**, 971-974 (2000).
[12] J. M. Graybeal and M. R. Beasley, Localization and interaction effects in ultrathin amorphous superconducting films, Phys. Rev. B **29**, 4167-4169 (1984); also J. M. Graybeal, Ph.D. dissertation, (Stanford University, 1985) unpublished.
[13] A. O. Caldeira and A. J. Leggett, Influence of dissipation on quantum tunneling in macroscopic systems, Phys. Rev. Lett. **46**, 211-214 (1981).
[14] A. Schmid, Diffusion and localization in a dissipative quantum system, Phys. Rev. Lett. **51**, 1506-1509 (1983).
[15] J. S. Penttila, U. Parts, P. J. Hakonen, M. A. Paalanen and E. B. Sonin, "Superconductor-insulator transition" in a single Josphson junction, Phys. Rev. Lett. **82**, 1004-1007 (1999).
[16] A. D. Zaikin, D. S. Golubev, A. van Otterlo and G. T. Zimanyi, Quantum phase slips and transport in ultrathin superconducting wires, Phys. Rev. Lett. **78**, 1552-1555 (1997); also A. D. Zaikin, D. S. Golubev, A. van Otterlo and G. T. Zimanyi, Quantum fluctuations and dissipation in thin superconducting wires, Usp. Fiz. Nauk **168**, 244-248 (1998).
[17] E. Demler, (2000), to be published.
[18] J. E. Mooij and G. Schön, Propagating plasma mode in thin superconducting filaments, Phys. Rev. Lett. **55**, 114-117 (1985).
[19] S. A. Bulgadaev, Phase diagram of a dissipative quantum system, JETP Lett. **39**, 315-319 (1984).
[20] M. Tinkham, *Introduction to Superconductivity*, 2nd ed., (McGraw-Hill, New York, 1996) p. 99.

Chapter 24

SUPERCONDUCTIVITY IN ULTRA-SMALL GRAINS: INTRODUCTION TO RICHARDSON'S EXACT SOLUTION

J. von Delft and F. Braun

Institut für Theoretische Festkörperphysik, Universität Karlsruhe
76128 Karlsruhe, Germany

Abstract Studies of pairing correlations in ultrasmall metallic grains have commonly been based on a simple reduced BCS-model describing the scattering of pairs of electrons between discrete energy levels that come in time-reversed pairs. This model has an exact solution, worked out by Richardson in the context of nuclear physics in the 1960s. Here we give a tutorial introduction to his solution, and use it to check the quality of various previous treatments of this model.

1. INTRODUCTION

Recent experiments by Ralph, Black and Tinkham, involving the observation of a spectroscopic gap indicative of pairing correlations in ultrasmall Al grains [1], have inspired a number of theoretical [2]-[11] studies of how superconducting pairing correlations in such grains are affected by reducing the grains' size, or, equivalently, by increasing its mean level spacing $d \propto \mathrm{Vol}^{-1}$ until it exceeds the bulk gap Δ. In the earliest of these, a grand-canonical (g.c.) BCS approach [2, 3, 4] was applied to a reduced BCS Hamiltonian for uniformly spaced, spin-degenerate levels; it suggested that pairing correlations, as measured by the condensation energy E^C, vanish abruptly once d exceeds a critical level spacing d^c that depends on the parity (0 or 1) of the number of electrons on the grain, being smaller for odd grains ($d_1^c \simeq 0.89\Delta$) than even grains ($d_0^c \simeq 3.6\Delta$). A series of more sophisticated canonical approaches (summarized in Section 3. below) confirmed the parity dependence of pairing correlations, but established [6]-[11] that the abrupt vanishing of

I. O. Kulik and R. Ellialtioğlu (eds.),
Quantum Mesoscopic Phenomena and Mesoscopic Devices in Microelectronics, 361–370.

pairing correlations at d^c is an artifact of g.c. treatments: pairing correlations do persist, in the form of so-called fluctuations, to arbitrarily large level spacings, and the crossover between the bulk superconducting (SC) regime ($d \ll \Delta$) and the fluctuation-dominated (FD) regime ($d \gg \Delta$) is completely smooth [10]. Nevertheless, these two regimes are qualitatively very different [9, 10]: the condensation energy, *e.g.*, is an extensive function of volume in the former and almost intensive in the latter, and pairing correlations are quite strongly localized around the Fermi energy ε_F, or more spread out in energy, respectively.

After the appearance of all these works, we became aware that the reduced BCS Hamiltonian on which they are based actually has an exact solution. It was published by R. W. Richardson in the context of nuclear physics (where it is known as the "picket-fence model"), in a series of papers between 1963 and 1977 [12]-[20] which until very recently seem to have completely escaped the attention of the condensed matter community. In this work, we (i) give a tutorial introduction (with no pretense of rigor) to his solution, and (ii) compare the results of various previously-used approximations against the benchmark set by the exact solution, in order to gauge their reliability for related problems for which no exact solutions exist [21, 22].

2. RICHARDSON'S EXACT SOLUTION

2.1 Reduced BCS Model

Ultrasmall superconducting grains are commonly described [2]-[11] by a reduced BCS model,

$$H = \sum_{j\sigma} \varepsilon_j c^\dagger_{j\sigma} c_{j\sigma} - g \sum_{ij} c^\dagger_{i+} c^\dagger_{i-} c_{j-} c_{j+}, \tag{24.1}$$

for a set S of N_S pairs of time-reversed states $|j, \pm\rangle$ labeled by a discrete index $j = 1, \ldots, N_S$, with energies ε_j and coupling $g = \lambda\, d$, where d is the mean level spacing and λ a dimensionless coupling constant. Unbeknownst to the authors that have studied this model recently, Richardson had long ago solved it exactly, for an arbitrary set of levels ε_j (degenerate levels are allowed, but are to be distinguished by distinct j-labels, *i.e.* they have $\varepsilon_i = \varepsilon_j$ for $i \neq j$).

The first step is to note that *singly-occupied* levels do not participate in the pairscattering described by H, and by the Pauli principle remain "blocked" [23] to such pairscattering; the labels of such levels are therefore good quantum numbers. A general eigenstate of H thus has the form

$$|n, B\rangle = \prod_{i \in B} c^\dagger_{i\sigma} |\Psi_n\rangle_U , \tag{24.2}$$

$$|\Psi_n\rangle_U = \sum_{j_1,\ldots,j_n}^{U} \psi(j_1, \ldots, j_n) \prod_{\nu=1}^{n} b^\dagger_{j_\nu=1} |0\rangle . \tag{24.3}$$

This describes $N = 2n + b$ electrons, b of which sit in a set B of singly-occupied, blocked levels, thereby contributing $\mathcal{E}_B = \sum_{i \in B} \varepsilon_i$ to the eigenenergy, while the remaining n pairs of electrons, created by the pair operators $b^\dagger_j = c^\dagger_{j+} c^\dagger_{j-}$, are distributed among the remaining set $U = S \backslash B$ of $N_U = N_S - b$ *unblocked* levels, with wave function $\psi(j_1, \ldots, j_n)$ ($\sum_j^U \equiv \sum_{j \notin B}$ denotes a sum over all *unblocked* levels). The dynamics of these pairs is governed by

$$H_U = \sum_{ij}^{U} (2\varepsilon_j \delta_{ij} - g)\, b^\dagger_i b_j , \tag{24.4}$$

and writing the eigenenergy of $|n, B\rangle$ as $\mathcal{E}_n + \mathcal{E}_B$, the state $|\Psi_n\rangle_U$ satisfies

$$H_U |\Psi_n\rangle_U = \mathcal{E}_n |\Psi_n\rangle_U , \qquad \sum_j^U b^\dagger_j b_j |\Psi_n\rangle_U = n |\Psi_n\rangle_U . \tag{24.5}$$

Diagonalizing H_U would be trivial if the b's were true bosons. However, they are not, and in the subspace spanned by the set U of all non-singly-occupied levels, instead satisfy the "hard-core boson" relations,

$$b^{\dagger 2}_j = 0, \quad [b_j, b^\dagger_{j'}] = \delta_{jj'}(1 - 2b^\dagger_j b_j), \quad [b^\dagger_j b_j, b^\dagger_{j'}] = \delta_{jj'} b^\dagger_j , \tag{24.6}$$

which reflect the Pauli principle for the fermions they are constructed from. In particular, $b^{\dagger 2}_j = 0$ implies that only those terms in (24.3) are non-zero for which the indices $j_1, \ldots j_n$ are all distinct.

In his original publications [12, 13, 14], Richardson derived a Schrödinger equation for $\psi(j_1, \ldots, j_n)$ and showed that its exact solution was simply a generalization of the form that $\psi(j_1, \ldots, j_n)$ would have had if the b's had been true (not hard-core) bosons. With the benefit of hindsight, we shall here follow an alternative, somewhat shorter root, also due to Richardson [24]: we first consider the related but much simpler case of *true* bosons and write down the generic form of its eigenstates; we then clarify why this form fails to produce eigenstates of the *hard-core* boson Hamiltonian; and having identified the reason for the failure, we show that (remarkably) only a slight generalization is needed to repair it and to obtain the sought-after hard-core-boson eigenstates.

2.2 True Bosons

Let $\tilde{b}_j$ denote a set of true bosons (*i.e.* $[\tilde{b}_j, \tilde{b}^\dagger_{j'}] = \delta_{jj'}$), governed by a Hamiltonian $\tilde{H}_U$ of precisely the form (24.4), with $b_j \to \tilde{b}_j$. This problem, being quadratic, can be solved straightforwardly by any number of methods. The solution is as follows: $\tilde{H}_U$ can be written as

$$\tilde{H}_U = \sum_J \tilde{E}_J \tilde{B}^\dagger_J \tilde{B}_J + \text{const.} \tag{24.7}$$

where the new bosons $\tilde{B}^\dagger_J$ (with normalization constants C_J) are given by

$$\tilde{B}^\dagger_J = gC_J \sum_j^U \frac{\tilde{b}^\dagger_j}{2\varepsilon_j - \tilde{E}_J}, \qquad \frac{1}{(gC_J)^2} = \sum_j^U \frac{1}{(2\varepsilon_j - \tilde{E}_J)^2}, \tag{24.8}$$

and the boson eigenenergies $\tilde{E}_J$ are the roots of the eigenvalue equation

$$1 - \sum_j^U \frac{g}{2\varepsilon_j - \tilde{E}_J} = 0\,. \tag{24.9}$$

This is an equation of order N_U in $\tilde{E}_J$. It thus has N_U roots, so that the label J runs from 1 to N_U. As the coupling g is turned to 0, each E_J smoothly evolves to one of the bare eigenenergies ε_j. A general n-boson eigenstate of $\tilde{H}_U$ and its eigenenergy $\tilde{\mathcal{E}}_n$ thus have the form

$$|\tilde{\Psi}_n\rangle_U = \prod_{\nu=1}^n \tilde{B}^\dagger_{J_\nu}|0\rangle\,, \qquad \tilde{\mathcal{E}}_n = \sum_{\nu=1}^n \tilde{E}_{J_\nu}\,, \tag{24.10}$$

where the n indices $J_1, \ldots, J_n$ that characterize this state need not all be distinct, since the $B^\dagger_J$ are true bosons.

2.3 Complications Arising For Hard-Core Bosons

Let us now return to the hard-core boson Hamiltonian H_U. Its eigenstates will obviously *not* be identical to the true-boson eigenstates just discussed, since matters are changed considerably by the hard-core properties of b_j. To find out exactly *what* changes they produce, it is very instructive to take an Ansatz for $|\Psi_n\rangle_U$ similar to (24.10) (but suppressing the normalization constants and taking all J_ν to be distinct), namely

$$|\Psi_n\rangle_U = \prod_{\nu=1}^n B^\dagger_{J_\nu}|0\rangle\,, \qquad \text{with} \qquad B^\dagger_J = \sum_j^U \frac{b^\dagger_j}{2\varepsilon_j - E_J}, \tag{24.11}$$

and to check explicitly whether or not it could be an eigenstate of H_U, *i.e.* to check under what conditions $(H_U - \mathcal{E}_n)|\Psi_n\rangle_U$ would equal zero, where $\mathcal{E}_n = \sum_\nu^n E_{J_\nu}$. To this end, we commute H_U to the right past all the $B^\dagger_{J_\nu}$ operators in $|\Psi_n\rangle_U$, using

$$\left[H_U, \prod_{\nu=1}^{n} B^\dagger_{J_\nu}\right] = \sum_{\nu=1}^{n} \left\{ \left(\prod_{\eta=1}^{\nu-1} B^\dagger_{J_\eta}\right) [H_U, B^\dagger_{J_\nu}] \left(\prod_{\mu=\nu+1}^{n} B^\dagger_{J_\mu}\right)\right\}. \quad (24.12)$$

To evaluate the commutators appearing here, we write H_U as

$$H_U = \sum_j^U 2\varepsilon_j b^\dagger_j b_j \; - \; g B^\dagger_0 B_0 \,, \qquad \text{where} \;\; B^\dagger_0 = \sum_j^U b^\dagger_j \,, \quad (24.13)$$

and use the following relations:

$$[b^\dagger_j b_j, B^\dagger_J] = \frac{b^\dagger_j}{2\varepsilon_j - E_J} \,, \qquad [B_0, B^\dagger_J] = \sum_j^U \frac{1 - 2b^\dagger_j b_j}{2\varepsilon_j - E_J} \,, \quad (24.14)$$

$$[H_U, B^\dagger_J] = E_J B^\dagger_J \; + \; B^\dagger_0 \left[1 - g \sum_j^U \frac{1 - 2b^\dagger_j b_j}{2\varepsilon_j - E_J}\right] . \quad (24.15)$$

Inserting these into (24.12) and using $H_U|0\rangle = 0$ and $\mathcal{E}_n = \sum_\nu^n E_{J_\nu}$, we find

$$\begin{aligned} H_U|\Psi_n\rangle_U = \mathcal{E}_n|\Psi_n\rangle_U &+ \sum_{\nu=1}^{n}\left[1 - \sum_j^U \frac{g}{2\varepsilon_j - E_{J_\nu}}\right] B^\dagger_0 \left(\prod_{\eta=1(\neq\nu)}^{n} B^\dagger_{J_\eta}\right)|0\rangle \\ &+ \sum_{\nu=1}^{n}\left\{\left(\prod_{\eta=1}^{\nu-1} B^\dagger_{J_\eta}\right)\left[\sum_j^U \frac{2gB^\dagger_0\, b^\dagger_j b_j}{2\varepsilon_j - E_{J_\nu}}\right]\left(\prod_{\mu=\nu+1}^{n} B^\dagger_{J_\mu}\right)\right\}|0\rangle \,. \quad (24.16)\end{aligned}$$

Now, suppose we do the same calculation for true instead of hard-core bosons (*i.e.* run through the same steps, but place a ˜ on H_U, b_j, E_J and $\mathcal{E}_n$). Then the second line of (24.16) would be absent (because the $b^\dagger_j b_j$ terms in the second of Eqs. (24.6) and (24.14) and in (24.15) would be absent); and the first line of (24.16) would imply that $(\tilde{H}_U - \tilde{\mathcal{E}}_n)|\tilde{\Psi}_n\rangle_U = 0$ provided that the term in square brackets vanishes, which is nothing but the condition that the $\tilde{E}_J$ satisfy the the true-boson eigenvalue equation of (24.9)! In other words, we have just verified explicitly that all true-boson states of the form (24.10) are indeed eigenstates of $\tilde{H}_U$, provided that the $\tilde{E}_J$ satisfy (24.9). Moreover, we have identified the term in second line of (24.16) as the extra complication that arises for hard-core bosons.

2.4 The Cure: A Generalized Eigenvalue Equation

Fortunately, this extra complication is tractable: first, we note that

$$\left[\sum_j^U \frac{2gB_0^\dagger b_j^\dagger b_j}{2\varepsilon_j - E_{J_\nu}}, B_{J_\mu}^\dagger\right] = \sum_j^U \frac{2gB_0^\dagger}{2\varepsilon_j - E_{J_\nu}} \frac{b_j^\dagger}{2\varepsilon_j - E_{J_\mu}} = 2gB_0^\dagger \frac{B_{J_\nu}^\dagger - B_{J_\mu}^\dagger}{E_{J_\nu} - E_{J_\mu}}. \tag{24.17}$$

The rightmost expression follows via a partial fraction expansion, and remarkably, contains only $B^\dagger$ operators and no more $b_j^\dagger b_j$s. This enables us to eliminate the $b_j^\dagger b_j$s from the second line of (24.16), by rewriting it as follows (we commute its term in square brackets to the right, using a relation similar to (24.12), but with the commutator (24.17) instead of $[H_U, B_{J_\mu}^\dagger]$):

$$\begin{aligned}
&\sum_{\nu=1}^n \left\{\left(\prod_{\eta=1}^{\nu-1} B_{J_\eta}^\dagger\right) \sum_{\mu=\nu+1}^n \left\{\left(\prod_{\eta'=\nu+1}^{\mu-1} B_{J_{\eta'}}^\dagger\right)\left[2gB_0^\dagger \frac{B_{J_\nu}^\dagger - B_{J_\mu}^\dagger}{E_{J_\nu} - E_{J_\mu}}\right]\left(\prod_{\mu'=\mu+1}^n B_{J_{\mu'}}^\dagger\right)\right\}\right\}|0\rangle \\
&\quad = \sum_{\mu=1}^n \left[\sum_{\nu=1}^{\mu-1} \frac{2g}{E_{J_\nu} - E_{J_\mu}}\right] B_0^\dagger \left(\prod_{\eta=1(\neq\mu)}^n B_{J_\eta}^\dagger\right)|0\rangle \\
&\qquad - \sum_{\nu=1}^n \left[\sum_{\mu=\nu+1}^n \frac{2g}{E_{J_\nu} - E_{J_\mu}}\right] B_0^\dagger \left(\prod_{\eta=1(\neq\nu)}^n B_{J_\eta}^\dagger\right)|0\rangle \\
&\quad = \sum_{\nu=1}^n \left[\sum_{\mu=1(\neq\nu)}^n \frac{2g}{E_{J_\mu} - E_{J_\nu}}\right] B_0^\dagger \left(\prod_{\eta=1(\neq\nu)}^n B_{J_\mu}^\dagger\right)|0\rangle\,.
\end{aligned} \tag{24.18}$$

(The last line follows by renaming the dummy indices $\nu \leftrightarrow \mu$ in the second line.) Substituting (24.18) for the second line of (24.16), we conclude that $(H_U - \mathcal{E}_n)|\Psi_n\rangle_U$ will be equal to zero, provided that

$$1 - \sum_j^U \frac{g}{2\varepsilon_j - E_{J_\nu}} + \sum_{\mu=1(\neq\nu)}^n \frac{2g}{E_{J_\mu} - E_{J_\nu}} = 0\,, \qquad \text{for} \quad \nu = 1, \ldots, n\,. \tag{24.19}$$

This consitutes a set of n coupled equations for the n parameters $E_{J_1}, \ldots,$ E_{J_n}, which may be thought of as self-consistently-determined pair energies. Eq. (24.19) can be regarded as a generalization of the true-boson eigenvalue equation (24.9), and was originally derived by Richardson by solving the Schrödinger equation for the wave-function $\psi(j_1, \ldots, j_n)$ of (24.3). It is truly remarkable that the exact eigenstates of a complicated many-body problem can be constructed by such a simple generalization of the solution of a quadratic (*i.e.* non-interacting) true-boson Hamiltonian!

Below we shall always assume the ε_j's to be all distinct. Then there exists a simple relation between the bare pair energies $2\varepsilon_j$ and the solutions of (24.19): as g is reduced to 0, it follows by inspection that each solution $\{E_{J_1}, \ldots, E_{J_n}\}$ reduces smoothly to a certain set of n bare pair energies, say $\{2\varepsilon_{j_1}, \ldots, 2\varepsilon_{j_n}\}$. Correspondingly, the state $|\Psi_n\rangle_U \equiv |J_1, \ldots J_n\rangle_U$ of (24.11) reduces smoothly to the state $|j_1, \ldots j_n\rangle_U \equiv \prod_{\nu=1}^{n} b_{j_\nu}^\dagger |0\rangle$ (up to a normalization factor not shown here). Thus there is a one-to-one correspondence between the set of all states $\{|J_1, \ldots, J_n\rangle_U\}$ and the set of all states $\{|j_1, \ldots j_n\rangle_U\}$. Since the latter constitute a complete eigenbasis for the n-pair Hilbert space defined on the set of unblocked levels U, the former do too.

2.5 Ground State

For a given set of blocked levels B, the lowest-lying of all states $|n, B\rangle$, say $|n, B\rangle_G$, is obtained by using that particular solution $E_{J_1}, \ldots E_{J_n}$ for which the total "pair energy" $\mathcal{E}_n$ takes its lowest possible value (as g is increased, some of the E_Js become complex; however, they always occur in complex conjugate pairs, so that $\mathcal{E}_n$ remains real [17]).

The lowest-lying of all eigenstates with n pairs and b blocked levels, say $|n, b\rangle_G$ with energy $\mathcal{E}_b^G(n)$, is that $|n, B\rangle_G$ for which the blocked levels in B are all as close as possible to ε_F, the Fermi energy of the uncorrelated N-electron Fermi sea $|F_N\rangle$. The E_{J_ν} for the ground state $|n, b\rangle_G$ coincide at $g = 0$ with the lowest n energies $2\varepsilon_j$ ($j = 1, \ldots, n$), and smoothly evolve toward lower values as g is turned on. This fact can be exploited during the numerical solution of (24.19), which can be simplified by first making some algebraic transformations, discussed in detail in [15], that render the equations less singular.

2.6 General Comments

Since the exact solution provides us with wave functions, it is in principle straightforward to calculate arbitrary correlation functions. Some such correlators are discussed by Richardson in [16, 17], who showed that they can be expressed in terms of certain determinants that are most conveniently calculated numerically. Moreover, it is natural to ask whether in the bulk limit, the standard BCS results can be extracted from the exact solution. Indeed they can, as Richardson showed in [20], by interpreting the problem of solving (24.19) for the E_{J_ν} as a problem in two-dimensional electrostatics. Exploiting this analogy, he showed that in the bulk limit ($N_S \to \infty$ at fixed $N_S d$), Eqs. (24.19) reduce to the well-known BCS gap equation and the BCS equation for the chemical potential, and the condensation energy $\mathcal{E}_0^C(n)$ (defined in Eq. (24.20) below) to its BCS result, namely $-\Delta^2/2d$.

3. COMPARISON WITH OTHER APPROACHES

We now apply the exact solution to check the quality of results previously obtained by various other methods. Most previous works [2, 3, 4], [6]-[10] studied a half-filled band with fixed width $2\omega_D$ of uniformly-spaced levels (*i.e.* $\varepsilon_j = j\,d$), containing $N = 2n + b$ electrons. Then the level spacing is $d = 2\omega_D/N$ and in the limit $d \to 0$ the bulk gap is $\Delta = \omega_D \sinh(1/\lambda)^{-1}$. Following [9], we take $\lambda = 0.224$ throughout this paper. To study the SC/FD crossover, two types of quantities were typically calculated as functions of increasing d/Δ, which mimics decreasing grain size: the even and odd ($b = 0, 1$) condensation energies

$$E_b^C(n) = \mathcal{E}_b^G(n) - \langle F_N|H|F_N\rangle \; ; \qquad (24.20)$$

and a parity parameter introduced by Matveev and Larkin (ML) [6] to characterize the even-odd ground state energy difference,

$$\Delta^{\mathrm{ML}}(n) = \mathcal{E}_1^G(n) - [\mathcal{E}_0^G(n) + \mathcal{E}_0^G(n+1)]/2\,. \qquad (24.21)$$

Following the initial g.c. studies [2]-[6], the first canonical study was that of Mastellone, Falci and Fazio (MFF) [7], who used Lanczos exact diagonalization (with $n \leq 12$) and a scaling argument to probe the crossover regime. Berger and Halperin (BH) [8] showed that essentially the same results could be achieved with $n \leq 6$ by first reducing the bandwidth and renormalizing λ, thus significantly reducing the calculational effort involved. To access larger systems and fully recover the bulk limit, fixed-n projected variational BCS wavefunctions (PBCS) were used in [9] (for $n \leq 600$); significant improvements over the latter results, in particular in the crossover regime, were subsequently achieved in [10] using the density matrix renormalization group (DMRG) (with $n \leq 400$). Finally, Dukelsky and Schuck [11] showed that a self-consistent RPA approach, that in principle can be extended to finite temperatures, describes the FD regime rather well (though not as well as the DMRG).

To check the quality of the above methods, we [21, 22] computed $E_b^C(n)$ and $\Delta^{\mathrm{ML}}(n)$ using Richardson's solution (Fig. 24.1). The exact results (a) quantitatively agree, for $d \to 0$, with the leading $-\Delta^2/2d$ behavior for $E_b^C(n)$ obtained in the g.c. BCS approach [2, 3, 4], which in this sense is exact in the bulk limit, corrections being of order d^0; (b) confirm that a completely smooth [10] crossover occurs around the scale $d \simeq \Delta$ at which the g.c. BCS approach breaks down; (c) show that

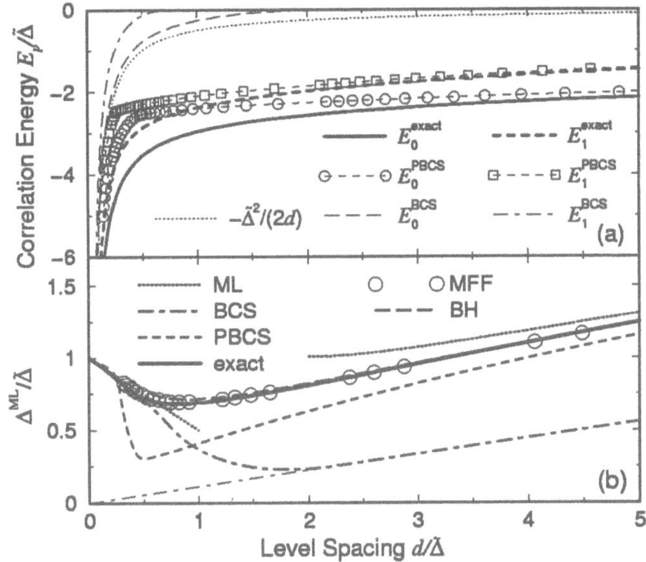

Figure 24.1 (*a*) The even and odd ($b = 0, 1$) condensation energies E_b^C of Eq. (24.20), calculated with BCS, PBCS and exact wave functions, as functions of $d/\Delta = 2\sinh(1/\lambda)/(2n + b)$, for $\lambda = 0.224$. For comparison the dotted line gives the "bulk" result $E_0^{\text{bulk}} = -\Delta^2/(2d)$. (*b*) Comparison of the parity parameters Δ^{ML} [6] of Eq. (24.21) obtained by various authors: ML's analytical result (dotted lines) [$\Delta(1 - d/2\Delta)$ for $d \ll \Delta$, and $d/2 \log(ad/\Delta)$ for $d \gg \Delta$, with $a = 1.35$ adjusted to give asymptotic agreement with the exact result]; grand-canonical BCS approach (dash-dotted line) [the naive perturbative result $\lambda d/2$ is continued to the origin]; PBCS approach (short-dashed line); Richardson's exact solution (thick solid line); exact diagonalization and scaling by MFF (open circles) and BH (long-dashed line).

the PBCS crossover [9] is qualitatively correct, but not quantitatively, being somewhat too abrupt; (d) are reproduced remarkably well by the approaches of MFF [7] and BH [8]; (e) are fully reproduced by the DMRG of [10] with a relative error of $< 10^{-4}$ for $n \leq 400$; our figures don't show DMRG curves, since they are indistinguishable from the exact ones and are discussed in detail in [10].

4. CONCLUSIONS

The main conclusion we can draw from these comparisons is that the two approaches based on renormalization group ideas work very well: the DMRG is essentially exact for this model, but the band-width rescaling method of BH also gives remarkably (though not quite as) good results with rather less effort. In contrast, the PBCS approach is rather unreliable in the crossover region.

Acknowledgements

The derivation of the exact solution shown above, which is shorter and perhaps somewhat more direct than the original published derivation, was invented by R. W. Richardson too; we thank him for a private communication suggesting this route. We also thank Moshe Schechter for helpful comments on the manuscript.

References

[1] D. C. Ralph, C. T. Black and M. Tinkham, Phys. Rev. Lett. **76**, 688 (1996); 78, 4087 (1997).
[2] J. von Delft *et al.*, Phys. Rev. Lett. **77**, 3189 (1996).
[3] F. Braun *et al.*, Phys. Rev. Lett. **79**, 921 (1997).
[4] F. Braun and J. von Delft, Phys. Rev. B **59**, 9527 (1999).
[5] R. A. Smith and V. Ambegaokar, Phys. Rev. Lett. **77**, 4962 (1996).
[6] K. A. Matveev and A. I. Larkin, Phys. Rev. Lett. **78**, 3749 (1997).
[7] A. Mastellone, G. Falci and R. Fazio, Phys. Rev. Lett. **80**, 4542 (1998).
[8] S. D. Berger and B. I. Halperin, Phys. Rev. B **58**, 5213 (1998).
[9] F. Braun and J. von Delft, Phys. Rev. Lett. **81**, 4712 (1998).
[10] J. Dukelsky and G. Sierra, Phys. Rev. Lett. **83**, 172 (1999); and cond-mat/9906166.
[11] J. Dukelsky and P. Schuck, to appear in Phys. Lett. B.
[12] R. W. Richardson, Phys. Lett. **3**, 277 (1963).
[13] R. W. Richardson, Phys. Lett. **5**, 82 (1963).
[14] R. W. Richardson and N. Sherman, Nucl. Phys. **52**, 221 (1964).
[15] R. W. Richardson, Phys. Lett. **14**, 325 (1965).
[16] R. W. Richardson, J. Math. Phys. **6**, 1034 (1965).
[17] R. W. Richardson, Phys. Rev. **141**, 949 (1966).
[18] R. W. Richardson, Phys. Rev. **144**, 874 (1966).
[19] R. W. Richardson, Phys. Rev. **159**, 792 (1966).
[20] R. W. Richardson, J. Math. Phys. **18**, 1802 (1977).
[21] F. Braun, Ph.D. thesis, Karlsruhe University (1999); F. Braun and J. von Delft, Advances in Solid State Physics, (Ed. B. Kramer), p. 341, Vieweg, Braunschweig (1999).
[22] G. Sierra, J. Dukelsky, G. G. Dussel, J. von Delft and F. Braun, cond-mat/9909015.
[23] V. G. Soloviev, Mat. Fys. Skrif. Kong. Dan. Vid. Selsk. **1**, 1 (1961).
[24] R. W. Richardson, private communication (1999).

Chapter 25

SUPERCONDUCTIVITY IN ULTRASMALL METALLIC PARTICLES

H. Boyaci, Z. Gedik and I. O. Kulik
Department of Physics, Bilkent University
Bilkent 06533 Ankara, Turkey

Abstract Recent single electron transport experiments in nanometer size samples renewed the question about the lower limits of the size of superconductors, and the crossover from superconducting to normal state. In order to give answers to these questions, a pairing Hamiltonian for fixed number of particles is studied including the degeneracy of levels around the Fermi energy. For d-fold degenerate states we find that the ratio of two successive parity parameters Δ_p is nearly $1 + 1/d$.

1. INTRODUCTION

Back in the year of 1959, Anderson [1] proposed that for a small metallic particle superconductivity should disappear as the mean level spacing δ becomes of the order of bulk gap Δ. Since the level spacing is related to the size of the material as $\delta \sim 1/\mathrm{Vol}$, according to Anderson's criterion, superconductivity would disappear in ultrasmall grains.

Interest in superconductivity in ultrasmall grains recently renewed with a series of experiments by Black, Ralph and Tinkham (BRT) [2, 3] (and more recently by Davidović and Tinkham [4, 5]). BRT accomplished in fabricating a single Al particle of nanometer size connected to two separate metal leads by tunnel junctions. They obtain the current-voltage ($I - V$) curve with discrete steps corresponding to tunneling via individual electronic states in the sample, providing the first spectroscopic measurement of these states. BRT observe that the spectroscopic gap parameter vanishes as the size of the sample decreases. For estimated level spacing $\delta \sim 0.02$ meV (corresponding sample size is $r \sim 10$ nm) a gap is observed, while for $\delta \sim 0.7$ meV ($r \sim 2.5$ nm) gap disap-

I. O. Kulik and R. Ellialtıoğlu (eds.),
Quantum Mesoscopic Phenomena and Mesoscopic Devices in Microelectronics, 371–380.

pears. Since $\Delta \sim 0.34$ meV for Al, BRT conclude that their experimental results are in qualitative agreement with Anderson's criterion. BRT also observed that the gap parameter persists for smaller samples with even number of electrons than those with odd number of electrons.

These experiments raised questions about the crossover from superconducting to normal state in ultrasmall grains with level spacing $\delta \sim \Delta$. Standard BCS theory gives a good description of the phenomenon of superconductivity for large samples. However one should expect that the quantum fluctuations of the order parameter grows as δ reaches Δ. Matveev and Larkin (ML)[6] show that the corrections to the mean field results which are small in large grains ($\delta \ll \Delta$), become important in the opposite limit ($\delta \gg \Delta$). ML [6] introduce a parameter for parity effect:

$$\begin{aligned} \Delta_p &= E_g^{2n+1} - \frac{1}{2}\left(E_g^{2n} + E_g^{2n+2}\right) , \\ \tilde{\Delta}_p &= -E_g^{2n} + \frac{1}{2}\left(E_g^{2n+1} + E_g^{2n-1}\right) . \end{aligned} \tag{25.1}$$

Although with standard BCS calculations it vanishes, ML show that, if the quantum fluctuations are properly taken into account, the parity parameter does not vanish for $\delta \gg \Delta$. They obtain the following asymptotic results

$$\begin{aligned} \frac{\Delta_p}{\Delta} &= 1 - \frac{\delta}{2\Delta} , \quad \frac{\delta}{\Delta} \ll 1, \\ \frac{\Delta_p}{\Delta} &= \frac{\delta}{\Delta}\frac{1}{2\ln\frac{\delta}{\Delta}}, \quad \frac{\delta}{\Delta} \gg 1 , \end{aligned} \tag{25.2}$$

In scope of these asymptotic results, ML conclude Δ_p/Δ has a minimum about $\delta \sim \Delta$. Note that this value corresponds to the crossover in question, that is transition from superconducting to normal state.

Mastellone, Falci and Fazio [7], and Berger and Halperin [8] solve the problem numerically by exact diagonalization. Both groups obtain similar results suggesting a minimum in Δ_p/Δ for $\delta \sim \Delta$, in agreement with ML's predictions. Braun and von Delft [9] approach the problem within a fixed-N picture of superconductivity. Instead of grand canonical ensemble, they solve the problem by using a canonical ensemble. Their results confirm the minimum predicted by ML.

2. THE MODEL

Although it is supposed that these ultrasmall samples are irregular in shape, it has been argued [10] that spatial symmetry may exist no matter how small the sample is. In case such a symmetry exists, for

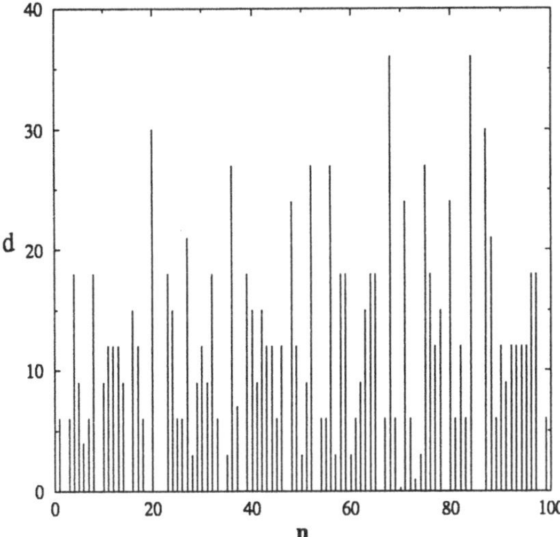

Figure 25.1 Degeneracy of energy levels for a parabolic dispersion, where $E_{\mathbf{n}} = \frac{\hbar^2\pi^2}{2mL^2}(n_1^2 + n_2^2 + n_3^2)$. The number of solutions d satisfying $n^2 = n_1^2 + n_2^2 + n_3^2$ for a range of energies about the Fermi energy is shown as vertical lines for each channel.

a parabolic dispersion, degeneracy is of the order of $k_F L$ where k_F is the Fermi momentum and L is the particle size, and typical distance between levels is of the order of $\hbar^2/mL^2$. Fig. 25.1 shows the degeneracy of energy levels in such a parabolic dispersion. If we assume that there is no spatial symmetry in the grain, then the only degeneracy is due to the time reversal symmetry. In order to understand the effect of time reversal symmetry, let us consider the standard BCS theory. For a grain where eigenstates are labeled by crystal momentum $\mathbf{k}$, time reversed states are $|\mathbf{k}\downarrow\rangle$ and $|-\mathbf{k}\uparrow\rangle$. Note that there is another similar but different pair between $|\mathbf{k}\uparrow\rangle$ and $|-\mathbf{k}\downarrow\rangle$. Since in usual BCS reduced Hamiltonian there is a summation over $\mathbf{k}$, both pairs are properly taken into account in calculations. However, when we sum over energy levels rather than the individual states we must be careful in including both pairs [11]. Nevertheless, the model without double degeneracy can still be considered to describe superconductivity in systems with real wave functions e.g. one dimensional infinite quantum well.

We address the question of ultrasmall superconducting grains within a pairing Hamiltonian

$$H = \sum_{f,\sigma} \epsilon_f c^\dagger_{f,\sigma} c_{f,\sigma} - g \sum_{f,f' \in S} c^\dagger_{f,\sigma} c_{f,\sigma} c^\dagger_{f',-\sigma} c_{f',-\sigma}, \tag{25.3}$$

where $c^\dagger_{f,\sigma}$ and $c_{f,\sigma}$ are fermion creation and annihilation operators which satisfy the anti-commutation relation

$$[c_{f,\sigma}, c_{f',\sigma'}]_+ = \delta_{ff'}\delta_{\sigma\sigma'}, \tag{25.4}$$

and f denotes the single particle quantum numbers including degeneracy, ϵ_f denotes the single particle energy levels, $\sigma = \pm 1$ denotes the states with up and down spin, which are conjugate with respect to time reversal symmetry, g is the pair-pair coupling term. The second summation is over a convenient set of levels S. For the BCS model this set is the collection of levels lying within a shell which has a width of $2\omega_D$ about the Fermi level. Hence, in the second sum we impose this restriction and consider only the states

$$f, f' \in S = \{-n_c, ..., n_c\}, \tag{25.5}$$

where $n_c = [\omega_D/\delta]$ (where [...] denotes integer part of the argument).

We write above model of many-fermion system (25.3) as an Hamiltonian of fermion pairs interacting via pairing forces in second quantized form

$$H = \sum_f 2\epsilon_f \hat{N}_f - g \sum_{f,f'\in S} b^\dagger_f b_{f'}, \tag{25.6}$$

where

$$\hat{N}_f = \frac{1}{2}\left(c^\dagger_{f,\sigma}c_{f,\sigma} + c^\dagger_{f,-\sigma}c_{f,-\sigma}\right), \tag{25.7}$$

and

$$b_f = c_{f,-\sigma}c_{f,\sigma}. \tag{25.8}$$

Since the second term of H defines interaction between pairs only, unpaired particles do not interact. Therefore, singly occupied levels are taken out from the set S. This is the so-called "blocking effect". Moreover, with the shift of chemical potential, levels included in S will also change. Fig. 25.2 shows both of these effects.

We first calculate ground state energies for three successive states with number of particles being equal to l, $l+1$ and $l+2$. Since we also consider degeneracy in the system, we obtain $2 \times d$ (d being the level degeneracy) different Δ_p values. We use the following labeling scheme

$$\Delta_p^{(m)} = (-1)^m \left[E_g^{(2N+m-1)} - \frac{1}{2}\left(E_g^{(2N+m-2)} + E_g^{(2N+m)}\right)\right], \tag{25.9}$$

where $N = n_c d$ (total number of levels within the shell below Fermi energy and above). Hence, for example, for $m = 1$

$$\Delta_p^{(1)} = -E_g^{(2N)} + \frac{1}{2}\left(E_g^{(2N-1)} + E_g^{(2N+1)}\right), \tag{25.10}$$

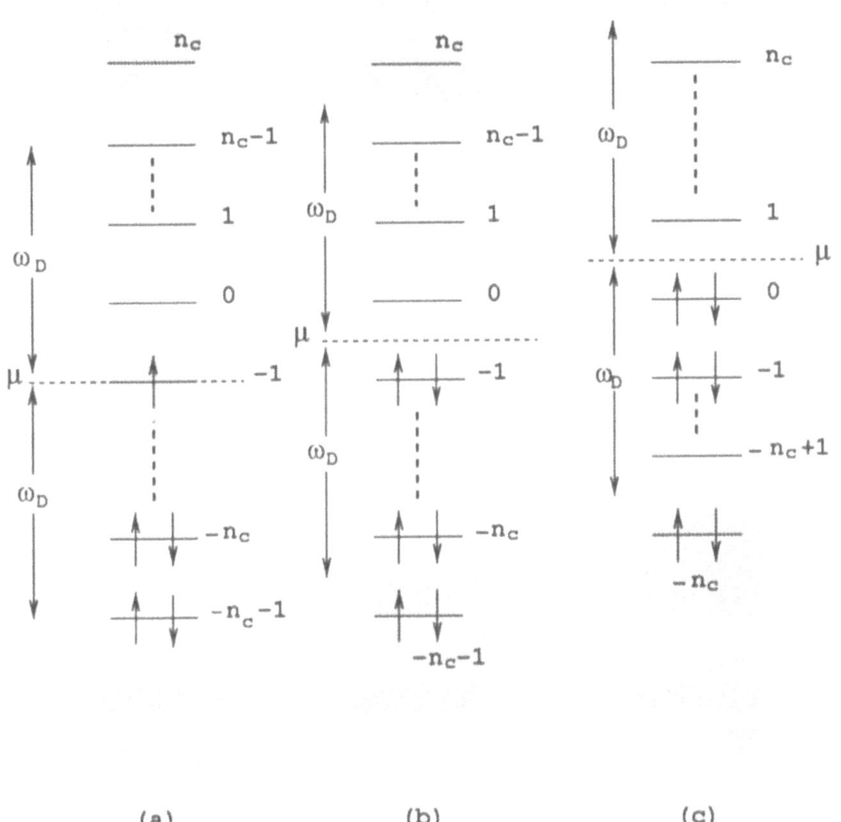

Figure 25.2 Levels included in the set S. The pair-pair interaction is assumed to be restricted to pairs with energy within the $2\omega_D$ shell about Fermi level. When Fermi level shifts, the levels which should be included in the set S change. In addition to this, singly occupied levels are taken out from the set S, which is the so called "blocking effect". Here, the shaded regions correspond to the levels which are occupied by non-interacting particles

which is schematically presented in Fig. 25.3.

The problem of determining the eigenstates of the pairing-force Hamiltonian was solved by R. W. Richardson and N. Sherman in 1964 [12]. However, this solution was forgotten for a long time despite of its importance in application to BCS theory of superconductivity. And only

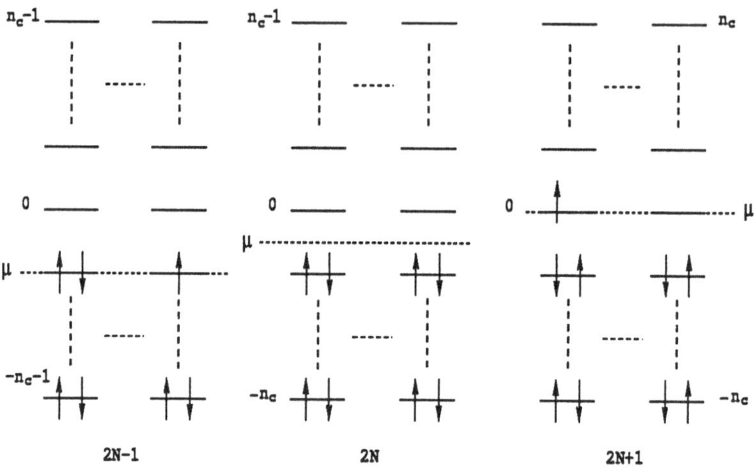

Figure 25.3 Three consecutive configurations for different number of electrons. Here, the three consecutive configurations whose ground state energies are used to calculate $\Delta_p^{(1)}$, are shown schematically. Note the shift of chemical potential for different number of electrons. Below each configuration, number of electrons are written for clarity.

recently, it has been "re-discovered" by the condensed matter community [13].

Ground state energy of the model is given by

$$E = E_1 + \ldots + E_N, \tag{25.11}$$

where the pair parameters E_i are obtained from the following coupled system of non-linear equations

$$\frac{d}{\lambda\delta} + 2\sum_{j\neq i}^{N} \frac{1}{E_j - E_i} - \sum_{n=1}^{M} \frac{\Omega(n)}{2\epsilon_n - E_i} = 0; \qquad i = 1, \ldots N, \tag{25.12}$$

where we introduced the dimensionless coupling parameter $\lambda = gd/\delta$. Here $\Omega(n)$ is the pair degeneracy of the level corresponding to energy $\epsilon_n = n\delta$, N is the total number of pairs, and M is the total number of levels in the set S. The roots E_i of (25.12) are required to be distinct. However, the domain of validity of the solution can be extended by letting E_i's to be complex [14]. Complex roots E_i occur in complex conjugate pairs. This preserves the reality of E which is the sum of all roots (25.11). Such complex conjugate roots of the system also preserves the reality of the wave function for the model [12]. Nevertheless, the existence of complex roots of (25.12) depend upon the state of the system and can not easily be treated in a general way.

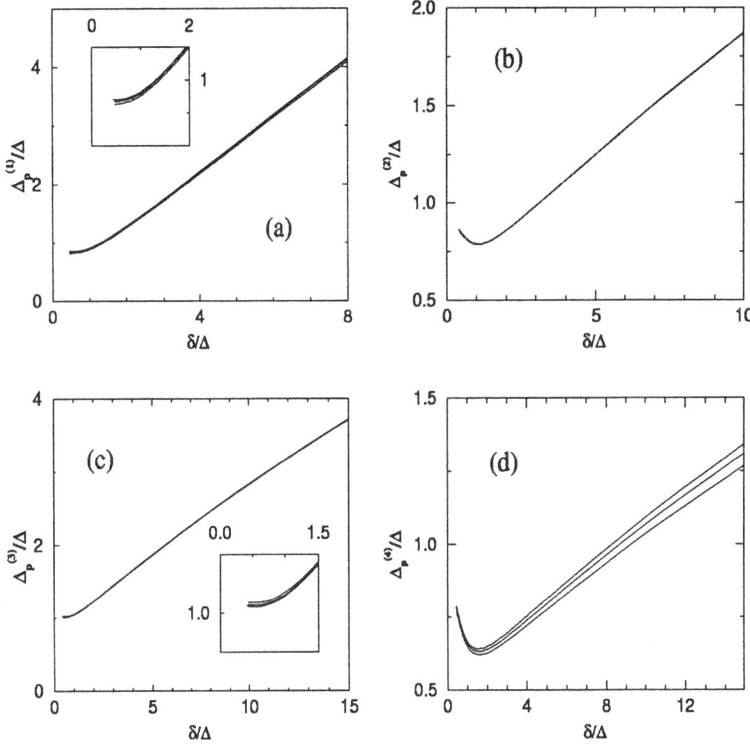

Figure 25.4 Dependence of parity effect parameter upon level spacing for doubly degenerate energy levels. $M = 41, 51, 61$ from top to bottom in (c), while from bottom to top in (a), (b) and (d).

3. RESULTS

We plot $\Delta_p^{(m)}/\Delta$ versus δ/Δ for double degeneracy in Fig. 25.4. For higher degeneracy, such as $d = 4$, we obtain similar results. It is seen that in the region $\delta \sim \Delta$, the curves have different behaviors. While $\Delta_p^{(odd)}/\Delta$ decreases steadily towards 1, $\Delta_p^{(even)}/\Delta$ makes a minimum at $\delta \sim \Delta$. A similar behavior is observed for the non-degenerate case [11].

As it has been mentioned earlier that for certain energy spectra and more generally for lattice symmetries, energy levels are strongly degenerate near the Fermi energy. Eq. 25.12 has an analytic solution for a single $d-$fold degenerate level[12]. In this case with our notation $n_c = 0$ and the ground state energy measured from the Fermi level is given by

$$E^{(2n)} = 2n\epsilon - \frac{\lambda\delta}{d}n[d - n + 1],$$

$$E^{(2n+1)} = 2n\epsilon - \frac{\lambda\delta}{d} n[(d-1) - n + 1], \tag{25.13}$$

where n is the number of hardcore bose particles, and ϵ is the energy measured from the Fermi level. Blocking effect is taken into consideration in the second equation of (25.13). We calculate the following parity parameters

$$\Delta_p^{(even)} = E^{(2n+1)} - \frac{1}{2}[E^{(2n)} + E^{(2n+2)}]$$
$$\Delta_p^{(odd)} = -E^{(2n)} + \frac{1}{2}[E^{(2n-1)} + E^{(2n+1)}]. \tag{25.14}$$

By substituting (25.13) into (25.14), we obtain

$$\Delta_p^{(even)} = \frac{\lambda\delta}{2}, \quad \Delta_p^{(odd)} = \frac{\lambda\delta}{2} + \frac{\lambda\delta}{2d}, \tag{25.15}$$

so that

$$\frac{\Delta_p^{(odd)}}{\Delta_p^{(even)}} = 1 + \frac{1}{d}. \tag{25.16}$$

These results suggest that the parity parameter remains non-zero for an infinitely degenerate single level. Comparing Figs. 25.4-(b) and (c), for example at $\delta/\Delta = 0.99$, we find that $\Delta_p^{(3)}/\Delta_p^{(2)} = 1.04/0.78 \sim 1.33$ which is quite close to 1.50 that we would obtain from (25.16). Moreover for some larger δ/Δ value ($\delta/\Delta = 10.36$), we obtain $\Delta_p^{(3)}/\Delta_p^{(2)} = 2.90/1.91 \sim 1.51$ which is even closer to corresponding value for single level spectrum. Table 25.1 shows $\Delta_p^{(odd)}/\Delta_p^{(even)}$ ratios for 2-fold and 4-fold degenerate cases. Note that the ratio (25.16) is applicable for three

Table 25.1 $\Delta_p^{(odd)}/\Delta_p^{(even)}$ ratios. As δ/Δ increases, the ratio $\Delta_p^{(odd)}/\Delta_p^{(even)}$ approaches the value $1+1/d$ as can be predicted by using the solution due to Richardson and Sherman, for d-fold degenerate single level.

	At $\frac{\delta}{\Delta} \sim 1$	At $\frac{\delta}{\Delta} \sim 10$
2-fold degeneracy; $d = 2$		
$\frac{\Delta_p^{(odd)}}{\Delta_p^{(even)}} = 1 + \frac{1}{d} = 1.5$	$\frac{\Delta_p^{(3)}}{\Delta_p^{(2)}} \sim 1.33$	$\frac{\Delta_p^{(3)}}{\Delta_p^{(2)}} \sim 1.51$
4-fold degeneracy; $d = 4$		
$\frac{\Delta_p^{(odd)}}{\Delta_p^{(even)}} = 1 + \frac{1}{d} = 1.25$	$\frac{\Delta_p^{(3)}}{\Delta_p^{(4)}} \sim 1.17$	$\frac{\Delta_p^{(3)}}{\Delta_p^{(4)}} \sim 1.24$

successive configurations where chemical potential does *not* shift. Thus, we conclude that, existence of degeneracy can be observed in experiments via the ratio of Δ_p values. If there is large difference in all successive Δ_p, then this is most probably a sign of non-degeneracy. However, if there is not such a difference for any Δ_p values, this can be interpreted as a sign of degeneracy. Moreover, level degeneracy can be predicted by observing these Δ_p's.

Acknowledgements

This work was partially supported by the Scientific and Technical Research Council of Turkey (TUBITAK), under grant No. TBAG 1736, and by the National Research Council of Italy, under the Research and Training Program for the Third Mediterranean Countries.

References

[1] P. W. Anderson, Theory of dirty superconductors, J. Phys. Chem. Solids **11**, 26 (1959).

[2] D. C. Ralph, C. T. Black, and M. Tinkham, Spectroscopic measurements of discrete electronic states in single metal particles, Phys. Rev. Lett. **74**, 3241 (1995).

[3] D. C. Ralph, C. T. Black, and M. Tinkham, Gate-voltage studies of discrete electronic states in aluminum nanoparticles, Phys. Rev. Lett. **78**, 4087 (1997).

[4] D. Davidović and M. Tinkham, Unconventional clustering of discrete energy levels in an ultrasmall Au grain, Appl. Phys. Lett. **73**, 3959 (1998).

[5] D. Davidović and M. Tinkham, Spectroscopy, interactions, and level splittings in Au nano-particles, cond-matt/990543 (1999).

[6] K. A. Matveev and A. I. Larkin, Parity effect in ground state energies of ultrasmall superconducting grains, Phys. Rev. Lett. **78**, 3749 (1997).

[7] A. Mastellone, G. Falci and Rosario Fazio, A small superconducting grain in the canonical ensemble, Phys. Rev. Lett. **80**, 4542 (1998). [cond-matt/9801179]

[8] S. D. Berger and B. I. Halperin, Parity effect in a small superconducting particle, Phys. Rev. B **58**, 5213 (1998). [cond-matt/9801286]

[9] F. Braun and J. von Delft, Superconductivity in ultrasmall metallic grains, Phys. Rev. B **59**, 9527 (1999).

[10] U. Landman, Clusters, dots, dot-molecules and Wigner crystallization, NATO Advanced Study Institute on "*Quantum Mesoscopic Phenomena and Mesoscopic Devices in Microelectronics*", Ankara/Antalya, 1999.

[11] H. Boyaci, Z. Gedik, and I. O. Kulik, Superconductivity in mesoscopic metal particles: The role of degeneracy, cond-mat/9909386.
[12] R. W. Richardson and N. Sherman, Exact eigenstates of the pairing-force Hamiltonian, Nuclear Physics **52**, 221 (1964).
[13] F. Braun and J. von Delft, Fixed-N Superconductivity: The Exact Crossover from the Bulk to the Few-Electron Limit, Advances in Solid State Physics **39**, 341 (Vieweg Braunschweig/Wiesbaden 1999). [cond-matt/9907402]
[14] R. W. Richardson, Numerical study of the 8-32-particle eigenstates of the pairing Hamiltonian, Phys. Rev. **141**, 949 (1966).

Chapter 26

TUNNELING SPECTROSCOPY OF METALLIC QUANTUM DOTS

M. Tinkham
Harvard University
Cambridge, MA 02138, USA

1. INTRODUCTION

A quantum dot is an isolated piece of conducting material which forms a small enough "box" that the confinement of electrons within it leads to resolvable discrete quantum energy levels, as opposed to the continuum of energies in a sample of macroscopic size. Here we specifically address metallic quantum dots produced by deposition of a thin, granular film onto an insulating substrate, from which a single selected grain is connected to two electrical leads by high-resistance, low-capacitance tunnel junctions. If these leads have tunneling resistances well in excess of the quantum resistance $R_Q = h/e^2 \simeq 23$ kΩ, the number of electrons on the grain is a good quantum number. This number can be changed one at a time by tunneling processes through the junctions, and the equilibrium number also can be controlled by an additional gate electrode which is coupled only electrostatically to the grain under study. In such a system, it is possible to carry out tunneling spectroscopy measurements which directly reveal the structure of the energy eigenvalues of the electrons in a small metallic grain which typically contains a few thousand conduction electrons. Such measurements were first carried out a few years ago by Ralph, Black, and myself on nanograins of Al [1]-[4]. More recent measurements have extended this work to nanoparticles of the heavy metal Au [5, 6] and to alloys of Al and Au [7]. This more recent work has pointed up the need to go beyond the simple single-electron model, which was used for an initial understanding of the spectra, to include both the Coulomb interaction between electrons and the spin-orbit interaction, as well as the role of nonequilibrium electronic populations

I. O. Kulik and R. Ellialtioğlu (eds.),
Quantum Mesoscopic Phenomena and Mesoscopic Devices in Microelectronics, 381–396.

and the Thouless energy, in order to gain an understanding of all the observations. In this paper, I shall review how such spectra can be observed and the evolving progress that has been made in interpreting these observations.

2. THEORETICAL OVERVIEW

If electrons had no charge, the most important characteristic energies would be simply the Fermi energy $E_F \simeq 5$ eV and the spacing between discrete electronic energy levels, namely $\delta \simeq E_F/N$, where N is the total number of conduction electrons in the grain. In the grains under study, the diameter is typically 5-10 nm, so that the number of electrons is typically 10^3 to 10^4, and so $\delta \simeq 0.5 - 5$ meV. These levels might be expected to yield resolvable tunneling features so long as the measurements are made at temperatures such that $kT \ll \delta$, because the characteristic width of the Fermi function derivative df/dE used to probe the tunnel spectrum is $\sim 3.5\ k_BT$. However, the true situation is more complicated.

The simple picture is modified in crucial ways by the existence of electron charge. The most straightforward effect is the classical electrostatic energy required to add (or subtract) an electron from an originally neutral grain. This charging energy is usually approximated by $E_c = e^2/2C_\Sigma$, the expression for the macroscopic charging energy of the capacitance between the grain and the environment. This C_Σ is usually dominated by $C_1 + C_2$, the sum of the capacitances of the two tunnel junctions coupling to the grain. This capacitance is proportional to the area of the tunnel junctions, and so typically scales with the surface area of these ultrasmall grains, while the density of electronic energy levels scales with the volume of the grain. For (approximately hemispherical) grains of the size of interest here, the Coulomb energy is typically an order of magnitude larger than the level spacing. Consequently, it is possible to distinguish the spectra of the grain with different numbers of electrons. It is also necessary to make this distinction, because, given the interelectronic Coulomb interactions beyond the mean field approximation, the spectra for $N \pm 1$ electrons cannot be derived from the spectrum for N electrons. In fact the energy of the "last" electron also depends on *which* other single-electron states are already occupied, as well as how many. As pointed out by Agam *et al.* [8], this fact, together with the existence of non-equilibrium populations when the tunneling rate is comparable to or exceeds the relaxation rate, can cause a single-particle energy level to show up in the tunneling spectrum as a cluster of closely spaced peaks, each associated with a different nonequilibrium occupation of other single-particle states.

The discussion of the preceding paragraph is still oversimplified, because it considers the shifts in the energies of tunneling resonances caused by interactions, but not the effect of the interactions in *broadening* the resonances by inducing transitions between different *many-body* states many of which can be accessed by tunneling into a given *quasiparticle* state. This issue was treated by Sivan, *et al.* [9, 10] in interpreting their experiments on semiconductor quantum dots. They found that as one considers states further above the Fermi sea ground state, the states become broader in energy, and merge into a continuum above an energy approximated by the Thouless energy $E_T = \hbar/\tau_T$. Here $\tau_T$is roughly the time required for a semiclassical electron to diffuse through the sample so as to determine its size and shape. That is, they found that the number of resolved energy levels should be *finite* even at T=0, and only of order E_T/δ, typically ~10 in their samples. This theoretical analysis was developed further by Altshuler *et al.* [11], who predicted that as one considered levels further and further above the system ground state, one should first find completely sharp levels, then clustered levels and levels of finite width, and finally (at $\sim E_T$) a continuous structureless tunneling density of states. Rather clear evidence for this progression has been found in the recent experiments of Davidovic and Tinkham [6] which will be reported below.

To complete this brief overview, we now mention the consequences of electrons having spin as well as charge. At the simplest level, this makes each single-particle orbital level doubly degenerate with respect to spin. At a somewhat deeper level, spin-orbit coupling mixes the two spin-states for a given orbital level to form a degenerate Kramers doublet state with a g-value which can be considerably reduced from the free-electron value of g=2. For example, in Au, which has high atomic number and hence strong spin-orbit coupling, the measured g-values are typically ~0.3-0.4, while in the low-Z metal Al, the measured g is typically ~1.98. Further complexity arises because of the spin-dependent "exchange" energy, which, for example, couples pairs of electrons in different orbital states, giving rise to singlet, triplet, and even higher multiplicity states for the system.

We now describe how samples are prepared and measured, and how an understanding of the observed spectra is approached by a series of successively improved approximations.

3. SAMPLE PREPARATION

In the deposition of thin granular metal films, the grains nucleate at a certain center-to-center separation (e.g., ~12 nm in the work of Davidovic [5]) which is determined by such parameters as surface diffusion rates and surface tension. As more metal is deposited, the grains grow in thickness and diameter, retaining the same separation. In preparing samples for the experiments on tunnel spectra of metallic nanograins, the deposition process is stopped before the individual grains form a percolating network. Instead, they form an irregular array of grains of similar but not identical size, and with a shape that is roughly hemispherical (although perhaps facetted), but tends to a pancake shape as the grains grow in area after a certain thickness is achieved. The challenge in the experiment is to find a way to make electrical tunnel contact with only a *single* grain from this array.

The pioneering experiments of Ralph *et al.* [1]-[4] were carried out using a novel configuration based on first creating a nanometer scale hole in a SiN membrane (by a process developed at Cornell University by Ralls *et al.* [12]), and using this hole to select a single grain for tunnel contact. A schematic diagram of the fabrication procedure is shown in Fig. 26.1.

In the more recent experiments of Davidovic and Tinkham [6], successive evaporations at carefully controlled angles, without breaking vacuum, are used to arrange that only a single grain is in tunnel contact with both the lower and upper Al electrode. [In some samples, it is possible that 2 or 3 grains are contacted, but because of the exponential dependence of tunnel resistance on barrier thickness, the current

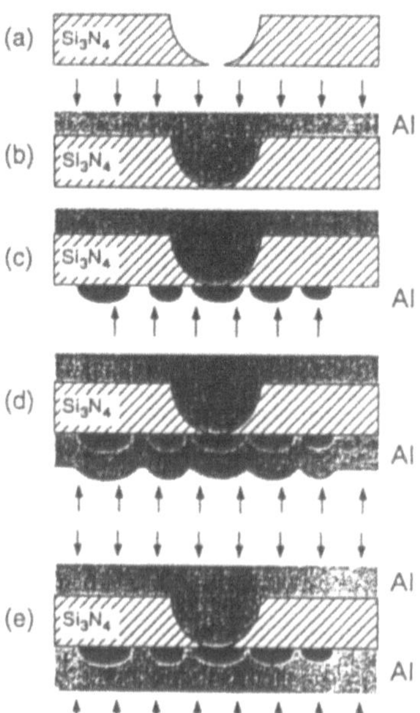

Figure 26.1 Device fabrication procedure used by Ralph and Black. Starting with a nm-scale hole etched in a 50 nm thick Si_3N_4 membrane (*a*), the device is defined by a series of metal depositions. In (*b*), a layer of Al is deposited into the bowl side of the membrane. After an oxidation step to form a tunnel barrier below the hole, a 15–20 Å layer of Al is deposited onto the flat side of the membrane (*c*), forming a layer of nm-scale grains. These particles are oxidized and then covered with a base electrode of Al (*d*). Finally, thick (1500 Å) Al layers are deposited on both sides of the device (*e*) for making contacts. (After Ref. [16])

through a particular single grain probably dominates. In either configuration, the dominance of tunneling through a single grain can be confirmed by carefully observing whether the shift in the positions of features in the measured $I-V$ curves when the superconducting energy gap in the Al electrodes is destroyed by a magnetic field corresponds to a unique junction capacitance ratio.]

The above description barely hints at the difficulty of actually preparing "good" samples. A good sample is one with no stray leakage conductances and especially one with a minimum amount of "charge noise" stemming from random motions of electronic charges within the dielectrics surrounding the metallic elements. These considerations motivate the dominant use of Al electrodes and Al_2O_3 tunnel barriers in these experiments. These materials are notably free of weakly conductive off-stoichiometry oxides, such as are found with metals such as Nb, and of possible additional complications from antiferromagnetism in materials such as Cr and Cr Oxides.

4. PARAMETER EVALUATION BY COULOMB STAIRCASE MEASUREMENTS

Before one can quantitatively interpret the spectra of resolved energy levels in a grain, one needs to know the capacitance and resistance of the tunnel junctions by which electrical contact is made. These parameters can be inferred by fitting the $I-V$ characteristic of the device measured at 4 K, where individual energy levels are not resolved, to the predictions of the classical "orthodox theory" of Averin and Likharev [13]. In this theory, for given tunnel resistances, transition rates are determined by differences in system energies calculated by classical electrostatics for states with various numbers of excess (or deficiency) charges on the grain, and as a function of the applied bias voltage V (and gate voltage V_g, if a gate electrode is used). With these ingredients, one can set up a "master equation", and solve to find the self-consistent steady-state current through the device as a function of the bias voltage. The junction capacitances C_1 and C_2 and (with less accuracy) the junction resistances R_1 and R_2 can be determined by fitting the data with $I-V$ curves simulated in this way for various parameter values.

The $I-V$ curves found in this way consist of straight line segments, which change slope (and/or value) at regularly spaced voltage values, at which either the equilibrium number of electrons on the grain changes by one, or at which some new cycle of charging and discharging becomes energetically possible. An example of such a "Coulomb staircase" is shown in Fig. 26.2. As can be seen from the identification of a number

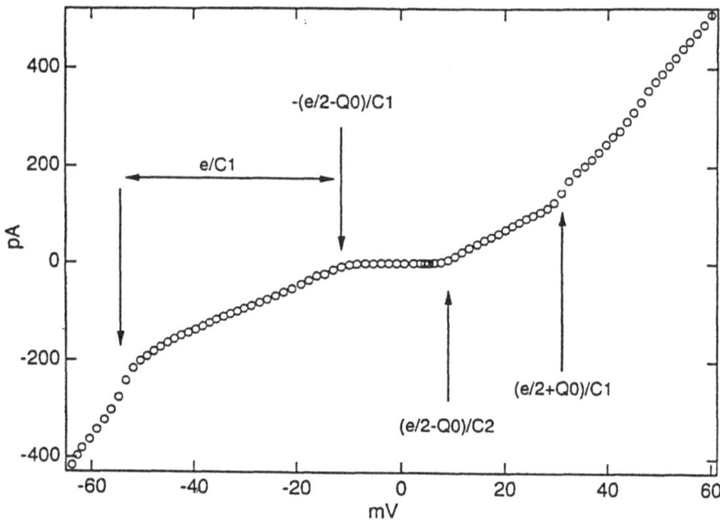

Figure 26.2 $I-V$ curve of typical Au-grain device at 4.2 K, showing the Coulomb staircase, with individual levels *not* resolved. The parameters determined by fitting these data to the orthodox theory are C_1=4 aF, C_2=5 aF, R_1=110 MΩ, R_2=24 MΩ, and Q_0=0.25e. (After Ref. [6])

of "corners" on the $I-V$ curve, the values of the two junction capacitances and of the offset charge Q_o can be inferred directly from the data. The sum of the two tunnel resistances is also readily determined from the overall slope of the $I-V$ curve, but the ratio of the two resistances can only be estimated from quantitative simulations. For details on this procedure, see, for example, Hanna and Tinkham [14]. The offset charge Q_o characterizes the effect of miscellaneous stray charges trapped in the surrounding dielectric. Unfortunately, Q_o is subject to random shifts, typically with a $1/f$ spectrum, which shows up as "charge noise" superposed on the data. This noise is a major problem in measurements of tunnel spectra. Fortunately, an occasional sample is unusually quiet, allowing high quality data to be obtained, typically after waiting a few days at low temperature for charge relaxation to occur. Without actually deriving the characteristic "corner" voltages in the $I-V$ curve, we can make an important point which enters into the interpretation of all data in these single-electron tunneling devices. When a voltage V is applied between the two electrodes, it is split by capacitive voltage divider action, with fractions $C_2/(C_1+C_2)$ and $C_1/(C_1+C_2)$ appearing across C_1 and C_2, respectively. Accordingly, the energy available to transfer an electron from an electrode to the grain when a voltage V is applied to the device will be eV times the one of these ratios which refers to the junction through which the tunneling process occurs. Thus, it is essential to know the ratio of C_1/C_2 in order to be able to correctly convert measured *voltage* intervals into *energy* level differences. For example, the accuracy with which g-values for spin splittings can be determined is limited by the accuracy with which this ratio is known. The role of Q_o in the complete formulas quoted in Fig. 26.2 is to reflect the fact that net charges in the environment, coupled electrostatically to the grain, shift the position of minimum electrostatic energy from zero charge to a (in general, fractional) charge Q_o. Given a non-zero Q_o (which is a signed quantity), clearly there is no reason for the $I-V$ curve to be simply antisymmetric about V=0, and the asymmetry is further complicated by the differences in capacitance and resistance between the two tunnel junctions. Thus, the $I-V$ curves for positive and negative voltages may show a certain resemblance to each other, but will not be exact inversion images.

5. DISCRETE LEVEL SPECTRA IN NORMAL GRAINS

The first successful observations of resolved discrete energy levels in metallic nanoparticles were carried out in the configuration shown in Fig. 26.1 by Ralph *et al.* [1] using aluminum particles and aluminum electrodes, with aluminum oxide tunnel barriers. Although bulk aluminum is superconducting, these first samples were in the normal state because the superconducting size effect (see Chapter 23) prevents superconductivity in sufficiently small particles. Representative data are shown in Fig. 26.3. Instead of the smooth onset of current above the Coulomb blockade (as in Fig. 26.2), the low temperature data in Fig. 26.3 show that the onset of current occurs in a series of discrete steps. Each step occurs when the applied voltage becomes sufficient to make it possible for an electron to tunnel from the continuum of energy states in one electrode (at the Fermi energy) into an empty additional, higher energy discrete quantum state in the island. Thereafter an electron must tunnel out of the island into the other electrode to complete the cycle, before another electron can enter the grain. These $I-V$ features are shown more clearly by plotting the derivative dI/dV, which displays a peak for each successive tunneling resonance.

Insofar as the energy state in the grain is much sharper than k_BT, the tunneling resonance should have the shape of df/dE, the derivative of the Fermi function. This is a bell-shaped curve with full width at half maximum of $\sim$3.5 k_BT, where T is the temperature of the electrons

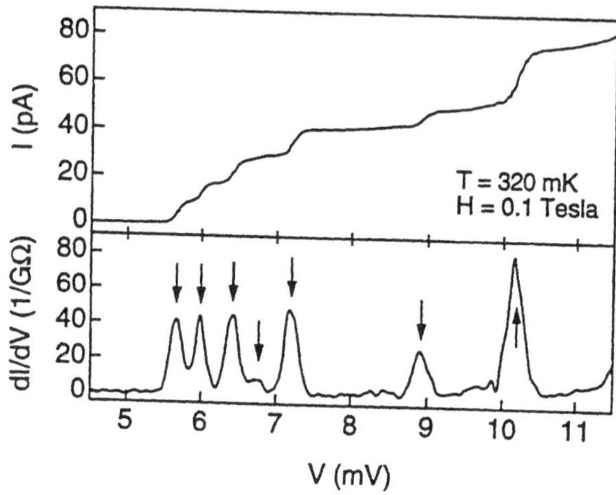

Figure 26.3 $I-V$ and dI/dV curves of an Al-grain device at 320 mK. The $I-V$ curve shows a sequence of irregularly-spaced steps, corresponding to tunneling via discrete electronic states in the nm-scale Al particle. Each step is converted to a peak in the dI/dV curve. The arrows indicate the positions of individual discrete levels. (After Ref. [16])

in the electrode. Experimentally, one finds semi-quantitative agreement with this prediction, except that the resonance widths bottom out at about 50-100 mK, indicating that the electron temperatures do not follow the dilution refrigerator mixing chamber temperature down below such temperatures. Also, the minimum resonance widths were found to increase with increase in the resonance energy, suggesting an additional energy-dependent intrinsic level width from lifetime considerations. We shall return to this issue in Section 6.

When a magnetic field H is applied to the sample, this produces a Zeeman splitting which lifts the spin degeneracy of the orbital levels. In the early experiments on Al grains, the g-value governing the splitting of the tunneling resonances was usually found to be 2 ± 0.05, near the free electron value of 2, as might be expected for a low-Z element for which spin-orbit coupling is weak. In a few samples, however, significantly lower g-values ($\sim$1.8) were found, and these spectra also showed avoided crossings of the positions of resonance peaks corresponding to up and down spins [2]. It appears likely that both these unusual features are due to the presence of a few high-Z impurities in these particular grains.

This idea has been tested recently by Salinas *et al.* [7], who found similar effects in Al grains with a deliberate admixture of ~4% of Au. Much more dramatic departures from $g = 2$ are found in the recent experiments of Davidovic and Tinkham [6] on grains of pure gold, a high-Z element. These experiments show g-values of ~0.3-0.4, indicating that the spin-orbit coupling in Au strongly mixes spin-up and spin-down states in the process of forming a Kramers doublet of time-reversed states which also diagonalizes the Hamiltonian.

It is worth pointing out that one can normally distinguish grains with even and odd numbers of electrons by the nature of the splitting of the tunneling resonances in a magnetic field. If the grain initially contains an even number of electrons, filling paired spin-up/down states, then an additional electron tunneling into the grain can enter any empty higher level with either spin direction, giving resonances that split in a magnetic field. On the other hand, if the number of electrons before tunneling is odd, then the lowest empty state will be the high energy spin orientation of the singly-occupied level. Since this state is uniquely determined, there is no splitting in a magnetic field, only an increase in energy linear in H, proportional to its g-value. This interpretation is based on the assumption that the relevant energy levels can be thought of as those of non-interacting electrons, just filling shells in order of ascending energy. Although this is a useful starting point, it can not readily explain some observations, and a wider range of possibilities will be considered in the next two sections.

6. EFFECTS OF ELECTRON INTERACTIONS: CLUSTERS AND FINITE LEVEL WIDTHS

We now review the experimental observations and theoretical developments which take us beyond the crude picture in which the energy state of the many-electron system is approximated by the occupation of a specified set of quasiparticle eigenstates, with simply additive energies as in basic Fermi liquid theory.

An important step beyond this approach was made by Agam *et al.* [8]. They pointed out the need to take account of cross terms in the Fermi liquid energy expansion. That is, the energy associated with a particular quasiparticle (or Fermi liquid excitation) will depend to some extent upon what other excitations are present as well. Thus, instead of a single energy value associated with tunneling into a quasiparticle state, there will be a cluster of possible values, depending on what other quasiparticle states are excited. This possibility would not arise if one considered only equilibrium states at $T \approx 0$, since only the lowest energy configuration of excitations would be present. But Agam *et al.* pointed out that if successive tunneling events took place more quickly than relaxation from previous events took place, there would be a certain probability of nonequilibrium occupation numbers, and hence of several possible energies for a given quasiparticle excitation. An important consequence of this model in its simplest form is that the lowest tunneling resonance should remain single, because a second tunnel event into a given level could not take place until the level had been vacated by a relaxation process, and the lowest excited level could not be relaxed except to the unique ground state of the island. This model was very successful in accounting for observations made by Ralph *et al.* [2]. Indeed, in many samples the lowest resonance appeared to be single, while higher resonances took the form of clusters of subresonances. Moreover, the cluster concept was able to explain in a natural way why the density of observed resonant peaks (particularly at somewhat higher energies) was considerably larger than the density expected from the classic density of single-particle states in a grain of given size, an observation that had been quite worrisome and puzzling when first made. Essentially, we were seeing eigenenergies of the *system*, not of single particles.

The approach of Agam *et al.*, was essentially a perturbative one, taking account of higher-order Fermi liquid corrections to excitation *energies*, based on interelectronic interactions. Somewhat earlier, Sivan *et al.* [9, 10], had applied a perturbative approach, based on earlier work by Altshuler and Aronov [15] to estimate the level *width* which results from the lifetime limitation due to the interelectronic interactions. Their conclusion was that the level widths should increase with excitation energy E, as expected from general considerations, and become as large as the level spacing when $E \sim E_T$, the Thouless energy. [This energy is of order $\hbar/\tau_T$, where $\tau_T$is the time required for an electron to sample the volume of the grain. For a sample of size L with diffusive transport, this is quite a well defined time $\sim L^2/D$, where D is the diffusion constant. For particles with ballistic transport and diffuse surface scattering, τ_T is less well defined but thought to be similar to the diffusive case, except that the sample dimension would set the effective mean free path.] This result implies that above E_T, the excitations, although still well-defined, are broad enough to blend together into a continuum. Careful analysis of experimental measurements by Sivan *et al.* [9, 10], on semiconductor quantum dots supported this conclusion. For example, an autocorrelation analysis of the tunneling data suggested that only ~10 discrete levels were found for a semiconductor system containing thousands of electrons.

The next major step in developing the theory of quasiparticle lifetimes was a *non*perturbative treatment by Altshuler *et al.* [11]. Their approach was to map the problem of lifetimes limited by electron-electron interactions into a problem of localization in the Fock space of wavefunctions, analogous to the problem of Anderson localization of wavefunctions in coordinate space. From this perspective, localized and delocalized regimes correspond to quasiparticle resonances of zero and finite width, respectively. According to their analysis, the quasiparticle spectrum of a quantum dot should fall into a series of four regimes with increasing excitation energy. The borders of these regimes are most conveniently described in terms of the dimensionless conductance $\mathbf{g} = E_T/\delta$, a parameter whose value is typically only of order 5-10 in these small grains. In terms of this parameter, the quasiparticle states are predicted to be sharp and single up to an energy $\sim (\mathbf{g}/\ln \mathbf{g})^{1/2}\delta$, then sharp but clustered up to $\sim (\mathbf{g})^{1/2}\delta$, then broad but resolvable up to $\sim \mathbf{g}\delta$, and finally forming an unresolved continuum above the Thouless energy $E_T = \mathbf{g}\delta$. This summary does not do justice to the subtlety of the theory. Moreover, these simplified results should be used with caution, since the parameters E_T and $\mathbf{g}$ can only be estimated, either from the data or *a priori* for a (presumably) ballistic grain. Nonetheless, these results provide a framework for interpreting data from tunnel spectroscopy.

The early data of Ralph *et al.*, were in qualitative agreement with the basic prediction that with increasing energy, the tunneling resonances should broaden and show a tendency to clustering and blending into an unresolved continuum. However, since these data were taken before this detailed theory was available, careful measurements were not made over a sufficient range of energies to provide a real test.

This situation has been remedied to a considerable degree by the recent data of Davidovic [6] shown in Fig. 26.4 for three different samples. For samples 1 and 2, the complete progression from resolved narrow resonances to an effectively uniform tunneling density of states is displayed, while for sample 3, one sees the progression from sharp isolated peaks to clusters, but the data do not extend to high enough voltage to reach the continuum. For the first two samples our estimates of the Thouless energy (37 and 75 meV) agree within a factor of 2 with the voltage of the transition to a uniform continuum density of states, while for sample 3 the estimated value (40 meV) is above the range of available data, and hence consistent with it, since the data do not extend far enough to reveal the start of the continuum. Taken all together, these data display rather directly the progression of regimes that are predicted by the theory, but further work will be required to test the more quantitative aspects of the theory by studies of a greater variety of samples.

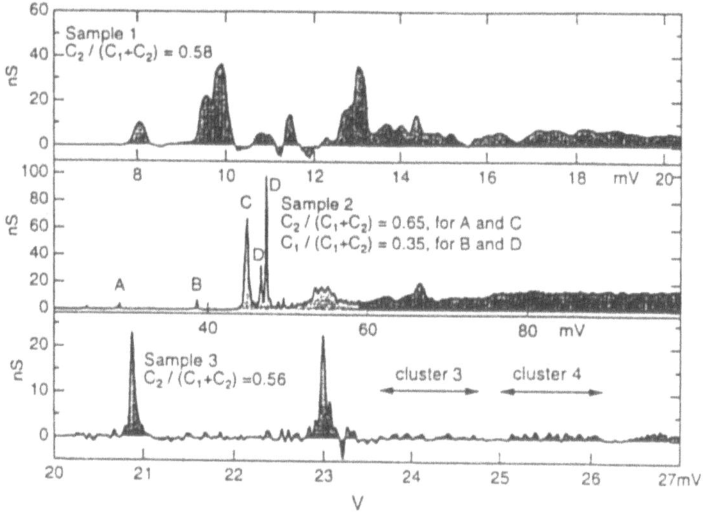

Figure 26.4 Excitation spectra in three different Au samples, at T=30 mK (and H=1 T to suppress superconductivity in the Al leads). In order to convert from voltages to particle eigenenergies, voltages must be multiplied by the appropriate capacitance-ratio factor, indicated for each graph. (After Ref. [6])

The theory basically ignores the Coulomb blockade charging energy, treating it as just a (large) additive constant for all the excited states for any given charge state of the grain. But all tunneling experiments involve a tunneling cycle, where one adds an electron followed by an electron leaving to the other electrode. In this case, the full energy eV ($\gg \delta$) is supplied by the bias source in each cycle, and a substantial fraction of this energy is deposited as hole and electron excitations in the grain, awaiting relaxation. This raises the question of what energy to compare with theory, the total energy supplied (eV), or only the part of this in excess of the Coulomb charging energy. In the above numerical comparison, we have compared the total energy eV at which the continuum is reached with the estimated Thouless energy. If we had subtracted out the Coulomb energy, the comparison would have been much less satisfactory. This result is qualitatively consistent with the earlier work of Ralph *et al.* [3] in which it was found that the tunneling resonances became sharper when a gate electrode was used to minimize the Coulomb blockade voltage by adjusting the offset charge Q_o.

7. EFFECTS OF SHELL STRUCTURE, SPIN-ORBIT, AND EXCHANGE INTERACTIONS

In building up the periodic table of ordinary atoms, one fills the successive shells and subshells in order of ascending energy: 1s, 2s, 2p, 3s, *etc.* until the desired number of electrons are accommodated. If this occurs at a point when all shells either are completely filled or completely empty, this leads to a non-degenerate ground state. However, for all other cases, there will be a partially filled shell, which implies a choice of the specific orbital or spin states that are occupied, all at the same total energy. This degeneracy is lifted by interelectronic interactions, so that a particular many-electron state will be lowest. This is the province of the famous Hund Rules, the first of which stipulates that the state of maximum total spin S that can be formed without elevating any electrons to higher subshells will have the lowest energy. Physically, this is a consequence of the correlation of motion of electrons with parallel spins to avoid each other in space that is enforced by the antisymmetry of the wavefunction under exchange of electrons. This correlated motion acts to reduce their mutual Coulomb repulsion energy. These effects are often referred to as "exchange" energies, and they are of the same order as other interelectronic Coulomb energies. In a metallic quantum dot, we assume that the symmetry is low enough that each orbital has a distinct energy, so that there are no spatial degeneracies to split. Still, if the interelectronic interactions are comparable with, or larger than, the level spacing, these exchange energies may still play a role in controlling the order of filling of system energy levels. For example, they might make a triplet (S=1) combination of two electrons in two different orbital states lie lower than a singlet (S=0) state, even if it requires occupation of a slightly higher energy orbital. This qualitative idea has been carried to a more quantitative level by Brouwer, Oreg, and Halperin [17]. Taking account of the statistical distribution of level spacings, they found that the probability of S=1 exceeds the probability of S=0 even when the ratio of interaction energy to mean level spacing is as small as 0.4. There is some experimental evidence from a proliferation of low-lying excitations that such higher spin states are observed [18]. In other words, although a simple picture of sequential filling of non-interacting single-particle energy levels can explain many spectra, it is not capable of explaining all observed spectra. An even greater deviation from the simple picture was found by Gueron *et al.* [19] in studying nanoparticles of ferromagnetic Co, in which a substantial fraction of the electrons were polarized, as in macroscopic samples. This contrasts with the case considered in [17] in which the net spin is found to be of order unity even in larger grains.

In conventional atoms, where orbital angular momentum L and spin angular momentum S can both be good quantum numbers, the relativistic spin-orbit coupling energy determines the so-called fine structure of the energy levels. These levels have well defined values of the total angular momentum J, as well as of L and S. In a magnetic field, the $(2J+1)$ levels associated with a given J split as described by the familiar Lande g-factor: $g = 1 + [J(J+1) + S(S+1) - L(L+1)]/2J(J+1)$. In a metallic quantum dot with less symmetry, orbital angular momentum is not a good quantum number, and no such simple result is possible. Nonetheless spin-orbit coupling *does* reduce the effective g-value below the free-electron value of 2 by mixing together states with spin up and spin down [20]. If the total angular momentum is half integral, these composite states exist as time-reversed pairs of states, forming a so-called Kramers doublet. In a magnetic field, the energies of these pairs of states split apart just like the spin up and down energies of a free electron, except that the splitting is described by a non-obvious g-value. This g-value can be much less than 2 if the spin-orbit coupling mixing the spin states is strong, as it is for a high-Z element like Au. For example, Davidovic *et al.* [6] find g-values in the range 0.3-0.4 in pure Au grains, while Salinas *et al.* [7] find g-values of $\sim$1.7-1.8 in Al grains containing a few percent of Au. These values are to be compared with the g-values of 1.98$\pm$0.03 found for pure Al grains [3] and the g-value 2.1 reported from spin-resonance measurements on pure bulk Au [21].

Summing up this final section, recent experiments [6, 7] together with the earlier data of Black [16] show that a model of filling successive single-electron orbitals, with alternating spin up and spin down, is too simplistic to account for all the data. Not only does spin-orbit coupling mix up and down spins to reduce the g-value, but interelectronic coupling energies often make the available energy level structure depend on the number of electrons present to an extent that can not be described in the simple scheme based on non-interacting electrons. Clearly more experimental and theoretical work is needed to build a complete understanding of these phenomena.

Acknowledgements

The author is happy to acknowledge the crucial contributions of his coworkers D. C. Ralph, C. T. Black, and D. Davidovic, to the work reported here. The permission to include previously unpublished figures from the Ph.D. thesis of C. T. Black is acknowledged with thanks. The financial support of this research by NSF grants DMR-97-01487, DMR 98-09363, and PHY98-71810, and by ONR grant N00014-96-0108 is gratefully acknowledged.

References

[1] D. C. Ralph, C. T. Black and M. Tinkham, Spectroscopic Measurements of Discrete Electronic States in Single Metal Particles, Phys. Rev. Lett. **74**, 3241-3244 (1995).

[2] D. C. Ralph, C. T. Black and M. Tinkham, Studies of ElectronEnergy Levels in Single Metal Particles, Physica B **218**, 258 (1996).

[3] D.C. Ralph, C. T. Black and M. Tinkham, Gate Voltage Studies of Discrete Electronic States in Al Nanoparticles, Phys. Rev. Lett. **78**, 4087-90 (1997).

[4] C. T. Black, D. C. Ralph and M. Tinkham, Spectroscopy of the Superconducting Gap in Individual Nanometer-Scale Aluminum Particles, Phys. Rev. Lett. **76**, 688-691 (1996).

[5] D. Davidovic and M. Tinkham, Coulomb Blockade and Discrete Energy Levels in Au Nanoparticles, Appl. Phys. Lett. **73**, 3959-3961 (1998).

[6] D. Davidovic and M. Tinkham, Spectroscopy, Interactions, and Level Splitting in Au Nanoparticles, Phys. Rev. Lett. **83**, 1644-1647 (1999).

[7] D. G. Salinas, S. Gueron, D. C. Ralph, C. T. Black and M. Tinkham, Effects of Spin-Orbit Interactions on Tunneling via Discrete Energy Levels in Metal Nanoparticles, Phys. Rev., (1999) to appear.

[8] O. Agam, N. S. Wingreen, B. L. Altshuler, D. C. Ralph and M. Tinkham, Chaos, Interactions, and Nonequilibrium Effects in the Tunneling Resonance Spectra of Ultrasmall Metallic Particles, Phys. Rev. Lett. **78**, 1956-1959 (1997).

[9] U. Sivan, F. P. Milliken, K. Milkove, S. Rishton, Y. Lee, J. M. Hong, V. Boegli, D. Kern and M. DeFranza, Spectroscopy, Electron-Electron Interaction, and Level Statistics in a Disordered Quantum Dot, Europhys. Lett. **25**, 605-611 (1994).

[10] U. Sivan, Y. Imry and A. G. Aronov, Quasi-Particle Lifetime in a Quantum Dot, Europhys. Lett. **28**, 115-120 (1994).

[11] B. L. Altshuler, Y. Gefen, A. Kamenev and L. S. Levitov, Quasiparticle Lifetime in a Finite System: A Nonperturbative Approach, Phys. Rev. Lett. **78**, 2803 (1997).

[12] K. S. Ralls, R. A. Buhrman and R. C. Tiberio, Fabrication of Thin-Film Metal Nanobridges, Appl. Phys. Lett. **55**, 2459-2461 (1989).

[13] D. V. Averin and K. K. Likharev, Single Electronics: A Correlated Transfer of Single Electrons and Cooper Pairs in Systems of Small Tunnel Junctions, in B. L. Altshuler, P. A. Lee, and R. A. Webb (eds.), *Mesoscopic Phenomena in Solids*, (Elsevier, NY, 1991) pp. 173-271.

[14] A. E. Hanna and M. Tinkham, Variation of the Coulomb Staircase in a Two-Junction System by Fractional Electron Charge, Phys. Rev. B **44**, 5919-5922 (1991).

[15] B. L. Altshuler and A. G. Aronov, Damping of One-electron Excitations inMetals, Pis'ma Zh. Eksp. Teor. Fiz. **30**, 514 (1979) [Sov. Phys. JETP Lett. **30**, 482 (1979)].

[16] C. T. Black, Tunneling Spectroscopy of Nanometer-Scale Metal Partaicles, Ph.D. Dissertation, (Physics Dept., Harvard University 1996).

[17] P. W. Brouwer, Y. Oreg and B. I. Halperin, Mesoscopic Fluctuations of the Ground State Spin of a Small Metal Particle, Phys. Rev. B **60**, R13977-R13980 (1999).

[18] D. Davidovic and M. Tinkham, (2000), to be published.

[19] S. Gueron, M. M. Deshmukh, E. B. Myers and D. C. Ralph, Tunneling via Individual Electronic States in Ferromagnetic Nanoparticles, cond-mat./9904248 (1999).

[20] W. P. Halperin, Quantum Size Effect in Metal Particles, Revs. Mod. Phys. **58**, 533-606 (1986).

[21] P. Monod and A. Janossy, Conduction Electron Spin Resonance in Gold, J. Low Temp. Phys. **26**, 311-316 (1977).

VII

QUANTUM COMPUTATION

Chapter 27

QUANTUM COMPUTATION AND SPIN ELECTRONICS

D. P. DiVincenzo
IBM Research Division, T. J. Watson Research Center
PO Box 218, Yorktown Heights, NY 10598 USA

G. Burkard, D. Loss, and E. V. Sukhorukov
Dept. of Physics and Astronomy, University of Basel,
Klingelbergstrasse 82, CH-4056 Basel, Switzerland

Abstract In this chapter we explore the connection between mesoscopic physics and quantum computing. After giving a bibliography providing a general introduction to the subject of quantum information processing, we review the various approaches that are being considered for the experimental implementation of quantum computing and quantum communication in atomic physics, quantum optics, nuclear magnetic resonance, superconductivity, and, especially, normal-electron solid state physics. We discuss five criteria for the realization of a quantum computer and consider the implications that these criteria have for quantum computation using the spin states of single-electron quantum dots. Finally, we consider the transport of quantum information via the motion of individual electrons in mesoscopic structures; specific transport and noise measurements in coupled quantum dot geometries for detecting and characterizing electron-state entanglement are analyzed.

1. BRIEF SURVEY OF THE HISTORY OF QUANTUM COMPUTING

The story of why quantum computing and quantum communication are theoretically interesting and important has been told in innumerable places before, and we will just point the reader to some of those. The "prehistory" of quantum computing (up to 1994) consists of the

I. O. Kulik and R. Ellialtioğlu (eds.),
Quantum Mesoscopic Phenomena and Mesoscopic Devices in Microelectronics, 399–428.

tinkerings of a small number of visionaries on the question of how data could be processed if bits could be put into quantum superpositions of states. The idea of a quantum gate was introduced, the basic possibilities of quantum algorithms were set forth, quantum communication (in the form of quantum cryptography) was well developed, and some rudimentary ideas of how physical systems could be made to implement quantum computing were considered. Actually, it is humbling to reflect on how few people it took (no more than about 15, we would say) to launch a set of basic concepts which have now come to occupy the attention of some hundreds of researchers.

Here is a bibliography of recent and not-so-recent review articles which would bring the reader up to speed on quantum computation and related areas. Those interested in seeing how complete the "prehistoric" perspective was should consult Ekert's influential paper at Atomic Physics 14 [1], which helped to make atomic physicists the earliest participants in the attempt to build a quantum computer. Another early but influential review which is a very good source for the Shor algorithm is [2]. An early general overview is [3], and an early simple discussion of quantum gate constructions is [4]. More recent general overviews are [5] and [6]. [7] and [8] give a set of ideas for the future direction of quantum communication and information theory. Many surveys have considered the question of how quantum computers might be implemented in actual physical devices; for the solid-state type of implementations which we will discuss below, we might mention [9] besides some of the other articles already cited.

2. CREATING THE QUANTUM COMPUTER

Of course, the reason for this contribution to appear in a school on mesoscopic quantum phenomena is that we hope, ultimately, that out of mesoscopic physics will emerge the capability actually to build a working, scalable quantum computer. Microelectronic mesoscopic devices will inevitably emerge, of course, but whether they will be usable for quantum computation will depend on whether they manage to satisfy a very specific set of requirements.

2.1 Five criteria for building a quantum computer

These requirements can be boiled down to a list of five criteria. We have written frequently about these five criteria before [8, 9], but they are worth considering in detail here again, since we have found that these criteria provide a unifying framework for these investigations which encompass an astonishingly broad and rich range of fundamental physics.

Further investigations have endowed this simple list of five with more and more richness and interest, and show how exciting the building of a quantum computer will be from the point of view of novel and basic physics.

Before getting into the serious work, we find it amusing to state the criteria stripped of all physics language and posed purely as a set of requirements for building a computer. In this light our five points are extremely trivial:

1. The machine should have a collection of bits.

2. It should be possible to set all the memory bits to 0 before the start of each computation.

3. The error rate should be sufficiently low.

4. It must be possible to perform elementary logic operations between pairs of bits.

5. Reliable output of the final result should be possible.

It has to be admitted that no great brainwork is required to arrive at this set of basic attributes which a computer ought to possess. But let us take the next step and translate these into a physics language, to specify what it really takes to achieve these very basic and simple requirements in a quantum setting:

1. A physical system with a collection of well characterized quantum two-level systems (qubits) is needed. Each qubit should be separately identifiable and externally addressable. The dynamics of the system should be under sufficient control that these qubits are never excited into any third level. It should be possible to add qubits at will.

2. It should be possible to, with high accuracy, completely decouple the qubits from one another, and it should be possible to start an experiment by placing each qubit in its lower (0) state.

3. The decoherence time of these qubits should be long, ultimately up to 10^4 times longer than the "clock time" (see the next requirement).

4. Logic operations should be doable. This involves having the one-body Hamiltonian of each qubit under independent and precise control (this gives the one-bit gates). Two-body Hamiltonians involving nearby qubits should also be capable of being turned on

and off under external control (these are the two-bit gates). In a typical operation, the one- or two-body Hamiltonian will be turned on smoothly from zero to some value and then turned off again, all within one clock cycle; the integral of this pulse should be controllable to again about one part in 10^4.

5. Projective quantum measurements on the qubits must be doable. It is useful, but not absolutely necessary, for these measurements to be doable fast, within a few clock cycles. It is also useful, but not necessary, for the measurement to have high quantum efficiency (say 50%). If the quantum efficiency is many orders of magnitude lower, then the quantum computation must be done in an "ensemble" style, in which many identical quantum computers are running simultaneously. As with all the other requirements, these measurements absolutely must be qubit specific.

The reader will notice that there is a lot more to be said about these criteria as physics requirements than as computer requirements. Indeed, we will see in the examples we review below that these criteria involve not just innovations in materials preparation (for numbers 1 and 3), not just novel device design and fabrication (for number 4), not just new, precision high-speed electronics at the nanometer scale (number 4), not just unprecedented capabilities for ultrasensitive metrology (number 5), but all of these at once!

2.2 Potential realizations

There is a remarkably long list of physical systems that have been proposed, and are under active experimental investigation, for the creation of a quantum computer. This list is remarkable not only for its length but also for its diversity; it seems that virtually every area of quantum physics has a candidate. (This is not quite true: we know of no proposed quantum computer emerging from high-energy particle physics.)

Since non-solid state implementations are not the focus of this article, we will merely mention the many approaches in this category and give a bibliography: In atomic physics, the original proposal by Cirac and Zoller [10] considered the internal states of ions as the qubits, coupled by the common vibrational mode of the ions in the trap. Later this group [11] introduced a variant in which the coupling is provided by a common cavity electromagnetic mode. This work has inspired several analogous solid state proposals which we will mention below. Many other variants of the cavity quantum electrodynamics schemes have been proposed; in some of these the qubit could be transmitted from one subprocessor to another as a single photon (in an optical fiber, say) [12]; these schemes are intriguing in that they offer an idea of how to do distributed quantum computing and distributed decision problems.

A very recent development from atomic physics, neutral-atom optical lattices, has been proposed to provide the necessary tools to build a quantum computer [13]. Here, the ability to move trapped atoms in a state-dependent way by changing the (classical) phase of the trapping laser beams is exploited to obtain the necessary conditional dynamics to do two-bit gates. There are many ideas for mixing and matching these various atomic-physics proposals.

Another large area of research has involved bulk NMR of small molecules containing sets of spin-1/2 nuclei in solution [14, 15]. This represents an "ensemble" approach, exploiting the ability to read out the result of quantum computation with only low quantum efficiency measurements (criterion 5 above). Unfortunately these schemes presently do not satisfy criterion 2 above, but in other respects it represents a possible approach to a scalable quantum computer.

An "almost" solid state approach, which we will mention just because it illustrates that quantum computing proposals are coming from just about every area of physics these days, is one proposed by Platzman and coworkers [16] involving two dimensional crystalline layers of electrons that can be trapped near the surface of liquid He. We are not knowledgeable enough to comment on this proposal in any detail (it seems to involve elements of the quantum dot proposals we will discuss in a moment, plus some ideas from the atomic physics proposals), but it is interesting to see contact made with another, evidently quite well developed, area of research in quantum physics.

3. SOLID STATE PROPOSALS

There have been almost as many proposals for solid state implementations of quantum computers as all the other proposals put together. We think there are some clear reasons for this: we believe that solid state physics is the most versatile branch of physics, in that almost any phenomenon possible in physics can be embodied in a correctly designed condensed matter system. Even rather esoteric properties of field theories which are of interest in high-energy physics have useful realizations, for example, in the fractional quantum Hall effect. A related reason is that solid state physics, being so closely allied with computer technology, has exhibited great versatility over the years in the creation of artificial structures and devices which are tailored to show a great variety of desired physical effects. This has been exploited very powerfully to produce ever more capable computational devices. It would be natural to extrapolate to say that this versatility will extend to the creation of solid state quantum computers as well; the plethora of proposals would indicate that this is indeed true, although time will tell whether any of these proposals will actually provide a successful route to a quantum computer!

3.1 The spintronic quantum dot proposal

We will first go through, in some detail, the proposals that we [17] have made for the use of coupled quantum dot arrays for quantum computation; we will also discuss several closely related proposals such as the "Kane model" [18, 19]. Our proposal may rightly be termed a "spintronic" proposal, as it attempts to use as much as possible the tools of standard electronics (control via voltage gates, readout by current detection) to accomplish quantum computing; but our proposal departs from conventional electronics in that it uses the *spin* of the electrons as the basic information carrier.

We think that **Criterion 1** above is most naturally satisfied by a genuine spin-1/2 system, which by its nature has a doubly-degenerate ground state to serve as a qubit. As we will discuss, the prospects for long coherence times are better for this qubit than for many other "pseudo-spin" degrees of freedom that might be chosen. (But we will mention several other possible ones being considered at the end of this section.)

We have to specify how one obtains the individually identifiable and addressable qubits of Criterion 1. Our scheme for doing this involves an extension of well-demonstrated techniques in the area of mesoscopic quantum-dot physics. In particular, we are interested in quantum dots created by lateral confinement in a two-dimensional electron gas. The structure is illustrated in Fig. 27.1. To have qubits in this structure (one per quantum dot) using the electron spin, it is necessary for the electron number to be controlled precisely; this is accomplished using the well-documented Coulomb blockade effect. The electron number must be fixed at an odd value in order for the spin quantum number to be 1/2. This can be achieved if the electron number can be controllably tuned down to one (technically difficult in structures in which the Coulomb blockade is probed by transport measurements, quite feasible in structures where the Coulomb blockade is sensed capacitively); alternatively, a consensus seems to be developing that a dot with a spin quantum number of 1/2 gives a characteristic zero-bias anomaly in transport which signals a Kondo effect. This approach will have to be used with caution, however–the spin quantum number of a dot may readily switch from 1/2

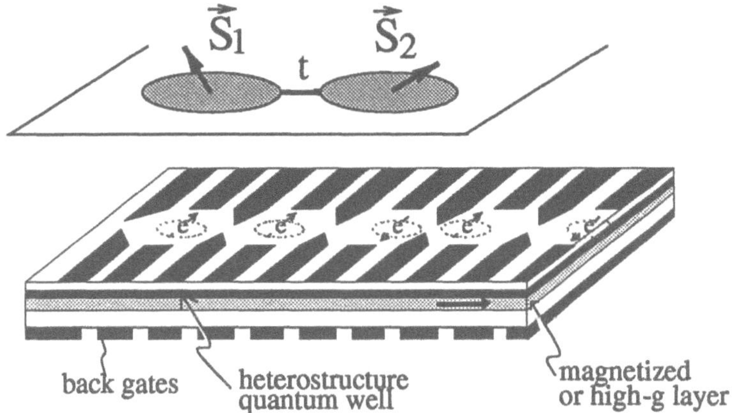

Figure 27.1 Above: Two coupled quantum dots, between which electrons can tunnel with amplitude t. This tunneling leads to an effective spin interaction $J \sim t^2/U$ (where U is the on-site Coulomb repulsion) between the excess spins $\mathbf{S}_1$ and $\mathbf{S}_2$ in the dots, which can be controlled by a number of external parameters. In principle, any type of tunnel-coupled confined structure is a candidate for the "spintronics" quantum computing proposal. *E.g.*, the dots can be defined electrically in a two-dimensional electron gas (as suggested by this drawing), or they can be vertically coupled dots, or even atoms, *etc.* Below: Concept for quantum-dot array device.

to 3/2 as a function of the dot shape [20]; it is of course essential that this switching does not occur between the configuration in which a qubit is characterized and that in which it is used.

Criterion 1 also requires that the next higher lying state should never (or hardly ever) be occupied; this requires that dots be sufficiently small that the next level is inaccessible to thermal excitation or non-adiabatic excitation. It appears that this requirement will be quite readily satisfied for dots of the 10s of nanometer sizes that are already studied, at normal (l-He) cryogenic temperatures. As for extendibility, as our Fig. 27.1 suggests this is "simply" a matter of making an array of dots, either one or two dimensional. Of course, this cartoon does not do justice to the actual complexity of the multilevel wiring layout required to address a large array of qubits; we hope we are justified in considering this an "engineering" matter, although by no means a trivial one.

In the other spin-qubit scheme we have considered [19], the "quantum dot" is provided by a single donor impurity atom lying near the surface of a Si-Ge heterostructure. Such a system is a kind of natural one-electron quantum dot; the hydrogenic impurity potential can normally only bind a single conduction-band electron (sometimes it can bind two: see Criterion 5 below). The characteristics of this qubit (its g-factor, orbital size, and so forth) are more nearly unique than in the lateral quantum dot proposal; this could ultimately be either an a advantage or a disadvantage. Also, placing individual impurities at specified locations near the surface of a semiconductor is a very daunting technical challenge, although perhaps one that the technological world will be needing to face in any case in the coming nanoelectronic age. In general, Criterion 1 poses very serious challenges in the art of material preparation and device fabrication, as our discussion should make clear.

Criterion 2 is a relatively straightforward one, but involves additional device-fabrication considerations. In order for the quantum-dot or donor-impurity qubit to start in a stable all-0 state, it is sufficient to place the spins in a several-Tesla magnetic field at liquid-He temperatures; then the probability of occupying the all-down ground state is sufficiently high. In order for this state to be stable upon removal of the magnetic field, we require a probably more difficult requirement: the qubits must be, to rather high accuracy (to be discussed further in the next two criteria), decoupled from one another. The device geometry is chosen so that, in the "resting" state, the "point contact" voltage probes (the pairs of electrodes between neighboring quantum dots) can be set to a high repulsive voltage so that the overlap between neighboring quantum-dot orbitals is negligible. Turning this voltage to low can increase this overlap by orders of magnitude, and this is an essential computational step (see Criterion 4).

In the donor-impurity scheme, top electrodes control the near-surface band bending of the semiconductor; in the flat-band condition, the donor hydrogenic orbitals are compact enough that the overlap between neighbors is negligible. By suitable band bending, the electron orbitals can be made much more delocalized, leading to orders of magnitude more spin-spin interaction.

Criterion 3 probably involves the must fundamental quantum physics, in that it relates to the issues of the decoherence of quantum systems and the transition between quantum and classical behavior. Of course, a lot of attention has been devoted in fundamental mesoscopics research to characterizing and understanding the decoherence of electrons in small structures. We remind the reader, however, that most of what has been probed (say in weak localization studies or the Aharonov-Bohm effect) is the *orbital* coherence of electron states, that is, the preservation of the relative phase of superpositions of spatial states of the electron (*e.g.*, in the upper or lower arm of an Aharonov-Bohm ring). The coherence times seen in these investigations are almost completely irrelevant to the *spin* coherence times which are important in our quantum computer proposal. There is some relation between the two if there are strong spin orbit effects, but our intention is that conditions and materials should be chosen such that these effects are weak.

Under these circumstances the spin coherence times (the time over which the phase of a superposition of spin-up and spin-down states is preserved) can be completely different from the charge coherence times, and in fact it is known that they can be orders of magnitude longer. This was actually one of our prime motivations for proposing spin rather than charge as the qubit in these structures. The experimental measurement of this kind of coherence is not so familiar in mesoscopic physics, but fortunately it is very familiar in the area of spectroscopy. The measurement to probe spin coherence is essentially equivalent of the characterization of the so-called T_2 time in spin resonance.

In fact, we suggest that the quantification of spin coherence in the quantum dot quantum computer will follow a line very familiar from the many recent experiments of Awschalom, Kikkawa, and collaborators [21] that have observed spin precession in a variety of semiconductor materials. We expect that these experiments will set up a large array of identical quantum dots; the dots don't have to be fully "wired up" for quantum computation, but they should otherwise have the same device structure as in the quantum computer (because we expect that the coherence times should be very device- and structure-specific). The spins will be set in a superposition of up and down in a magnetic field, and the decay of free induction as these spins precess in the field will be observed. Conventional "spin echo" tricks should be used to eliminate the effects of residual inhomogeneities. Note that this measurement only requires low quantum efficiency, so it is not nearly so difficult as the quantum measurements which we will discuss in Criterion 5.

Awschalom and coworkers have already done many such measurements on a variety of semiconductor systems, and sees decoherence times up to hundreds of nanoseconds in some structures [21]. For donor impurities in silicon, traditional electron spin resonance measurements have seen T_2 times for the P donor spin up to hundreds of microseconds. As we see decoherence times in the range from microseconds to milliseconds as acceptable for quantum computation, these results are very encouraging; but they are not conclusive. Decoherence times, depending on the details of the quantum degrees of freedom in the environment that the qubit can become entangled with, are expected to be very sensitive to the details of the makeup of the physical device. Since none of the experiments have been done on an actual quantum computing structure as we envision it, the existing results cannot be viewed as conclusive. Because of this sensitivity to details, theory can only give general guidance about the mechanisms and dependencies to be looked for, but cannot make reliable *a priori* predictions of the decoherence times.

In fact there are further complications in store: we know theoretically that decoherence is not actually fully characterized by a single rate; in fact, a whole set of numbers is needed to fully characterize the decoherence process (12 in principle for individual qubits), and no experiment has been set up yet to completely measure this space of parameters, although the theory of these measurements is available. Even worse, decoherence effects will in principle be modified by the act of performing quantum computation (during gate operation, decoherence is occurring in a coupled qubit system [17]). We believe that the full characterization of decoherence will involve ongoing iteration between theory and experiment, and will probably be inseparable from the act of building a reliable quantum computer.

Finally, the decoherence time τ_ϕ by itself is not a figure of merit of a quantum computer proposal — the amount of coherent computation which can be performed depends on the ratio τ_ϕ/τ_s where τ_s denotes the switching time (τ_s^{-1} is the "clock frequency" of the quantum computer).

Criterion 4 also requires an extensive discussion of the physics of the proposed qubit device. It is necessary to identify simple, reliable mechanisms by which specified qubits can be subjected to one-body (one-bit gates) and two-body (two-bit gates) Hamiltonians which can be turned on and off in time. Almost all the complexity of the proposed devices arises from the need to achieve these capabilities.

In the quantum dot array structure, the two-bit gates are obtained by a controlled lowering of the potential barrier produced by the "point contact" gates between neighboring quantum dots. When this barrier is lowered, the two electrons are brought together, forming, temporarily, an artificial hydrogen molecule. The effective spin-spin interaction that this produces should, to high accuracy, be given by a Heisenberg interaction $J\vec{S}_i \cdot \vec{S}_{i+1}$, where the exchange coupling should be tunable up to about 0.1 meV [22], with the "off" value being many orders of magnitude lower than this. The exponentially strong suppression of the coupling J in the "off" state of the gate is essential, because it assures that no correlated errors occur due to the switching mechanism. Corrections to the Heisenberg form should arise only from relativistic effects (spin-orbit coupling); although spin-orbit corrections to the band parameters like the g-factor are fairly large in GaAs, the residual relativistic effects on the low-angular momentum effective-mass conduction band states can be estimated to be negligibly small [9](c).

For laterally tunnel-coupled quantum dots in a two-dimensional electron system it was found that besides electrical gating, an external magnetic field can be used to switch on and off the spin-spin interaction [22]. Recently, there has been great interest in vertically tunnel-coupled dot structures, both in etched vertical columnar heterostructures [23] and in double-layer self-assembled quantum-dot structures [24]. We have analyzed the spin-interaction in such vertically coupled quantum dots [25], and have found, in addition to the ones known from laterally coupled dots, a new mechanism which allows the external control of the spin interaction. In this scenario, two coupled quantum dots of different size are subject to an external electric field. The field shifts the big dot by a larger distance than the small dot, therefore leading to an effective increase in the inter-dot distance, which causes an exponential suppression of the spin exchange coupling.

The proposed two-bit gate mechanism in the donor-impurity quantum computer is conceptually almost identical to the proposals above. By weakening the binding of the electron to the impurity by band bending, the orbitals can be made to spread out so that the overlap of neighboring orbitals becomes appreciable; the exchange physics, and the expected effective Hamiltonian, is the same as the quantum dot case.

It is known that universal, fault tolerant quantum computation can be obtained with the logic gates obtainable from this nearest neighbor exchange interaction [26]. For dots of tens of nanometer size, a smooth (*i.e.* adiabatic) turning on and off the Hamiltonian on a time scale of 10s or 100s of picoseconds would be desirable. (The smoothness is required so that higher-lying states of the dot are not unintentionally excited [22].) This is technically feasible, although it requires very high-bandwidth control (10-100 GHz) of the voltages on each individual electrode (Fig. 27.1) of the structure, which is a daunting job of microwave engineering.

Especially daunting is the theoretical precision requirement, which is that the integrated strength of the exchange Hamiltonian should be controlled to about one part in 10^4. This number, based on the analysis of the efficacy of error correction techniques in quantum computation, may come down as better error correction schemes are devised. Also, if this number were "only" 10^2 in an experiment, many interesting studies of quantum computation could still be performed. As an example, we have described a minimal experimental test for quantum error correction involving (at least) three coupled quantum dots [27]. Time will tell what the ultimate technical requirements will be. A fortunate fact about GaAs quantum dot structures, holding out the hope that precision manipulation will be possible, is the observation that, when treated correctly, quantum dot structures show essentially no "charge switching" effects which are the origin of traditional $1/f$ noise in transport

[28]. Note that uncontrolled charge switching is not nearly so great a problem for spin qubits as for charge qubits, since this switching does not couple directly to the spin degree of freedom. But, since the second order effects of charge motion could change the strength of the exchange coupling J by more than a part in 10^{-4}, the ability to suppress $1/f$ effects will be very important for switching in quantum computation.

One-qubit gates must also be considered, and these involve rather different physics. Theoretically the requirement is very simple for a spin-1/2 qubit: it must be possible to subject a specified qubit to a (real or effective) magnetic field of specified direction and strength. We have offered many suggestions previously [17, 22] on how this requirement may be met: by the application of real, localized magnetic fields using a scanned magnetic particle or nanoscale electric currents; by the use of a magnetized dot or magnetized barrier material that the electron can be inserted in and out of by electric gating; by the judicious choice of g-factor-modulated materials [8]. We have performed some detailed analysis of this last mechanism recently, and it is also the preferred mechanism in the Si-Ge heterostructure scheme (Si and Ge have very different g-factors), so we will discuss this recent work here.

Due to spin-orbit coupling, the Landé g-factor in bulk semiconductor materials differs from the free-electron value $g_0 = 2.0023$ and ranges from large negative to large positive numbers for various materials. In confined structures such as quantum wells, wires, and dots, the g-factor is modified with respect to the bulk material and sensitive to an external bias voltage [29]. Here, we study the simpler case of a layered structure in which the effective g-factor of electrons is varied by electrically shifting their equilibrium position from one layer (with g-factor g_1) to another (with another g-factor $g_2 \neq g_1$). For simplicity, we use the bulk g-factors of the layer materials, an approximation which becomes increasingly inaccurate as the layers become thinner [30].

We consider a quantum well (*e.g.* AlGaAs-GaAs-AlGaAs), in which some fraction y of the Ga atoms are replaced by In atoms in the upper half of the heterostructure (we have used $y = 0.1$). The sequence of layers in the heterostructure is then $Ga_{1-x}Al_xAs$-GaAs-$Ga_{1-y}In_yAs$-$Ga_{1-x-y}Al_xIn_yAs$, where x denotes the Al content in the barriers (typically around 30%). Changing the vertical position of the electrons in the quantum well via top or back gates permits control of the effective g-factor for the corresponding electrons: If the electron is mostly in a pure GaAs environment, then its effective g-factor will be around the GaAs bulk value ($g_{\mathrm{GaAs}} = -0.44$) whereas if the electron is in the InGaAs region, the g-factor will be somewhere between the GaAs and the InAs values ($g_{\mathrm{InAs}} = -15$). We have analyzed the problem of a single electron in such a structure, neglecting screening due to surrounding electrons. This procedure is justified, since we are interested in isolated electrons located in quantum dots. In a quantum well with a high electron density, however, many-body effects should be taken into account.

We have solved the one-dimensional problem,

$$\left[-\frac{d}{dz}\frac{\hbar^2}{2m(z)}\frac{d}{dz} + V(z)\right]\Psi(z) = E\Psi(z), \qquad (27.1)$$

of a single electron with a spatially varying effective mass $m(z)$ numerically by discretizing it in real space and subsequently performing exact diagonalization. The potential $V(z)$ describes the quantum well (conduction band offset ΔE_c=270 meV) and the electric field E in growth direction. For the effective masses and g-factors of the various layers we have linearly interpolated between the GaAs, AlAs, and InAs values. The resulting effective g-factor was calculated by averaging the g-factor over the electronic ground-state wavefunction,

$$g_{\text{eff}} = \int dz g(z)|\Psi(z)|^2. \qquad (27.2)$$

In Fig. 27.2, we plot the effective g-factor g_{eff} versus the electric field for a quantum well which is w=10 nm wide. For the barrier thickness we have assumed w_B=10 nm. At moderate electric fields, g_{eff} interpolates roughly between the GaAs and $Ga_{1-y}In_yAs$ g-factors. If the electric energy $eEw_B = eU_B$ becomes larger than the barrier ΔE_c, we observe a vertical deconfinement of the electrons. In our plot (Fig. 27.2) the electric deconfinement is clearly seen as a jump of the effective g-factor to the barrier material value at $E = \pm\Delta E_c/ew_B = \pm 27$ mV/nm. The electric field required for a substantial change in g_{eff} is of the order of 10 mV/nm, corresponding to a voltage of 100 mV, which is about one order

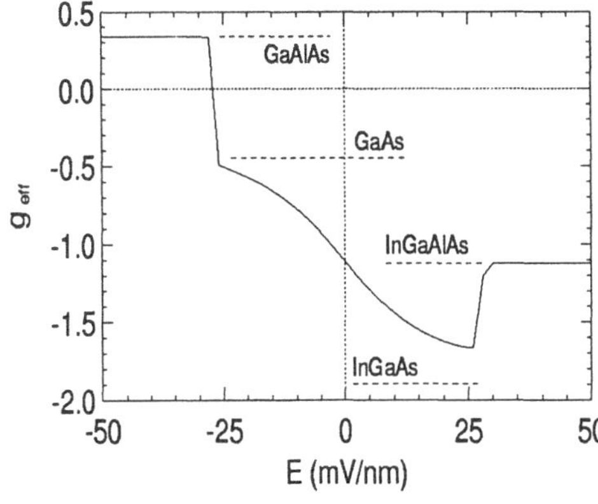

Figure 27.2 The effective g-factor g_{eff} of electrons confined in a $Ga_{1-x}Al_xAs$ - GaAs - $Ga_{1-y}In_yAs$ - $Ga_{1-x-y}Al_xIn_yAs$ heterostructure (x=0.3, y=0.1) as a function of the applied electric field E in growth direction. The widths of the quantum well and the barriers are $w = w_B$=10 nm. The g-factors which are used for the materials are indicated with horizontal lines.

of magnitude smaller than the band gap (1.5 eV for GaAs at T=0).

Since the g-factor is spatially varying, the Zeeman coupling influences the electron wavefunction, which in principle could lead to a non-linear spin splitting $\Delta E(B)$. For the above materials and parameters however, we find numerically that the splitting is almost exactly linear, $\Delta E(B) \simeq g_{\text{eff}}\mu_B B$. The irrelevance of the Zeeman coupling for the orbital wavefunction is due to the fact that the typical electronic kinetic energy is at least two orders of magnitude larger than the typical Zeeman energy.

The described quantum well can host an array of electrostatically defined quantum dots, containing a single excess electron (and thus a single spin 1/2) each. In order to carry out a single-qubit operation on one of the spins, the whole system is placed into a homogeneous magnetic field. By changing the voltage at the electric gate on top of a single quantum dot, the effective g-factor g_{eff} for the spin in this quantum dot can be changed by about $\Delta g_{\text{eff}} \approx 1$ with respect to the g-factor of all remaining spins. This leads to a relative rotation about the direction of **B** by an angle of roughly $\phi = \Delta g_{\text{eff}}\mu_B B\tau/2\hbar$. The typical switching time τ for a $\phi = \pi/2$ rotation using a field of 1 T is then approximately $\tau \approx 2\phi\hbar/\Delta g_{\text{eff}}\mu_B B \approx 30\,\text{ps}$. Controlling the top gate at $\tau^{-1} \approx 30\,\text{GHz}$ seems very challenging; we emphasize however that the single-qubit operation can be done much more slowly (a lower limit is provided by the spin dephasing time). The switching can be slowed down either by choosing a smaller Δg_{eff} or by replacing ϕ by $\phi + 2\pi n$ where n is an integer.

Actually, we should note that there is a substantial degree of flexibility in how universal quantum computation is achieved. We have noted in our original work [17] that switchable effective magnetic fields on the dots are not needed for the implementation of one bit gates, if there are some dots which have a higher static magnetic field, either because they are magnetized or because of the presence of a fixed magnetic field gradient. Then, one-qubit operations can be effected by swapping the qubit onto the magnetized dot, then swapping if off again once the desired interaction with the magnetic field has occurred.

In yet another variant along these lines, Bacon *et al.* [31] have very recently shown that, at the price of increasing the number of quantum-dot spins required to represent each qubit, all computation can be done with exchange interactions alone, without the need for any local magnetic fields, except during the measurement operation. These workers define a logical qubit as the two-level system of the singlet states of four spins, in which

$$\begin{aligned} |0_L\rangle &= |S\rangle \otimes |S\rangle, \\ |1_L\rangle &= \frac{1}{\sqrt{3}}[|T_+\rangle \otimes |T_-\rangle - |T_0\rangle \otimes |T_0\rangle + |T_-\rangle \otimes |T_+\rangle]. \end{aligned} \tag{27.3}$$

Here $|S\rangle$ is the singlet state of two spins and $|T_{+,-,0}\rangle$ are the three triplet states of two spins. The initial preparation of $|0_L\rangle$ is easy; introduce a strong exchange interaction between pairs of spins (*e.g.*, on the "even" bonds but not on the "odd" bonds of a one-dimensional chain); the ground state of this Hamiltonian is the desired state. Ref. [31] shows that all necessary one- and two-qubit gates between these logical qubits can be done by sequences of exchange interactions only. (More work should be done to make these sequences explicit and short.) This is possible because the exchange interactions can move the quantum state vector anywhere within the total singlet (total spin=0) sector. Ref. [31] also notes that the measurement of the logical qubit (anticipating Criterion 5 below) can be done if the first two spins in the four-spin block are measured in the z bases and the next two are measured in the x basis.

We have recently noted that even this potentially inconvenient requirement for two different measurement bases can be eliminated. Only z measurements are needed, if it is assumed that other spins, initialized in the $|0\rangle$ state (*not* $|0_L\rangle$), are also available, which is possible if some spins are initially cooled in a uniform external field without being exchange-coupled to their neighbors. Then, the measurement protocol of Fig. 27.3 suffices.

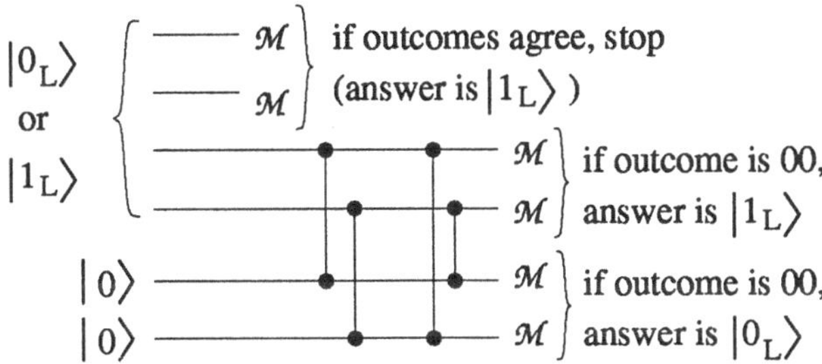

Figure 27.3 Two ancilla spins set to the $|0\rangle$ state suffice to measure the logical qubit $|0_L, 1_L\rangle$. First, the first two spins of the logical qubit are measured in the z basis. If the two measurement outcomes agree, then the procedure is finished, the outcome is $|1_L\rangle$. Otherwise, the ancilla must be used; first, the four two-bit gates are performed in the sequence indicated. The gate is the square-root-of-swap; see [17, 27]. After this circuit, all four remaining spins are measured in the z basis. If the top two measurements give 00, the outcome is $|1_L\rangle$; if the bottom two measurements give 00, the outcome is $|0_L\rangle$. No other measurement results are possible.

This little digression about coded qubits illustrates a more general theme here: the theory indicates that many tradeoffs are possible, in this case between the number of spins needed and the complexity of the required gate operations. Each specific experimental case needs to be carefully assessed to see what tradeoffs are possible, and how they can be optimized.

We think that **Criterion 5** hold the prospect for some early successes in this long road we have laid out to achieving a quantum computer, successes that will advance some fundamentally interesting agendas in solid state physics. This criterion can be simply posed: how do we measure the state of the spin of a single electron in a solid? Actually there is a lot of flexibility in this requirement, in that high quantum efficiency is not in principle needed: if the quantum efficiency is 1%, say, then the computer can still give a reliable output if the qubit to be read is first copied, say, 200 times, then each of these qubits is subjected to this low-efficiency measurement. (Note that this does not violate the no-cloning theorem, because the extra qubits act as "copies" only in the basis specified for the measurement). But there are various reasons why high quantum efficiency is desirable: obviously, the number of extra qubits needed is less, but also, it is desirable for error correction (although not absolutely necessary) that the whole measurement be complete, and the measurement outcome available, in a time much shorter (by a factor of 100, say) than the decoherence time.

Indeed, we think that there is a real prospect that such fast, high-sensitivity measurements can be achieved for spins. This is so despite the fact that direct, magnetometric measurements of the spin via its magnetic moment have, after many years of effort [32], so far failed to achieve anything close to single-spin sensitivity; it is for this reason that we are pessimistic about a magnetic-force-microscope based quantum computer [33]. We feel that a very promising general strategy involves first converting the spin degree of freedom of the electron, which is very hard to measure, into a charge degree of freedom, which is measurable at the single-electron level by well-established electrometric techniques. There are several possible ways that this could be achieved; we have previously discussed [17, 34] a scheme in which a partially transparent barrier (say, the bottom wall of the quantum dot) has a barrier height which is spin-dependent, either because of a large g-factor or an exchange splitting of the barrier. Then, the measurement would consist of applying a bias voltage to the dot so that the electron is pressed up against this barrier; if the spin is up, say, the barrier height is low, so the electron has a high chance of tunneling and being detected by an electrometer circuit underneath; while if the spin is down, the barrier height is high, tunneling does not take place, and the electron is not detected.

In the donor-impurity scheme we can use a similar approach, which uses some interesting physics of the donor impurity that was pointed out by Kane [18]. The phosphorus impurity in Si has a stable singly charged state (two electron state), but only if the relative spin of the two electrons is a singlet. If we wish to measure whether the spins on two neighboring Ps are in a singlet or triplet (not identical to a single spin measurement, but almost equally effective in satisfying Criterion 5), a bias voltage is applied between the two Ps so that the electron will tunnel from one P to another if a final state is available (the singlet) but not otherwise. Then, again, electrometry can be used to determine whether the P has become charged or not.

3.2 Superconducting Qubits

We will not review the superconducting proposals in detail, but we will give a brief idea of how the five criteria are to be satisfied and indicate to the reader where he can find more information on these developments. The proposals fall into two broad classes: ones in which qubits involve the charge degree of freedom of the superconductor, and ones where the qubits are embodied by the flux states in a superconducting structure.

The charge models [35, 36] use as their basic structure a pair of small superconducting islands separated by a Josephson junction. The isolation of the islands causes the Cooper-pair number on the islands to be almost a good quantum number; a qubit can be defined as one in which a Cooper pair is resident on either the left or the right island. Changing the voltages on the islands and, possibly, the strength of the effective Josephson coupling, provide, in principle, sufficient control to do one-bit gates, although it is not clear whether the necessary parameters can be controlled with sufficient precision, and whether "$1/f$" phenomena (random switching of impurities near the device) can be adequately suppressed. Two-bit gate action is obtained by electrostatic coupling between islands [35] or coupling via the quantum states of an LC circuit [36]. This coupling is not formally scalable to large numbers of qubits, but it would be feasible for modest (10) numbers of qubits. Readout would be by single-electron electrometry, which is reasonably well understood. The standard phenomenological model of Josephson circuits suggest that the decoherence rates will be low enough that gate operations can be performed.

The flux models come in a number of varieties [37, 38]; the approach favored by the experimentalists [11] involves a low-Tc (Al or Nb) SQUID circuit which, classically, has two degenerate low energy states which differ slightly in their flux configuration; from the quantum point of view, this is a double-well potential problem with two nearly degenerate ground states, which functions as the qubit. The observation of single-qubit control is then equivalent to the "MQC" phenomenon which has been sought in these structures for many years. Controlled inductive coupling between SQUIDs is proposed for two-qubit gate action. Again, solutions for the five requirements are all proposed; it is possibly worrisome that "extrinsic" effects resulting from the large-scale motion of magnetic flux which might cause decoherence, *e.g.*, coupling to stray spins in the substrate, have apparently not been evaluated.

3.3 Optical

We would finally like to give a very brief mention of another general line of attack that has some promise for solving some of the problems contained in the five criteria: the use of solid-state optical physics. We have not studied these schemes in great detail, but it is clear that they have the prospect of providing some of the best solutions to the problems, for example, of single-quantum measurement. It is also clear that optical expertise is of great value in the characterization of the quantum behavior of spins; all the recent determinations of decoherence times of spins in semiconductors [21] were performed using optical techniques.

So far, many of these proposals are not fully worked out; for the most part, they conceptually follow the quantum-optics proposal involving trapped atoms in a small optical cavity of Pellizzari *et al.* [11]. For example, the proposal of Brun and Wang [40] uses a very familiar object from quantum optics, the whispering-gallery optical modes of a silica microsphere, to couple quantum dots. Other optical cavities with reasonably high Q are available in solid state physics: in the proposal of Imamoglu and coworkers [41], the microdisk optical cavity, a common structure in laser research, is studied; in this structure the optical cavity is formed by a pair of circular mirrors created by the deposition of multiple layers of different III-V semiconductor materials (typically) with different dielectric constants. The mode volume is typically also occupied by another III-V semiconducting material, but it can also contain a collection of quantum dots. In a manner related to [11] (and also to the more recent proposal [42]), the quantized cavity modes provide a means of turning on and off two-bit interactions between individual spins in the quantum dots (again, the qubit of choice). Single-qubit operations are accomplished by near-resonant two-beam Raman transitions obtained by directing (classical) light of the right frequency at particular quantum dots. The decoherence properties are estimated to be marginal given the current state of the art of these microcavities, but this technology is expected to continue to improve.

This and other proposals are at a concrete enough stage that experiments can begin to probe the ability to satisfy some of the five criteria. Many related approaches are possible, and we only mention the electric-dipole transition qubits proposed by Sherwin *et al.* [43] (see also [44]), the use of transitions in optical hole burning materials [45], and the use of the spectroscopy of excitons in quantum wells in III-V materials [46]. We don't want to predict where any of these approaches will lead us, but they all have the possibility of future success.

4. QUANTUM COMMUNICATION WITH ELECTRONS

The essential resources for quantum communication [47] are EPR (Einstein-Podolsky-Rosen) pairs [48]—pairwise entangled qubits—the members of which are shared between two parties ("Alice and Bob"). These parties are located at different places and their goal for instance is to communicate with each other in an absolutely secure way (which is not possible with classical means only). The prime example of an EPR pair considered here is the singlet state formed by two electron spins. The intrinsic non-locality of these states gives rise to striking phenomena such as violations of Bell inequalities and has a number of possible applications in quantum information such as quantum teleportation, quantum key distribution, and entanglement purification. The non-locality of EPR pairs has been experimentally investigated for photons [49, 50], but not yet for *massive* particles such as electrons, let alone in a solid state environment. This is so because it is difficult to first produce and to then detect entanglement of electrons in a controlled way. In the following we review two scenarios we have recently proposed [51, 52] where the entanglement of electrons (once produced *e.g.* as described in the previous sections) can be detected in mesoscopic transport and noise measurements. One goal of the following discussion is to show that there exists an interesting connection between the field of quantum communication and the field of transport theory in electronic nanostructures. Another goal is to show that the investigation and concrete tests of fundamental phenomena such as quantum non-locality for electrons are within experimental reach within the not-so-distant future. Such phenomena go beyond the standard single-particle interference effects which have been well studied in mesoscopic systems over the last decade or so. Instead, they involve genuine two-particle effects where, due to strong correlations leading to entanglement, the quantum phases of two identical particles interfere with each other in a constructive or destructive way. This two-particle interference manifests itself in observable Aharonov-Bohm phase oscillations in the electric transport current and in the non-equilibrium current-current correlations.

4.1 Probing Entanglement Of Electrons In A Double Dot

We consider a double-dot (DD) system which contains two metallic leads which are in equilibrium with associated reservoirs kept at the chemical potentials $\mu_{1,2}$. Each lead is weakly coupled to *both* dots with tunneling amplitudes Γ, and these leads are probes where the currents $I_{1,2}$ are measured. Note that the DD system is put in parallel in contrast to the standard situation where the coupled dots are put in series (*i.e.* lead1-dot1-dot2-lead2). The quantum dots contain one (excess) electron each, and are coupled to each other by the tunneling amplitude t, which leads to a level splitting [17, 22] $J = E_{\rm t} - E_{\rm s} \sim 4t^2/U$ in the DD, with U being the single-dot Coulomb repulsion energy, and $E_{\rm s/t}$ are the singlet/triplet energies. We recall that for two electrons in the DD (and for weak magnetic fields) the ground state is given by a spin singlet. For convenience we count the chemical potentials μ_i from $E_{\rm s}$.

The tunneling Hamiltonian [53] reads $H = H_0 + V$, where $H_0 = H_{\rm D} + H_1 + H_2$ with $H_{\rm D}$ describing the DD and $H_{1,2}$ the leads (assumed to be Fermi liquids). The tunneling between leads and dots is described by $V = V_1 + V_2$, where

$$V_n = \Gamma \sum_s \left[D_{n,s}^\dagger c_{n,s} + c_{n,s}^\dagger D_{n,s} \right] , \quad D_{n,s} = e^{\pm i\varphi/4} d_{1,s} + e^{\mp i\varphi/4} d_{2,s} \ , \tag{27.4}$$

and where $c_{n,s}$ and $d_{n,s}$, $n = 1, 2$, annihilate electrons with spin s in the $n^{\rm th}$ lead and in the $n^{\rm th}$ dot, respectively. The Peierls phase φ in the hopping amplitude accounts for an Aharonov-Bohm (AB) or Berry phase (see below) in the presence of a magnetic field. The upper sign belongs to lead 1 and the lower to lead 2. The average current through the DD system is $I = \langle I_2 \rangle$ with $I_n = ie\Gamma \sum_s \left[D_{n,s}^\dagger c_{n,s} - c_{n,s}^\dagger D_{n,s} \right]$.

We consider now the Coulomb blockade (CB) regime where we can neglect double (or higher) occupancy in each dot for all transitions including virtual ones, *i.e.* we require $\mu_{1,2} < U$. Further we assume that $\mu_{1,2} > J, k_{\rm B}T$ to avoid resonances which might change the DD state. Γ is assumed to be weak (*i.e.* $J > 2\pi\nu_{\rm t}\Gamma^2$, where $\nu_{\rm t}$ is the tunneling density of states of the leads) so that the state of the DD is not perturbed; this will allow us to retain only the first non-vanishing contribution in

Γ to I. In analogy to the single-dot case [54, 55] we refer to the above CB regime as the cotunneling regime. In leading order, the cotunneling current involves the tunneling of one electron from the DD to, say, lead 1 and of a second electron from lead 2 to the DD. However, due to the weak coupling Γ, the DD will have returned to its equilibrium state *before* the next electron passes through it. We focus on the regime, $|\mu_1 - \mu_2| > J$, where elastic and inelastic cotunneling occurs, with singlet and triplet contributions being different. In this regime we can neglect the dynamics generated by J compared to the one generated by the bias, and we finally obtain (for $k_B T < \mu_{1,2}$)

$$I = e\pi\nu_t^2\Gamma^4 C(\varphi)\frac{\mu_1 - \mu_2}{\mu_1\mu_2} \;, \tag{27.5}$$

$$C(\varphi) = \sum_{s,s'} \left[\langle d_{1s'}^\dagger d_{1s} d_{1s}^\dagger d_{1s'}\rangle + \cos\varphi \langle d_{1s'}^\dagger d_{1s} d_{2s}^\dagger d_{2s'}\rangle\right] \;. \tag{27.6}$$

Eq. (27.5) shows that the cotunneling current depends on the properties of the ground state of the DD through the coherence factor $C(\varphi)$ given in (27.6). The first term in C is the contribution from the topologically trivial tunneling path which runs from lead 1 through, say, dot 1 to lead 2 and the same path back. The second term (phase-coherent part) in C is the ground state amplitude of the exchange of electron 1 with electron 2 via leads 1 and 2 such that a closed loop is formed enclosing an area A. Thus, in the presence of a magnetic field B, an AB phase factor $\varphi = ABe/h$ is acquired.

Next, we evaluate $C(\varphi)$ explicitly in the singlet-triplet basis. Note that only the singlet $|S\rangle$ and the triplet $|T_0\rangle$ are entangled EPR pairs while the remaining triplets $|T_+\rangle = |\uparrow\uparrow\rangle$, and $|T_-\rangle = |\downarrow\downarrow\rangle$ are not (they factorize). Assuming that the DD is in one of these states we obtain

$$C(\varphi) = \begin{cases} 2 - \cos\varphi\,, & \text{for the singlet,} \\ 2 + \cos\varphi\,, & \text{for all triplets.} \end{cases} \tag{27.7}$$

Thus, we see that the singlet and the triplets contribute with *opposite sign to the phase-coherent part of the current.* One has to distinguish, however, carefully the entangled from the non-entangled states. The phase-coherent part of the entangled states is a genuine *two-particle* interference effect, while the one of the product states cannot be distinguished from a phase-coherent *single-particle* interference effect. Indeed, this follows from the observation that the phase-coherent part in C factorizes for the product states $T_\pm$ while it does not do so for the entangled states S, T_0. Also, for states such as $|\uparrow\downarrow\rangle$ the coherent part of C vanishes, showing that two different (and fixed) spin states cannot lead to a phase-coherent contribution since we *know* which electron goes in which part of the loop. Finally we note that due to the AB phase the role of the singlet and triplets can be interchanged, which is to say that we can continually transmutate the statistics of the entangled pairs S, T_0 from fermionic to bosonic (like in anyons): the symmetric orbital wave function of the singlet S goes into an antisymmetric one at half a flux quantum, and vice versa for the triplet T_0.

The amplitude of the AB oscillations is a direct measure of the phase coherence of the entanglement, while the period via the enclosed area $A = h/eB_0$ gives a direct measure of the non-locality of the EPR pairs, with B_0 being the field at which $\varphi = 1$. Thus, the measurement of the AB amplitude will provide us with an entanglement dephasing length, which tells us how far we can spatially separate two electrons from each other in a conductor (in the presence of many other electrons, spin-orbit interaction, spin-impurities, *etc.*) before the entanglement in the total spin state is lost. No doubt, it would be highly desirable to obtain experimental information about this length scale since this will allow us to assess if and under which conditions quantum communication (which makes essential use of separated EPR pairs) will be possible in mesoscopic structures.

The triplets themselves can be further distinguished by applying a directionally inhomogeneous magnetic field (around the loop) producing a Berry phase Φ^{B} [56], which is positive (negative) for the triplet $m = 1(-1)$, while it vanishes for the EPR pairs S, T_0. Thus, we will eventually see beating in the AB oscillations due to the positive (negative) shift of the AB phase Φ by the Berry phase, $\varphi = \Phi \pm \Phi^{\mathrm{B}}$. We finally note that the closed AB-loop can actually be made as large as the dephasing length by using wave guides forming a loop with leads attached to it. Thus, a moderately weak field can be applied to produce the AB oscillations with negligible effect of the orbital state of the DD.

We discuss now the spectral density (noise) of the current cross-correlations, $S(\omega) = \int dt\ e^{i\omega t}\,\mathrm{Re}\,\langle \delta I_2(t)\,\delta I_1(0)\rangle$. Under the same assumptions as before (cotunneling regime), we obtain for the zero-frequency noise, its Poissonian value, *i.e.* $S(0) = -e|I|$. This shows that the Fano factor (noise-to-current ratio) is universal and the current and its cross-correlations contain the same information. For finite frequencies in the regime $|\mu_1 - \mu_2| > J$ and at T=0, we find

$$S(\omega) = (e\pi\nu_t\Gamma^2)^2 C(\varphi)\left[X_\omega + X^*_{-\omega}\right] ,$$

where

$$\mathrm{Im}X_\omega = \left[\theta(\mu_1 - \omega) - \theta(\mu_2 - \omega)\right]/2\omega, \tag{27.8}$$

$$
\begin{aligned}
\mathrm{Re}X_\omega &= \frac{1}{2\pi\omega}\,\mathrm{sign}(\mu_1-\mu_2+\omega)\ln\left|\frac{(\mu_1+\omega)(\mu_2-\omega)}{\mu_1\mu_2}\right| \\
&-\frac{1}{2\pi\omega}\left[\theta(\omega-\mu_1)\ln\left|\frac{\mu_2-\omega}{\mu_2}\right|+\theta(\omega-\mu_2)\ln\left|\frac{\mu_1-\omega}{\mu_1}\right|\right]. \qquad (27.9)
\end{aligned}
$$

The noise again depends on the phase-coherence factor C with the same properties as discussed before. Here, $\mathrm{Re}S(\omega)$ is even in ω, while $\mathrm{Im}S(\omega)$ is non–zero (for finite frequencies) and odd, in contrast to single-barrier junctions, where $\mathrm{Im}S(\omega)$ vanishes, since $\delta I_1 = -\delta I_2$ for all times. At small bias $\Delta\mu = \mu_1 - \mu_2 \ll \mu = (\mu_1+\mu_2)/2$, the odd part, $\mathrm{Im}S(\omega)$, given in (27.8) exhibits two narrow peaks at $\omega = \pm\mu$, which lead to slowly decaying oscillations in time, $S_{odd}(t) = \pi\nu_t^2\Gamma^4 C(\varphi)\sin(\Delta\mu t/2)\sin(\mu t)/\mu t$. These oscillations can be interpreted as a temporary charge-imbalance on the DD during an uncertainty time $\sim \mu^{-1}$, which results from the cotunneling of electrons and an associated time shift (induced by a finite ω) between incoming and outgoing currents.

There are a few obvious generalizations to the material presented so far: (1) multi-dot and multiterminal set-ups which implement n-particle entanglement, a prime example being the 3-particle entangled GHZ states *etc.*; (2) variations of the geometries such as the phase-coherent transport from additional "feeding leads" into dots 1 and 2. Such a set-up corresponds topologically to a scattering experiment in which we can arrange for scattering of unentangled electrons (as considered previously in noise studies [57]) but now also of entangled ones. In the latter case we get a non-trivial Fano factor [52] due to antibunching (triplets) and bunching (singlet) effects in the noise [8]: see below. (3) We can replace leads 1 and 2 each by quantum dots which are connected to the double-dot by spin-selective tunneling devices [58] (such spin-filters would allow us to measure spin via charge [17]). Such or similar set-ups would be needed to measure all spin correlations contained in the electronic EPR pairs and thus to test Bell inequalities for electrons in a solid state environment.

4.2 Noise Of Entangled Electrons: Bunching And Antibunching

In this section we discuss a related but alternative scenario in which entanglement of electrons can be measured through a bunching and antibunching behavior in the noise of conductors [8, 52]. The basic idea is rather simple and well known from the scattering theory of two identical particles [59, 60]. In the center-of-mass system the differential scattering cross-section can be expressed in terms of the scattering amplitude $f(\theta)$

and scattering angle θ [60],

$$\begin{aligned}\sigma(\theta) &= |f(\theta) \pm f(\pi - \theta)|^2 \\ &= |f(\theta)|^2 + |f(\pi - \theta)|^2 \pm 2\mathrm{Re} f^*(\theta) f(\pi - \theta).\end{aligned} \quad (27.10)$$

The first two terms in the second equation are the "classical" contributions which are obtained if the particles were distinguishable, whereas the third term results from the indistinguishability which gives rise to constructive (destructive) *two-particle interference effects.* Here the plus sign applies for spin-1/2 particles in the singlet state (described by a symmetric orbital wave function), while the minus sign applies for their triplet states (described by an antisymmetric orbital wave function). The very same two-particle interference mechanism which is responsible for the enhancement/reduction of the scattering cross section σ near $\theta = \pi/2$ leads to a bunching/antibunching behavior in the statistics [61].

We have previously described in detail how two electron spins can be deterministically entangled by weakly coupling two nearby quantum dots, each of which contains one single (excess) electron [17, 22]. The recently investigated coupling between electrons which are trapped by surface acoustic waves on a semiconductor surface [62] might provide another possibility of producing EPR pairs in a solid-state environment. Generalizing the above two-particle scattering experiment to a mesoscopic system, we have discussed an experimental set-up by which the entanglement of electrons (moving in the presence of a Fermi sea) can be detected in measurements of the current correlations (noise) [8, 52]. For this purpose we employ a beam splitter which has the property that electrons fed into its two incoming leads have a finite amplitude to be interchanged (without mutual interaction) before they leave through the two outgoing leads. In our case, the electrons are entangled before they enter the beam splitter. The quantity of interest is then the noise measured in the outgoing leads of the beam splitter. It is well-known that particles with symmetric wave functions show bunching behavior [57] in the noise, whereas particles with antisymmetric wave functions show antibunching behavior. The latter situation is the one considered recently for electrons in the normal state of mesoscopic transport systems both in theory [63, 64] and in experiments [65, 66]. However, since the noise is produced by the charge degrees of freedom we can expect [8] that in the absence of spin scattering processes the noise is sensitive to the symmetry (singlet or triplet) of only the *orbital part* of the wave function. We have verified this expectation explicitly, by extending the standard scattering matrix approach for transport in mesoscopic systems [63] to a situation with entanglement [52].

The electron current operator in lead α of a multiterminal conductor is

$$I_\alpha(t) = \frac{e}{h\nu} \sum_{E,E',\sigma} \left[a^\dagger_{\alpha\sigma}(E)a_{\alpha\sigma}(E') - b^\dagger_{\alpha\sigma}(E)b_{\alpha\sigma}(E')\right] e^{i(E-E')t/\hbar}, \tag{27.11}$$

where $a^\dagger_{\alpha\sigma}(E)$ creates an incoming electron in lead α with spin σ and energy E, and the operators $b_{\alpha\sigma}$ for the outgoing electrons are related to the operators a_α for the incident electrons via $s_{\alpha\beta}$, the (spin- and energy-independent) scattering matrix, $b_{\alpha\sigma}(E) = \sum_\beta s_{\alpha\beta}a_{\beta\sigma}(E)$. Note that since we are dealing with discrete energy states here, we normalize the operators $a_\alpha(E)$ such that $\left\{a_{\alpha\sigma}(E), a_{\beta\sigma'}(E')^\dagger\right\} = \delta_{\sigma\sigma'}\delta_{\alpha\beta}\delta_{E,E'}/\nu$, where the Kronecker symbol $\delta_{E,E'}$ equals 1 if $E = E'$ and 0 otherwise, and ν stands for the density of states in the leads. We also assume that each lead consists of only a single quantum channel; the generalization to leads with several channels is straightforward but is not needed here.

We evaluate the spectral density for the current fluctuations $\delta I_\alpha = I_\alpha - \langle I_\alpha \rangle$ between the leads α and β,

$$S_{\alpha\beta}(\omega) = \lim_{T\to\infty} \frac{h\nu}{T} \int_0^T dt\; e^{i\omega t}\langle\Psi|\delta I_\alpha(t)\delta I_\beta(0)|\Psi\rangle, \tag{27.12}$$

for the entangled incident state

$$|\Psi\rangle = |\pm\rangle = \frac{1}{\sqrt{2}}\left(a^\dagger_{2\downarrow}(\epsilon_2)a^\dagger_{1\uparrow}(\epsilon_1) \pm a^\dagger_{2\uparrow}(\epsilon_2)a^\dagger_{1\downarrow}(\epsilon_1)\right)|0\rangle. \tag{27.13}$$

The state $|-\rangle$ is the spin singlet, $|S\rangle$, while $|+\rangle$ denotes one of the spin triplets $|T_{0,\pm}\rangle$; in the following we will present a calculation of the noise for $|+\rangle = |T_0\rangle$, i.e. the triplet with $m_z = 0$. Evaluating the matrix elements we obtain the current correlation between the leads α and β,

$$S_{\alpha\beta}(0) = \frac{e^2}{h\nu}\left[\sum_{\gamma\delta}{}' A^\alpha_{\gamma\delta}A^\beta_{\delta\gamma} \mp \delta_{\epsilon_1,\epsilon_2}\left(A^\alpha_{12}A^\beta_{21} + A^\alpha_{21}A^\beta_{12}\right)\right], \tag{27.14}$$

where $A^\alpha_{\beta\gamma} = \delta_{\alpha\beta}\delta_{\alpha\gamma} - s^*_{\alpha\beta}s_{\alpha\gamma}$, and $\sum'_{\gamma\delta}$ denotes the sum over $\gamma = 1, 2$ and all $\delta \neq \gamma$, and where again the upper (lower) sign refers to triplets (singlets).

We apply these formulas now to our scattering set-up involving a beam splitter with four attached leads (leads 1 and 2 incoming, leads 3 and 4 outgoing) described by the single-particle scattering matrix elements, $s_{31} = s_{42} = r$, and $s_{41} = s_{32} = t$, where r and t denote the reflection and transmission amplitudes at the beam splitter, respectively. We assume that there is no backscattering, $s_{12} = s_{34} = s_{\alpha\alpha} = 0$. The unitarity of the s-matrix implies $|r|^2 + |t|^2 = 1$. The final result for the noise correlations for the incident state $|\pm\rangle$ is then[67],

$$S_{33}(0) = S_{44}(0) = 2\frac{e^2}{h\nu}T\,(1-T)\,(1 \mp \delta_{\epsilon_1,\epsilon_2})\,, \tag{27.15}$$

$$S_{34}(0) = 2\frac{e^2}{h\nu}\mathrm{Re}\left[r^{*2}t^2\right](1 \mp \delta_{\epsilon_1,\epsilon_2})\,, \tag{27.16}$$

where $T = |t|^2$ is the probability for transmission through the beam splitter. The calculation for the remaining two triplet states $|+\rangle = |T_\pm\rangle = |\uparrow\uparrow\rangle, |\downarrow\downarrow\rangle$ yields the same results Eqs. (27.15) and (27.16) (upper sign). For the average current in lead α we obtain $|\langle I_\alpha\rangle| = e/h\nu$, with no difference between singlets and triplets. Then, the Fano factor $F = S_{\alpha\alpha}(0)/\,|\langle I_\alpha\rangle|$ takes the form

$$F = 2eT(1-T)\,(1 \mp \delta_{\epsilon_1,\epsilon_2})\,, \tag{27.17}$$

and correspondingly for the cross correlations. This result implies that if two electrons with the same energies, $\epsilon_1 = \epsilon_2$, in the singlet state $|s\rangle = |-\rangle$ are injected into lead 1 and lead 2, respectively, then the zero frequency noise is *enhanced* by a factor of two, $F = 4eT(1-T)$, compared to the shot noise of uncorrelated particles, $F = 2eT(1-T)$. This enhancement of noise is due to *bunching* of electrons in the outgoing leads, caused by the symmetric orbital wavefunction of the spin singlet $|s\rangle$. On the other hand, the triplet states $|+\rangle = |T_{0,\pm}\rangle$ exhibit an *anti-bunching* effect, leading to a complete suppression of the zero-frequency noise, $S_{\alpha\alpha}(0) = 0$. The noise enhancement for the singlet $|S\rangle$ is a unique signature for entanglement (there exists no unentangled state with the same symmetry), therefore entanglement can be observed by measuring the noise power of a mesoscopic conductor. The triplets can be further distinguished from each other if we can measure the spin of the two electrons in the outgoing leads, or if we insert spin-selective tunneling devices [58] into leads 3, 4 which would filter a certain spin polarization.

Note that above results remain unchanged if we consider states $|\pm\rangle$ which are created above a Fermi sea. We have shown elsewhere [8] that the entanglement of two electrons propagating in a Fermi sea gets reduced by the quasiparticle weight z_F (for each lead one factor) due to the presence of interacting electrons. In the metallic regime z_F assumes typically some finite value [68], and thus as long as spin scattering processes are small the above description for non-interacting electrons remains valid.

5. CONCLUSION

We hope that workers in mesoscopic physics will find this brief survey of recent theoretical developments in quantum computation stimulating. As we continue to learn about quantum computing and quantum communication, we see more and more connections with present-day experimental physics. Quantum computing is not just a mathematical abstraction, it changes our outlook on a variety of fundamental issues in mesoscopics: on the desirability of having long coherence times in mesoscopic structures, on the role of precise time-dependent control of these structures for manipulating the interaction of electron states, on the need to develop high quantum efficiency measurements for spin and other single-quantum properties. Quantum computing and communication clearly have a fascinating role to play in some far-future technologies; we hope that we have illustrated how they can also play a role in the direction of fundamental physics research today.

Acknowledgements

We would like to thank K. Ensslin for kindly providing us with essential material parameters, and A. Chiolero for advising us on our numerical method for the g-factor calculation. DPD is grateful for funding under grant ARO DAAG55-98-C-0041. GB, DL, and EVS acknowledge the funding from the Swiss National Science Foundation.

References

[1] A. Ekert, "Quantum Computation," in *Atomic Physics 14*, 14th International Conference on Atomic Physics, Boulder, CO, 1994 (AIP Conference Proceedings **323**, AIP Press, New York, 1995), eds. D. J. Wineland, C. E. Wieman, and S. J. Smith, p. 450; see http://eve.physics.ox.ac.uk/NewWeb/Publications/oldftp.htm.

[2] A. Ekert and R. Jozsa, Rev. Mod. Phys. **68**, 733 (1996), and Ref. [1].

[3] C. H. Bennett, Physics Today **48** (10), 24 (1995).

[4] D. P. DiVincenzo, Proc. R. Soc. London A **454**, 261 (1998); quant-ph/9705009.

[5] A. Barenco, Contemp. Phys. **37**, 375 (1996).

[6] A. Steane, Rep. Prog. Phys. **61**, 117 (1998).

[7] C. H. Bennett and P. W. Shor, IEEE Trans. Info. Theory **44**, 2724 (1998).

[8] D. P. DiVincenzo and D. Loss, J. Magn. Mag. Matl. **200**, 202 (1999); cond-mat/9901137.

[9] (a) D. P. DiVincenzo, in *Mesoscopic Electron Transport*, eds. L. Sohn, L. Kouwenhoven, and G. Schön (Vol. 345, NATO ASI Series

E, Kluwer, 1997), p. 657 (cond-mat/9612126); (b) D. P. DiVincenzo, Science **270**, 255 (1995); (c) D. P. DiVincenzo and D. Loss, Superlattices and Microstructures **23**, 419 (1998).
[10] J. I. Cirac and P. Zoller, Phys. Rev. Lett. **74**, 4091 (1995).
[11] T. Pellizzari, S. A. Gardiner, J. I. Cirac, and P. Zoller, Phys. Rev. Lett. **75**, 3788 (1997).
[12] S. J. van Enk, J. I. Cirac, and P. Zoller, Phys. Rev. Lett. **78**, 4293 (1997).
[13] G. K. Brennan *et al.*, Phys. Rev. Lett. **82**, 1060 (1999), quant-ph/9806021; D. Jaksch *et al.*, Phys. Rev. Lett. **82**, 1975 (1999), quant-ph/9810087.
[14] I. L. Chuang, N. A. Gershenfeld and M. Kubinec, Phys. Rev. Lett. **80**, 3408 (1998).
[15] D. Cory, A. Fahmy, and T. Havel, Proc. Natl. Acad. Sci. USA **94**, 1634 (1997).
[16] P. M. Platzman and M. I. Dykman, Science **284**, 1967 (1999).
[17] D. Loss and D. P. DiVincenzo, Phys. Rev. A **57**, 120 (1998); cond-mat/9701055.
[18] B. Kane, Nature **393**, 133 (1998).
[19] R. Vrijen *et al.*, "Electron spin resonance transistors for quantum computing in silicon-germanium heterostructures," submitted to Phys. Rev. A; quant-ph/9905096.
[20] P. W. Brouwer *et al.*, cond-mat/9907148; H. U. Baranger *et al.*, cond-mat/9907151; but, see P. Jacquod and A. D. Stone, cond-mat/9909313.
[21] J. M. Kikkawa and D. D. Awschalom, Phys. Rev. Lett. **80**, 4313 (1998).
[22] G. Burkard, D. Loss and D. P. DiVincenzo, Phys. Rev. B **59**, 2070 (1999); cond-mat/9808026.
[23] D. G. Austing, T. Honda, K. Muraki, Y. Tokura and S. Tarucha, Physica B **249-251**, 206 (1998).
[24] R. J. Luyken, A. Lorke, M. Haslinger, B. T. Miller, M. Fricke, J. P. Kotthaus, G. Medeiros-Ribiero and P. M. Petroff, preprint.
[25] G. Burkard, G. Seelig and D. Loss, cond-mat/9910105.
[26] D. Gottesman, "Fault-Tolerant Quantum Computation with Local Gates", quant-ph/9903099.
[27] G. Burkard, D. Loss, D. P. DiVincenzo, and J. A. Smolin, Phys. Rev. B **60**, 11404 (1999); cond-mat/9905230.
[28] L. Kouwenhoven and C. Marcus, private communcation.
[29] E. L. Ivchenko, A. A. Kiselev and M. Willander, Solid State Comm. **102**, 375 (1997).

[30] A. A. Kiselev, E. L. Ivchenko and U. Rössler, Phys. Rev. B **58**, 16353 (1998).
[31] D. Bacon, J. Kempe, D. A. Lidar and K. B. Whaley, "Universal fault-tolerant computation on decoherence-free subspaces", quant-ph/9909058.
[32] K. Wago, D. Botkin, C. S. Yannoni and D. Rugar, Phys. Rev. B **57**, 1108 (1998), and references therein.
[33] G. Berman *et al.*, quant-ph/9909033.
[34] D. P. DiVincenzo, J. Appl. Phys. **85**, 4785 (1999); cond-mat/9810295.
[35] D. Averin, Solid State Commun. **105**, 659 (1998).
[36] Y. Makhlin *et al.*, Nature **398**, 305 (1999).
[37] L. B. Ioffe, V. B. Geshkenbein, M. V. Feigel'man, A. L. Fauchère and G. Blatter, Nature **398**, 679 (1999).
[38] A. M. Zagoskin, cond-mat/9903170; A. Blais and A. M. Zagoskin, quant-ph/9905043.
[39] J. E. Mooij, T. P. Orlando, L. Levitov, L. Tian, C. H. van der Wal and S. Lloyd, Science **285**, 1036 (1999).
[40] T. Brun and H. Wang, "Coupling nanocrystals to a high-Q silica microsphere: entanglement in quantum dots via photon exchange", quant-ph/9906025.
[41] A. Imamoglu, D. D. Awschalom, G. Burkard, D. P. DiVincenzo, D. Loss, M. Sherwin and A. Small, Phys. Rev. Lett. **83**, 4204 (1999); quant-ph/9904096.
[42] A. Sorensen and K. Molmer, Phys. Rev. Lett. **82**, 1971 (1999).
[43] M. Sherwin, A. Imamoglu and Thomas Montroy, "Quantum computation with quantum dots and terahertz cavity quantum electrodynamics", quant-ph/9903065.
[44] G. D. Sanders *et al.*, "An optically driven quantum dot quantum computer", quant-ph/9909070.
[45] S. M. Shahriar *et al.*, unpublished.
[46] D. Steele and D. Gammon, unpublished.
[47] C. H. Bennett and G. Brassard, in *Proceedings of the IEEE International Conference on Computers, Systems and Signal Processing, Bangalore, India* (IEEE, New York, 1984), p. 175.
[48] A. Einstein, B. Podolsky, N. Rosen, Phys. Rev. **47**, 777 (1935).
[49] A. Aspect, J. Dalibard, G. Roger, Phys. Rev. Lett. **49**, 1804 (1982); W. Tittel *et al.*, Phys. Rev. Lett. **81**, 3563 (1998).
[50] D. Bouwmeester *et al.*, Nature **390**, 575 (1997); D. Boschi *et al.*, Phys. Rev. Lett. **80**, 1121 (1998).
[51] D. Loss and E. Sukhorukov, cond-mat/9907129.

[52] G. Burkard, D. Loss and E. V. Sukhorukov, cond-mat/9906071.
[53] G. D. Mahan, *Many-Particle Physics*, 2nd Ed. (Plenum, New York, 1993).
[54] D. V. Averin and Yu. V. Nazarov, in *Single Charge Tunneling*, eds. H. Grabert and M. H. Devoret, NATO ASI Series B: Physics Vol. 294, Plenum Press, New York, 1992.
[55] J. König, H. Schoeller and G. Schön, Phys. Rev. Lett. **78**, 4482 (1997).
[56] D. Loss and P. Goldbart, Phys. Rev. B **45**, 13544 (1992).
[57] R. Hanbury Brown and R. Q. Twiss, Nature (London) **177**, 27 (1956).
[58] G. A. Prinz, Science **282** 1660 (1998).
[59] R. P. Feynman, R. B. Leighton and M. Sands, *The Feynman Lectures* (Addison-Wesley, Reading, MA, 1965), Vol. 3.
[60] L. E. Ballentine, *Quantum Mechanics*, pp. 352, Prentice Hall, New Jersey, 1990.
[61] R. Loudon, Phys. Rev. A **58**, 4904 (1998).
[62] C. H. W. Barnes, private communication.
[63] M. Büttiker, Phys. Rev. Lett. **65**, 2901 (1990); Phys. Rev. B**46**, 12485 (1992).
[64] Th. Martin and R. Landauer, Phys. Rev. B **45**, 1742 (1992)
[65] R. C. Liu *et al.*, Nature **391**, 263 (1998); M. Henny *et al.*, Science **284**, 296 (1999); W. D. Oliver *et al.*, Science **284**, 299 (1999).
[66] For a positive sign in the noise cross correlations due to the boson-like properties of Cooper pairs see, J. Torrès, T. Martin, cond-mat/9906012.
[67] For finite frequencies, we obtain the noise power $S_{\alpha\alpha}(\omega) = (2e^2/h\nu)[(1-\delta_{\omega,0}) + 2T(1-T)(\delta_{\omega,0} \mp \delta_{\omega,\epsilon_1-\epsilon_2})]$.
[68] For instance, in metals such as bulk Cu the quasiparticle weight becomes, within the RPA approximation, $z_F = 0.77$ [69], while for a GaAs 2DEG we find (also within RPA) $z_F = 1 - r_s(1/2 + 1/\pi) = 0.66$ for the GaAs interaction parameter $r_s = 0.61$ (the details of the calculation will be given elsewhere).
[69] T. M. Rice, Ann. Phys. **31** 100 (1965).

Chapter 28

DECOHERENCE OF THE SUPERCONDUCTING PERSISTENT CURRENT QUBIT

L. Tian[1], L. S. Levitov[1], C. H. van der Wal[4], J. E. Mooij[2,4], T. P. Orlando[2], S. Lloyd[3], C. J. P. M. Harmans[4] and J. J. Mazo[2,5]

[1]*Department of Physics, Center for Material Science & Engineering,*

[2]*Department of Electrical Engineering and Computer Science,*

[3]*Department of Mechanical Engineering, Massachusetts Institute of Technology;*

[4]*Department of Applied Physics and Delft Institute for Microelectronics and Submicron Technologies, Delft University of Technology;*

[5]*Department de Física de la Mataeria Condensada, Universidad de Zaragoza*

Abstract Decoherence of a solid state based qubit can be caused by coupling to microscopic degrees of freedom in the solid. We lay out a simple theory and use it to estimate decoherence for a recently proposed superconducting persistent current design. All considered sources of decoherence are found to be quite weak, leading to a high quality factor for this qubit.

1. INTRODUCTION

The power of quantum logic [1] depends on the degree of coherence of the qubit dynamics [2, 3]. The so-called "quality factor" of the qubit, the number of quantum operations performed during the qubit coherence time, should be at least 10^4 for the quantum computer to allow for quantum error correction [4]. Decoherence is an especially vital issue in solid state qubit designs, due to many kinds of low energy excitations in the solid state environment that may couple to qubit states and cause dephasing.

In this article we discuss and estimate **some of the main sources of** decoherence in the superconducting persistent current qubit proposed recently [3]. The approach will be presented in a way making it easy to

I. O. Kulik and R. Ellialtioğlu (eds.),
Quantum Mesoscopic Phenomena and Mesoscopic Devices in Microelectronics, 429–438.

generalize it to other systems. We emphasize those decoherence mechanisms that illustrate this approach, and briefly summarize the results of other mechanisms.

The circuit [3] consists of three small Josephson junctions which are connected in series, forming a loop, as shown in Fig. 28.1. The charging energy of the qubits $E_C = e^2/2C_{1,2}$ is $\sim$ 100 times smaller than the Josephson energy $E_J = \hbar I_0/2e$, where I_0 is the qubit Josephson critical current. The junctions discussed in [3] are 200 nm by 400 nm, and $E_J \approx$ 200 GHz.

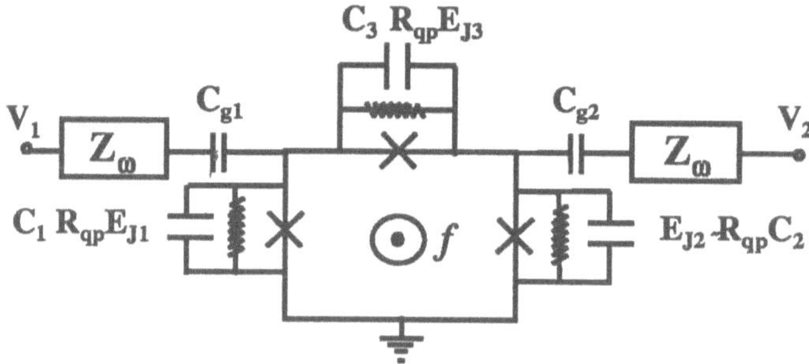

Figure 28.1 Schematic qubit design [3] consisting of three Josephson junctions connected as shown. Josephson energy of one of the junctions (number 3 in the figure) is adjustable by varying the flux in the SQUID loop. The impedances Z_ω model electromagnetic environment coupled to the qubit via gate capacitances $C_{g(1,2)}$. Shunt resistors model quasiparticle subgap resistance effect.

Qubit is realized by two lowest energy states of the system corresponding to opposite circulating currents in the loop. The energy splitting of these states $\varepsilon_0 \approx 10$ GHz is controlled by the external magnetic field flux f, the barrier height is $\simeq$ 35 GHz and the tunneling amplitude between the two states is $t \approx 1$ GHz. The Hamiltonian derived in [3] for the two lowest energy levels of the qubit has the form

$$\mathcal{H}_0 = \begin{pmatrix} -\varepsilon_0/2 & t(q_1, q_2) \\ t^*(q_1, q_2) & \varepsilon_0/2 \end{pmatrix}, \quad (28.1)$$

where $t(q_1, q_2)$ is a periodic function of gate charges $q_{1,2}$. In the tight binding approximation [3], $t(q_1, q_2) = t_1 + t_2 e^{-i\pi q_1/e} + t_2 e^{i\pi q_2/e}$, where t_1 is the amplitude of tunneling between the nearest energy minima and t_2 is the tunneling between the next nearest neighbor minima in the model [3]. Both t_1 and t_2 depend on the energy barrier height and width exponentially. With the parameters of our qubit design, $t_2/t_1 < 10^{-3}$, the effect of fluctuations of $q_{1,2}$ should be small.

Below we consider a number of decoherence effects which seem to be most relevant for the design [3], trying to keep the approach general enough, so that it can be applied to other designs.

2. BASIC APPROACH

We start with a Hamiltonian of a qubit coupled to environmental degrees of freedom in the solid: $\mathcal{H}_{\text{total}} = \mathcal{H}_{\text{Q}}(\vec{\sigma}) + \mathcal{H}_{\text{bath}}(\{\xi_\alpha\})$, where $\mathcal{H}_{\text{Q}} = \mathcal{H}_0 + \mathcal{H}_{coupling}$:

$$\mathcal{H}_{\text{Q}} = \frac{\hbar}{2}\left(\vec{\Delta}(t) + \vec{\eta}(t)\right)\cdot\vec{\sigma}\ , \qquad \vec{\eta} = \sum_\alpha \vec{A}_\alpha \hat{\xi}_\alpha\ , \tag{28.2}$$

where $\vec{\sigma} = (\sigma_x, \sigma_y, \sigma_z)$ is the vector of Pauli matrices acting on the qubit states, the vector $\vec{\Delta}$ represents external control, and $\vec{\eta}$ is noise due to coupling to the bath variables ξ_α. In (28.1), $\Delta_z = -\varepsilon_0$, $\Delta_x - i\Delta_y = t(q_1, q_2)$.

The degrees of freedom that may decohere qubit dynamics are:

- charge fluctuations in the gates coupling qubit states to other states;
- quasiparticles in the superconductor giving rise to subgap resistance;
- nuclear spins in the solid creating fluctuating magnetic fields;
- electromagnetic radiation causing damping of Rabi oscillations;
- coupling between qubits affecting operation of an individual qubit.

In all cases except the last one, the qubit is coupled to a macroscopic number of degrees of freedom $N \gg 1$ with about the same strength A_α to each. In such a situation, the qubit decoherence rate is much larger than the characteristic individual coupling frequency $A_\alpha/\hbar$. This means that dephasing happens on a shorter time scale than it would have taken to create an entangled state of the qubit and one particular element of the bath. In other words, on the decoherence time scale each element of the bath remains in its initial state with probability $1 - O(1/N)$, and it is only due to a large number of relevant degrees of freedom N that the state of the qubit is significantly affected on this time scale.

This observation makes the analysis quite simple, especially because the condition $N \gg 1$ allows one to replace generally noncommuting quantum variables $\hat{\xi}_\alpha(t)$ by bosonic fields $\eta_{x,y,z}(t)$ fluctuating in time. (Because at large N the commutators $[\eta_i(t), \eta_j(t)]$ are well approximated by c−numbers.) As a result, the problem becomes equivalent to that of longitudinal and transverse spin relaxation times T_1 and T_2 in NMR, corresponding to the noise $\eta_i(t)$ either flipping the qubit spin, or contributing a random phase to the qubit states evolution, respectively. Thus we can use the standard Debye–Bloch theory of relaxation in two-level systems.

To adapt this theory to our problem, we assume, without loss of generality, that $\vec{\Delta}(t) \parallel \hat{z}$ and is constant as a function of time. Then one can eliminate the term $\frac{1}{2}\vec{\Delta}\cdot\vec{\sigma}$ by going to the frame rotating around the z−axis with the Larmor frequency $\Delta = |\vec{\Delta}|$. In the rotating frame the Hamiltonian (28.2) becomes:

$$\widetilde{\mathcal{H}}_{\mathrm{Q}} = \frac{\hbar}{2}\left(\eta_{||}(t)\sigma_z + e^{-i\Delta t}\eta_\perp(t)\sigma_+ + e^{i\Delta t}\eta_\perp^*(t)\sigma_-\right) , \qquad (28.3)$$

where $\eta_{||}(t)$ and $\eta_\perp(t)$ correspond to components of vector $\vec{\eta}(t)$ in (28.2) parallel and perpendicular to $\vec{\Delta}$, respectively.

The time evolution due to noise $\vec{\eta}(t)$ is given by the evolution operator $T\exp\left(-i\int\widetilde{\mathcal{H}}_{\mathrm{Q}}(t')dt'\right)$ written in the rotating Larmor basis. However, for a simple estimate below we ignore noncommutativity of different parts of the Hamiltonian (28.3), and consider a c-number phase factor instead of an operator exponent.

Then the decoherence can be characterized using the function

$$R(t) = \max\left[\langle\phi_{||}^2(t)\rangle,\ \langle|\phi_\perp(t)|^2\rangle\right] , \qquad (28.4)$$

where $\langle ... \rangle$ stands for ensemble average, and

$$\phi_{||}(t) = \int_0^t \eta_{||}(t')dt' , \qquad \phi_\perp(t) = \int_0^t e^{-i\Delta t'}\eta_\perp(t')dt' \qquad (28.5)$$

The function $R(t)$ grows with time, and one can take as a measure of decoherence the time τ for which $R(\tau) \simeq 1$. There are several assumptions implicit in this criterion. First, we ignore noncommutativity of different terms in (28.3), which is legitimate at short times, when $R(t) \ll 1$. Second, we include in (28.4) the zero-point fluctuations of $\hat{\xi}_\alpha(t)$. The issue of decoherence due to zero-point motion in some cases can be subtle. However, since including the zero-point fluctuations in $R(t)$ can only overestimate the rate of loosing coherence, one expects the criterion $R(\tau) \simeq 1$ to still give a good lower bound on decoherence time.

Finally, we note that (28.4) contains statistical average over an ensemble of bath realizations. Hence care needs to be taken in the interpretation of τ when the bath is "frozen" into a particular configuration so that the ensemble averaging does not apply. In this situation one has to distinguish between decohering individual qubit dynamics and averaged dynamics of a qubit array. An example of such a situation is provided by the problem of coupling to the nuclear spins, a system with long relaxation times.

Since $\vec{\eta} = \sum\limits_{\alpha} \vec{A}_\alpha \hat{\xi}_\alpha(t)$, it is the time evolution of $\hat{\xi}_\alpha(t)$ defined by $\mathcal{H}_{\text{bath}}$ that is what eventually leads to decoherence. One can express quantities of interest in terms of the noise spectrum of the components of $\vec{\eta}$:

$$\langle \phi_{\|}^2(t) \rangle = \int d\omega \frac{|1 - e^{i\omega t}|^2}{2\pi\omega^2} \langle \eta_{\|}(-\omega) \eta_{\|}(\omega) \rangle \quad (28.6)$$

$$\langle |\phi_{\perp}(t)|^2 \rangle = \int d\omega \frac{|1 - e^{i\omega t}|^2}{2\pi\omega^2} \langle \eta_{\perp}(-\omega - \Delta) \eta_{\perp}(\omega + \Delta) \rangle \quad (28.7)$$

In thermal equilibrium, by virtue of the Fluctuation–Dissipation theorem, the noise spectrum in the RHS of (28.6) and (28.7) can be expressed in terms of the out-of-phase part of an appropriate susceptibility.

3. ESTIMATES FOR PARTICULAR MECHANISMS

Here we discuss the above listed decoherence mechanisms and use the expressions (28.6) and (28.7) to estimate the corresponding decoherence times. We start with the effect of **charge fluctuations on the gates** due to electromagnetic coupling to the environment modeled by an external impedance Z_ω (see Fig. 28.1), taken below to be of order of 400 Ω, the vacuum impedance.

The dependence of the qubit Hamiltonian on the gate charges $q_{1,2}$ is given by (28.1), where $q_{1,2}$ vary in time in response to the fluctuations of gate voltages, $\delta q_{1,2} \approx C_g \delta V_{g(1,2)}$, where the gate capacitance is much smaller than the junction capacitance: $C_g \ll C_{1,2}$. The gate voltage fluctuations are given by the Nyquist formula: $\langle \delta V_g(-\omega) \delta V_g(\omega) \rangle = 2 Z_\omega \hbar\omega \coth \hbar\omega / kT$.

In our design, $|t(q_1, q_2)| \ll \varepsilon_0$, and therefore fluctuations of $q_{1,2}$ generate primarily transverse noise $\eta_\perp$ in (28.3), $\eta_\perp(t) \simeq (2\pi/\hbar e) t_2 C_g \delta V_g(t)$. In this case, according to (28.7), we are interested in the noise spectrum of δV_g shifted by the Larmor frequency Δ. Our typical $\Delta \simeq 10\,\text{GHz}$ is much larger than the temperature $k_B T/h = 1\,\text{GHz}$ at $T = 50\,\text{mK}$, and thus one has $\omega \simeq \Delta \gg kT/\hbar$ in the Nyquist formula.

The Nyquist spectrum is very broad compared to Larmor frequency and other relevant frequency scales, and thus in (28.7) we can just use the $\omega = \Delta$ value of the noise power. Evaluating $\int |(1 - e^{i\omega t})/\omega|^2 d\omega = 2\pi t$, we obtain

$$R(t) = \langle |\phi_\perp(t)|^2 \rangle = \frac{2t}{\hbar} \left(\frac{2\pi}{e} t_2 C_g \right)^2 \Delta Z_{\omega=\Delta} \tag{28.8}$$

Rewriting this expression as $R(t) = t/\tau$, we estimate the decoherence time as

$$\tau = \Delta^{-1} \frac{\hbar}{2e^2} Z_{\omega=\Delta}^{-1} \left(\frac{e^2}{2\pi C_g t_2} \right)^2 \tag{28.9}$$

where $\hbar/2e^2 \simeq 4\,\mathrm{k}\Omega$. In the qubit design $e^2/2C_g \simeq 100\,\mathrm{GHz}$, and $t_2 \simeq 1\,\mathrm{MHz}$ when $t_2/t_1 \leq 10^{-3}$. With these numbers, one has $\tau = 0.1\,\mathrm{s}$.

The next effect we consider is dephasing due to **quasiparticles on superconducting islands**. At finite temperature, quasiparticles are thermally activated above the superconducting gap Δ_0, and their density is $\sim \exp(-\Delta_0/kT)$. The contribution of quasiparticles to the Josephson junction dynamics can be modeled as a shunt resistor, as shown in Fig. 28.1. The corresponding *subgap resistance* is inversely proportional to the quasiparticle density, and thus increases exponentially at small temperatures: $R_{\mathrm{qp}} \approx R_n \exp \Delta_0/kT$, where R_n is the normal state resistance of the junction. For Josephson current $I_0 = 0.2\,\mu\mathrm{A}$, $R_n \approx 1.3\,\mathrm{k}\Omega$. At low temperatures the subgap resistance is quite high, and thus difficult to measure [5]. For estimates below we take $R_{\mathrm{qp}} = 10^{11}\ \Omega$ which is much smaller than what follows from the exponential dependence for $T = 50\,\mathrm{mK}$.

The main effect of the subgap resistance in the shunt resistor model is generating normal current fluctuations which couple to the phase on the junction. The Hamiltonian describing this effect is

$$\mathcal{H}^{\mathrm{qp}}_{\mathrm{coupling}} = \sum_i \frac{\hbar}{2e} \varphi_i I_i^{\mathrm{qp}}(t) \ , \tag{28.10}$$

where $i = 1, 2, 3$ labels Josephson junctions. Projecting (28.10) to the two qubit states, one obtains the Hamiltonian (28.2) with $\eta_z(t) = I_i^{\mathrm{qp}}(t)/e$, $\eta_{x,y} = 0$.

The noise spectrum of the quasiparticle current is given by Nyquist formula:

$$\langle I^{\mathrm{qp}}(-\omega) I^{\mathrm{qp}}(\omega) \rangle = 2R_{\mathrm{qp}}^{-1} \hbar\omega \coth(\hbar\omega/kT) \tag{28.11}$$

After rotating the basis and transforming the problem to the form (28.3) we have $\eta_\perp(t) \simeq (t_1/\varepsilon_0)\eta_{||}(t)$, where $\eta_{||}(t) \simeq I_i^{\mathrm{qp}}(t)/e$ since $t_1 \ll \varepsilon_0$.

The analysis of $\langle|\phi_\perp(t)|^2\rangle$ and $\langle|\phi_{||}(t)|^2\rangle$ is similar to that described above for charge fluctuations on the gates, and one obtains $R_\perp(t) = 2t(t_1/\varepsilon_0)^2\hbar\Delta/(e^2R_{\rm qp})$, and $R_{||}(t) = 2t\,kT/(e^2R_{\rm qp})$ which gives

$$\tau = \min\left[\tau_\perp, \tau_{||}\right] = \min\left[\frac{e^2R_{\rm qp}}{2\hbar\Delta}\left(\frac{\varepsilon_0}{t_1}\right)^2, \frac{e^2R_{\rm qp}}{2kT}\right] \qquad (28.12)$$

Taking $R_{qp} = 10^{11}\,\Omega$, $T = 50\,\mathrm{mK}$, and $\varepsilon_0/t_1 = 100$, the decoherence times are $\tau_{||} = 1\,\mathrm{ms}$ and $\tau_\perp = 10\,\mathrm{ms}$.

The decoherence effect of **nuclear spins** on the qubit is due to their magnetic field flux coupling to the qubit inductance. Alternatively, this coupling can be viewed as Zeeman energy of nuclear spins in the magnetic field $\vec{B}(r)$ due to the qubit. The two states of the qubit have opposite currents, and produce magnetic field of opposite sign. The corresponding term in (28.2) is

$$\mathcal{H}_{\rm coupling} = -\sigma_z \sum_{r=r_i} \mu\vec{B}(r)\cdot\vec{s}(r) \qquad (28.13)$$

where r_i are positions of nuclei, μ is nuclear magnetic moment and $\hat{s}(r_i)$ are spin operators.

Nuclei are in thermal equilibrium, and their spin fluctuations can be related to the longitudinal relaxation time T_1 by the Fluctuation-Dissipation theorem. Assuming that different spins are uncorrelated, one has

$$\langle s_\omega(r)s_{-\omega}(r)\rangle = 2k_BT\frac{\chi''(\omega)}{\omega} = \frac{2k_BT_1\chi_0}{1+\omega^2T_1^2}\,, \qquad (28.14)$$

where $\chi_0 = 1/k_BT$ is static spin susceptibility.

The spectrum (28.14) has a very narrow width set by the long relaxation time T_1. This width is much less then k_BT and Δ. As a result, only longitudinal fluctuations $\eta_{||}$ survive in (28.6) and (28.7). One has

$$\langle\phi_{||}^2(t)\rangle = \int d\omega\frac{|1-e^{i\omega t}|^2}{2\pi\hbar^2\omega^2}\sum_{r=r_i}\mu^2B^2(r)\langle s_\omega(r)s_{-\omega}(r)\rangle\,. \qquad (28.15)$$

Plugging the spectrum (28.14) in (28.15) and integrating, one obtains

$$R(t) = \frac{T_1}{\tau_0^2}\left(|t| - T_1 + T_1e^{-|t|/T_1}\right),$$

$$\tau_0 = \left(\int\frac{2\mu^2}{\hbar^2}n(r)B^2(r)d^3r\right)^{-1/2}, \qquad (28.16)$$

where $n(r)$ is the nuclei concentration. The **ensembled-averaged** decoherence time that defined by $R(\tau) \simeq 1$ is then estimated as:

$$\tau = \begin{cases} \tau_0 & \text{for} \quad T_1 > \tau_0 \\ \tau_0^2/T_1 & \text{for} \quad T_1 < \tau_0 \end{cases} \tag{28.17}$$

In superconducting Al, nuclear spin relaxation time is strongly varying with temperature: $T_1 \simeq (300/T\,[\mathrm{K}])e^{\Delta/k_B T}\,\mathrm{s}$. At $T = 50\,\mathrm{mK}$, the time T_1 is of order of minutes, which exceeds all time scales relevant for qubit operation. To estimate τ_0, we use the magneton $\mu \simeq e\hbar/Mc$, where M is proton mass, and $\int B^2(r)d^3r \simeq 10^{-5}\Phi_0^2/w$, where $w \simeq 0.5\,\mu\mathrm{m}$ is the thickness of Al wires in the circuit, and $\Phi_0 = hc/2e$ is the flux quantum. The resulting $\tau_0 \simeq 3 \times 10^{-8}\,\mathrm{s} \ll T_1$.

According to (28.17), one apparently obtains a worryingly short time $\tau = \tau_0$. However, we note that this result corresponds to ensemble averaging, and one should be careful in applying it to an individual qubit.

The physical picture is that the nuclei spin configuration stays the same over times $\leq T_1$. At such times the perturbation $\vec{\eta}$ due to spins has essentially no time dependence, and so nuclei can be viewed as sources of random *static* magnetic field. The fluxes of this field induced on the qubits depend on initial conditions, and are uncorrelated for different qubits. Typical value of this flux corresponds to the change in Larmor frequency of order of $\delta\Delta \simeq \tau_0^{-1} \simeq 30\,\mathrm{MHz}$.

To summarize, for an individual qubit the effect of nuclear spins is equivalent to a random detuning caused by random change in Δ. For an ensemble of qubits, there will be a distribution of Larmor frequencies of width $\delta\Delta \simeq 30\,\mathrm{MHz}$, even if all qubits are identical. However, since the qubit phase can be kept coherent within a time $\leq T_1$, an indirect observation of Rabi oscillations is still possible by using the so-called "spin-echo technique."

A similar theory can be employed to estimate the effect due to magnetic impurities. The main difference is that for impurity spins the relaxation time T_1 is typically much shorter than for nuclear spins. If T_1 becomes comparable to the qubit operation time, the ensemble averaged quantities will describe a *real* dephasing of an individual qubit, rather than effects of inhomogeneous broadening, like for nuclear spins.

4. OTHER MECHANISMS

Some sources of decoherence are not amenable to the basic approach considered above, such as radiation losses which we estimate to have $\tau \simeq 10^3\,\mathrm{s}$.

Another such source of decoherence is caused by the magnetic dipole interaction between the qubits. This **interaction between qubits** is described by

$$\mathcal{H}_{\text{coupling}} = \sum_{i,j} \hbar\lambda_{ij}\sigma_z^{(i)} \otimes \sigma_z^{(j)} \ , \qquad \hbar\lambda_{ij} \approx \frac{\mu_i\mu_j}{|r_i - r_j|^3} \tag{28.18}$$

This interaction is strongest for nearest neighbors. For a square lattice of qubits with the spacing $R = 10\,\mu\text{m}$, one has the nearest neighbor coupling $\lambda \simeq 6\,\text{kHz}$. The corresponding decoherence time $\tau = \lambda^{-1} \simeq 0.2\,\text{ms}$ is relatively short.

Several alterations of the design can be implemented to reduce the effect of qubit-qubit interaction. One can arrange qubits in pairs with opposite sign of circulating currents. This will eliminate dipole moment of a pair, and reduce coupling between different pairs to a somewhat weaker quadrupole interaction. The same result can be achieved by using a superconducting base plane, in which magnetic dipoles will be imaged by dipoles of opposite sign, which will partially cancel the qubit-qubit coupling. Also, one can detune Larmor frequencies of neighboring qubits, moving them apart by more than λ, which will make couplings (28.18) off-resonant and reduce their effect.

Unwanted coupling between qubits is a common problem in quantum computers. Sophisticated decoupling techniques that have been developed for NMR designs [6], could equally be used here. The idea is to apply a sequence of single bit operations that effectively average out the coupling Hamiltonian over time. Such methods would also be effective for reducing the coupling to the environment [7]. These techniques are fully compatible with quantum computation and could be used to lengthen significantly the effective coherence times.

5. SUMMARY

Our analysis shows that for the qubit design [3] the decoherence time is limited by qubit-qubit coupling. By using methods discussed above the decoherence time can be made at least 1 ms which for $f_{\text{Rabi}} = 100\,\text{MHz}$ gives a quality factor of 10^5, passing the criterion for quantum error correction.

In addition to the effects we discussed, some other decoherence sources are worth attention, such as low frequency charge fluctuations resulting from electron hopping on impurities in the semiconductor and charge configuration switching near the gates [8]. These effects cause $1/f$ noise in electron transport, and may contribute to decoherence at low frequencies. Also, we left out the effect of the *ac* field coupling the two low energy states of the qubit to higher energy states. Results of our numerical simulations of the coupling matrix in the qubit [3] show that Rabi oscillations can be observed even in the presence of the *ac* excitation mixing the states (to be published elsewhere).

Acknowledgements

This work is supported by ARO grant DAAG55-98-1-0369, NSF Award 67436000IRG, Stichting voor Fundamenteel Onderzoek der Materie and the New Energy and Industrial Technology Development Organization.

References

[1] S. Lloyd, Science **261**. 1589 (1993); A. Barenco *et al.*, Phys. Rev. A **52**, 3457 (1995).

[2] J. I. Cirac and P. Zoller, Phys. Rev. Lett. **74**, 4091 (1995); C. Monroe *et al.*, Phys. Rev. Lett. **75**, 4714 (1995); J. M. Raimond *et al.*, Phys. Rev. Lett. **79**, 1964 (1997); N. A. Gershenfeld, I. L. Chuang, Science **275**, 350 (1997); D. Loss and D. P. DiVincenzo, Phys. Rev. A**59**, 120 (1998). A. Shnirman *et al.*, Phys. Rev. Lett. **79**, 2371 (1997); Yu. Makhlin *et al.*, Nature **398**, 305 (1999); L. B. Ioffe *et al.*, Nature **398**, 679 (1999).

[3] J. E. Mooij *et al.*, Science **285** , 1036 (1999); T.P. Orlando *et al.*, Phys. Rev. B**60**, 15398 (1999).

[4] P. Shor, *Proceeding of the 37th Annual Symposium on the Foundations of Computer Science*, 56, IEEE Computer Society Press, Los Alamos, 1996; D. P. DiVincenzo and P. W. Shor, Phys. Rev. Lett.**77**, 3260 (1996); E. Knill and R. Laflamme, Phys. Rev. A**55**, 900 (1997); A. Steane, *Proceedings of the Royal Society of London A*452, 2551 (1996).

[5] J. R. Kirtley *et al.*, Phys. Rev. Lett. **61**, 2372 (1988); A. T. Johnson *et al.*, Phys. Rev. Lett. **65**, 1263, (1990); C. H. van der Wal and J. E. Mooij, J. Superconductivity **12**, 807 (1999).

[6] D. G. Cory *et al.*, Proc. Natl. Acad. Sci. **94**, 1634-1639, 1997, I. L. Chuang, N. A. Gershenfeld, Science **275**, p. 350, 1997

[7] L. Viola, S. Lloyd, Phys. Rev. A**58**, 2733 (1998); L. Viola *et al.*, Phys. Rev. Lett. **82**, 2417 (1999).

[8] T. Henning *et al.*, Eur. Phys. J. B **8**. 627 (1999); V. A. Krupenin *et al.*, J. Appl. Phys. **84**, 3212 (1998); N. Zimmerman *et al.*, Phys. Rev. B **56**, 7675 (1997); E. H. Visscher *et al*, Appl. Phys. Lett. **66**, 305 (1995).

Chapter 29

QUANTUM COMPUTING AND JOSEPHSON JUNCTION CIRCUITS

Y. Makhlin[1,2], G. Schön[1] and A. Shnirman[1]
[1]*Institut für Theoretische Festkörperphysik, Universität Karlsruhe D-76128 Germany*
[2]*Landau Institute for Theoretical Physics Kosygin st. 2, Moscow, Russia*

Abstract Among possible realizations of quantum computing systems nano-electronic circuits appear very promising as they can be easily scaled up to large numbers of qubits and integrated in electronic circuits. We discuss the proposal to use Josephson junctions in the charge regime as quantum bits. They combine the intrinsic coherence of superconducting systems with control possibilities of single-electron circuits. Elementary quantum logic operations can be performed by means of voltage and current pulses. The system can stay phase coherent for a long time which allows many elementary steps before the coherence is lost. To read out the result of computation a quantum measurement is needed which can be accomplished by coupling a single-electron transistor to the qubit.

1. INTRODUCTION

The idea to use quantum systems for information processing was discussed since the beginning of the 1980s. The interest to quantum computing was revived recently, after Shor found that quantum computers can factorize large integers exponentially faster than known classical algorithms allow. Further applications of quantum computers include fast database search and simulation of quantum dynamics. Apart from algorithms, a growing number of applications are developed such as quantum communication techniques, quantum cryptography *etc.* Theory of quantum information processing has achieved tremendous progress in recent years (see Ref. [1] for introduction) and many proposals have been

I. O. Kulik and R. Ellialtioğlu (eds.),
Quantum Mesoscopic Phenomena and Mesoscopic Devices in Microelectronics, 439–446.

made for realization of quantum bits. The best studied ones involve ion traps, NMR in the liquid state and optical systems. Among other realizations nano-electronic devices, and Josephson circuits in particular, appear promising since they can be easily scaled up to large numbers of qubits and integrated in electronic circuits.

Here we briefly introduce the idea of quantum computation concentrating on the requirements to a physical system which serves as a quantum computer. We discuss these requirements using a model system of spin-1/2 particles. Then we demonstrate how these requirements can be satisfied in Josephson junction circuits.

2. A MODEL QUANTUM COMPUTER

To realize a quantum computer a physical system needs to satisfy a number of criteria [2]. First of all, it should serve as a set of two-state systems isolated from the environment. The Hilbert space of the whole system factorizes into a tensor product of elementary 2-dimensional subspaces. This makes 2^N dimensions for N qubits.

Unlike 'classical' bits qubits can exist not only in logic states 0 and 1 but also in superpositions of these states $a\,|0\rangle + b\,|1\rangle$. When a computation is being performed, the qubits are prepared in some well-defined initial states (*e.g.*, all qubits in the state $|0\rangle$). Then a quantum algorithm is applied to the content of the qubit register which is a series of elementary operations. Each of these operations (logic gates) affects the state of only one or two qubits similar to elementary logic gates in 'classical' computation such as AND, OR, NOT. Finally, the result of the computation should be read out of the register. Below we discuss the details of this process in a model system of spin-1/2 particles which can be thought of as qubits.

Consider a system of N spin-1/2 particles with the following Hamiltonian:

$$\mathcal{H} = -\sum_{i=1}^{N} \vec{H}^i(t)\hat{\vec{\sigma}}^i + \sum_{i\neq j} J^{ij}(t)\hat{\sigma}_+^i\hat{\sigma}_-^j + \mathcal{H}_{\text{meas}}(t) + \mathcal{H}_{\text{env}} \qquad (29.1)$$

The first term describes an external magnetic field which can be controlled, individually for each spin. The second term describes spin-spin interactions which are also controlled separately for each pair of qubits. Here $\hat{\sigma}_\pm \equiv \hat{\sigma}_x \pm i\hat{\sigma}_y$ are Pauli matrices. These first two terms can be used to control the Hamiltonian and therefore the quantum dynamics of the system. To read out the result of manipulations, coupling to the meter is needed, described by $\mathcal{H}_{\text{meas}}(t)$. The coupling is switched on only at the final read-out stage so that the meter does not influence the coherent dynamics of the qubits. Uncontrollable degrees of freedom of the environment are inevitably coupled to the spins (the term $\mathcal{H}_{\rm env}$). They influence the quantum spin system, producing computational errors. This coupling to the environment has to be sufficiently weak so that many elementary logic operations can be performed before the influence of the environment becomes significant.

To prepare the initial state of the system we switch on large magnetic field $H_z^i \gg k_{\rm B}T$ for all the qubits while $H_x^i = H_y^i = J^{ij} = 0$. Then after a long relaxation time the system ends up in the ground state, $|\downarrow\downarrow \ldots \downarrow\rangle$.

If we keep all $\vec{H}^i = 0$ and $J^{ij} = 0$ the system does not evolve ($\mathcal{H}_{\rm env}$ can be neglected on short times). To manipulate a particular spin i one can switch on H_x^i for a finite time τ. Integrating the Schrödinger equation we find that a unitary operator $\exp(i\alpha\hat{\sigma}_x^i) = \begin{pmatrix} \cos\alpha & i\sin\alpha \\ i\sin\alpha & \cos\alpha \end{pmatrix}$ is applied to the quantum state. Depending on $\alpha \equiv H_x^i\tau/\hbar$ it produces a spin-flip, for $\alpha = \pi/2$, or a superposition of spin-up and spin-down states, $\frac{1}{\sqrt{2}}(|\uparrow\rangle + i\,|\downarrow\rangle)$ for $\alpha = \pi/4$ if applied to the initial state.

Another single-bit operation can be performed by switching on H_z^i which produces a phase shift $\exp(i\beta\hat{\sigma}_z^i)$ by a controlled angle $\beta \equiv H_z^i\tau/\hbar$. *Any* unitary 2×2 transformation of the state of a qubit can be realized as a sequence of x- and z-rotations. The third, y-component is not needed.

To perform a two-bit gate on qubits (spins) i and j we switch on the coupling $J^{ij} \neq 0$ for time τ. This results in the unitary transformation

$$\begin{pmatrix} 1 & 0 & 0 & 0 \\ 0 & \cos\gamma & i\sin\gamma & 0 \\ 0 & i\sin\gamma & \cos\gamma & 0 \\ 0 & 0 & 0 & 1 \end{pmatrix} \tag{29.2}$$

in the basis $|\uparrow_i\uparrow_j\rangle\,, |\uparrow_i\downarrow_j\rangle\,, |\downarrow_i\uparrow_j\rangle\,, |\downarrow_i\downarrow_j\rangle$, with $\gamma = J^{ij}\tau/\hbar$. For $\gamma = \pi/2$ the operation is a spin-swap exchanging quantum states of two spins, while $\gamma = \pi/4$ yields a 'square-root swap'. The latter transforms the product state $|\uparrow_i\downarrow_j\rangle$ into an entangled state $\frac{1}{\sqrt{2}}(|\uparrow_i\downarrow_j\rangle + i\,|\downarrow_i\uparrow_j\rangle)$. These gates, together with single-bit operations discussed above, are sufficient for *any* quantum computation (*cf.* [3]).

The techniques discussed above are very much reminiscent of those used in NMR. Similarly to NMR experiments the quantum dynamics can also be driven by resonant ac-pulses of the fields $\vec{H}^i$ and J^{ij} (*cf.* [5]).

3. JOSEPHSON JUNCTION QUANTUM BITS

The simplest design of a Josephson qubit [5] is shown in Fig. 29.1*a*. It is a superconducting electron box which consists of a Josephson junc-

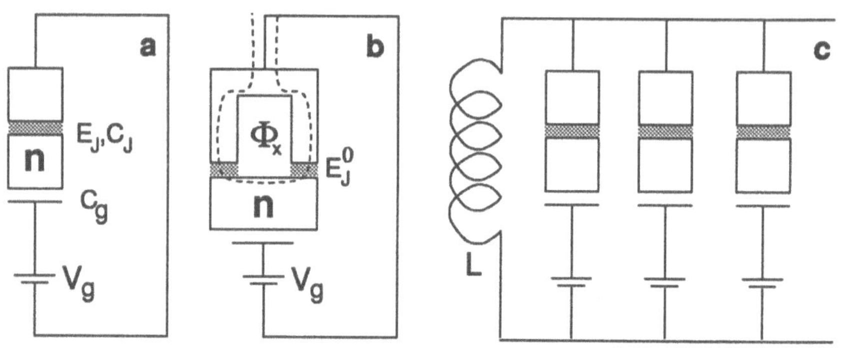

Figure 29.1 *a*) Josephson junction qubit. *b*) A qubit with SQUID-controlled Josephson coupling. *c*) Coupled qubits.

tion, with the Josephson energy E_J and capacitance C_J, biased by a voltage source V_g through a gate capacitor C_g. If the superconducting gap is large enough, at low temperatures single-electron tunneling is suppressed [4] and only even-parity are allowed. The relevant conjugated variables are the number n of extra Cooper pairs on the superconducting island and the phase difference φ across the junction. The dynamics of the circuit is described by the Hamiltonian

$$\mathcal{H} = \frac{(2ne - C_g V_g)^2}{2(C_J + C_g)} - E_J \cos\varphi\,, \tag{29.3}$$

where $e^{i\varphi} = \sum_n |n+1\rangle\langle n|$. The eigenenergies of the system in the regime when the charging energy, of order $E_C = e^2/2(C_J + C_g)$, dominates over the Josephson coupling E_J, are plotted in Fig. 29.2 as a function of the bias voltage V_g. Away from voltages which are multiples of e/C_g, eigenstates are very close to charge states of the system. At low temperatures $k_B T \ll E_C$ the system is in the charge state corresponding to the lowest parabola. However, near the voltages $V_{deg} = (2n+1)e/C_g$ the two lowest states, with n and $n+1$ Cooper pairs, are almost degen-

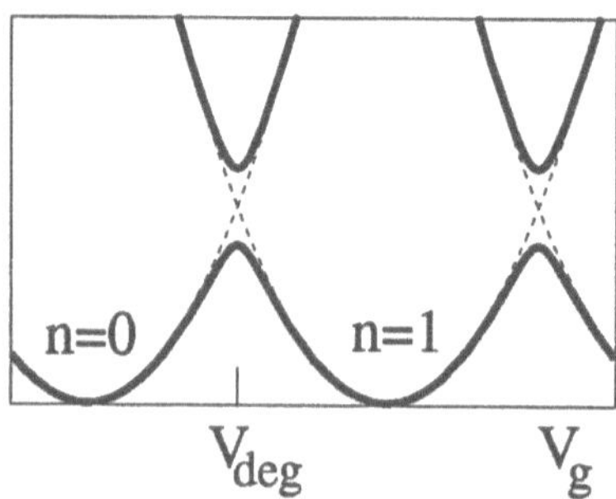

Figure 29.2 Eigenenergies of the superconducting box for different numbers n of Cooper pairs and different bias voltages V_g.

erate and the Josephson coupling mixes them strongly. In this regime the system reduces to a two-state one and can serve as a quantum bit. The Hamiltonian of the qubit is

$$\mathcal{H} = \frac{1}{2}\delta E_{\text{ch}}(V_{\text{g}})\,\hat{\sigma}_z - \frac{1}{2}E_{\text{J}}\,\hat{\sigma}_x \tag{29.4}$$

in the charge basis $|\uparrow\rangle \equiv |n\rangle$, $|\downarrow\rangle \equiv |n+1\rangle$. Here $\delta E_{\text{ch}} = 2e\frac{C_{\text{qb}}}{C_{\text{J}}}(V_{\text{g}} - V_{\text{deg}})$ is the difference in the charging energies between two relevant charge states, and the capacitance of the qubit in the circuit is $C_{\text{qb}}^{-1} = C_{\text{J}}^{-1} + C_{\text{g}}^{-1}$.

Only z-component of the Hamiltonian (29.4) is tunable. To achieve control over x-component one can replace the Josephson junction with a dc-SQUID [6] as shown in Fig. 29.1*b*. If an externally supplied magnetic flux Φ_{x} penetrates the loop of the SQUID, it serves as a Josephson junction [4] with the Josephson coupling $E_{\text{J}} = 2E_{\text{J}}^0\cos(2\pi\Phi_{\text{x}}/\Phi_0)$, where E_{J}^0 is the bare coupling in the junctions. Single-bit logic gates can be performed by voltage and flux pulses. The typical time span of single-bit operations is of order $\hbar/E_{\text{J}}$, $\hbar/\delta E_{\text{ch}}$.

4. COUPLED QUBITS

To couple Josephson qubits in a controlled way we join them in parallel to a common inductor L, see Fig. 29.1*c*. The dynamics of the qubits are coupled to the oscillations in the LC-circuit formed by the inductor and capacitances of all the qubits in parallel. If the oscillations are very fast compared to qubits' typical time scales, $\omega_{LC} = (NLC_{\text{qb}})^{-1/2} \gg \delta E_{\text{ch}}, E_{\text{J}}$, the oscillatory degree of freedom can be integrated out producing an effective coupling between the qubits [5]. To understand the physics of the coupling let us consider the limit of small L (weak coupling) and use the adiabatic approximation. The current through (the voltage source of) the qubit i can be expressed as $I^i = (2e/\hbar)E_{\text{J}}\sin\varphi = (eE_{\text{J}}/\hbar)\hat{\sigma}_y^i$, and the inductor is biased by the current $I = \sum_i I^i$. The Hamiltonian of the fast degree of freedom, the LC-circuit, is $\frac{q^2}{2NC_{\text{qb}}} + \frac{\Phi^2}{2L} - I\Phi = \frac{q^2}{2NC_{\text{qb}}} + \frac{(\Phi - LI)^2}{2L} - \frac{LI^2}{2}$ in terms of the flux Φ through the inductor and conjugated charge q. The oscillator is always in its ground state, with the energy given by the ground state energy of the first two terms, $\frac{1}{2}\hbar\omega_{LC}$, plus the last term $-\frac{L}{2}\left(\sum_i I^i\right)^2$ which produces the interqubit coupling. Introducing $E_L = \frac{\Phi_0^2}{\pi^2 L}\left(\frac{C_{\text{qb}}}{C_{\text{J}}}\right)^2$ we rewrite the interaction as

$$\mathcal{H}_{\text{int}} = -\sum_{i<j}\frac{E_{\text{J}}^i E_{\text{J}}^j}{E_L}\hat{\sigma}_y^i\hat{\sigma}_y^j\,, \tag{29.5}$$

where $E_J^i \equiv E_J(\Phi_x^i)$ are controlled by the external fluxes Φ_x^i. Using this coupling we can perform all logic operations. Single-bit operations can be performed on one qubit at a time, keeping the interaction off. When a two-bit gate is needed, two Josephson couplings are switched on, yielding the total Hamiltonian $\mathcal{H} = -\frac{1}{2}\left(E_J^1\hat{\sigma}_x^1 + E_J^2\hat{\sigma}_x^2 + \frac{E_J^1 E_J^2}{E_L}\hat{\sigma}_y^1\hat{\sigma}_y^2\right)$. Although different from the interaction discussed in Section 2., this coupling also allows to perform a two-bit gate which together with single-bit ones forms a universal set. The typical time span of the two-bit gate is of order $\hbar E_L/E_J^2$.

5. ENVIRONMENT AND DEPHASING

The most important sources of dephasing for Josephson qubits are electromagnetic fluctuations in the external circuit and background charge fluctuations. To estimate the effect of the former one can model them by an effective impedance of the voltage source. A two-state system coupled to environment was discussed in the context of the spin-boson model [7]. Using those considerations we find [5, 6] that the effect of the environment on the qubit is two-fold: it induces random fluctuations of the qubit's eigenlevels and incoherent transitions between them. As a result, in the eigenbasis of (29.4) off-diagonal elements of the reduced 2×2 density matrix of the qubit decay to zero and the diagonal ones relax to their thermal equilibrium values. The corresponding time scales are given by

$$\tau_\varphi = 2\tau_{\text{rel}} = \frac{1}{2\pi}\frac{R_K}{R}\left(\frac{C_J}{C_{qb}}\right)^2 \frac{\hbar}{E_J}\tanh\frac{E_J}{2k_B T}\,, \tag{29.6}$$

at the degeneracy point, $V_g = V_{\text{deg}}$, while far away from degeneracy, at $\delta E_{\text{ch}} \gg E_J$, relaxation becomes very slow and the dephasing time is $\tau_\varphi = \frac{1}{8\pi}\frac{R_K}{R}\left(\frac{C_J}{C_{qb}}\right)^2 \frac{\hbar}{k_B T}$. Here $R_K \approx 26$ kΩ is the quantum resistance, and R is the resistance in the voltage circuit. The time scale of a single-bit gate is of order $\hbar/E_J$. Hence, for $R \sim 50\ \Omega$, relatively small gate capacitance $C_g/C_J \approx 0.1$ and the optimal temperature $k_B T \sim E_J/2$, one can perform about 10^4 operations before the coherence is destroyed.

6. DISCUSSION

To read out the information from a quantum computer we need a quantum measurement. For Josephson qubits discussed above it can be accomplished by coupling a single-electron-transistor capacitively to the qubit. During the computation the SET is kept in the off-state, with no additional dephasing effects on the qubit. To perform the measurement the transport voltage is switched on inducing dissipative current through the SET. Monitoring the current one can extract information about the quantum state of the qubit [10].

Recent experiments have proven that the superconducting box can be described as a two-state quantum system. Nakamura *et al.* [8] have observed coherent oscillations between two charge states demonstrating an elementary single-bit operation (x-rotation). It was the first observation of quantum coherence in macroscopic systems [9]. Experiments with Josephson junctions in the phase dominated regime, $E_J \gg E_C$, are also underway [11].

Further experimental progress in the realization of controlled interqubit coupling and optimization of the measurement process should enable applications of the circuits as well as investigation of fundamental properties of quantum mechanical systems.

References

[1] D. P. DiVincenzo, Quantum computation, Science **269**, 255–261 (1995); A. Ekert and R. Jozsa, Quantum computation and Shor's factoring algorithm, Rev. Mod. Phys. **68**, 733–753 (1996); A. Barenco, Quantum Physics and Computers, *Contemp. Phys.* **37**, 375–389 (1996); D. Aharonov, Quantum computation, in D. Stauffer (ed.) *Annual Reviews of Computational Physics VI*, (World Scientific, Singapore, 1999).

[2] D. P. DiVincenzo, Topics in quantum computers, in L. Kowenhoven, G. Schön, and L. Sohn (eds.), *Mesoscopic Electron Transport*, (Kluwer Ac. Publ., Dordrecht 1996).

[3] A. Barenco *et al.*, Elementary gates for quantum computation, Phys. Rev. A **52**, 3457–3467 (1995).

[4] M. Tinkham, *Introduction to Superconductivity*, (McGraw-Hill, New York 1996).

[5] A. Shnirman, G. Schön and Z. Hermon, Quantum manipulations of small Josephson junctions, Phys. Rev. Lett. **79**, 2371–2374 (1997).

[6] Yu. Makhlin, G. Schön and A. Shnirman, Josephson junction qubits with controlled couplings, *Nature* **398**, 305–307 (1999).

[7] U. Weiss, *Quantum dissipative systems*, (World Scientific, Singapore, 1993).

[8] Y. Nakamura, Yu. A. Pashkin and J. S. Tsai, Coherent control of macroscopic quantum states in a single-Cooper-pair box, Nature **398**, 786–788 (1999).

[9] A. J. Leggett, Quantum mechanics at the macroscopic level, in J. Souletie, J. Vannimenus, R. Stora (eds.) *Chance and Matter*, (Elsevier Sci. Publ., 1987) pp. 395–506.

[10] A. Shnirman and G. Schön, Quantum measurements performed with a single-electron transistor. Phys. Rev. B **57**, 15400–15407 (1998).

[11] J. E. Mooij *et al.*, Josephson persistent-current qubit, *Science* **285**, 1036–1039 (1999).

VIII

NANO-ELECTRONICS

Chapter 30

RECENT ADVANCES IN NANOTECHNOLOGY: AN OVERVIEW

R. Ellialtıoğlu
Department of Physics, Bilkent University
06533 Bilkent, Ankara, Turkey

Advanced materials can be obtained by changing the chemical composition of the components. The bulk properties of such materials can be predicted with great accuracy. On the other hand, controlling the size and shape of the components in nano scale also leads to new advanced materials. However, these nanostructures would have properties quite different from the bulk properties of the same material systems due to quantum size effects and to surface reconstructions. To this end, one may consider the field of nanotechnology as the laboratory of quantum effects. The building blocks of nanotechnology, so-called isolated nanostructures, may be created by synthesis, by forming clusters, by self-assembly, by modifying molecular structures already available in nature, or by initiating local reactions using extremely fine probes. As a further step down the line, isolated nanostructures may be assembled into much larger functional complexes. But, some of the properties of the isolated nanostructures may be lost in the process. So, it is extremely difficult to design such complexes to function in a predictable way even if the properties of the building blocks are precisely known.

Although the first step, *i.e.* to investigate new effects and phenomena related to many types of isolated nanostructures and nano devices, and to develop methods of fabricating them, is very appealing, the second step, the *manufacturing* of functional *nano entities* is more challenging in the sense that the real impact of nanotechnology to the society will be seen when the latter is achieved. However, it should not be considered as a near term goal.

I. O. Kulik and R. Ellialtioğlu (eds.),
Quantum Mesoscopic Phenomena and Mesoscopic Devices in Microelectronics, 449–456.

Nanoscience is an inter-disciplinary field that benefits from the tools and techniques developed by each discipline separately, although each one possesses different perspectives in nanotechnology. As a multi-disciplinary field nanoscience finds growing research areas in condensed matter physics, materials science, electrical engineering, solid state chemistry, and molecular biology. In fact, nanotechnology became a trans-disciplinary field dictating the needs and the directions of research and development. For instance, a biologist would have never thought of a DNA strand to be used as a conducting wire in an electronic circuit, neither a physicist nor an electrical engineer for that matter. Hence, the field of nanoscience and nanotechnology requires a new breed of scientists and engineers who can work across disciplines, in order for rapid progress. Furthermore, with the budgets reaching a billion dollar mark, one can say that nanoscience is already a mega-science.

The main objectives are to nanostructure materials for novel performance, to characterize them with higher resolution and sensitivity, to understand the mechanisms responsible for the observed novel properties, and to design and manufacture new devices from nanoscale building blocks with improved properties and functionalities.

Computation and modeling of nanostructures is essential for predicting the properties and accurately simulating the functional behavior of nanoscale devices as the sizes vary from atomic to mesoscopic, and even to macroscopic dimensions. Structural, chemical and thermal stability of nanostructures and devices is of most concern in applications where the environments have changing chemistry and temperature. Due to very high specific surface areas of nanostructures (especially nanoparticles) and their local surface chemistry which play critical roles in the interaction between the nanoscale building blocks, it is important to control their size, composition and morphology, as well as their spatial distribution, size distribution and composition variation when embedded in a matrix. In some applications, such as the quantum computer architecture, there is no room for an error in the realization of the designed circuit layout, where not only the switching elements are to be nanodevices, but the interconnects as well. So, another goal is to develop and improve enabling technologies for the manufacturing of functionally large systems that might need to use bottom-up approach, depending upon the stringency of the required properties.

With the recent advances in fabrication and characterization techniques towards higher sensitivity and higher resolution, the promises of nanotechnology now seem more plausible. The ability to control over the composition and the thickness of thin films grown on a substrate with extreme precision using epitaxial techniques, and the ability to shrink

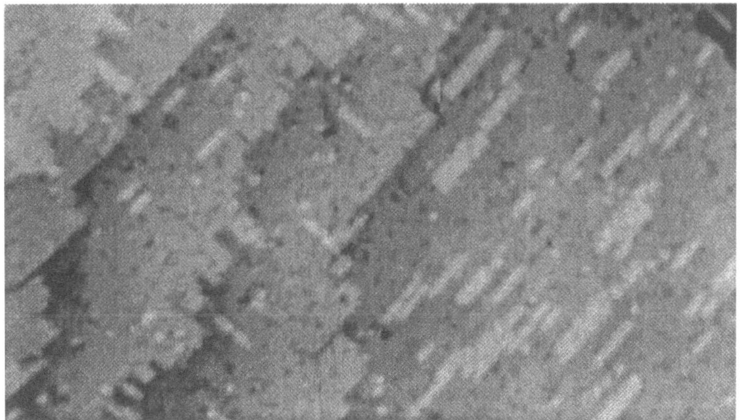

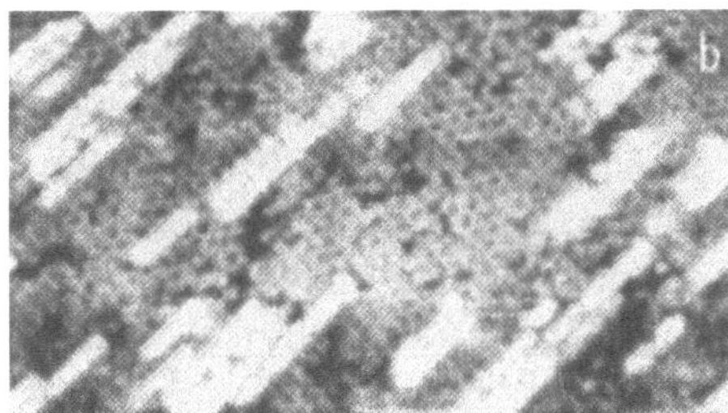

Figure 30.1 (*a*) Room temperature STM image of a vicinal Si(001) (2×1) surface with 0.11 ML Si grown at T_s = 300 °C (image size ~1020×1020 Å). After Ref. [2]. (*b*) 0.25 ML $Si_{.36}Ge_{.64}$ grown on Si(001) at T_s = 310 °C (image size ~340 ×230 Å). After Ref. [3].

the lateral dimensions down to the order of 10 nm using e-beam and X-ray lithography has already made it possible for more than a decade to fabricate structures and devices in the nanoscale. However, the most important tool discovered in this period is the Scanning Tunneling Microscope (STM) [1] and many of its derivatives, which are being employed in all fields of nanoscience and nanotechnology. STM has been the work horse in the investigations of surface reconstructions, epitaxial growth kinetics, surface diffusion, adsorption, *etc.* An example of such studies was carried out in author's laboratory, where Oral and Ellialtioglu [3] investigated the initial stages of epitaxial growth in a UHV chamber equipped with a home-built STM. They studied kinetics of growth of Si and SiGe on a vicinal Si(001) surface at various substrate temperatures, T_s. Fig. 30.1*a* shows the STM images of 0.11 monolayer (ML) Si grown, where the dimer rows were formed orthogonal to the underlying dimer rows due to 2×1 reconstruction. Since there is no strain in homoepitaxy, the symmetric nature of dimers are clear. However, when Ge is incorporated in the growth, buckling of dimers occurs, as shown in Fig. 30.1*b*. The strain in the overlayer changes the reconstruction of the Si underlayer from 2×1 to 4×2. They also observed 2×n reconstructions at intermediate temperatures. At higher substrate temperatures, *i.e.* around the actual growth temperature, the vacancies and defects are healed for a few monolayer growth.

The importance of Scanning Probe Microscopies (SPM) are not only because of their ability to give direct information on the topographical, spectroscopic, or any other respective properties of the material surface in atomic or near atomic resolution, but also because of their utilization in atomic scale manipulation of nanostructures and nano-devices.

Eigler [4] used STM, for the first time, as a tool to move the atoms on a surface and position them as designed. In this and similar works that followed [5], the so-called vertical manipulation have been carried out at cryogenic temperatures, where the atoms were picked up at one place, moved to another place and deposited there. Lateral manipulation, *i.e.*, pulling, pushing and sliding the species with the STM tip, was used in order to demonstrate the repositioning of a molecules adsorbed on a surface *at room temperature* [6, 7]. Recently, Fishlock *et al.* [8] maneuvered a single Br atom on a Cu(001) surface at room temperature. They discovered that the triggering mechanism for the atom motion is not the electric field or the proximity of tip and surface, but rather the tunneling current density. Fig. 30.2 shows the STM images before and after the displacement of the marked atom by five lattice positions.

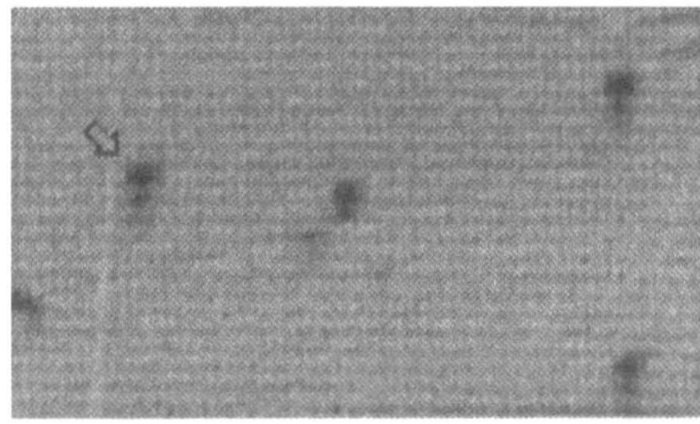

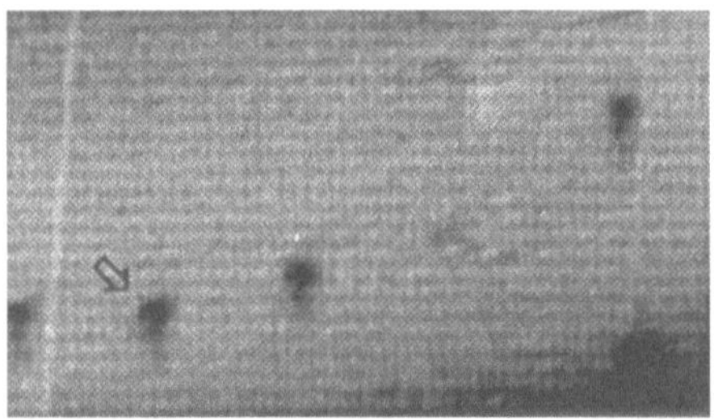

Figure 30.2 Controlled positioning of a single Br atom on Cu(001) surface at room temperature. The lower image is taken before, and the upper image after, a manipulation stroke on the marked atom. Both images have been taken at 1 nA and 10 mV; the manipulation has been at 3 nA. After Ref. [8].

Meyer *et al.* [9] removed a single atom from the step edge of Cu(211) using yet another type of manipulation without touching the atom but through a step roughening mechanism as a result of tip surface interaction [10]. Recently, H. J. Lee and W. Ho [11] used STM to bond a single CO molecule to a single Fe atom, followed by another CO molecule bonding to the same Fe atom to form $Fe(CO)_2$. Then they used the STM to characterize the bonds that formed, by measuring the single molecule vibrational modes.

The discovery of carbon nanotubes [12], has opened a new horizon in nanotechnology research. A very important property of carbon nanotubes, in addition to their high strength and high thermal conductivity, is that they can be either semiconducting or metallic depending on the chirality and their diameter. This makes the carbon nanotube a possible candidate for a building block of a molecular computer. They have been used to assemble nanotube circuits with AFM [13], to build transistors out of single nanotubes [14, 15], and have been synthesized into arrays of reconfigurable nanoswitches for memory and logic elements [16]. Furthermore, the utilization of carbon nanotubes includes applications such as AFM or STM tip [17], templates for a nanotube of an inorganic oxide, *e.g.* zirconia or yttria-stabilized zirconia [18], nanotweezers [19] where the tweezer ends close due to elastic deformation when a small voltage is applied to the nanotubes.

Organic molecules are also used for the purpose of building molecular devices and eventually molecular computers. Reed and Tour [20] have synthesized molecules to fabricate molecular switches and memory devices using self-assembly techniques, which is a method of preparation that allows the fabrication of identical nanostructures in large numbers.

Self-assembly is a process in which microscopic phase separation or surface segregation occurs due to competing molecular interactions between the component materials resulting in a hierarchy of bond lengths in the final nanostructured material; hence, the shape. One example to the utilization of this method is the nanofabrication of self-organized quantum dots using strained layer epitaxy. To name a few, arrays of InAs quantum dots on GaAs [21], Ge quantum dots on Si [22], SiGe on Si [23] have been grown, and such self assembled quantum dots have been incorporated into microelectronic devices, *e.g.* semiconductor laser structures [24].

Another method of preparation which is utilized in solid state chemistry is colloidal synthesis of nanocrystals. This has proven to be an enabling technology to produce nanoparticulate dispersions and coatings that finds many applications already in the market; for a review, see Ref. [25].

In biology, gene splicing, cloning, overexpression in bacterial production systems are being practiced routinely. Borrowing these techniques and modifying the biological materials like DNA, proteins, enzymes, *etc.*, unfolds another path to building functional nanostructures, in addition to top-down (lithography) and bottom-up approaches. An example to artificial biological devices is a protein motor F_1-ATPase [26], which is powered by ATP (Adenosine Tri-Phosphate). Thus, it can be embedded into the body and may, for instance, be utilized in controlled drug delivery in the circulatory system or in other cellular transport applications.

One of the most challenging goal of nanoscience is to develop a quantum computer. Major candidates for a new processor (possibly a cellular automata type) architecture are the resonant tunneling transistor (RTT), the single electron transistor (SET), and the quantum dot (QD). For an extended review of quantum computation, see Chapter 27 of this volume by D. P. DiVincenzo.

In the area of modeling and simulation of nanostructures, the research activities extend from electronic structure and/or energetics of an adatom to the design of a nanomechanical manipulator composed of tens of thousands of atoms. The number of atoms that nanostructures contain is treatable considering the computing power available today. A recent study by Kılıc *et al.* [27], proposes a model for quantum heterostructures with variable (and reversible) electronic properties, that can be realized on a single wall carbon nanotube (SWNT). They calculated the electron density $|\Psi_{i,\mathbf{k}}(j)|^2$ using a transferable tight-binding model. Their result for A_8B_8 superlattice generated from the (7,0) zigzag nanotube is shown in Fig. 30.3. For this particular nanotube, distortion to elliptic cross section reduces the band gap. The confinement of states at the band edge indicates a multiple quantum well structure (MQWS) behavior. They argued that these kinds of behavior can be exploited to produce interesting quantum structures and devices on a SWNT, such

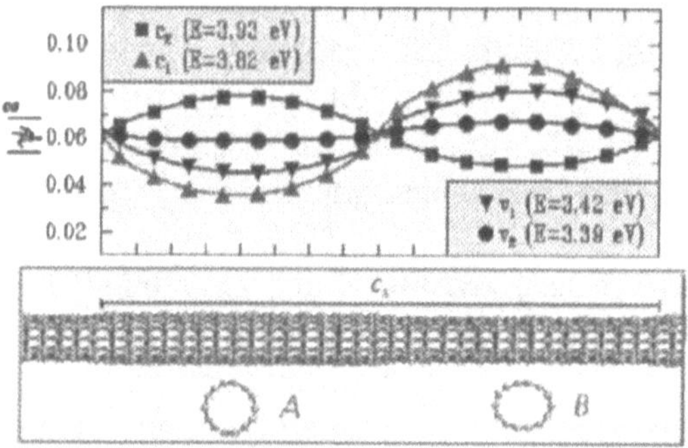

Figure 30.3 Upper panel: the electron density $|\Psi_{i,\mathbf{k}}(j)|^2$ integrated at the cell j of the supercell for first and second states at the edge of the conduction band (c_1 and c_2), and first and second states at the edge of the valence band (v_1 and v_2). Lower panel: The supercell of A_8B_8 superlattice generated from the (7,0) SWNT. c_s is the lattice parameter. Region A is undeformed whereas region B is deformed to have elliptic cross section.

as MQWS, resonant tunneling double barrier structures, rectifying junctions, and variable nanoresistors.

In summary, the progress in nanoscience and nanotechnology research during the last decade was extremely rapid in approaching the objectives of nanotechnology, *i.e.*, better understand the properties of isolated nanostructures, design nanostructures to have desired properties using appropriate computing tools, make nanostructures with atomic-level precision, mass produce such structures inexpensively. But, the next decade promises much more rapid and intensive developments towards the far reaching goals of the nanotechnology era offering revolutionary applications with direct impact to society.

This chapter attempts to review the recent advances with examples covering only a small cross-section of the demonstrations and achievements in nanoscience and nanotechnology. Since the progress is so rapid and the field is so broad that this review is by no means complete and will be out-dated soon. However, it is important to recognize that the research in the field of quantum mesoscopic physics will continue to be the driving power in understanding the structure-property relations of the nanostructures and to lead to the engineering of novel materials and devices with unprecedented properties.

References

[1] G. Binnig, H. Rohrer, Ch. Gerber and E. Weibel, Phys. Rev. Lett. **49**, 57 (1982).

[2] A. Oral, Ph.D. Thesis (Bilkent University, 1994).

[3] A. Oral and R. Ellialtioglu, Surf. Sci. **323**, 295 (1995).

[4] D. M. Eigler, E. K. Schweizer, Nature (London) **344**, 524 (1990);

[5] M. F. Crommie, C. P. Lutz, D. M. Eigler, Science **262**, 218-220 (1993); M. F. Crommie, C. P. Lutz, D. M. Eigler, Nature **363**, 524-527 (1993); M. F. Crommie, C. P. Lutz, D. M. Eigler, E. J. Heller, Surface Review and Letters **2**, 127-137 (1995); M. T. Cuberes, R. R. Schlitter and J. K. Gimzewski, Appl. Phys. Lett. **69**, 3016 (1996).

[6] T. A. Jung, R. R. Schlitter, J. K. Gimzewski, H. Tang and C. Joachim, Science **271**, 181 (1996).

[7] L. Bartels, G. Meyer and K. H. Rieder, Phys. Rev. Lett. **79**, 697 (1997);

[8] T. W. Fishlock, A. Oral, R. G. Egdell and J. B. Pethica, Nature **404** 743 (2000).

[9] G. Meyer, L. Bartels, S. Zöphel, E. Henze and K. H. Rieder, Phys. Rev. Lett. **78**, 1512 (1997).

[10] J. J. Schulz, R. Koch and K. H. Rieder, Phys. Rev. Lett. **84**, 4597 (2000).

[11] H. J. Lee and W. Ho, Science **286**, 1719 (1999).

[12] T. Guo, P. Nikolaev, A. Thess, D. T. Colbert and R. E. Smalley, Chem. Phys. Lett. **243**, 49 (1995).

[13] J. Lefebvre, J. F. Lynch, M. Llaguno, M. Radosavljevic and A. T. Johnson, Appl. Phys. Lett. **75**, 3014 (1999).

[14] S. J. Tans, A. R. M. Verschueren and C. Dekker, Nature **393**, 49 (1998).

[15] R. Martel, T. Schmidt, H. R. Shea, T. Hertel and Ph. Avouris, Appl. Phys. Lett. **73**, 2447 (1998).

[16] C. P. Collier, E. W. Wong, M. Belohradsky, F. J. Raymo, J. F. Stoddart, P. J. Kuekes, R. S. Williams, and J. R. Heath, Science **285**, 391 (1999).

[17] H. Dai, J. H. Hafner, A. G. Rinzler, D. T. Colbert and R. R. Smalley, Nature **384**, 147 (1996).

[18] C. N. R. Rao, B. C. Satishkumar and A. Govindaraj, Chem. Commun. 1581 (1997).

[19] P. Kim and C. M. Lieber, Science **286**, 2148 (1999).

[20] C. Zhou, M. R. Deshpande, M. A. Reed and J. M. Tour, Appl. Phys. Lett. **71**, 611 (1997); M. A. Reed, C. Zhou, C. J. Muller, T. P. Burgin, and J. M. Tour, Science **278**, 252 (1997).

[21] R. Mirin, A. Gossard and J. Bowers, Elect. Lett. **32**, 1732-34 (1996).

[22] Y.-W. Mo, D. E. Savage, B. S. Swartzentruber and M. Lagally, Phys. Rev. Lett. **65**, 1020 (1990).

[23] R. M. Tromp, F. M. Ross and M. C. Reuter, Phys. Rev. Lett. **84**,4641 (2000).

[24] D. Bimberg, N. Kirstaedter, N. N. Ledentsov, Zh. I. Alferov, P. S. Kopev and V. M. Ustinov, IEEE J. of Selected Topics in Quant. Electronics **3**, 196-205 (1998).

[25] J. Mendel, Dispersions and Coatings, in *Nanostructure Science and Technology*, eds., R. W. Siegel, E. Hu and M. C. Roco, (WTEC, Loyola College, Maryland, 1999), pp. 35-47.

[26] H. Noji, R. Yasuda, M. Yoshida and K. Kinosita, Jr., Nature **386**, 299 (1997).

[27] Ç. Kılıç, S. Ciraci, O. Gülseren and T. Yildirim, Phys. Rev. Lett. (submitted, 2000).

Chapter 31

QUANTUM ELECTRON OPTICS AND ITS APPLICATIONS

W. D. Oliver, R. C. Liu, J. Kim, X. Maitre, L. Di Carlo and Y. Yamamoto
Departments of Electrical Engineering and Applied Physics
Ginzton Laboratory, Stanford, CA 94305

1. INTRODUCTION

With the development of two-dimensional electron gas (2DEG) substrates and submicron lithography techniques, it has become possible to fabricate mesoscopic devices with length scales smaller than the inelastic and elastic scattering lengths of electrons at cryogenic temperatures. In these ballistic devices, the wave-nature of the electron can be probed through dc conductance measurements, proportional to the first-order correlation function of the wavefunction amplitude. Electron focusing, diffraction, and Aharonov-Bohm interference experiments are examples of classical optical phenomena clearly observed in mesoscopic electron systems [1].

The study of higher-order correlation functions, or *quantum optical* phenomena, of electrons and composite particles in mesoscopic systems opens the field of quantum electron optics. Identical quantum particles are inherently indistinguishable, and this oftentimes leads to nonclassical behaviors manifest in higher-order correlation functions. Since the 1950's, probing higher-order correlation functions in photon and atom-cavity systems has led to a more complete understanding of fundamental quantum statistical [2, 3] and quantum mechanical phenomena resulting from entanglement [4, 5, 6], including violations of Bell's inequality [7, 8, 9], quantum nondemolition measurements [10, 11], and teleportation [12, 13, 14]. Recent experiments have demonstrated that quantum optical effects can also be observed clearly in mesoscopic electron systems [15]-[21].

I. O. Kulik and R. Ellialtioğlu (eds.),
Quantum Mesoscopic Phenomena and Mesoscopic Devices in Microelectronics, 457–466.

In this paper, we present two recent experiments that use current fluctuation measurements to probe the second-order electron correlation function (fourth-order in wavefunction amplitude): an intensity interferometry experiment called the Hanbury Brown and Twiss (HBT) experiment [2, 21], and an electron collision experiment [19]. Finally, we discuss how these two experiments can be used to characterize the unique behavior of Bell's state entangled electrons in a proposed electron bunching experiment [22, 23, 24]. Throughout the paper, we consider only ballistic systems, in which the inelastic phonon scattering and the elastic ionized impurity scattering lengths are much longer than the characteristic size of the system at cryogenic temperatures (typically 1.5 K in our experiments). The screening length (typically $\lambda_{sc} \approx 5$ nm) is assumed to be much smaller than the Fermi wavelength (typically $\lambda_F \approx 40$ nm) so that Coulomb interactions can be neglected. We assume ideal thermal reservoirs, independent transport channels, and transmission probabilities independent of the applied bias voltage. This approach directly follows the coherent scattering formalism [25, 26].

2. EQUILIBRIUM AND NON-EQUILIBRIUM NOISE

There are three primary contributions to the noise in the mesoscopic systems considered here: thermal reservoir, single-particle partition, and two-particle collision noise. At thermal equilibrium, a reservoir exhibits noise due to the quantum statistical nature in which particles occupy an available mode [25, 26]. In a thermal boson reservoir, additional particles are emitted into a given mode through a spontaneous and stimulated emission process yielding particle number fluctuations $\langle \Delta N_k{}^2 \rangle_B = \langle N_k \rangle (1 + \langle N_k \rangle)$, where $\langle N_k \rangle = f_B(E_k)$ is the Bose-Einstein distribution [2]. A thermal fermion reservoir will emit particles into a given mode through a phase-space filling process, a manifestation of the Pauli exclusion principle, yielding $\langle \Delta N_k{}^2 \rangle_F = \langle N_k \rangle (1 - \langle N_k \rangle)$ where $\langle N_k \rangle = f_F(E_k)$ is the Fermi-Dirac distribution [27]. In the low-degeneracy limit, $\langle N_k \rangle \ll 1$, both fermions and bosons carry Poisson-like statistics, $\langle \Delta N_k{}^2 \rangle \approx \langle N_k \rangle$. Since many classical phenomena also carry Poisson-like statistics, it is convenient to use Poisson statistics as a reference point to characterize noise processes. We define the Fano factor, $F = \langle \Delta N_k{}^2 \rangle / \langle N_k \rangle$, where $F_C = 1$ (Poisson distributed noise) for a classical reservoir, $F_B = 1 + \langle N_k \rangle > 1$ (super-Poisson distributed noise) for a thermal boson reservoir, and $F_F = 1 - \langle N_k \rangle$, $0 \leq F_F < 1$ (sub-Poisson distributed noise) for a thermal fermion reservoir.

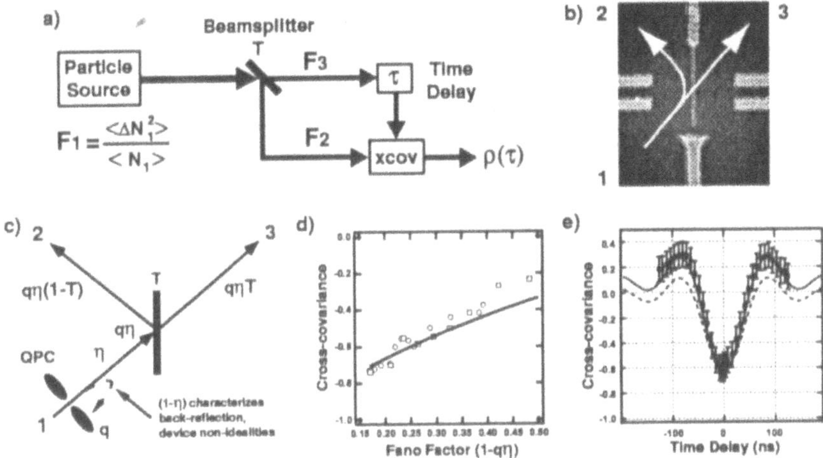

Figure 31.1 *a*) Cross-covariance schematic. The Fano factor of the source, F_1, can be determined from the cross-covariance, $\rho(\tau)$. *b*) Device used for the HBT-type intensity interferometer for electrons. Arrows indicate electron entry and partitioning at the electron beamsplitter. *c*) Probability model for HBT-type intensity interferometer. q is the QPC transmission probability, η accounts for device non-idealities, and T is the beamsplitter transmission probability. *d*) Cross-covariance *vs.* input Fano factor with $T \approx 0.5$ and $\eta \approx 0.83$. The cross-covariance decreases towards -1 as the Fano factor is reduced towards zero, indicating a quiet source. The solid line is a theoretical trace using Eq. (31.2). *e*) Cross-covariance *vs.* delay time at $F = 0.23$. The cross-covariance features a sinc-like behavior due to the finite detection bandwidth (2-10 MHz). A simulation yields the solid line for the actual 2-10 MHz bandwidth, while the dotted line represents a 0-10 MHz lowpass configuration for comparison.

Under non-equilibrium conditions when there is a net flux of particles from a source, partition noise can arise due to the random deletion of the input flux and is fundamentally a single-particle effect. Non-equilibrium collision noise, fundamentally a two-particle effect, is discussed in Section 4.. In Fig. 31.1*a*, we consider the partitioning of particles from a source with Fano factor F_1 at a beam splitter with transmission probability T. The average number of particles transmitted to output 3 is $\langle N_3 \rangle = T\langle N_1 \rangle$. The variance in particle number at output 3 is given by the Burgess variance theorem, $\langle \Delta N_3^2 \rangle = \langle \Delta N_1^2 \rangle T^2 + \langle N_1 \rangle T(1-T)$ in which the first term is the attenuated source noise and the second term is the partition noise [28]. It follows that the Fano factor at output 3 is $F_3 = TF_1 + (1-T)$. Note that this mesoscopic beamsplitter also models a single channel through a quantum point contact (QPC). Between the $(n-1)$ and n'th conductance plateau, there are $(n-1)$ channels (beamsplitters) transmitting with unity probability, and the n'th channel (beamsplitter) transmitting with probability T_n (all beamsplitters in parallel). In this case, for a quiet source ($F_1 = 0$), it can be shown that ${F_3}^{(n)} = T_n(1 - T_n)/[T_n + (n-1)]$ at the output of the n-channel QPC [15, 16, 19].

3. HANBURY BROWN AND TWISS-TYPE INTENSITY INTERFEROMETER

In principle, one can measure the current noise of a particle source (reservoir, beamsplitter, QPC, *etc.,*) by placing a noise detector immediately after it. This is usually done in mesoscopic systems using a square-law device followed by an averager to get a quantity proportional to the time average of the squared current. However, a careful calibration is needed to interpret the resulting value, because the transfer function accounting for the noise detection circuit is usually not known *a priori*. One possibility is to put a known noise source in parallel with one's device, but this may hamper the operation of the device. Another possibility is to characterize the detection system without the device, but then adding the device later may change the transfer function.

Another approach is to physically place a beamsplitter after the noise source and calculate the cross-covariance of the output fluxes as shown in Fig. 31.1*a*. The normalized cross-covariance is defined as

$$\rho(\tau) = \frac{\langle \Delta N_2(t) \Delta N_3(t+\tau) \rangle}{\langle \Delta N_2^2 \rangle^{\frac{1}{2}} \langle \Delta N_3^2 \rangle^{\frac{1}{2}}}, \tag{31.1}$$

where τ is the relative delay time between the beamsplitter outputs. By definition, this is the cross-correlation of the output fluctuations ΔN_2 and ΔN_3. One can uniquely determine the Fano factor of a noise source from the cross-covariance for a $T = 0.5$ beamsplitter with zero delay, $\rho(\tau = 0) = (F_1 - 1)/(F_1 + 1)$. In this case, the cross-covariance is positive for $F_1 > 1$ (super-Poisson distributed noise), negative for $0 \leq F_1 < 1$ (sub-Poisson distributed noise), and zero for $F_1 = 1$ (classical, Poisson distributed noise). The advantage of this approach is that, for matched electronics between the two outputs, the transfer function of the experimental detection system is normalized out of the expression. The positive cross-covariance for photons from a thermal photon reservoir was demonstrated more than forty years ago by Hanbury Brown and Twiss [2]. This intensity interferometry technique has subsequently become an important tool for probing the statistics of photons generated by various types of sources [29, 30, 31].

We recently demonstrated intensity interferometry with electrons in a 2DEG system shown in Fig. 31.1*b* [21]. Schottky gates and an etched trench define a four-port device with a narrow, tunable electron beamsplitter. An equivalent transmission probability model is shown in Fig. 31.1*c*. The QPC serves as the Fermi-degenerate, single-mode, electron source. In this model, the QPC is ideal with transmission probability $q \in [0,1]$. Non-idealities, such as coherent back-reflection from the electron beamsplitter, are characterized by the conditional transmission probability, η, which can be experimentally determined. Using this probability model, one can derive an analytic expression for the cross-covariance,

$$\rho(\tau=0) = \frac{F-1}{\sqrt{F+\frac{T}{1-T}}\sqrt{F+\frac{1-T}{T}}} = -\left[\frac{q\eta T \cdot q\eta \bar{T}}{(1-q\eta T)(1-q\eta \bar{T})}\right]^{\frac{1}{2}} \quad (31.2)$$

where $F = (1 - q\eta)$ is the Fano factor of the source (including the reservoir, QPC transmission probability, and losses) as seen from the beamsplitter, and $\overline{T} \equiv (1-T)$. Experimentally, one can vary the Fano factor of the source, F, by tuning the QPC transmission probability, q. The cross-covariance as a function of the Fano factor is plotted in Fig. 31.1*d*. As the transmission, q, through the QPC is increased, the Fano factor, F, is decreased, and the input electron flux to the beam splitter carries less (current-normalized) noise. The experimental cross-covariance coefficient approaches -1 as F decreases, in close agreement with the analytical trace calculated using Eq. (31.2). Fig. 31.1*e* shows the cross-covariance as a function of the relative delay time, τ, between the two output arms for $F = 0.23$. The sinc-like oscillatory behavior and side-lobe dc-offset are due to the detection bandwidth of our measurement system, 2-10 MHz. The solid line is a zero-parameter fit from a simulation which accounts for the actual detection system including this bandwidth. For comparison, the dashed line is from a simulation using a low pass bandwidth 0-10 MHz which removes the side-lobe dc-offset.

4. QUANTUM INTERFERENCE IN ELECTRON COLLISION

The Hanbury Brown and Twiss-type intensity interferometry experiment is a useful tool to probe the noise of a given system. If that system is known to be a degenerate thermal reservoir, then one can also conclude the type of particle, fermion or boson, from the cross-covariance. However, this is not true for arbitrary noise sources or reservoirs in the non-degenerate limit. For example, a single-photon turnstile device is an engineered source of quiet, single-file photons [32]. An HBT-type measurement of this source would yield a *negative* cross-covariance since the source is quiet, *i.e.*, $F_1 = 0$. To determine directly the statistical nature of quantum particles, one should investigate the two-particle contributions to the noise. Such terms arise in the collision of particles [3, 19].

The symmetrization postulate for quantum particles can lead to non-classical interferences affecting the collision of two quantum particles at a beamsplitter [25, 26]. In Figs. 31.2*a* and *b*, the two-particle scattering amplitudes are calculated for two identical particles from different inputs incident on a $T = 0.5$ beamsplitter. In Fig. 31.2*a*, the detection of one particle in each output, the anti-bunched (1,1) state, is analyzed. The "direct" term interferes quantum mechanically with the physically indistinguishable "exchange" term in which the roles of the particles are reversed. This interference is destructive for bosons and constructive for fermions. The analogous calculation for two particles in a single output, the bunched (2,0) state, is shown in Fig. 31.2*b*. In contrast, two

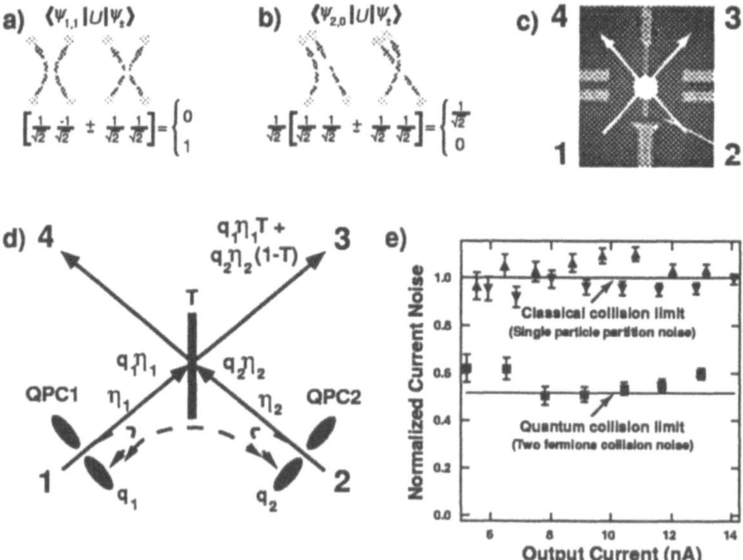

Figure 31.2 *a-b*) Quantum interference in the two-particle scattering amplitudes. The input two-particle bosonic (+) or fermionic (−) wavefunction, $|\psi_{\pm}\rangle$, is scattered by the beam splitter's unitary transformation, U, and projected onto the measurement outcome of *a*) one particle in each output ($|\psi_{1,1}\rangle$), or *b*) two particles in an output ($|\psi_{2,0}\rangle$). *c*) Device used for electron collision experiment. Arrows indicate electron entry, collision, and partitioning at the electron beamsplitter. *d*) Probability model for electron collision. q_1, q_2 are the QPC transmission probabilities, η_1, η_2 account for device non-idealities, and T is the beamsplitter transmission probability. *e*) Normalized current noise as a function of the output current. Triangles represent the noise when either input is biased alone. The theoretically expected fermionic collision noise based on the transmissions (lower solid line) agrees well with the measured electron collision noise (squares).

classical particles would independently scatter into a statistical mixture of the anti-bunched and bunched outputs. In all three cases, the average particle flux at an output is one. However, the Fano factor after collision distinguishes the three cases: $F_3 = 1/2$ for classical particles, $F_3 = 1$ for bosons, and $F_3 = 0$ for fermions.

The device and corresponding model for the electron collision experiment are shown in Figs. 31.2*c* and *d*. Quantum point contacts at the inputs are biased at unity transmission probability, $q_1 = q_2 = 1$, to supply quiet streams of identical particles to the beamsplitter. Non-idealities, such as coherent back-reflection and loss to the opposite inputs, are characterized by η_1 and η_2, and one can use conditional probability arguments to account for the quantum mechanical interactions at the beam splitter. To characterize the noise at an output after collision, we use the output Fano factor of a single-particle partition noise experiment (single-sided input experiment). This is because the output of a single-particle partition noise experiment has the same statistics as the output of a collision experiment using classical particles. The output Fano factor in the electron collision case normalized by the output Fano factor for the single-particle partition (classical collision) case is

$$\frac{F_{electron}}{F_{classical}} = \frac{1 - q\eta}{1 - q\eta T} \tag{31.3}$$

where it is assumed for simplicity that $q_1 = q_2 \equiv q$, $\eta_1 = \eta_2 \equiv \eta$, and the beamsplitter behaves identically as seen from each input. The experimental results are shown in Fig. 31.2*e*. The normalized Fano factor in the collision case is approximately 0.56, representing a significant noise reduction compared with the single-particle partition (classical collision) case. With $q_1 = q_2 = 1$, $T = 1/2$, and using $\eta_1 \approx \eta_2 \approx 0.65$ (as measured independently), the simple model predicts a normalized Fano factor of 0.52, in good agreement with the experimental results. It should be noted that in the dissipative transport limit, partition noise suppression and the recovery of the generalized Johnson-Nyquist noise is due to the frequent elastic/inelastic scattering of electrons and the Pauli exclusion principle [33].

5. ELECTRON ENTANGLEMENT

The HBT and collision experiments can be used as tools to investigate fundamental quantum mechanics, for example to characterize entangled states and to test Bell's inequality [22]. Here we consider an electron "bunching" and "anti-bunching" experiment utilizing electrons entangled in specific spin-singlet Bell's states [22, 23, 24]. As shown in Fig. 31.3, we assume the existence of an electron entangler, which emits

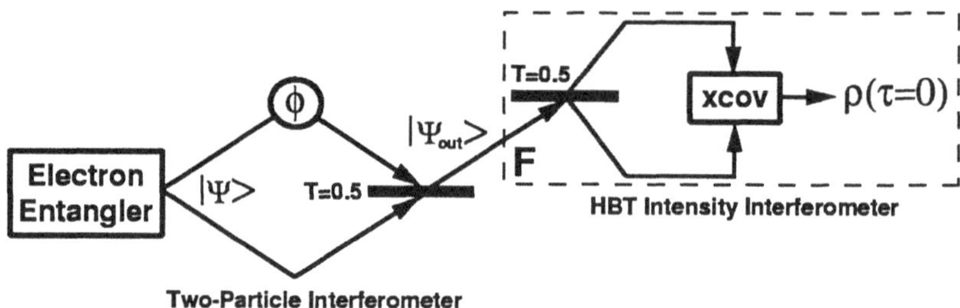

Figure 31.3 Proposed electron bunching and anti-bunching experiment. A two-electron entangler emits electrons in the spin-singlet state into a two-particle interferometer. A phase shift, ϕ, is applied to spin down electrons in the upper arm. The output Fano factor after collision is measured using an HBT-type intensity interferometer.

electrons in the following Bell's state into a two-particle interferometer:

$$|\Psi\rangle = \frac{1}{2}\left(|L_1U_2\rangle + |U_1L_2\rangle\right) \otimes \left(|\uparrow_1\downarrow_2\rangle - |\downarrow_1\uparrow_2\rangle\right), \qquad (31.4)$$

where 1 and 2 denote the particle label, L and U denote the spatial wavefunctions with translation to the lower and upper arms of the interferometer, and $\uparrow$ and $\downarrow$ are spins up and down. The spatial wavefunction is symmetric and the spin wavefunction is anti-symmetric, making the total wave function, $|\Psi\rangle$, anti-symmetric under particle exchange as it must be for electrons. We now assume a means to phase shift selectively any spin-down electron which passes through the upper arm. The electrons are then collided at a $T = 0.5$ beamsplitter to yield the following output state:

$$\begin{aligned} |\Psi_{out}\rangle &= \frac{1}{4}[-i(e^{i\phi}+1)(|L_1L_2\rangle + |U_1U_2\rangle) \otimes (|\uparrow_1\downarrow_2\rangle - |\downarrow_1\uparrow_2\rangle) \\ &+(e^{i\phi}-1)(|L_1U_2\rangle + |U_1L_2\rangle) \otimes (|\uparrow_1\downarrow_2\rangle - |\downarrow_1\uparrow_2\rangle)]. \qquad (31.5) \end{aligned}$$

The two-electron output can be shifted from bunching ($|L_1L_2\rangle$ and $|U_1U_2\rangle$) to anti-bunching ($|L_1U_2\rangle$ and $|U_1L_2\rangle$) states as a function of ϕ. The Fano factor at either output as a function of ϕ is

$$F(\phi) = \frac{1+\rho(\tau=0,\phi)}{1-\rho(\tau=0,\phi)} = \frac{1}{2}(1+\cos\phi), \qquad (31.6)$$

and can be measured using HBT-type intensity interferometry. Note that two-particle bunching corresponds to $F = 1$ (twice the Fano factor of classical partition noise, $F = 1/2$), while anti-bunching corresponds to $F = 0$. Conditional probability models like those used in Sections 3. and 4. can be used to characterize the non-idealities in the electron entangler, collision, and HBT stages.

6. SUMMARY

We presented two quantum electron optics experiments, intensity interferometry and electron collision, and proposed a test of electron entanglement based on these techniques. Investigating quantum optical effects in mesoscopic systems will lead to a more fundamental understanding of the quantum statistics and quantum mechanics of electrons, as well as particles in other systems, for example, the composite particles in the fractional quantum Hall regime [17, 18], Cooper pairs in superconductors [34, 35], and excitons in semiconductors [36].

Acknowledgements

The authors gratefully acknowledge S. Machida for his contributions to the cross-correlation circuit, S. Tarucha for his contributions to the beam-splitter device, and support from the Quantum Entanglement Project (ICORP, JST), MURI, and JSEP. W. D. O. gratefully acknowledges additional support from MURI and the NDSEG Fellowship Program.

References

[1] C. W. J. Beenakker and H. van Houten, *Solid State Physics vol. 44*, Academic Press, San Diego (1991).
[2] R. Hanbury Brown and R. Q. Twiss, Nature **178**, 1447 (1956).
[3] C. K. Hong *et al.*, Phys. Rev. Lett. **59**, 2044 (1987).
[4] A. Einstein *et al.*, Phys. Rev. **47**, 777 (1935).
[5] N. Bohr and L. Rosenfeld, Phys. Rev. **48**, 669 (1938).
[6] D. Bohm, *Quantum Theory*, Constable, London (1954).
[7] J. S. Bell, Physics **1**, 195 (1964).
[8] A. Aspect *et al.*, Phys. Rev. Lett. **47**, 460 (1981).
[9] G. Weihs *et al.*, Phys. Rev. Lett. **81**, 5039 (1998).
[10] P. Grangier *et al.*, Nature **396**, 537 (1998).
[11] G. Nogues *et al.*, Nature **400**, 239 (1999).
[12] D. Bouwmeester *et al.*, Nature **390**, 575 (1997).
[13] D. Boschi *et al.*, Phys. Rev. Lett. **80**, 1121 (1998).
[14] A. Furusawa *et al.*, Science **282**, 706 (1998).
[15] M. Reznikov *et al.*, Phys. Rev. Lett. **75**, 3340 (1995).
[16] A. Kumar *et al.*, Phys. Rev. Lett. **76**, 2778 (1996).
[17] L. Saminadayar *et al.*, Phys. Rev. Lett. **79**, 2526 (1997).
[18] R. de Picciotto *et al.*, Nature **389**, 162 (1997).
[19] R. C. Liu *et al.*, Nature **391**, 263 (1998).
[20] M. Henny *et al.*, Science **284**, 296 (1999).
[21] W. D. Oliver *et al.*, Science **284**, 299 (1999).

[22] X. Maître *et al.*, to be published in Physica E: Low Dim. Sys. and Nanostruct.
[23] M. Michler *et al.*, Phys. Rev. A **53**, R1209 (1996).
[24] G. Burkard *et al.*, cond-mat/9906071 (1999).
[25] M. Büttiker, Phys. Rev. B **54**, 12485 (1992).
[26] T. Martin and R. Landauer, Phys. Rev. B **45**, 1742 (1992).
[27] E. M. Purcell, Nature **178**, 1449 (1956).
[28] R. E. Burgess, Discussions of the Faraday Society **28**, 151 (1959).
[29] G. A. Rebka, Jr. *et al.*, Nature **180**, 1035 (1957).
[30] H. J. Kimble *et al.* Phys. Rev. Lett. **39**, 691 (1977).
[31] F. Diedrich and H. Walther, Phys. Rev. Lett. **58**, 203 (1987).
[32] J. Kim *et al.*, Nature **397**, 500 (1999).
[33] R. C. Liu and Y. Yamamoto, Phys. Rev. B **50**, 17411 (1994).
[34] J. C. Cuevas *et al.*, Phys. Rev. Lett. **82**, 4086 (1999).
[35] J. Torrès and Th. Martin, cond-mat/9906012 (1999).
[36] A. Imamoğlu and R. J. Ram, Phys. Lett. A **214**, 193 (1996).

Chapter 32

PHYSICS AND APPLICATIONS OF PHOTONIC CRYSTALS

B. Temelkuran[†], M. Bayindir and E. Ozbay
Department of Physics, Bilkent University
Bilkent, 06533 Ankara, Turkey

Abstract Photonic crystals are three-dimensional periodic structures having the property of reflecting the electromagnetic (EM) waves in all directions, for a certain range of frequencies. Defects or cavities around the same geometry can also be built by means of adding or removing material. The EM fields in such cavities are usually enhanced, and by placing active devices in such cavities, one can make the device benefit from the wavelength selectivity and large enhancement of the resonant EM field within the cavity. In this work, we have demonstrated the resonant cavity enhanced (RCE) effect by placing microwave detectors in defect structures built around dielectric based photonic crystals. A power enhancement factor of 3450 was measured for planar cavity structures. The tuning bandwidth of the RCE detector extends from 10.5 to 12.8 GHz. We also used these defect structures to demonstrate waveguiding around layer-by-layer photonic crystals. An air gap introduced between two photonic crystal walls was used as waveguide. We observed full transmission of the EM waves through these planar waveguide structures within the frequency range of the photonic band gap. The dispersion relations obtained from the experiments were in good agreement with the predictions of our waveguide model. We also observed 35% transmission for the EM waves traveling through a sharp bend in an L-shaped waveguide carved inside the photonic crystal.

1. INTRODUCTION

The content of optics is the generation, propagation and detection of light. In the recent years, the term photonics has come into use, in analogy to electronics. Electronics deal with the behavior of electric-charge, while photonics concerns with the behavior of photons. The periodic-

I. O. Kulik and R. Ellialtioğlu (eds.),
Quantum Mesoscopic Phenomena and Mesoscopic Devices in Microelectronics, 467–478.

ity of atoms results in an energy band-gap for the electrons, where the electric-charge flow is forbidden. A decade ago, it was suggested that an artificially created periodic structures might result in a similar band gap for electromagnetic (EM) waves, where the propagation of the waves are inhibited in a certain range of frequencies in all directions [1, 2]. In analogy with electronic bandgaps in semiconductors, these structures are called photonic band gap (PBG) materials or photonic crystals [3, 4].

The initial interest in this area came from the proposal to use photonic crystals to control spontaneous emission in photonic devices [1]. However, the technological challenges restricted the experimental demonstrations and relevant applications of these crystals to millimeter wave and microwave frequencies [5]-[8]. Recently, a photonic crystal with a band gap at optical frequencies was reported [9, 10]. With this breakthrough, initially proposed applications like thresholdless semiconductor lasers [11] and single-mode light-emitting diodes [12, 13] became feasible.

One other important issue of the photonic crystals is that, just like the donor or acceptor states in an electronic crystal, breaking the periodicity of the crystal results in localization of the EM field within the defect volume [14]. With these properties, photonic crystals are novel structures that can be used to control the behaviour of light. Very recently, the two-dimensional bandgap laser was demonstrated. The cavity consisted one filled hole (a defect) in an otherwise periodic array of holes penetrating a light emitting, semiconducting film [15].

In this paper, we will present two important applications of photonic band gap materials, namely detector and waveguide. The former is the detection of EM wave, in which we will introduce a detector whose sensitivity and selectivity is significantly improved using photonic crystals. In the latter, EM wave propagates through a waveguide built around photonic crystals.

2. LAYER-BY-LAYER PHOTONIC CRYSTAL

In our experiments, we used a layer-by-layer structure [16, 17] which was constructed by using square-shaped alumina rods (0.32 cm $\times$ 0.32 cm $\times$ 15.25 cm) of refractive index 3.1 at 12 GHz. The stacking sequence repeats every four layers, which has the equivalent geometry of a face centered tetragonal (fct) lattice, corresponding to a single unit cell in the stacking direction. The crystal has a center to center separation of 1.12 cm, corresponding to a dielectric filling ratio of ~ 0.29 (Fig. 32.1a). The layer-by-layer photonic crystal is the first structure that was fabricated at optical frequencies [9, 10].

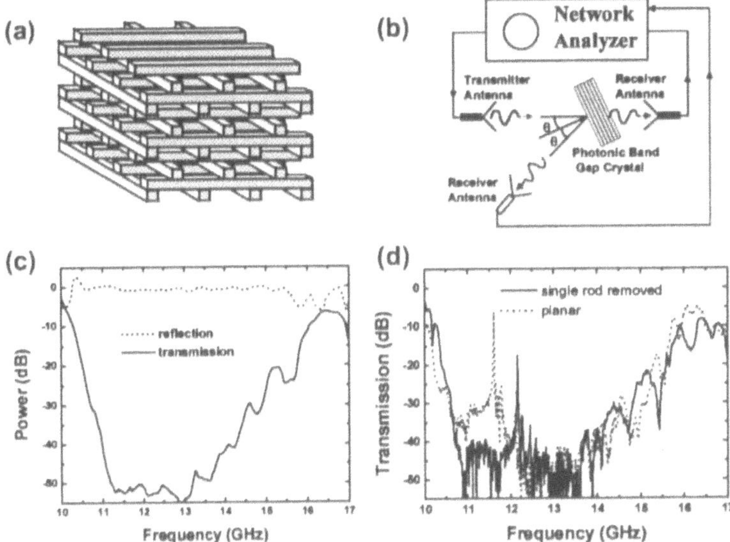

Figure 32.1 (*a*) Schematics of a three-dimensional layer-by-layer photonic crystal. (*b*) The experimental setup for measuring the transmission and reflection characteristics of the photonic crystal. (*c*) Transmission (solid line) and reflection (dotted line) profiles of 4 unit cell periodic structure along the stacking direction. (*d*) Transmission characteristics of a single rod removed (solid line) and planar (dotted line) defect structures.

We measured the transmission and reflection properties of the structure by using a Hewlett-Packard 8510C network analyzer. Standard gain horn antennas were used to transmit and receive the EM radiation (Fig. 32.1*b*). Surroundings of the setup were covered with absorbers resulting in a sensitivity around 70 dB. Fig. 32.1*c* shows the transmission (solid line) and reflection (dotted line) through a 4 unit cell crystal along the stacking direction. Almost all incident EM wave is reflected within the stop-band of the photonic crystal. The transmission is around -55 dB within the band gap, corresponding to 3.5 dB attenuation per layer. The transmission measurements performed at different angles and polarizations show that the three-dimensional stop band, which is referred as the photonic band gap, extends from 10.6 GHz to 12.7 GHz, which agrees well with the expectations of the theory [18].

Breaking the periodicity of the crystal results in evanescent modes within the PBG. We tested two types of such defect structures. Fig. 32.1*d* (solid line) shows the transmission spectrum of a 16 layer (4 unit cell) crystal with a single rod missing from the 8th layer. The resonant frequency of the defect mode is at 12.16 GHz with a Q-factor (quality factor defined as center frequency divided by the peak's full width at half-maximum) is 1380. We also create planar defects by separating the 8th and 9th layers of a 16 layer crystal. The defect frequency, which can be tuned by changing the width of the air gap, appears to be at 11.61 GHz for a separation of 8.6 mm (Fig. 32.1*d*, dotted line), with a Q-factor of 1570 [19].

3. RESONANT CAVITY ENHANCED DETECTORS

Defect structures built around the crystal were tested by putting them in the beam-path of the EM waves propagating along the stacking direction. A square law microwave detector was placed inside the defect volume of the photonic crystal, along with a monopole antenna. The monopole antenna was kept parallel to the polarization vector **e** of the incident EM wave in all measurements. The DC voltage on the microwave detector was used to measure the power of the EM field within the cavity. We also measured the enhanced field by feeding the output of the monopole antenna into the input port of the network analyzer. The monopole antenna was constructed by removing the shield around one end of a microwave coaxial cable. The exposed center conductor, which also acted as the receiver, was 2 mm long. The calibrated enhancement measurements were performed in the following manner. We first measured the enhanced EM field by the probe inside the cavity. While keeping the position of the probe fixed, we removed the crystal and repeated the same measurement. This single pass absorption data of the probe was then used for calibration of the first measurement.

We first investigated the planar defect structure we described in the previous section. Fig. 32.2*a* shows the enhancement characteristics of this defect structure with a separation width of 8.5 mm. The measurement was done by the network analyzer and the frequency was chosen to cover the photonic band gap of our crystal. We observed a power enhancement factor of 1600 at a defect frequency of 11.68 GHz with a Q-factor 900. We then measured the enhancement characteristics of the same defect structure (Fig. 32.1*b*), with a microwave detector, inserted inside the same cavity. An enhancement factor of 450 along with a Q-factor of 1100, were observed at the same defect frequency (Fig. 32.2*b*, solid line).

The discrepancy between two measured enhancement factors can be explained by modeling our structure as a Fabry-Perot cavity. The crystals on each side of the cavity are considered as the photonic mirrors of the Fabry-Perot cavity. The probe we used in our experiments was sim-

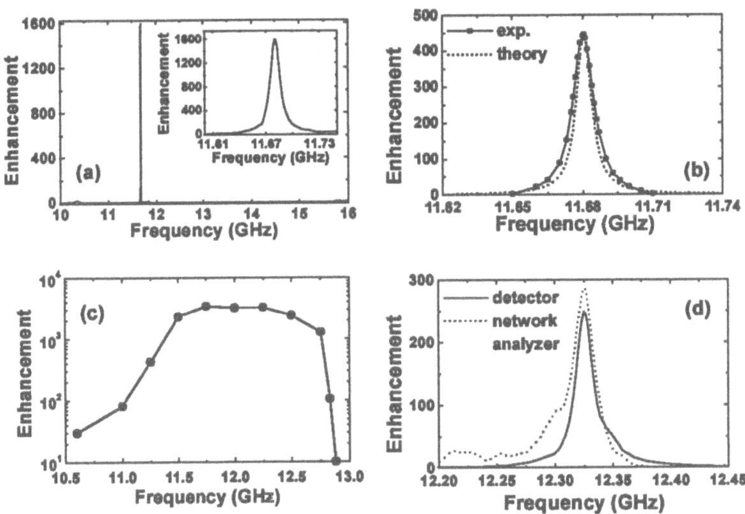

Figure 32.2 (*a*) Experimental enhancement factors obtained for a planar defect structure using the network analyzer. (*b*) Comparison of the experimental (solid line) and theoretical (dotted line) enhancement factors obtained for the RCE detector in the planar defect structure. (*c*) The power enhancement can be obtained at different resonant frequencies by changing the cavity width. This corresponds to a tuning band-width ranging from 10.5 to 12.8 GHz. (*d*) Enhancement characteristics of the box-like cavity measured by the network analyzer (dotted line) and the microwave detector (solid line).

ulated by an absorption region of thickness d, with a relative absorption coefficient (α). We can write the power enhancement factor η, which is defined as the ratio of the stored power inside the absorption layer, to the incident EM wave, for the absorption region within the Fabry-Perot cavity,

$$\eta = \frac{(1 + R_2 e^{-\alpha d})(1 - R_1)}{1 - 2\sqrt{R_1 R_2} e^{-\alpha d} \cos(2\beta L + \phi_1 + \phi_2) + R_1 R_2 e^{-2\alpha d}}, \tag{32.1}$$

where R_1 and R_2 are the reflectivities, ϕ_1 and ϕ_2 are the reflection phase of the mirrors of the cavity, β is the propagation constant of the EM wave in air, and L is the separation width of the cavity.

The above result is normalized with respect to the incident field absorbed by the detector in the absence of the crystal. The aforementioned planar defect structure have symmetric mirrors where $R = R_1 = R_2$. We used the measured transmission characteristics to obtain the reflectivities of our photonic mirrors. As the rods are made of high quality alumina with a very low absorption coefficient, the absorption in the crystal can be neglected [19]. At the defect frequency, the transmission of an 8-layer crystal was 30 dB below the incident EM wave. The reflectivity of the photonic mirrors was then obtained as $R = 1 - T$=0.999. The ideal case which maximizes η corresponds to $\alpha d = 0$, which gives a maximum enhancement factor of 2000. We then varied αd to obtain enhancement factors closer to our experimental measurements. For $\alpha d = 0.0001$, Eq. (32.1) yields an enhancement factor of 1600 (which corresponds to the value obtained from the network analyzer), while $\alpha d = 0.0011$ results in an enhancement factor of 450 (microwave detector). The increased absorption factor for the detector measurement can be explained by the relatively large volume size of the microwave detector compared to monopole antenna alone. Fig. 32.2*b* compares the measured (solid line) and simulated (dotted line) enhancements obtained for the RCE microwave detector within the planar defect structure. The theoretical Q-factor (1500) is comparable with the experimental Q-factor (1100).

The Fabry-Perot model suggests that η is maximized for the matching case $R_1 = R_2 e^{-2\alpha d}$ [20]. To increase the enhancement, we increased R_2 by adding one more unit cell (4 layers) to the mirror at the back. This result in an asymmetric planar cavity with a 2 unit cell thick front mirror, and a 3 unit cell thick back mirror. By varying the width of the planar cavity, we measured the enhancement factors at different resonant frequencies. As shown in Fig. 32.2*c*, the tuning bandwidth of the RCE detector extends from 10.5 GHz to 12.8 GHz. This tuning bandwidth of the RCE detector is in good agreement with the full photonic band gap (10.6-12.7 GHz) of the crystal [17]. As expected, the measured enhancement factors are relatively higher when compared with the symmetrical defect case. The maximum enhancement was measured to be 3450 at a defect frequency of 11.75 GHz. The theory predicted enhancement factors around 5500, which is higher than the measured values. The discrepancy can be explained by the finite size of the photonic crystal, which limits the power enhancement of the field within the cavity.

In order to obtain a defect that is localized in three dimensions, we modified a 16 layer crystal structure in the following manner. Part of the rods on the 8th and 9th layers were removed to obtain a rectangular prism-like cavity. The dimensions of the cavity was $4a \times 4a \times 2d$, where $a = 1.12$ cm was the center to center distance between parallel rods, and $d = 0.32$ cm was the thickness of the alumina rods. We measured the power enhancement characteristics of this structure using the method described earlier. Fig. 32.2*d* (dotted line) shows the measurement made by the network analyzer. An enhancement factor of 290, and a Q-factor of 540 were measured at a defect frequency of 12.32 GHz. We then used a microwave detector within the cavity to probe the EM field inside the localized defect. As shown in Fig. 32.2*d* (solid line), the maximum enhancement (245) occurred at the same frequency, along with a Q-factor of 680. Both measurements clearly indicate the resonant cavity enhancement for the localized defect.

4. WAVEGUIDE

We report our experimental results where we have observed waveguiding in photonic crystal structures [7]. The basic motivation in photonic crystal based waveguides aroused when the following properties of these crystals, which are essential for many applications, were considered. First, photonic crystals have the property of reflecting the EM waves within the band gap frequencies in all directions. Second, defect structures in which the EM wave is trapped, can be created by breaking the periodicity of the crystal. Combining these two properties, an opening carved all through an otherwise-perfect crystal (which resembles a continuous defect structure), may serve as a waveguide. Once the EM wave is coupled inside the guide, the trapped wave, which has no where else to go, is guided through the opening inside the crystal. This guiding mechanism is superior to traditional waveguides which rely on total internal reflection of the EM waves. The serious leakage problem for the EM waves traveling around tight corners in a traditional waveguide can be solved by using a photonic crystal based waveguide, and smaller scale optoelectronic integrated circuits can be successfully built [21, 22]. Fig. 32.3*a* shows the schematics of the measurement set up that was used in our experiments. We measured the transmission-phase and transmission-amplitude properties of the two different waveguide structures, namely a parallel-plate and an L-shaped. We constructed the parallel-plate type waveguide by using two separate 3 unit cell thick layer-by-layer photonic crystals. The crystals were brought together along the stacking direction with a separation width (d) between them, while keeping a mirror type of symmetry between the rods of the two crystals (see Fig. 32.3*a*). For the planar defect structure we have investigated in the previous sections, the propagation direction of the EM wave was perpendicular to the plane of the cavity. If the propagation direction is chosen to be parallel to the plane of the cavity, the structure will have the geometry of a parallel-plate waveguide. We expect the wave to be guided through the introduced air gap, starting from a cut-off frequency which depends on the width of the gap. The guiding is limited with the full band gap frequency range of the photonic crystal, for which the crystal has the property of reflecting the EM waves in all directions.

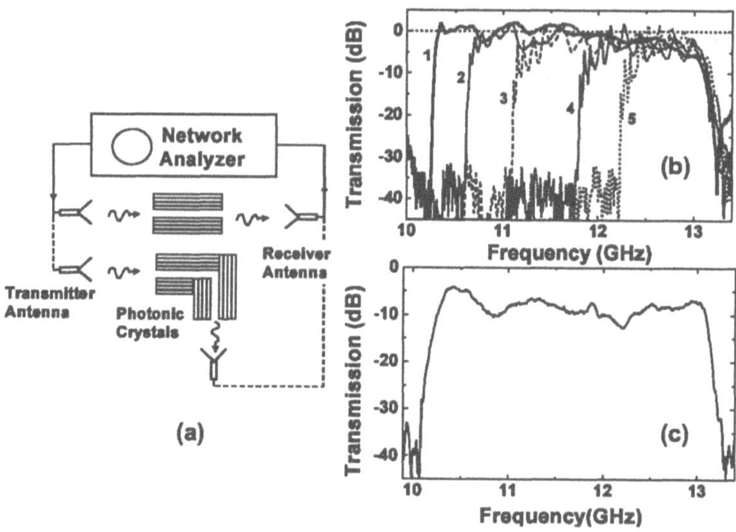

Figure 32.3 (*a*) Experimental setup used to investigate the parallel-plate (upper object) and L-shaped (lower object) waveguide structures. (*b*) Transmission amplitude measured from parallel-plate waveguides as the separation width of the waveguide is changed. The numbers given in the plot are assigned to width of the guides as (1) 18, (2) 16, (3) 14, (4) 12, (5) 10.5 mm. (*c*) Transmission characteristics of the L-shaped waveguide for $d = 20$ mm.

We tested this waveguiding argument, by measuring the transmission properties of these structures along the plane of the cavity. Fig. 32.3*b* shows the transmission properties of the waveguide structure for different separation widths. We observed full transmission (100%) of the EM waves along a certain frequency range. The waveguiding was first observed at a minimum separation width around 10 mm, and the cut-off shifted to lower frequency values as the width of the air gap was increased. Independent of the width of the cavity, the guiding was observed to vanish at a fixed upper cut-off frequency (13.2 GHz), which corresponds to the upper band-edge of the photonic band gap for the EM wave traveling perpendicular to the stacking direction. This is along our expectations as the crystals do not act as mirrors (in all directions) beyond the full band gap frequencies. The lower cut-off frequency is determined by the width of the cavity and corresponds to the resonant frequency of the Fabry-Perot resonator. This resonant frequency can easily be predicted by a Fabry-Perot defect model we have used in our earlier work [6].

As we have pointed earlier, photonic crystal based waveguides were predicted to have the property of guiding the wave through sharp bends

[3]. To demonstrate this effect, we constructed an L-shaped waveguide in the following manner. We coupled the output of the previously described planar waveguide structure, to the input of an other but identical waveguide making 90° with the first one, as shown in the second configuration of the set-up (see Fig. 32.3*a*). Each wall of the waveguide is a 2 unit cell photonic crystal. The width of the cavity is kept at a value of 2 cm, for which the frequency range of the waveguide will overlap with the full band gap of the crystals. Fig. 32.3*c* shows the transmission of the EM waves through the L-shaped waveguide. The maximum magnitude of the transmitted signal was 35% of the incident signal. The frequency range of the L-shaped waveguide again covers the full band gap frequencies of the photonic crystal. The exchange of the receiver and transmitter antennas did not affect the transmission characteristics. The relatively poor performance of the transmission magnitude can be further increased by a proper design of the bend [23]. These results show that photonic crystals can be used for various waveguide configurations.

We investigated the dispersion characteristics of the planar waveguide by measuring the phase difference of the transmitted wave introduced by the guide. This phase difference, ϕ_{trans}, can be written as $\phi_{\mathrm{trans}} = kL - k_z L$, where $k = 2\pi f/c$ is the free space wavevector, k_z is the component of the wavevector along the waveguide (see Fig. 32.4*a*), and L is the length of the waveguide. This can be used to find the normalized propagation constant, k_z/k, as a function of frequency,

$$\frac{k_z}{k} = 1 - \frac{\phi_{\mathrm{trans}}}{kL} = 1 - \frac{\phi_{\mathrm{trans}} c}{2\pi f L} \tag{32.2}$$

The dispersion relation calculated by this phase-measurement method is shown in Fig. 32.4*b* (solid lines) for different separation widths of the waveguide. The separation widths are chosen to be the same with those widths used in the transmission measurements as in Fig. 32.3*c*.

The dispersion relations can also be calculated by a parallel-plate waveguide model. Since the dielectric photonic crystal walls of the waveguide has a certain penetration depth that can be calculated using the reflection-phase information from the walls of the cavity, we can define an effective width for the waveguide. This approach was previously used to investigate the defect characteristics built around dielectric and metallic photonic crystals [6]. In the calculation of this effective penetration depth, one has to consider the angle dependence of the reflection phase, since the wave is considered to be bouncing between the walls of the waveguide at different angles for different frequencies. We measured the reflection phase of the EM waves from the walls of the cavity for the frequency range of the band gap, as a function of θ, the angle between

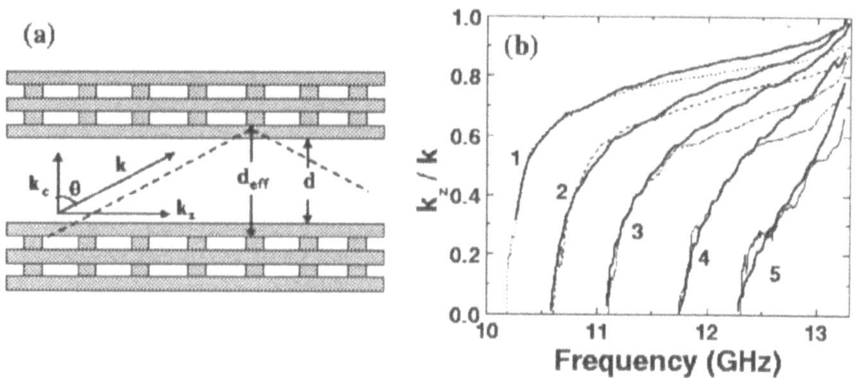

Figure 32.4 (*a*) The diagram of the wave vector for the propagating wave inside the photonic crystal based waveguide. (*b*) Comparison of predicted (solid lines) and calculated (dotted lines) dispersion diagrams for the waveguides with different separation widths (see Fig. 32.3*b*).

the wavevector $\vec{k}$ and its component along the stacking direction of the crystal, k_c, as shown in Fig. 32.4*a*. We calculated the effective width of the waveguide using the total phase contributions of both walls of the cavity, $\phi_{\text{ref}}(\theta, f)$,

$$d_{\text{eff}} = d + \frac{\phi_{\text{ref}}(\theta, f)}{2k}, \tag{32.3}$$

where d is the actual separation width of the waveguide. The corresponding propagation angle for each frequency is obtained from Eq. (32.1) as,

$$\theta = \arcsin\left(\frac{k_z}{k}\right) = \arcsin\left(1 - \frac{\phi_{\text{trans}} c}{2\pi f L}\right). \tag{32.4}$$

This angle information can be used in Eq. (32.3) to find an effective width of the guide at each frequency. The k_c component of the wavevector can be calculated as $k_c = 2\pi/\lambda_c$ where $\lambda_c = 2d_{\text{eff}}$ is the cut-off wavelength of the waveguide. The dispersion relation can now be expressed as

$$\frac{k_z}{k} = \frac{\sqrt{k^2 - k_c^2}}{k}. \tag{32.5}$$

Note that, since k_c is considered to be constant, after some frequency, the square-root becomes real, so that the waves after that cut-off frequency (defined by k_c) can propagate in the guide. For frequencies less than the cut-off frequency, k_z is imaginary, and such modes (evanescent modes) cannot propagate in the waveguide [24]. Fig. 32.4*b* compares the parallel-plate waveguide model dispersion relations (obtained from Eq. (32.5), dotted lines) with the dispersion relations obtained from the transmission phase measurements (using Eq. (32.1), solid lines). As can be seen from the plots, the results are in good agreement for different separation widths of the guide, except for the higher frequency regions of the waveguide. This discrepancy is mainly related to the inaccurate reflection phase information (due to experimental limitations) at higher incidence angles, $\theta > 70°$.

Recently, we have reported a new type of waveguiding which was based on the coupling of localized defect modes in layer-by-layer photonic crystals. Unlike the other types of waveguiding, photons propagate through neighboring defect sites via hopping [25].

5. CONCLUSION

In conclusion, we have investigated the transmission and defect characteristics of layer-by-layer photonic crystals and suggested two applications based on this structure. First, we suggest the possibility of using an embedded detector inside the crystal, as an RCE detector. By using smaller size photonic crystals and higher frequency detectors, the RCE effect can also be obtained at millimeter and far-infrared frequencies. These frequency selective RCE detectors have increased sensitivity and efficiency when compared to conventional detectors, and can be used for various applications. The second application is the guidance of the EM waves with 100% transmission, using photonic crystals. We have developed a parallel-plate waveguide model for our structures. The dispersion diagrams calculated using the transmitted phase measurements and by the waveguide model were in good agreement. We also observed 35% transmission for the EM waves traveling through a sharp bend in an L-shaped waveguide.

Acknowledgements

This work is supported by NATO Grant No. SfP971970, National Science Foundation Grant No. INT-9812322, and NATO-Collaborative Research Grant No. 950079. Ames Laboratory is operated for the U.S Department of Energy by Iowa State University under contract No. W-7405-Eng-82.

References

[1] E. Yablonovitch, Phys. Rev. Lett. **58**, 2059 (1987).

[2] S. John, Phys. Rev. Lett. **58**, 2486 (1987).

[3] J. D. Joannopoulos, R. D. Meade, and J. N. Winn, *Photonic Crystals, Molding the Flow of Light* (Princeton University Press, Princeton, NJ, 1995).

[4] For a recent review, see articles in *Photonic Band Gap Materials*, edited by C. M. Soukoulis (Kluwer, Dortrecht, 1996).
[5] M. C. Wanke, O. Lehmann, K. Muller, Q. Wen, and M. Stuke, Science **275**, 1284 (1997).
[6] B. Temelkuran, E. Ozbay, J. P. Kavanaugh, G. Tuttle, and K. M. Ho, Appl. Phys. Lett. **72**, 2376 (1998).
[7] B. Temelkuran and E. Ozbay, Appl. Phys. Lett. **74**, 486 (1999).
[8] B. Temelkuran, Mehmet Bayindir, E. Ozbay, R. Biswas, M. M. Sigalas, G. Tuttle, and K. M. Ho, J. Appl. Phys. **87**, 603 (2000).
[9] S. Y. Lin *et al.*, Nature (London) **394**, 251 (1998).
[10] J. G. Fleming and Shawn-Yu Lin, Opt. Lett. **24**, 49 (1999).
[11] P. R. Villenevue *et al.*, Appl. Phys. Lett. **67**, 167 (1995).
[12] P. L. Gourley, J. R. Wendt, G. A. Vawter, T. M. Brennan, and B. E. Hammons, Appl. Phys. Lett. **64**, 687 (1994).
[13] J. P. Dowling, M. Scalora, M. J. Bloemer, and C. M. Bowden, Appl. Phys. Lett. **75**, 1896 (1994).
[14] E. Yablonovitch, T. J. Gmitter, R. D. Meade, A. M. Rappe, K. D. Brommer, and J. D. Joannopoulos, Phys. Rev. Lett. **67**, 3380 (1991).
[15] O. Painter, R. K. Lee, A. Scherer, A. Yariv, J. D. O'Brien, P. D. Dapkus, and I. Kim, Science **284**, 1819 (1999); Physics Today, November 1999, p. 20.
[16] K. M. Ho, C. T. Chan, C. M. Soukoulis, R. Biswas, and M. Sigalas, Solid State Commun. **89**, 413 (1994).
[17] E. Ozbay, J. Opt. Soc. Am. B **13**, 1945 (1996).
[18] E. Ozbay, A. Abeyta, G. Tuttle, M. Tringides, R. Biswas, C. T. Chan, C. Soukoulis, and K. M. Ho, Phys. Rev. B **50**, 1945 (1994).
[19] E. Ozbay and B. Temelkuran, Appl. Phys. Lett. **69**, 743 (1996).
[20] M. Selim Unlu and S. Strite, J. Appl. Phys. **78**, (1995).
[21] A. Mekis, J. C. Chen, I. Kurland, S. Fan, P. R. Villeneuve, and J. D. Joannopoulos, Phys. Rev. Lett. **77**, 3787 (1996).
[22] Shawn-Yu Lin, E. Chow, V. Hietala, P. R. Villeneuve, and J. D. Joannopoulos, Science **282**, 274 (1998).
[23] M. M. Sigalas, R. Biswas, K. M. Ho, C. M. Soukoulis, and D. D. Crouch, Phys. Rev. B **60**, 4426 (1999).
[24] J. D. Jackson, *Classical Electrodynamics*, 2nd ed. (Wiley, New York, 1975).
[25] Mehmet Bayindir, B. Temelkuran, and E. Ozbay, Phys. Rev. Lett. **84**, 2140 (2000); Mehmet Bayindir, B. Temelkuran, and E. Ozbay, Phys. Rev. B **61**, xxxx (2000).

Chapter 33

CONDUCTANCE IN METALLIC SUBMICRON CROSS-JUNCTIONS

R. Ellialtıoğlu and İ. İ. Kaya[1]
Department of Physics, Bilkent University
06533 Bilkent, Ankara, Turkey

Abstract Cross conductance characteristics of a hot-spot transistor was presented. The observed nonlinearity was attributed to the reabsorption of non-equilibrium phonons emitted in the contact region by the injected electrons in the channels [1]. The temperature due to electron heating is proportional to the applied bias, $T \propto V$, which results in a substantial increase of the resistance in both channels. Acoustic coupling to the substrate was avoided by fabricating the devices as free-standing. The agreement with the theory is excellent.

1. INTRODUCTION

Recently, there have been various experiments studying the electron heating in metal wires [2]-[5], free-standing metal wires [6, 7, 8] and free-standing semiconductor wires [9] at low temperatures. Kulik, in his review paper in this book, discusses the metallic wires and point contacts in different size regimes, from atomic contacts to macroscopic structures. In this work, our interest has been focused on diffusive, clean, submicron metallic wires and junctions. Hot electrons injected in such structures relax their excess energy by electron-phonon scattering causing the lattice temperature to rise, and as a result a temperature dependent resistivity is added to the residual resistivity as given by the Matthiessen's rule, $\rho = \rho_r + \rho_{ph}(T)$. When $T \gtrsim \Theta_D$, where Θ_D is the Debye temperature, the phonon related resistivity varies linearly with

[1]Present address: Rowland Institute for Science, Cambridge, MA 02142 USA. e-mail: kaya@rowland.org.

I. O. Kulik and R. Ellialtioğlu (eds.),
Quantum Mesoscopic Phenomena and Mesoscopic Devices in Microelectronics, 479–484.

T, whereas for $T \ll \Theta_D$ it obeys the Bloch-Gruneissen T^5 law. Kulik suggested a formula [1] to combine the two temperature regions as $\rho = \rho_0 + \rho_1[1 + (T/\Theta_D)^5]^{1/5}$, which was modified [10] by introducing an experimental fitting parameter $\beta = 3200$ to yield

$$\rho = \rho_0 + \rho_1[1 + \beta(T/\Theta_D)^5]^{1/5}. \tag{33.1}$$

If hot electrons are injected into a metal film or a wire the excess energy will be readily absorbed by the phonons and the phonon distribution function will deviate from Planck distribution. Depending on the dimensions of the structure with respect to the dominant phonon wavelength one may have reabsorption of the non-equilibrium phonons by electrons, non-equilibrium phonon–thermal phonon scattering, or phonon–impurity scattering if the metal is not "clean" enough. In either case, the ratio of the thermal conductivity to the electronic conductivity is proportional to the temperature through the Wiedemann–Franz law, $\kappa/\sigma = \mathcal{L}T$, where $\mathcal{L}$ is the Lorenz constant, which deviates [11] at moderately low temperatures from the Sommerfeld value of $\mathcal{L}_o = (\pi^2/3)(k_B/e)^2$ which is valid for higher temperatures and also for very low temperatures. When an external voltage is applied to the ends of a metallic wire, the temperature at the narrowest part, if there is, or otherwise near the middle of the structure rises to considerably high values. This hot spot temperature is related to the applied voltage [1] by $eV = \gamma k_B T_{hs}$, where $\gamma = 2\pi/\sqrt{3} = 3.63$. Starting from a temperature distribution of the form of an inverted parabola [12]

$$T_e^2(x) = T_b^2 + \frac{3}{\pi^2}\left(\frac{e}{k_B}\right)^2 E^2\left(Lx - x^2\right) \tag{33.2}$$

one obtains the average temperature over the length L of the wire, as $\overline{T}_e \cong T_b/2 + (\pi\alpha/4)V$, where T_b is the bath temperature and $\alpha = e/\gamma k_B = 3200$ K/V. The integrated resistance of the wire in the linear temperature region, *i.e.*, $T \gtrsim \Theta_D$, is given by [10] $R = (L/A)(\rho_0 + \rho_1\beta^{1/5}\overline{T}_e/\Theta_D) \approx (L/A)\rho_1\beta^{1/5}\overline{T}_e/\Theta_D$, where A is the cross-sectional area of the wire. For small voltages of the order of ~ 10 mV and temperatures above $\sim \Theta_D/3$ one can replace $\overline{T}_e$ with T_b, and obtains the empirical parameter β from the slope of the R *vs* T_b curve. Similar expressions were obtained for the resistances of the two channels in a micro-junction structure, where the cross coupling between the channels were taken into account [10].

2. HOT-SPOT TRANSISTOR

Kulik [1] proposed a geometry of two metal wires crossing each other at right angle at a junction, predicting that the nonlinearity in the conductance of one channel can be controlled by the cross channel current; hence he coined the name "hot-spot transistor". He calculated the coupling between the channels by considering the reabsorption of nonequilibrium phonons generated at the crossing point. We observed a nonlinear conductance in such a structure in contact with the substrate, but the observed effect was not as pronounced as the theory predicted, due to the heat leak to the substrate material [13]. However, we later investigated the conductance changes in a free-standing [14] two-wire metallic cross junction as functions of the currents in both channels. The observed nonlinearity is in good agreement with Kulik's results.

3. EXPERIMENT

Small n-type Si pieces were cleaned using conventional methods. A 0.5 μm thick sacrificial SiO_x layer was deposited by Plasma Enhanced Chemical Vapor Deposition, followed by a thin layer of PMMA spun on the sample. Then 0.1 μm Au film was deposited by e-beam evaporation and lifted off after pattern definition by electron beam lithography. Large bonding pads of 0.2 μm thick Ti/Au were placed by means of optical lithography, after removing the residual resists with oxygen plasma ashing. Then, the samples were dipped into a 1:9 $HF:H_2O$ solution for 10 sec to remove the SiO_x layer underneath the metal so that the wires become free-standing. A scanning electron microscope picture of a typical device is shown in Fig. 33.1. Details of the sample preparation and measurement are given elsewhere [14]. The sample resistances were measured as a function of temperature from 8 K to 295 K. Since the current was kept small, < 3 mA, the resistance measurements were in complete equilibrium; hence, the temperature readings correspond to the bath (or lattice) temperature. Then, the cross conductance mea-

Figure 33.1 Scanning Electron Microscope view of a "hot-spot transistor". The gap between the free standing Au cross and the Si substrate is 0.5 μm. The wide leads and the cross were defined by e-beam lithography. The feature at the top right corner is the tip of the wider gold pad defined by optical lithography. Each wire is 0.5 μm wide and 9.3 μm long.

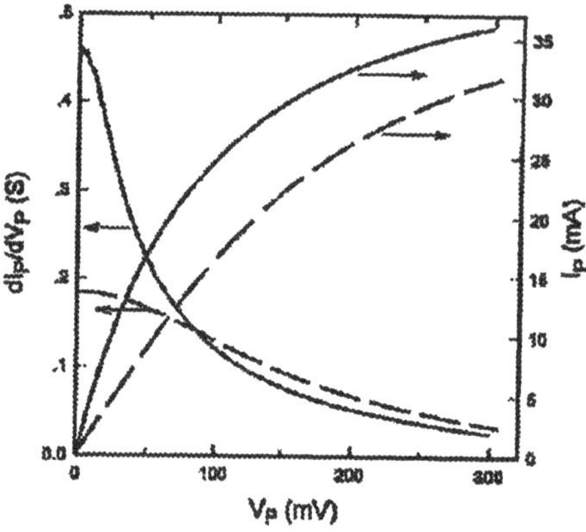

Figure 33.2 Probe current and probe conductance (differential) as functions of voltage drop across the channel with no current in the control channel at 8 K (solid line) and at room temperature (dashed line). Note that nonlinear conductance is observed also at room temperature. (After Ref. [14]).

surements were performed at a moderately low temperature (8 K), and also at room temperature. First we varied the current through one of the wires, call it the probe wire, and measured the the voltage developed across it, while maintaining a zero current through the second wire, the control channel. We observe a nonlinear behavior when we increase the current in the wire allowing hot electrons to be generated. Typical plots of I_p *vs.* V_p and dI_p/dV_p *vs.* V_p, for $I_c = 0$, are shown in Fig. 33.2. The solid curves are for the bath temperature at 8 K and the dashed curves are those of room temperature. The I/V characteristics clearly depart from Ohm's law, and the on-set of this departure is faster for the low temperature case, since the number of thermal phonons is much smaller than that at room temperature. Therefore, the effect of the non-equilibrium phonons, emitted by hot electrons at 8 K, for which kT<1 meV, becomes observable at lower electron currents, although electrons have lower energies for the same current at 8 K. Moreover, as expected, the differential conductance is higher initially at 8 K and the change is much steeper with respect to the room temperature curve. There is an excellent agreement with the calculated conductance characteristics of Ref. [1]. In the last part of the experiments the conductance of the probe wire was measured with respect to both probe and control currents, as shown in Fig. 33.3, where the measurements were taken at 8 K. When we fix the probe current to a finite value and vary the control current, we observe a decrease in the conductance of the probe channel, G_p. This is not a linear change, as can be seen from the figure. At I_p=1 mA, initially, the hot spot temperature, T_{hs}, is at bath temperature, and the probe conductance is large. As we increase I_c, T_{hs} increases, and accordingly the resistance of the probe wire increases; *i.e.*, G_p decreases. If we set I_p to a higher value, we start out with an intermediate lattice temperature, so the effect of I_c on G_p becomes less important. At

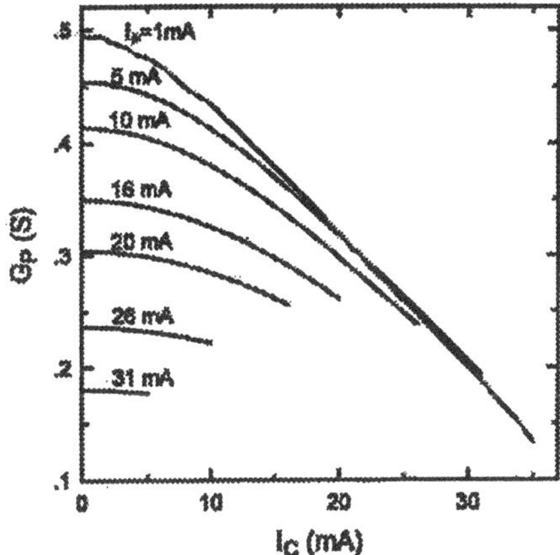

Figure 33.3 The effect of the control current on the probe channel differential conductance at various probe currents. Switching of I_c from 0 to 20 mA would result in a 28% reduction in the probe conductance at I_p=5 mA. (After Ref. [14])

I_p=31 mA, both wires are already hot and the variation of I_c does not yield much influence. In other words, the larger current dominates the conductance of the whole device. We repeated the experiment with samples of different sizes keeping the symmetry of the crosses; *i.e.*, $L_p = L_c$ and $W_p = W_c$. There were no systematic changes in the effective probe temperature characteristics with varying size, although slight departures were observed that seem to be within the experimental tolerance.

4. CONCLUSION

We verified experimentally the nonlinear conductance behavior of metallic submicrometer cross-junctions in the diffusive regime, in good agreement with the theoretical predictions. The strong coupling between the cross-channels was studied extensively, revealing that the larger of the two currents dominates the conductance of the whole device. Thus, one may consider this device as a "hot-spot transistor" with a transfer ratio less than unity. On the other hand, it can be used as a fast device with an estimated response time of $\sim 10^{-9}$ s, operating in the microwave region, for a hot-spot size of ~ 1 μm.

Acknowledgements

We are indebted to Professor Igor O. Kulik for attracting our attention to the problem, and for valuable discussions.

References

[1] I. O. Kulik, J. Appl. Phys. **76**, 1920 (1994).

[2] M. L. Roukes *et al.*, Phys. Rev. Lett. **55**, 422 (1985).

[3] P. M. Echternach, M. R. Thoman, C. M. Gould and H. M. Bozler, Phys. Rev. **B 46**, 10339 (1992).
[4] P. M. Echternach, M. E. Gershenson and H. M. Bozler, Phys. Rev. **B 47**, 13659 (1993).
[5] M. Kanskar, M. N. Wybourne and K. Johnson, Phys. Rev. **B 47**, 13769 (1993).
[6] M. Kanskar and M. N. Wybourne, Phys. Rev. Lett. **73**, 2123 (1994).
[7] J. F. diTusa, K.Lin, M. Park, M. S. Isaacson and J. M. Parpia, Phys. Rev. Lett. **68**, 678 (1992).
[8] J. F. diTusa, K.Lin, M. Park, M. S. Isaacson and J. M. Parpia, Phys. Rev. Lett. **68**, 1156 (1992).
[9] A. Potts, M. J. Kelly, D.J. Hasko, J. R. A. Cleaver, H. Ahmed, D. A. Ritchie, J. E. F. Frost and G. A. C. Jones, Semicond. Sci. Technol. **7**, B231 (1992).
[10] R. Ellialtıoğlu, Turkish J. Phys., submitted.
[11] Chambers, *Electrons in Metals and Semiconductors* (Chapman and Hall, Suffolk, 1990). Smith and H. H. Jensen, *Transport Phenomena* (Oxford University Press, New York, 1989).
[12] D. E. Prober, Phys. Rev. Lett. **75**, 3964 (1995). M. N. Wybourne and M. Kanskar, Phys. Rev. Lett. **75**, 3965 (1995). C. G. Smith and M. N. Wybourne, Solid State Commun. **57**, 411 (1986).
[13] İ. İ. Kaya, Hot Electron Interactions in Nanostructures, Ph.D. Thesis, Bilkent University, Ankara, 1997.
[14] İ. İ. Kaya and R. Ellialtıoğlu, Phys. Rev. B, submitted.
[15] N. W. Ashcroft and N. D. Mermin, *Solid State Physics* (Holt, Rineheart and Winston, New York, 1976).

Author Index

Citation Index

1-A. Halbritter, 2-M. Nita, 3-K. Flensberg, 4-L. Borda, 5-H. Bruus, 6-A. Örmeci, 7-X. Oriols, 8-, 9-S. Çıracı, 10-I. Shorubalko, 11-K.A. Matveev, 12-L. Glazman, 13-G. Goeppert, 14-C.W.J. Beenakker, 15-E.A. Pashitskii, 16-S.I. Shevchenko, 17-A.N, Omelyanchuk, 18-I.O. Kulik, 19-, 20-J. Goeres, 21-J.M. Van Ruitenbek, 22-I.K. Yanson, 23-, 24-, 25-D. Esteve, 26-, 27-J.J. Saenz, 28-E. Sheer, 29-A.M. Kosevich, 30-T. Reker, 31-A. Garcia-Martin, 32-E. Keçecioğlu, 33-J.C. Cuevas, 34-A. Clerk, 35-H. Boyacı, 36-, 37-Y.E. Losovik, 38-H.J.H. Smilde, 39-M. Tinkham, 40-, 41-K. Güven, 42-M. Schechter, 43-G. Ammendola, 44-H. Kroha, 45-R. Ellialtıoğlu, 46-Å. Ingerman, 47-Mrs. Tinkham, 48-R. Migliore, 49-P. Samuelsson, 50-B. Baelus, 51-B. Camarota, 52-M. Bonsager, 53-M. Pustilnik, 54-A.A. Nikolaeva, 55-J. Beyer, 56-F. Pierre, 57-W. Oliver, 58-E.N. Bogachek, 59-R. Cron, 60-A. Özpineci, 61-Z. Gedik, 62-Ö. Çakır, 63-M. Bayındır, 64-C. Yalabık, 65-Ç. Kılıç 66-A. Bek, 67-Mrs. Yalabık, 68-E.V. Sukhorukov, 69-K. Arutyunov, 70-, 71-L. Tian 72-, 73-A. Komnik, 74-G. Schön, 75-B.L. Altshuler, 76-C. Untied.

Zeitfracht Medien GmbH
Ferdinand-Jühlke-Straße 7
99095 Erfurt, Deutschland
produktsicherheit@kolibri360.de